AF840390

VOLUME ONE HUNDRED AND EIGHTEEN

ADVANCES IN PROTEIN CHEMISTRY AND STRUCTURAL BIOLOGY

Protein Misfolding

VOLUME ONE HUNDRED AND EIGHTEEN

ADVANCES IN
PROTEIN CHEMISTRY AND STRUCTURAL BIOLOGY

Protein Misfolding

Edited by

Rossen Donev
Head of Research
MicroPharm Ltd
Newcastle Emlyn, Carmarthenshire
United Kingdom

ELSEVIER

ACADEMIC PRESS
An imprint of Elsevier

Academic Press is an imprint of Elsevier
125 London Wall, London EC2Y 5AS, United Kingdom
525 B Street, Suite 1650, San Diego, CA 92101, United States
50 Hampshire Street, 5th Floor, Cambridge, MA 02139, United States
The Boulevard, Langford Lane, Kidlington, Oxford OX5 1GB, United Kingdom

First edition 2020

Copyright © 2020 Elsevier Inc. All rights reserved.

No part of this publication may be reproduced or transmitted in any form or by any means, electronic or mechanical, including photocopying, recording, or any information storage and retrieval system, without permission in writing from the publisher. Details on how to seek permission, further information about the Publisher's permissions policies and our arrangements with organizations such as the Copyright Clearance Center and the Copyright Licensing Agency, can be found at our website: www.elsevier.com/permissions.

This book and the individual contributions contained in it are protected under copyright by the Publisher (other than as may be noted herein).

Notices
Knowledge and best practice in this field are constantly changing. As new research and experience broaden our understanding, changes in research methods, professional practices, or medical treatment may become necessary.

Practitioners and researchers must always rely on their own experience and knowledge in evaluating and using any information, methods, compounds, or experiments described herein. In using such information or methods they should be mindful of their own safety and the safety of others, including parties for whom they have a professional responsibility.

To the fullest extent of the law, neither the Publisher nor the authors, contributors, or editors, assume any liability for any injury and/or damage to persons or property as a matter of products liability, negligence or otherwise, or from any use or operation of any methods, products, instructions, or ideas contained in the material herein.

ISBN: 978-0-12-817750-1
ISSN: 1876-1623

For information on all Academic Press publications visit our website at https://www.elsevier.com/books-and-journals

Publisher: Zoe Kruze
Acquisition Editor: Ashlie Jackman
Editorial Project Manager: Leticia M. Lima
Production Project Manager: Abdulla Sait
Cover Designer: Greg Harris

Typeset by TNQ Technologies

ELSEVIER Book Aid International Working together to grow libraries in developing countries
www.elsevier.com • www.bookaid.org

Contents

Contributors

Parvez Alam
Interdisciplinary Biotechnology Unit, Aligarh Muslim University, Aligarh, India

Sara Amidian
Department of Biochemistry, University of Alberta, Edmonton, Canada; Centre for Prions and Protein Folding Diseases, University of Alberta, Edmonton, Canada

Alan Bakalinsky
Department of Food Science and Technology, Oregon State University, Corvallis, Oregon, United States

Audrey M. Bernstein
SUNY Upstate Medical University, Department of Ophthalmology and Visual Sciences, Syracuse, NY, United States

Parbati Biswas
Department of Chemistry, University of Delhi, Delhi, India

John A. Carver
Research School of Chemistry, The Australian National University, Acton, ACT, Australia

Yraima Cordeiro
Faculty of Pharmacy, Federal University of Rio de Janeiro, Rio de Janeiro-RJ, Brazil

Andrea Curioni
Department of Agronomy, Food, Natural Resources, Animals and Environment (DAFNAE), University of Padua, Legnaro (PD), Italy; Centre for Research in Viticulture and Enology (CIRVE), Conegliano (TV), Italy

Guilherme A.P. de Oliveira
Programa de Biologia Estrutural, Instituto de Bioquímica Médica Leopoldo de Meis, Instituto Nacional de Biologia Estrutural e Bioimagem, Centro Nacional de Ressonância Magnética Nuclear Jiri Jonas, Universidade Federal do Rio de Janeiro, Rio de Janeiro, RJ, Brazil; Department of Biochemistry and Molecular Genetics, University of Virginia, Charlottesville, VA, United States

Jeffrey G. Dickhout
Department of Medicine, Hamilton Centre for Kidney Research, McMaster University and St. Joseph's Healthcare Hamilton, Hamilton, ON, Canada

Lyudmyla Dorosh
Department of Electrical and Computer Engineering, University of Alberta, Edmonton, Canada

Joseph C. Genereux
Department of Chemistry, University of California, Riverside, CA, United States

Carl Holt
Institute of Molecular, Cell and Systems Biology, University of Glasgow, Glasgow, United Kingdom

Claudia Honisch
Institute of Biomolecular Chemistry of CNR, Padua Unit, Padova, Italy; Department of Chemical Sciences, University of Padua, Padova, Italy

Rizwan Hasan Khan
Interdisciplinary Biotechnology Unit, Aligarh Muslim University, Aligarh, India

Elise A. Kikis
Biology Department, The University of the South, Sewanee, TN, United States

S. Imindu Liyanage
Krembil Research Institute, University Health Network, Toronto, ON, Canada

Nabeela Majid
Interdisciplinary Biotechnology Unit, Aligarh Muslim University, Aligarh, India

Sadia Malik
Interdisciplinary Biotechnology Unit, Aligarh Muslim University, Aligarh, India

Matteo Marangon
Department of Agronomy, Food, Natural Resources, Animals and Environment (DAFNAE), University of Padua, Legnaro (PD), Italy; Centre for Research in Viticulture and Enology (CIRVE), Conegliano (TV), Italy

Christiane Mendre
Institut de Génomique Fonctionnelle, CNRS, INSERM, Université de Montpellier, Montpellier, France

Bernard Mouillac
Institut de Génomique Fonctionnelle, CNRS, INSERM, Université de Montpellier, Montpellier, France

Samera Nademi
Department of Medicine, Hamilton Centre for Kidney Research, McMaster University and St. Joseph's Healthcare Hamilton, Hamilton, ON, Canada

Robert Ritch
Einhorn Clinical Research Center, New York Eye and Ear Infirmary of Mount Sinai, New York, NY, United States

Paolo Ruzza
Institute of Biomolecular Chemistry of CNR, Padua Unit, Padova, Italy

Gerold Schmitt-Ulms
Laboratory Medicine and Pathobiology, University of Toronto, Toronto, Canada

Mohammad Khursheed Siddiqi
Interdisciplinary Biotechnology Unit, Aligarh Muslim University, Aligarh, India

Jerson L. Silva
Programa de Biologia Estrutural, Instituto de Bioquímica Médica Leopoldo de Meis, Instituto Nacional de Biologia Estrutural e Bioimagem, Centro Nacional de Ressonância Magnética Nuclear Jiri Jonas, Universidade Federal do Rio de Janeiro, Rio de Janeiro, RJ, Brazil

Maria Stepanova
Department of Electrical and Computer Engineering, University of Alberta, Edmonton, Canada

Tuane C.R.G. Vieira
Programa de Biologia Estrutural, Instituto de Bioquímica Médica Leopoldo de Meis, Instituto Nacional de Biologia Estrutural e Bioimagem, Centro Nacional de Ressonância Magnética Nuclear Jiri Jonas, Universidade Federal do Rio de Janeiro, Rio de Janeiro, RJ, Brazil

Simone Vincenzi
Department of Agronomy, Food, Natural Resources, Animals and Environment (DAFNAE), University of Padua, Legnaro (PD), Italy; Centre for Research in Viticulture and Enology (CIRVE), Conegliano (TV), Italy

Donald F. Weaver
Krembil Research Institute, University Health Network, Toronto, ON, Canada; Departments of Medicine (Neurology), Chemistry and Pharmaceutical Sciences, University of Toronto, Toronto, ON, Canada

Holger Wille
Department of Biochemistry, University of Alberta, Edmonton, Canada; Centre for Prions and Protein Folding Diseases, University of Alberta, Edmonton, Canada; Neuroscience and Mental Health Institute, University of Alberta, Edmonton, Canada

J. Mario Wolosin
Eye and Vision Research Institute, Department of Ophthalmology, Icahn School of Medicine at Mount Sinai, New York, NY, United States

CHAPTER ONE

Theoretical and computational advances in protein misfolding

Parbati Biswas*
Department of Chemistry, University of Delhi, Delhi, India
*Corresponding author: E-mail: pbiswas@chemistry.du.ac.in

Contents

Abstract

Misfolded proteins escape the cellular quality control mechanism and fail to fold properly or remain correctly folded leading to a loss in their functional specificity. Thus misfolding of proteins cause a large number of very different diseases ranging from errors in metabolism to various types of complex neurodegenerative diseases. A theoretical and computational perspective of protein misfolding is presented with a special emphasis on its salient features, mechanism and consequences. These insights quantitatively analyze different determinants of misfolding, that may be applied to design disease specific molecular targets.

1. Introduction

Protein misfolding is a natural consequence of protein folding where the polymeric sequence of amino acids fold into a well-defined three dimensional functional native state in the cellular environment. For globular proteins, the native state represents the most stable structure under physiological conditions. Protein folding is vital for executing proper cellular

Advances in Protein Chemistry and Structural Biology, Volume 118
ISSN 1876-1623
https://doi.org/10.1016/bs.apcsb.2019.08.010
© 2020 Elsevier Inc.
All rights reserved.

functions. Folding of a polypeptide chain involves a directed search through the vast conformational space to reach the native state. The process of folding relies on the correct interactions involving a small subset of the residues of a protein. Intrinsic conformational fluctuations in a protein enable different residues, which are far apart in the polypeptide chain, to come in contact with one another. The native contacts stabilize the protein relative to the non-native ones, as the favorable enthalpy of such contacts, arising from the decrease in internal energy, balances the decreasing entropy and helps the protein to achieve the desired native conformation. Experimental and theoretical studies have established that most proteins fold via funnel-like rugged free-energy landscapes, that are encoded in the amino acid sequence of the protein, constrained by the rules of natural selection (Frauenfelder, Sligar, & Wolynes, 1991; Wolynes et al., 1995).

The energy landscape theory provides a statistical description of the thermodynamics and kinetics of protein folding. The folding energy landscape comprises all accessible protein conformations present in the folding pathway. Due to the restricted covalent connectivity of the protein backbone, various non-covalent interactions (Dill, 1990) present in the protein are not simultaneously optimized making the energy surface frustrated (Wolynes et al., 1995). Frustration in proteins arises from these competing interactions that are not satisfied simultaneously. The native state represents the minimally frustrated (Bryngelson & Wolynes, 1987; Ferreiro, Hegler, Komives, & Wolynes, 2007) conformation which lies at the global free energy minimum (Anfinsen, 1973). Under native conditions, a protein has a complex conformational manifold with innumerable near-native conformations residing at various local minima close to the native conformation, molding a dauntingly rugged landscape (Fig. 1). However, naturally occurring proteins are minimally frustrated and surmount this rugged energy terrain to fold reversibly to their unique native states.

The partly unfolded near-native conformations differ from the native fold only at some positions of the amino acid residues. While the native states of most globular proteins are well characterized, little is known about the rarely populated near-native ensemble. This is partly due to the fact that each of these individual conformations hardly affects the macroscopic properties and it is difficult to assess their respective contributions experimentally (Mello & Barrick, 2004). Recent experimental techniques like hydrogen exchange (Bai, Sosnick, Mayne, & Englander, 1995; Chamberlain, Handel, & Marqusee, 1996; Englander, 2000) and NMR relaxation (Volkman, Lipson, Wemmer, & Kern, 2001; Yang & Kay, 1996) and molecular dynamics

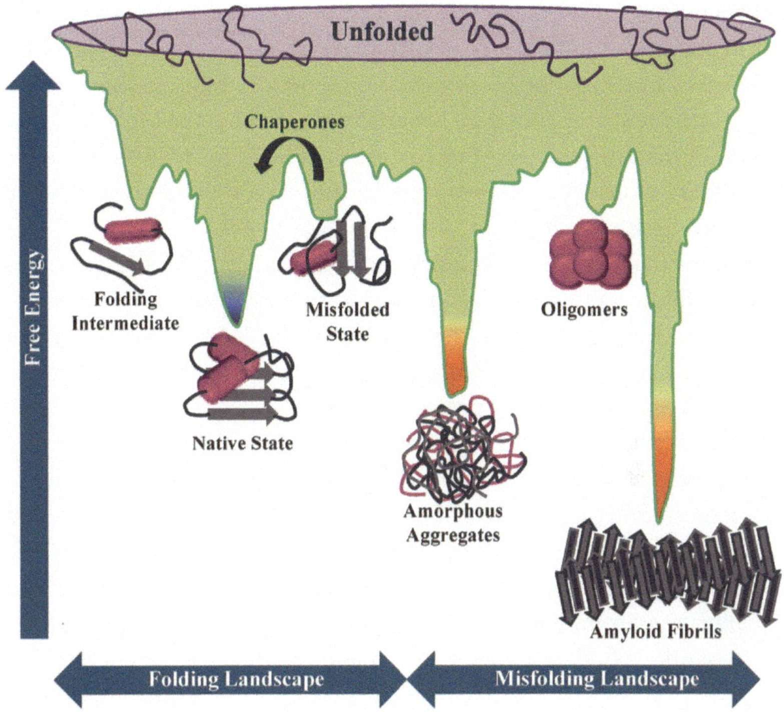

Fig. 1 A schematic representation of the energy landscape of protein folding, misfolding and aggregation.

(MDs) simulations (Alonso, DeArmond, Cohen, & Daggett, 2001; Voelz, Bowman, Beauchamp, & Pande, 2010) confirm the existence of such conformational states. The ensemble of partially folded conformations, which are thermodynamically distinct, but are comparable to the energy of the native state, may be transiently populated under native conditions (Tang & Dill, 1999).

Folding or refolding to such conformations has important implications in drastically altering protein functions, particularly responsible for allosteric regulation of enzymes and site-to-site communication in the cell. Characterizing the nature of these partially formed native-like structures constitutes the first step in understanding the relative stability of the misfolded states, mechanism of misfolding and the factors which promote misfolding. Partly or completely misfolded proteins are often involved in various cellular functions like signal transduction, translocation across membranes, transcription and cell cycle regulation. These near-native conformational states reside at various local minima in the free energy landscape, separated from the native state by a relatively low-energy barrier as shown in Fig. 1. Due to changes in solution conditions such as low pH, changes in temperature and pressure or an unwanted mutation in the amino acid sequence or any changes resulting from an inherited or acquired gene variation, a protein may fold into any of these conformations, instead of the native one. Such proteins are recognized as misfolded because of improper folding and are subsequently targeted to a degradation pathway.

The specificity of misfolding depends on the nature of the protein, the efficacy of folding and degradation mechanisms, cellular and environmental conditions and various interacting genetic factors. Mutations in the amino acid sequence also alter the folding ability by changing the folding energetics (Cohen & Kelly, 2003). Most proteins fold primarily in the cytoplasm, while others fold in the mitochondria or endoplasmic reticulum after translocation via membranes. The endoplasmic reticulum contains molecular chaperones, which assist in folding proteins either by non-specific binding to the protein or by catalyzing folding at later stages, while the respective proteins are evaluated by a stringent quality control mechanism before it is released in the extracellular environment (refer to Fig. 2). Eukaryotic cells possess an extensive network of components to maintain protein homeostasis (or

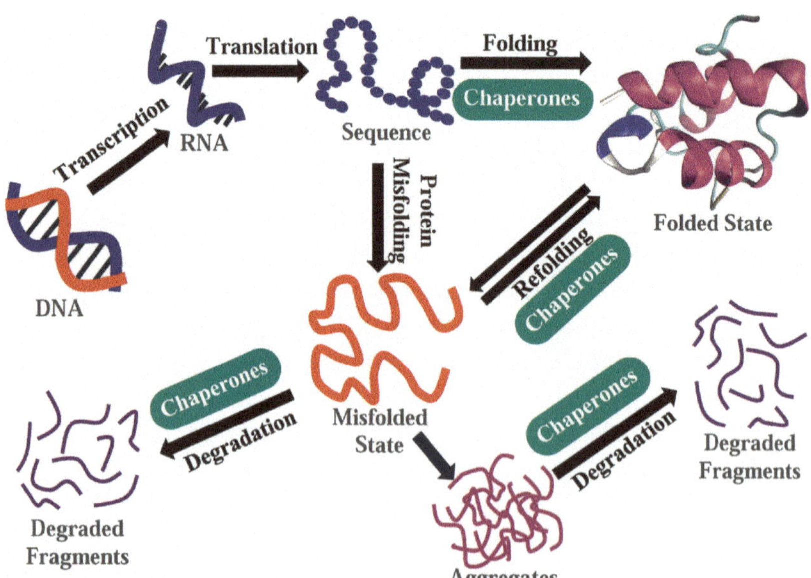

Fig. 2 Schematic diagram of the different stages in protein folding and misfolding.

proteostasis) that governs the cellular concentration, folding and localization of proteins (Labbadia & Morimoto, 2015), However, when the intracellular proteins or the newly synthesized proteins are unable to fold efficiently, the cell activates protein quality control system that facilitates the refolding of misfolded proteins or degradation of incorrectly folded proteins through molecular chaperones and proteasomes (Hipp, Park, & Hartl, 2014). Degradation in cells involve an alternative way of eliminating protein aggregates through molecular chaperones (see Fig. 2).

Typically, the long-lived proteins are degraded by lysosomes, while proteosomes are responsible for the degradation of short-lived proteins (Jackson & Hewitt, 2016). However, if the generation of the misfolded proteins exceeds the cellular degradative or refolding capacity causing an imbalance in the protein homeostasis (imbalance of correctly and incorrectly folded proteins), they assemble as aggregates. Thus, the successive failure to degrade misfolded proteins or dysfunction of the degradation pathways may result in aggregated assemblies that are associated with various human disorders, aging, perturbed cellular functions eventually leading to cell death (Koike et al., 2010). Thus a large number of very different diseases ranging from errors in metabolism to various types of complex neurodegenerative diseases may be attributed to protein misfolding (Dobson, 2002, 2003). Some of the common misfolding diseases include amyloidosis, Alzheimer's disease, Parkinson's disease, Creutzfeldt-Jakob disease, type II diabetes and some forms of cancer. While some maladies like cystic fibrosis result from the incorrect folding of the protein, others like amyloidoses, Parkinson's and Alzheimer's disease occur due to the deposition of specific proteins in some organs and tissues. Such misfolding diseases caused by the accumulation of soluble proteins outside the cell culminates in protein aggregation associated with toxic effects which, in some cases, lead to biochemical dysfunction and cell death (Cohen & Kelly, 2003; Selkoe, 1999).

2. Mechanism of protein misfolding

When proteins are exposed to specific environmental changes like high temperature and concentration, acidic or basic pH, elevated glucose levels or oxidative agents, their native structure is disrupted with a loss of functionality. Such a protein may be either misfolded or unfolded depending upon the number of non-native interactions between residues that differ from that of the native state. Most experimental studies suggest that protein misfolding follows 'nucleation growth' mechanism (Bitan et al., 2003; Scherzinger et al., 1999). The exposure of hydrophobic residues in an incompletely folded chain may induce inappropriate contacts with other molecules in the local environment. In such cases, transient non-native interactions may arise to bury these hydrophobic residues in the protein core. This process of aggregation is initiated by the exposed hydrophobic residues on the protein surface, β-sheet propensity and low net charge. These localized segments form oligomers as intermediates, which comprise either amorphous clusters or crystalline fibrils that act as precursors anchoring aggregation (refer to Fig. 3). The precursors may be generated either due to incomplete or incorrect proteolysis or a mutated protein, which generates an ensemble of oligomers that form aggregates (Herczenik & Gebbink, 2008). Often these aggregates possess well-defined crystalline structures comprising intermolecular β-sheets, known as amyloid fibrils.

Amyloid fibril formation occurs through nucleation-dependent polymerization (Harper & Lansbury, 1997; Soto, Estrada, & Castilla, 2006), which explains the formation of oligomers via a slow lag phase followed

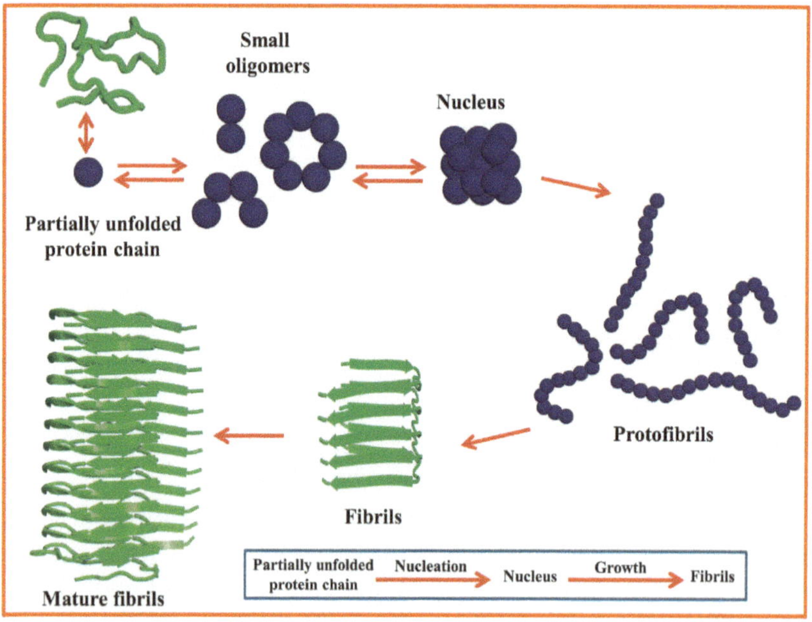

Fig. 3 Schematic diagram illustrating various stages of aggregation. *Modified from Fig. 2 of Morriss-Andrews, A., & Shea, J.-E. (2015). Computational studies of protein aggregation: Methods and applications.* Annual Review of Physical Chemistry, 66, *643–666.*

by polymerization. The templated assembly (Esler et al., 2000; Nguyen, Li, Stock, Straub, & Thirumalai, 2007) process comprises a series of unfavorable interactions between monomers in the slow lag phase to yield an oligomeric nucleus with minimal thermodynamic stability, which subsequently acts as a scaffold for the formation of larger aggregates. The initial association of monomers in the slow lag phase is energetically unfavorable, however once the oligomeric seed of a critical size is formed, rapid aggregation becomes thermodynamically favorable. The nucleation rate depends partly on the change in the free energy of nucleation, ΔG_{nuc}. However, the order of the overall nucleation reaction depends on each constituent second order association reactions, which may contribute cumulatively to the formation of the critical nucleus, since some of these individual reactions are expected to be rate controlling (Eigen, 1996). An oligomer of a critical size passes through a maximum value of ΔG_{nuc} to form a nucleus. The number of molecules in the critical nucleus depends on the monomer concentration, the equilibrium rate constant of monomer-monomer and monomer-fibril association and the protein solubility (Bader, Bamford, Zurdo, Luisi, & Dobson, 2006). The slow lag phase may be eliminated by seeding, where the preformed aggregates are added to the protein solution. In some cases, aggregation is triggered by the chemically modified forms of proteins that results from deamidation or oxidative stress (Nilsson, Driscoll, & Raleigh, 2002).

The conversion from the normal to the abnormal prion proteins in prion diseases usually follow this nucleation growth mechanism (Soto, 2012; Soto et al., 2006). Another model proposed for protein misfolding is the 'nucleated conformational conversion' model. The first step involves the conversion of the native monomer to a misfolded conformation that initiates self-assembly and forms molten oligomers (Serio et al., 2000). These oligomers then reorganize themselves to undergo structural changes yielding amyloid-like oligomers that act as the nucleus. This nucleus then triggers the formation of aggregates like other molten oligomers, which acquire amyloid conformations through the induced-fit mechanism (Serio et al., 2000), subsequently yielding higher order oligomers and fibrils. Unlike the nucleation growth mechanism, where the rate limiting step is the formation of the nucleus, the nucleated conformational conversion is characterized by the rapid formation of oligomers, where the rate limiting step involves the conversion of the misfolded oligomer into amyloid oligomers.

3. Simulations: atomistic and coarse grained models

Characterizing various structural and dynamical properties of misfolded proteins and aggregates is a major challenge of structural biology. Despite significant experimental advances in understanding the misfolding mechanism and fibril formation (Bemporad & Chiti, 2012; Williams & Pukala, 2013), other structural and dynamical aspects of aggregated proteins are often difficult to measure as experimental techniques are limited by the size of proteins and the time required to form aggregates. Computer simulations are also fraught with limitations with respect to the time and length scales associated with these processes. Nevertheless, they are particularly useful in probing the initial stages and the end-product of protein misfolding and aggregation in a detailed manner with fully atomistic protein models (Dupuis, Wu, Shea, & Bowers, 2011). A detailed study of the entire protein misfolding process, starting from small oligomers to the formation of large aggregates (or mature fibrils), is computationally too expensive to permit sufficient sampling at longer times.

Simulations of the detailed pathways of misfolding are mostly limited to small peptide segments (Matthes, Gapsys, & de Groot, 2012; Thirumalai, Reddy, & Straub, 2011) and have not been observed for large proteins due to high computational cost and sparse conformational sampling. However, it is necessary to simulate a sufficient number of complete protein chains to comprehend both mechanism of formation of oligomers in early stages and aggregation or fibril formation (Li, Uversky, & Fink, 2001). Several studies also point out the fact that the native structure predictions during the protein folding are reliable for shorter proteins rather than the longer ones due to restricted conformational sampling (Lindorff-Larsen, Piana, Dror, & Shaw, 2011; Nguyen, Maier, Huang, Perrone, & Simmerling, 2014; Shaw et al., 2007).

Atomistic simulation studies of predicting the native structures of long proteins with time scales varying between μs to hours, is a daunting task with present-day computational resources, as the folding time exponentially increases with the protein size (Naganathan & Muñoz, 2005). Thus, investigating protein misfolding and aggregation of full length proteins with a wide window of time scale, varying between *ms* to years, poses a formidable challenge for simulations, which is beyond the scope of existing computing facility. In this regard, coarse grained models (high resolution models with less details) compromise atomistic details (by neglecting some of the

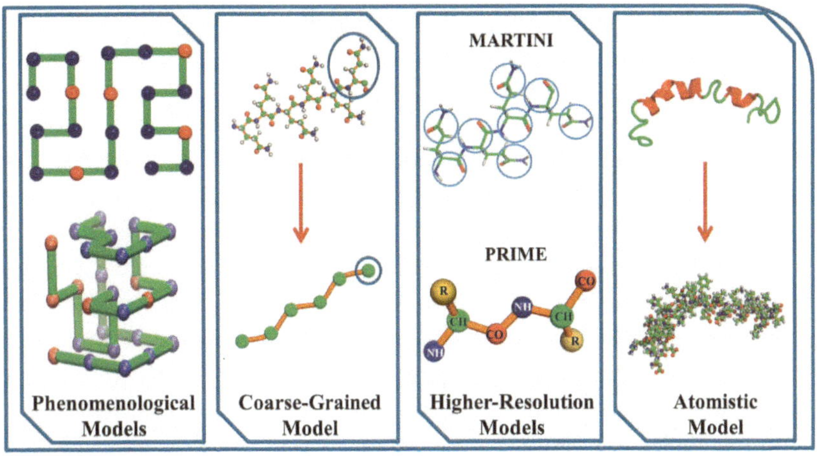

Fig. 4 Different coarse-grained and atomistic models in theory and simulations.

atomistic degrees of freedom) and accuracy at the expense of simplified calculations with a significantly faster convergence (Ingólfsson et al., 2014), with a deeper insight into the generic features of aggregation/fibril formation.

Coarse-grained models yield qualitatively similar folding free energy landscapes, as compared to those obtained through detailed atomistic simulations (Coluzza, Muller, & Frenkel, 2003; Shaw et al., 2010). However, coarse grained models are less accurate relative to the fully atomistic models, as most of these models use a combination of knowledge-based and physics-based potentials. However, such models allow the study of much larger systems by accessing a broad range of length and time scales that are inaccessible to the atomistic models. Different coarse-grained and atomistic models that are adapted in theory and simulations are schematically shown in Fig. 4. In following sections, some of the key findings in protein misfolding embodied in coarse grained and traditional all atom models are discussed.

4. Atomistic simulations of early-stages monomers, small oligomers and end products of aggregation

Since aggregation originates from monomers, understanding monomers and small oligomers in atomic details may reveal the mode of self-assembly of nontoxic monomers to form disease-causing toxic species. Structural analysis of the monomer, initiation of aggregation and oligomerization is typically achieved through enhanced sampling techniques such as

Replica Exchange Molecular Dynamics (REMD), simulated tempering and metadynamics. While REMD (Sugita & Okamoto, 1999) involves running multiple replicas of a system at different temperatures with periodic exchange of configurations among these replicas based on a Metropolis criterion, metadynamics (Laio & Parrinello, 2002) uses a history dependent bias, which compels the system to sample the rarely visted regions of the free energy landscape. Initial studies on protein misfolding were performed with implicit solvent models (Khandogin & Brooks, 2007; Lam, Teplow, Stanley, & Urbanc, 2008; Santini, Mousseau, & Derreumaux, 2004), on small protein segments, which included the classical MD simulations of the peptide segment 21−30 of amyloidβ (Aβ) peptide (Fawzi et al., 2008), Replica-Exchange Molecular Dynamics (REMDs) simulations of the fragment 15−28 of Aβ peptide (Baumketner, Krone, & Shea, 2008), REMD simulations of the 41 residue peptide with the non amyloid component region of α-synuclein with various force fields and solvent conditions (Yoon, Park, Jang, Lee, & Shin, 2008) and REMD simulations of the peptide fragment 21−31 of β_2-microglobulin (Nishino, Sugita, Yoda, & Okamoto, 2005). However, with the recent advances in computational algorithms and efficiency, atomistic simulations with explicit water models were used to explore the conformational changes and structural arrangements of the β-strand in amyloids (Carballo-Pacheco & Strodel, 2016; Tsai et al., 2005), as exemplified by multiple all-atom MD simulations of Aβ protein in explicit water (Lee & Ham, 2011) and REMD simulations of the prion protein (Baillod, Garrec, Colombo, Tavernelli, & Rothlisberger, 2012). A detailed account of the recent advances in the atomistic simulation studies of protein aggregation may be found in some current reviews (Carballo-Pacheco & Strodel, 2016; Nasica-Labouze et al., 2015).

All-atom REMD simulation studies of Aβ (Rosenman, Connors, Chen, Wang, & García, 2013) (see review (Nasica-Labouze et al., 2015) for details) and human Islet Amyloid Polypeptide (Dupuis, Wu, Shea, & Bowers, 2009; Qiao, Bowman, & Huang, 2013; Reddy et al., 2010) predicted β-hairpin structures of peptides that may act as a nucleus for self-assembly. The β-sheet enriched structure of the human islet amyloid polypeptide (*hIAPP*) was amply demonstrated through different REMD simulations, i) with *AMBERff*96 force field and the generalized-Born model *igb*5 (Dupuis et al., 2009), ii) *AMBERff*99*sb* force field with *TIP3P* water model (Qiao et al., 2013) and iii) *GROMOS*9653*a*6 force field with *SPC* water model (Reddy et al., 2010). However, classical MD and REMD simulations of the same protein performed with *OPLS-AA* force field with *SPC/ E* water model (Andrews & Winter 2011) and *ff03w* force field with *TIP4P/2005* water model (Miller, Zerze, & Mittal, 2013) respectively, hardly displayed any β-sheet content. This discrepancy may be attributed to the choice of force fields, which stabilize the secondary structures in proteins and hence are better optimized for globular proteins as compared to the misfolded and disordered proteins. Another view that emerged from the analysis of the free-energy landscape of the segment 25−35 of the Aβ peptide, depicts the participation of both α-helix and β-sheet in the formation of oligomers (Tran & Ha-Duong, 2015). Along similar lines, REMD simulation of Aβ peptides (Sgourakis, Yan, McCallum, Wang, & Garcia, 2007) suggested a strong correlation between the amyloid fibrils and the hydrogen bond forming propensity of the backbone polypeptide chain.

Prion diseases caused by the aggregation of the prion protein involves a conformational transition from the soluble cellular isoform PrP^C to the insoluble β rich PrP^{Sc}. REMD simulations of the monomeric *PrP* C-terminal domain reveals β enriched folds with exposed β-sheet edges and increased hydrophobic solvent accessible surface area (Baillod et al., 2012), which contradicts the earlier findings of a spiral (DeMarco & Daggett, 2004) and β-helix fibril (Govaerts, Wille, Prusiner, & Cohen, 2004; Wille et al., 2009). While some simulation studies suggested the disruption of antiparallel β-sheets of prion proteins under pathogenic conditions with parallel β-helix structure in the misfolded monomer (Barducci et al., 2006; Wille et al., 2002), other works established the evidence for the antiparallel β-sheet structure in oligomeric fibrils (Alonso et al., 2001; DeMarco & Daggett, 2004). Also the effect of mutations on disease related misfolding investigated by either classical MD simulations (Alonso et al., 2001; DeMarco & Daggett, 2004) or other advanced simulation methods such as REMD simulations (Sugita & Okamoto, 1999) and non-Markovian metadynamics (Barducci et al., 2006, 2011) revealed the stability of the β-sheet structure or the β-sheet motif in mouse prion protein and its mutant under pathological conditions.

Atomistic simulations are also employed to characterize the structural aspects of amyloid fibrils apart from exploring early stages of protein misfolding. Structural analysis of fibrils revealed a strand-loop-strand motif, as demonstrated by the work of Ma and Nussinov (2002) and Luca, Yau, Leapman, and Tycko (2007). Discrete and discontinuous molecular dynamics simulations probed the detailed mechanism of protofibril formation with concomitant conformational transition in Aβ and polyalanine peptides (Borreguero et al., 2005; Nguyen & Hall, 2006). Full atomistic simulations also analyzed the growth phase of fibril formation. While, a transition path sampling study (Schor, Vreede, & Bolhuis, 2012) proposed a dock-lock mechanism for fibril elongation (Esler et al., 2000), REMD simulations suggested that fibrils with parallel cross-β structures may act as template for the formation of β-rich conformations for monomeric Aβ yielding fibrillar aggregates (Xi, Li, & Wang, 2012).

Molecular dynamics simulations (Nguyen & Hall, 2006) established the presence of disordered oligomers in the early stages of protein misfolding where the fibrillar growth occurs in two ways; i) lateral growth by templated protofilament assembly through lateral addition of β-sheet layers and ii) by longitudinal elongation of β-sheet through addition of monomers to the extremities of fibrils. This work and the other related simulations (Bellesia & Shea, 2009; Kashchiev & Auer, 2010; Pellarin, Guarnera, & Caflisch, 2007) supported the monomer-addition elongation mechanism proposed by Straub et al. (Massi & Straub, 2001; Straub & Thirumalai, 2011). These simulations are also extensively used to explore various stages of protein misfolding, which include monitoring the conformational transitions (Alonso et al., 2001; DeMarco & Daggett, 2004), analyzing the effect of environmental factors such as change in pH, and temperature (Flöck, Colacino, Colombo, & Di Nola, 2006) and the effect of mutations (van der Kamp & Daggett, 2011) explicating the complex mechanism involved in protein misfolding. Langevin dynamics simulation with 'reaction path annealing' algorithm explored the effect of mutations on the kinetics of structural transitions in protein misfolding through the estimation of the energetic barrier of amyloid formation of the seven residue peptide *GNNQQNY* of yeast prion-like protein *Sup*35 (Lipfert, Franklin, Wu, & Doniach, 2005). The effect of pH on the fibril stability was analyzed for *hIAPP* which demonstrated that acidic conditions disfavored the formation of fibrils by disrupting the favorable electrostatic interactions and hydrogen bonding interactions (Wei et al., 2011).

An insight into the role of the environmental factors, such as change in pH, on protein misfolding was implemented via Constant pH molecular dynamics simulations (CpHMD), which directly captured the correlation between the conformation and the protonation states. Much of the work of Machuqueiro and Baptista (Machuqueiro and Baptista, 2006, 2009) accounted for the misfolding transition of proteins at low pH, which revealed a gradual decrease in the helical content with a subsequent increase in the average β-sheet structure as a function of the acidity of the medium. CpHMD simulations of human prion protein (Campos, Machuqueiro, &

Baptista, 2010) confirmed this observation. However, CpHMD simulations of human α-lactalbumin performed in acidic, neutral and basic pH (Bhattacharjee, Rani, & Biswas, 2013), displayed a complete disruption of the β-sheets in both acidic and basic pH, while, helices comprising the α-domains were moderately preserved in both acidic and alkaline medium as shown in Fig. 5. An examination of the conformational properties (shown in Table 1) revealed an increase in the size of the protein, as evident by an increased radius of gyration, decrease in the probability of the native contacts and enhanced mean square fluctuations in both acidic and basic medium. An analysis of the local and global conformations of the protein obtained from the calculation of mean fractal dimensions supported the emergence of the

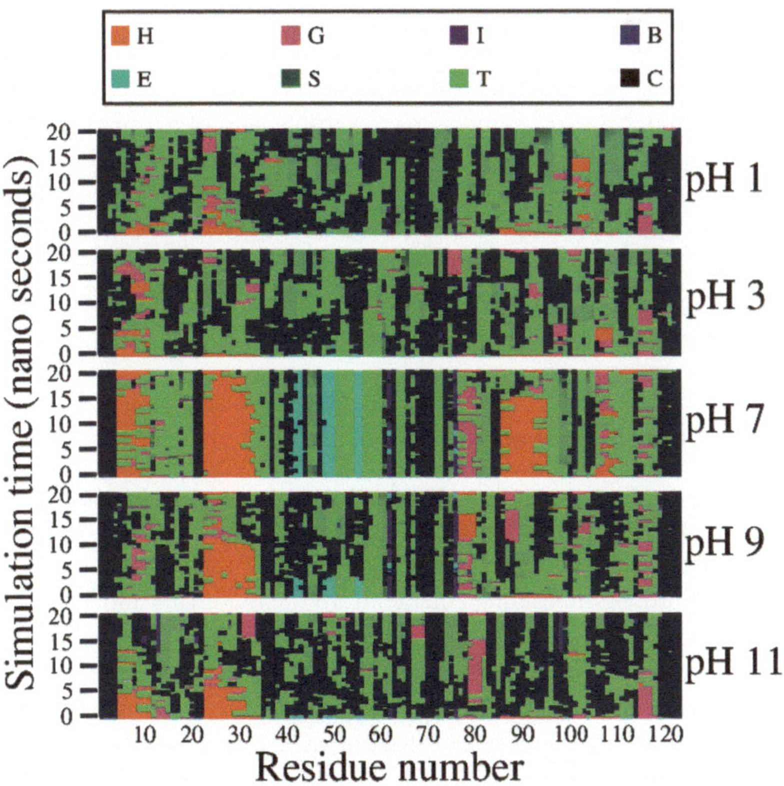

Fig. 5 Time evolution of secondary structures in α-lactalbumin during simulations at different pH. *Reprinted from Bhattacharjee, N., Rani, P., & Biswas, P. (2013). Capturing molten globule state of α-lactalbumin through constant ph molecular dynamics simulations.* The Journal of Chemical Physics, 138(9), *095101, with the permission of AIP Publishing.*

Table 1 Parameters describing the structure of α-lactalbumin at different pH: (i) All atoms Root-Mean-Square Deviation (RMSD) from the crystallographic structure, (ii) Radius of Gyration (R_G), (iii) Number of Native Contacts (NCs), (iv) Number of Total Contacts (TCs) and (v) Fractal dimension (D_1 and D_2).

			Native	Total	Fractal dimension	
pH	**RMSD (Å)**	**R_G (Å)**	**Contacts**	**Contacts**	**D_1 (M $\leq$ 10)**	**D_2 (M $>$ 10)**
1	13.59 (2.17)	17.17 (1.63)	568 (37)	821 (53)	1.39 (0.04)	1.57 (0.10)
3	12.45 (2.22)	15.16 (0.63)	553 (47)	848 (41)	1.39 (0.03)	1.63 (0.08)
7	4.43 (1.94)	14.27 (0.44)	797 (34)	900 (28)	1.42 (0.04)	1.62 (0.06)
9	9.60 (1.73)	15.88 (0.73)	631 (54)	825 (30)	1.39 (0.03)	1.63 (0.10)
11	11.44 (2.03)	15.44 (2.75)	548 (53)	835 (41)	1.38 (0.04)	1.75 (0.14)

Standard deviations from mean values are shown in parentheses.
Reprinted from Bhattacharjee, N., Rani, P., & Biswas, P. (2013). Capturing molten globule state of α-lactalbumin through constant ph molecular dynamics simulations. *The Journal of Chemical Physics, 138(9)*, 095101, with the permission of AIP Publishing.

molten globule state in both acidic and basic pH. Similar CpHMD simulations of prion proteins exhibited a reversible conformational transition induced by decreased pH ($pH = 2$) (Vila-Vicosa, Campos, Baptista, & Machuqueiro, 2012) which supported related experimental results (Bjorndahl et al., 2011). A series of MD simulations with explicit water models also captured the effect of pH on the misfolding of prion proteins (Alonso et al., 2001; Campos et al., 2010; Colombo, Meli, Morra, Gabizon, & Gasset, 2009; van der Kamp & Daggett, 2011).

5. Coarse grained models

Coarse grained models compromise the atomistic details while capturing the basic physical principles that characterizes protein misfolding and aggregation over longer time scales. Much of the present day understanding of protein misfolding emerges from the minimalist model proteins that are simplified representations of the polypeptide chain. These coarse-grained models provide viable alternatives that trade-off between accuracy and computational efficacy and hence are studied extensively by both analytical theory and simulations (Shakhnovich, Farztdinov, Gutin, & Karplus, 1991; Shakhnovich & Gutin, 1990; Tozzini, 2005; Yue et al., 1995). Such models unravel the thermodynamic and kinetic details of the misfolding process, mechanism of misfolding and the nature of the misfolded intermediates. Different types of coarse-grained models vary in the degree of resolution or parametrization; while the simplest models classify amino acids as either polar or hydrophobic, the sophisticated models use specific parameters for each of the twenty amino acids thereby distinguishing different types of amino acids. These models are frequently used to explain misfolding despite they are less accurate as compared to the fully atomistic models.

Highly coarse grained models considers each protein chain as a single unit, exemplified by the tube model (Auer, Meersman, Dobson, & Vendruscolo, 2008), cuboid model (Zhang & Muthukumar, 2009), orientable stick model (Irbäck et al., 2013) and rod-like model (Vácha & Frenkel, 2011). These simplest coarse grained models provided insights into the nucleation growth mechanism of protein misfolding. The nucleation growth mechanism was investigated via Monte Carlo simulations of a highly coarse-grained model protein (Irbäck et al., 2013). Another such example, the cuboid model, suggested two types of lag phases in the formation of amyloid fibrils, where the first lag phase represented the classical nucleation mechanism observed at high temperature or low concentration, and the second lag phase was caused by the slow Ostwald ripening observed at low temperature and intermediate concentrations. A similar mechanism in the lag phase was demonstrated in the Monte Carlo simulation study of Li, Klimov, Straub, & Thirumalai (2008), which explored the mechanism and kinetics of fibril formation for lattice proteins. The cuboid model and the tube model, justified the mechanism for elongation, where less stable protofibrils formed the primary fibrils. They also pointed out that the different stages of fibril formation conformed to the three components of the basic mechanism of fibril formation, i.e., nucleation and growth, templated assembly and nucleated conformational conversion.

Another simulation study employed the higher resolution tube model to probe the mechanistic details of oligomers. This work justified the self-templated nucleation mechanism with the initial formation of dynamic and disordered oligomers stabilized by hydrogen-bonding and hydrophobic interactions, followed by subsequent rearrangements in the oligomeric structure to yield β-rich fibrillar structures that acted as a template for further growth (Auer, Dobson, Vendruscolo, & Maritan, 2008, 2009). The tube model of polypeptide chains in conjunction with the generic hypothesis of amyloid formation implies that the formation of early disordered oligomers is favored with the increase in the concentration such that at low concentrations the peptide chains directly formed β-sheet structures, rather than the disordered oligomers.

Phenomenological models are the simplest coarse grained models that classify the amino acid residues as charged, polar and hydrophobic. Such models may be categorized as lattice and off-lattice models with low to medium resolution (Morriss-Andrews & Shea, 2015). Most of the existing lattice models are limited by their inability to account for the backbone-specific interactions that play an important role to understand the transition from oligomers to fibrils. The coarse grained off-lattice models simulate the transition from oligomers to aggregates (Matthes et al., 2012; Nguyen & Hall, 2004), but are computationally too expensive to study the detailed mechanism of misfolding. Lattice models are single bead models that represent the protein as a cubic lattice polymer, where each residue/structural unit occupies a lattice site (Li, Helling, Tang, & Wingreen, 1996). Off-lattice models of low to medium resolution include either two-beads-per-residue Caflisch model (Friedman & Caflisch, 2011; Magno, Pellarin, & Caflisch, 2012) or three-beads-per-residue (two for the backbone and one for the side chain) (Bellesia & Shea, 2007). Lattice and off-lattice models for folded and misfolded proteins are demonstrated in Fig. 6.

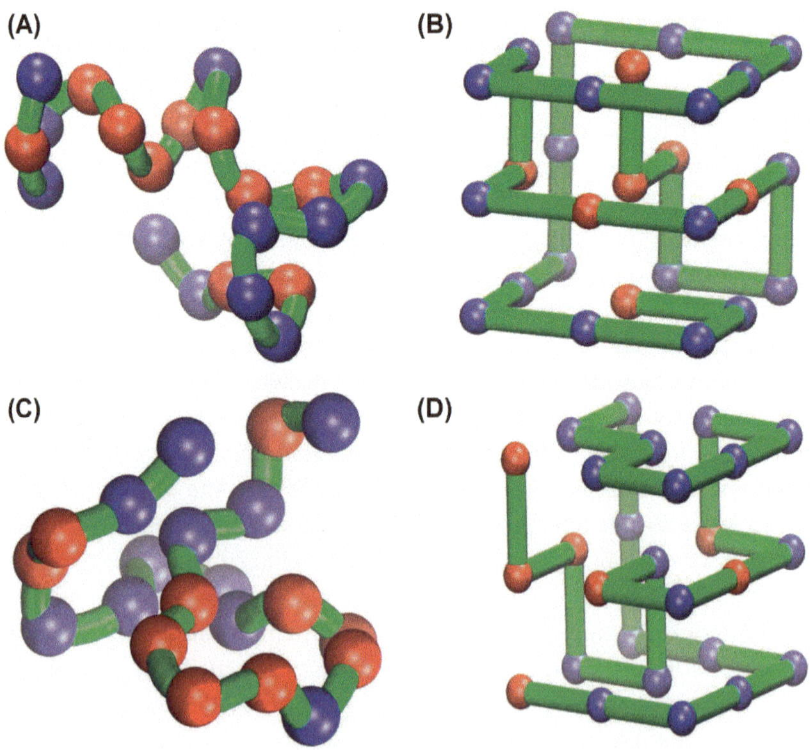

Fig. 6 The native structures of (A) 1*EDN* and (B) lattice protein. (C) and (D) are the respective misfolded structures. Blue beads represent polar residues and red beads represent hydrophobic residues. *Reproduced from Baruah, A. & Biswas, P. (2014).* Physical Chemistry Chemical Physics, 16, *13964–13973, with permission of the PCCP Owner Societies.*

Minimalist protein models with weak energetic frustration, represented by lattice heteropolymers with three different types of contacts, were designed by Metropolis Monte Carlo (MMC) algorithm with simulated annealing to probe multiple metastable native-like states (Locker and Hernandez, 2001, 2004). Results for the designed model generalized the rugged energy landscape picture for misfolded proteins, which are kinetically trapped in various long-lived conformations, that are inaccessible under typical folding conditions. Another work used a lattice model with the inclusion of hydrogen-bond and directionality of side chains (Abeln, Vendruscolo, Dobson, & Frenkel, 2014) to model β-strand formation in proteins. This study correlated the formation of amyloid and β-sheet to the amino acid composition of peptides, where the hydrophobic sequences favored the formation of such structures. A schematic energy landscape of the β-amyloid was constructed from a kinetic model to explore fibril association and reorganization (Massi & Straub, 2001). A steady state kinetic analysis combined with experimental data (Esler et al., 2000) evaluated the rate constants for the adsorption/desorption and deorganization/reorganization of the peptide and the fibril. Lattice models were also used to explore the nucleation and lag phases in misfolding. The generic features of the mechanism and kinetics of fibril formation as a function of temperature and the number of chains were captured through the simplified lattice models using exhaustive MC simulations (Li et al., 2008). This work established the validity of simple lattice models to identify the aggregate-prone conformations.

Various low resolution to medium resolution off-lattice coarse-grained models such as the three-bead model by Shea et al. (Bellesia & Shea, 2007, 2009) and the two bead model of Caflisch et al. (Pellarin et al., 2007) proposed that the proteins with low β-sheet propensities in the monomeric state generated higher number of prefibrillar species that is linked with toxicity. These works speculated that the peptides with lower β-sheet propensities produced more toxic end product in the misfolding process as compared to those with higher β-sheet propensities. Results of these simulation studies also suggested that the peptides with high β-sheet propensity directly assembled into ordered β-rich oligomers with the initial formation of a single β-sheet followed by the formation of small double-layered sheet. However, peptides with low β-sheet propensity initiated disordered oligomers, which internally rearranged into β-sheet structures (Morriss-Andrews & Shea, 2014). Different coarse-grained models explored the effects of temperature and concentration on protein misfolding. The number and size of fibrils as a function of temperature was investigated by Muthukumar et al. (Zhang & Muthukumar, 2009), where longer fibrils were formed with increase in temperature and fibril dissembly occurred at very high temperatures. An increase in the aggregation propensity of 27 polyglutamine peptides was observed as a function of chain length and concentration (Wang & Voth, 2010).

Detailed representations of proteins are required to investigate the chemical properties of protein misfolding, demanding the high resolution nonphenomenological models. High resolution nonphenomenological models include the MARTINI model (Monticelli et al., 2008) that maps four atoms to a single bead for protein and water respectively, the PRIME model (Cheon, Chang, & Hall, 2010) which was parametrized using learning algorithms to maximize the energetic gap between decoy and PDB structures and the OPEP model (Sterpone et al., 2014) that utilizes all-atom backbone with coarse-grained side chain with a high degree of chemical specificity. Coarse-grained MD simulation with the MARTINI model of amylin peptide depicted a temperature dependent fibril growth mechanism where low temperature facilitates elongation over the lateral growth (Sørensen, Periole, Skeby, Marrink, & Schiøtt, 2011). The MARTINI model in explicit water was used to portray the mechanism of aggregate formation in a system of 140 peptide amphiphiles with sequence *SLSLAAAEIKVAV* (Lee, Cho, & Schatz, 2012). The PRIME model in conjunction with discontinuous MD simulation is extensively used for the analysis of the kinetics of fibril formation for 48 tau fragment peptides, palindromic peptides from the Syrian hamster prion protein and the mouse prion protein and polyalanine peptides (Phelps & Hall, 2012; Wagoner, Cheon, Chang, & Hall, 2011, 2012). The OPEP model was used for the analysis of early stages of oligomer formation for the *GNNQQNY* and *NNQQ* peptides of the yeast prion Sup35 (Lu, Wei, & Derreumaux, 2012; Nasica-Labouze, Meli, Derreumaux, Colombo, & Mousseau, 2011).

Coarse-grained model proteins were also used to depict the intrinsic features of a given sequence, which determined its folding ability with respect to a unique native structure. A self-consistent mean field theory was developed to design misfolded sequences, that quantified the sequence properties in terms of a generalized foldability criterion. This energy based criterion characterized the sequence energy landscape in terms of folded, unfolded and misfolded conformational ensemble (Bhattacherjee & Biswas, 2011). For a particular foldability criterion, this theory identified folded, unfolded and misfolded ensemble of conformations of real and lattice proteins by

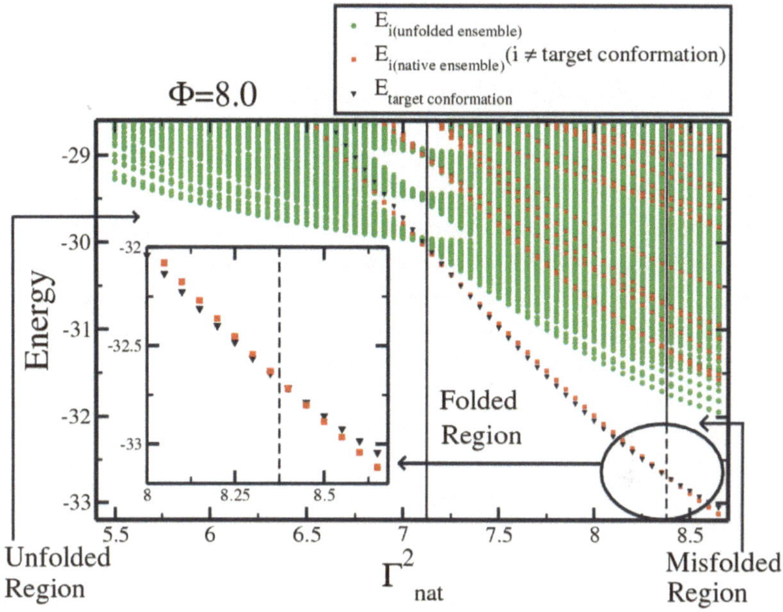

Fig. 7 Plot of the sequence energies of all conformations versus variance in the energy of the native state ensemble, Γ^2_{nat}, of 1WY3 for the native state ensemble with 55% similarity to the target conformation. The energy of the target conformation is denoted by black color, other conformations in the native state ensemble are represented by red squares. The unfolded ensemble of states is represented by green circles. Inset highlights the folding-misfolding transition. *Reprinted with permission from Bhattacherjee, A., & Biswas, P. (2011).* The Journal of Physical Chemistry B 115, *113–119. Copyright (2011) American Chemical Society.*

tuning the energy variance, Γ^2_{nat}, of the native state ensemble as shown in Fig. 7. This self-consistent field theory was also applied to investigate the role of conformational fluctuations in the native state and unfolded state ensemble in real and lattice proteins (Baruah, Bhattacherjee, & Biswas, 2012). Folded, unfolded or misfolded conformational ensemble of the designed sequences were modulated by the extent of conformational heterogeneity in the native state ensemble. Results portrayed that for increasingly heterogeneous native state ensembles, the protein compromised stability at the expense of increased structural flexibility and misfolded into any conformation other than the native one. This observation models protein evolution and predicts that conformationally flexible proteins opted by natural selection have reduced folding stability. The conformational fluctuations altered the surface accessibility of particular core or surface residues, which were either exposed to the surface or buried inside the protein core, that caused misfolding due to unfavorable interactions.

The specific roles of local and non-local interactions in protein misfolding were investigated through a Monte Carlo simulation based sequence design algorithm, where the backbone of proteins with different secondary structure content was modeled as a C_α chain (Kumar, Baruah, & Biswas, 2017). A statistical potential was developed from a selected dataset of proteins that comprised of hydrogen bonding, hydrophobic and electrostatic interactions. The relative weights of the local and nonlocal interactions were tuned to capture the essential features of folding, unfolding and misfolding of proteins as portrayed in Fig. 8 for a designed sequence of the 70-mer protein (PDB ID 5D14, SCOP $\alpha + \beta$). For a constant value of the weight of local interactions, the weight of nonlocal interactions was decreased stepwise. Most designed sequences selected expanded unfolded conformations below a critical value of the weight of nonlocal interactions. This implies that

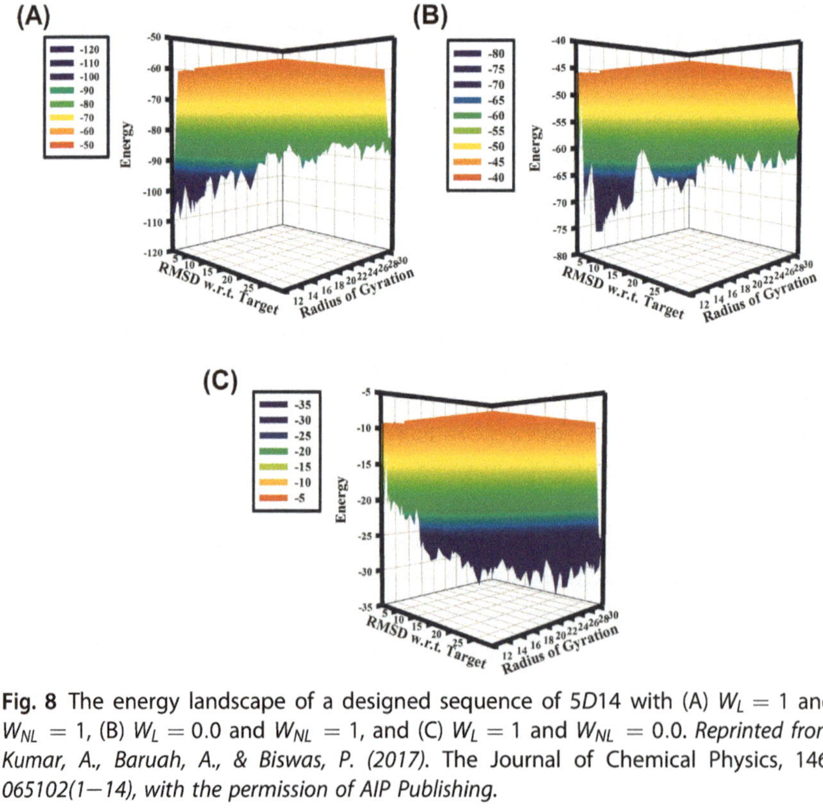

Fig. 8 The energy landscape of a designed sequence of 5*D*14 with (A) $W_L = 1$ and $W_{NL} = 1$, (B) $W_L = 0.0$ and $W_{NL} = 1$, and (C) $W_L = 1$ and $W_{NL} = 0.0$. *Reprinted from Kumar, A., Baruah, A., & Biswas, P. (2017).* The Journal of Chemical Physics, 146, *065102(1–14), with the permission of AIP Publishing.*

nonlocal interactions govern the fold-unfold structural transition, while local interactions determine the fold-misfold transition of proteins.

Simplified models also analyzed the effect of structural perturbations like mutations, that trigger protein misfolding, in the free energy landscape of proteins (Karanicolas and Brooks III, 2002; Stoycheva, Brooks, & Onuchic, 2004). The effect of mutation on the folding mechanism of protein *S*6 was characterized by Matysiak et al. (Matysiak & Clementi, 2006) who used the coarse-grained Gō like model to analyze the effect of such perturbations on the free energy landscape of *S*6 variant by mutating four residues of *S*6 protein. The findings established that four glutamic acid gatekeeper residues, i.e., 41, 42, 46 47, controlled the flexibility of the active site of *S*6 which regulated the balance between folding and misfolding. These residues minimized the non-native contacts and smoothened the free energy landscape. However, mutations of these four residues created an alternative funnel corresponding to the formation of partially misfolded states in the energy landscape (Otzen, Miron, Akke, & Oliveberg, 2004; Otzen & Oliveberg, 1999).

Random point mutations at specific sites drastically affect the stability and function of any protein, confirming that all sites in a protein are not equivalent. Furthermore, the design of rational mutations are complicated by the choice of the type and location of the mutated residues. A self-consistent mean field based theory was applied to probe the effect of site-directed point mutations in designing folded/misfolded sequences in both lattice and real proteins (Baruah & Biswas, 2014b). The respective roles of the core and surface residues were explored with respect to the structural stability of the protein. Results indicated that the core sites of the protein must possess a critical number of hydrophobic residues below which a protein misfolded to any conformation other than the native one. A clash-match calculation predicted the number of residue pairs in a sequence with favorable and unfavorable interactions due to site-directed point mutations as shown in Fig. 9. Mutations at sites with higher number of clashes between specific residue pairs generated misfolded proteins. A similar finding is depicted in Fig. 10, which was obtained from a Monte Carlo simulation study of a coarse-grained C_α chain model protein used to investigate the role of site-directed point mutations in 5D14, with 20 letter amino acid alphabet (Kumar & Biswas, 2019b).

Conformational flexibility is an important determinant for protein misfolding. This flexibility was measured in terms of the protein's propensity to fold into the native conformation through recurrent mutations of a specific residue-pair for a given foldability criterion (Baruah & Biswas, 2014a).

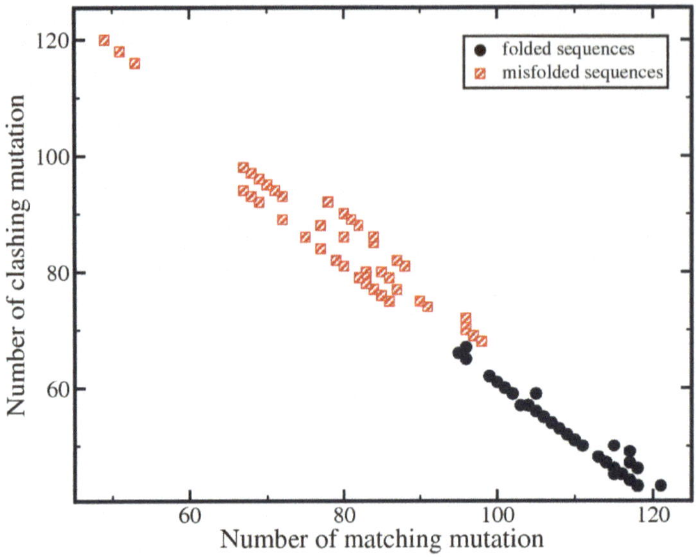

Fig. 9 Plot of the number of clashing mutations versus the number of matching mutations for folded and misfolded sequences. *Reproduced from A. Baruah, A. & Biswas, P. (2014).* Physical Chemistry Chemical Physics, 16, *13964–13973, with permission of the PCCP Owner Societies.*

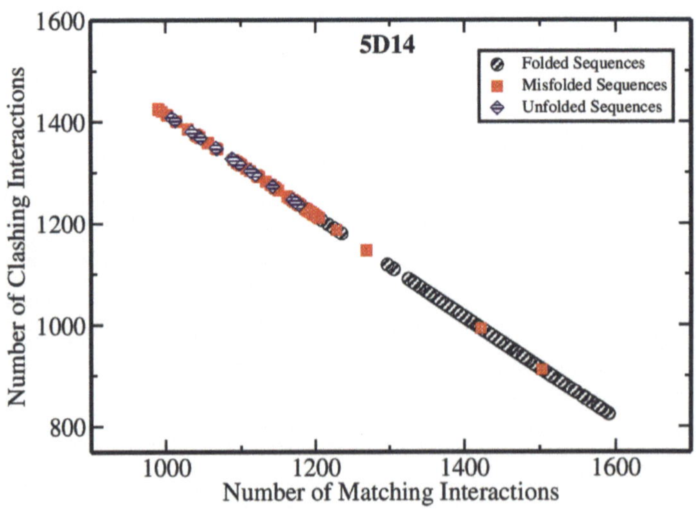

Fig. 10 Plot of the number of clashing mutations versus the number of matching mutations for the folded, misfolded and unfolded designed sequences of 5D14. *Reprinted from Kumar, A., & Biswas, P. (2019).* Proteins: Structure, Function, and Bioinformatics, 87, *760–773, with the permission of John Wiley & Sons, Inc.*

Conserved residues exhibited a higher preference of specific residues at the respective sites, which was reflected through a higher deviation from the random values of the residue probabilities. These residues may be mutated to alter the conformational stability and flexibility of any sequence. Experiments also support the fact that multiply mutated sequences, which retain all conserved residues contribute to a higher stability of a protein.

Correlated residue pairs significantly influence the stability and flexibility of a protein. A coarse-grained C_α chain backbone model was used to explore the effect of pair mutations in different globular proteins with varying secondary structures (Kumar & Biswas, 2019a). Metropolis MC simulations were employed to design sequences for the selected proteins, which were chosen for both correlated and random mutations. Selected sequences were subjected to i) mutations of most correlated residue pairs and ii) random pairwise mutations. The probability of misfolding recorded a significant increase for the correlated pair mutations as compared to the random mutations. as shown in Fig. 11. Pair mutations also affect the residue pairs that are in contact with the mutated residues, in addition to the most correlated ones. The results of this work may be used as an input for the targeted mutation experiments to identify conserved and coevolving residue pairs that are needed to preserve the stability and functional specificity of the protein.

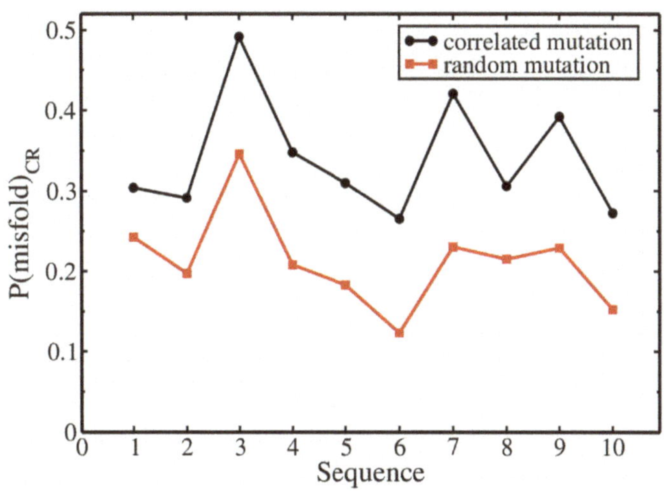

Fig. 11 Plot of the probability of misfolding for both correlated and random mutations. *Reprinted with permission from by Kumar, A., & Biswas, P. (2019).* The Journal of Physical Chemistry B, 123, *5069–5078. Copyright (2019) American Chemical Society.*

6. Conclusions

Protein misfolding originates from an aberrant folding behavior of proteins, which escape the stringent cellular quality control mechanism and fail to fold properly or retain the folded structure causing loss of function. However, the failure of this machinery causes an accumulation of the misfolded proteins, that include the toxic prefibrillar oligomers, which act as a template for formation of larger aggregates. These aggregates often crystallized into well-defined amyloid fibril structures. Misfolding of proteins usually trigger a wide spectrum of diseases, ranging from errors in metabolism to complex neurodegenerative and amyloidosis diseases. This chapter presents a theoretical and computational perspective of protein misfolding explained through different mechanisms, such as nucleation and growth, nucleated conformational conversion and the templated assembly for the formation of these pathogenic fibrils. In this context, various theoretical and simulation studies are discussed including the atomistic and coarse-grained models, that are suitable for proteins of varying sizes spanning a broad range of time scales. The insights provided by theory and simulations elucidate the underlying misfolding mechanism and quantitatively analyze different determinants of misfolding, that may be used to craft specific molecular targets for therapeutic interventions designed to combat the debilitating diseases.

Acknowledgments

The author gratefully acknowledges the financial support provided by DST-SERB, India (Project No.EMR/2016/006619) and DU-DST Purse Grant Phase-II (CD/2018/1638).

References

Abeln, S., Vendruscolo, M., Dobson, C. M., & Frenkel, D. (2014). A simple lattice model that captures protein folding, aggregation and amyloid formation. *PLoS One, 9*(1). e85185.

Alonso, D. O., DeArmond, S. J., Cohen, F. E., & Daggett, V. (2001). Mapping the early steps in the ph-induced conformational conversion of the prion protein. *Proceedings of the National Academy of Sciences, 98*(6), 2985–2989.

Andrews, M. N., & Winter, R. (2011). Comparing the structural properties of human and rat islet amyloid polypeptide by md computer simulations. *Biophysical Chemistry, 156*(1), 43–50.

Anfinsen, C. B. (1973). Principles that govern the folding of protein chains. *Science, 181*(4096), 223–230.

Auer, S., Dobson, C. M., Vendruscolo, M., & Maritan, A. (2008). Self-templated nucleation in peptide and protein aggregation. *Physical Review Letters, 101*(25), 258101.

Auer, S., Meersman, F., Dobson, C. M., & Vendruscolo, M. (2008). A generic mechanism of emergence of amyloid protofilaments from disordered oligomeric aggregates. *PLoS Computational Biology, 4*(11). e1000222.

Auer, S., Trovato, A., & Vendruscolo, M. (2009). A condensation-ordering mechanism in nanoparticle-catalyzed peptide aggregation. *PLoS Computational Biology, 5*(8). e1000458.

Bader, R., Bamford, R., Zurdo, J., Luisi, B. F., & Dobson, C. M. (2006). Probing the mechanism of amyloidogenesis through a tandem repeat of the pi3-sh3 domain suggests a generic model for protein aggregation and fibril formation. *Journal of Molecular Biology, 356*(1), 189–208.

Bai, Y., Sosnick, T. R., Mayne, L., & Englander, S. W. (1995). Protein folding intermediates: Native-state hydrogen exchange. *Science, 269*(5221), 192–197.

Baillod, P., Garrec, J., Colombo, M.-C., Tavernelli, I., & Rothlisberger, U. (2012). Enhanced sampling molecular dynamics identifies prpsc structures harboring a c-terminal β-core. *Biochemistry, 51*(49), 9891–9899.

Barducci, A., Bonomi, M., & Parrinello, M. (2011). Metadynamics. *Wiley Interdisciplinary Reviews: Computational Molecular Science, 1*(5), 826–843.

Barducci, A., Chelli, R., Procacci, P., Schettino, V., Gervasio, F. L., & Parrinello, M. (2006). Metadynamics simulation of prion protein: β-Structure stability and the early stages of misfolding. *Journal of the American Chemical Society, 128*(8), 2705–2710.

Baruah, A., Bhattacherjee, A., & Biswas, P. (2012). Role of conformational heterogeneity on protein misfolding. *Soft Matter, 8*(16), 4432–4440.

Baruah, A., & Biswas, P. (2014a). Designing sequences with varied flexibility and stability through pair mutations. *RSC Advances, 4*(16), 8031–8038.

Baruah, A., & Biswas, P. (2014b). The role of site-directed point mutations in protein misfolding. *Physical Chemistry Chemical Physics, 16*(27), 13964–13973.

Baumketner, A., Krone, M. G., & Shea, J.-E. (2008). Role of the familial Dutch mutation e22q in the folding and aggregation of the 15–28 fragment of the alzheimer amyloid-β protein. *Proceedings of the National Academy of Sciences, 105*(16), 6027–6032.

Bellesia, G., & Shea, J.-E. (2007). Self-assembly of β-sheet forming peptides into chiral fibrillar aggregates. *The Journal of Chemical Physics, 126*(24), 245104.

Bellesia, G., & Shea, J.-E. (2009). Effect of β-sheet propensity on peptide aggregation. *The Journal of Chemical Physics, 130*(14), 145103.

Bemporad, F., & Chiti, F. (2012). Protein misfolded oligomers: Experimental approaches, mechanism of formation, and structure-toxicity relationships. *Chemistry & Biology, 19*(3), 315–327.

Bhattacharjee, N., Rani, P., & Biswas, P. (2013). Capturing molten globule state of α-lactalbumin through constant ph molecular dynamics simulations. *The Journal of Chemical Physics, 138*(9), 095101.

Bhattacherjee, A., & Biswas, P. (2011). Designing misfolded proteins by energy landscaping. *The Journal of Physical Chemistry B, 115*(1), 113–119.

Bitan, G., Kirkitadze, M. D., Lomakin, A., Vollers, S. S., Benedek, G. B., & Teplow, D. B. (2003). Amyloid β-protein (aβ) assembly: Aβ40 and aβ42 oligomerize through distinct pathways. *Proceedings of the National Academy of Sciences, 100*(1), 330–335.

Bjorndahl, T. C., Zhou, G.-P., Liu, X., Perez-Pineiro, R., Semenchenko, V., Saleem, F., et al. (2011). Detailed biophysical characterization of the acid-induced prpc to prpβ conversion process. *Biochemistry, 50*(7), 1162–1173.

Borreguero, J. M., Urbanc, B., Lazo, N. D., Buldyrev, S. V., Teplow, D. B., & Stanley, H. E. (2005). Folding events in the 21-30 region of amyloid β-protein (aβ) studied in silico. *Proceedings of the National Academy of Sciences, 102*(17), 6015–6020.

Bryngelson, J. D., & Wolynes, P. G. (1987). Spin glasses and the statistical mechanics of protein folding. *Proceedings of the National Academy of Sciences, 84*(21), 7524–7528.

Campos, S. R., Machuqueiro, M., & Baptista, A. M. (2010). Constant-ph molecular dynamics simulations reveal a β-rich form of the human prion protein. *The Journal of Physical Chemistry B, 114*(39), 12692–12700.

Carballo-Pacheco, M., & Strodel, B. (2016). Advances in the simulation of protein aggregation at the atomistic scale. *The Journal of Physical Chemistry B, 120*(12), 2991–2999.

Chamberlain, A. K., Handel, T. M., & Marqusee, S. (1996). Detection of rare partially folded molecules in equilibrium with the native conformation of rnaseh. *Nature Structural Biology, 3*(9), 782–787.

Cheon, M., Chang, I., & Hall, C. K. (2010). Extending the prime model for protein aggregation to all 20 amino acids. *Proteins: Structure, Function, and Bioinformatics, 78*(14), 2950–2960.

Cohen, F. E., & Kelly, J. W. (2003). Therapeutic approaches to protein-misfolding diseases. *Nature, 426*(6968), 905–909.

Colombo, G., Meli, M., Morra, G., Gabizon, R., & Gasset, M. (2009). Methionine sulfoxides on prion protein helix-3 switch on the α-fold destabilization required for conversion. *PLoS One, 4*(1). e4296.

Coluzza, I., Muller, H., & Frenkel, D. (2003). Designing refoldable model molecules. *Physical Review E, 68*(4), 046703.

DeMarco, M. L., & Daggett, V. (2004). From conversion to aggregation: protofibril formation of the prion protein. *Proceedings of the National Academy of Sciences, 101*(8), 2293–2298.

Dill, K. A. (1990). Dominant forces in protein folding. *Biochemistry, 29*(31), 7133–7155.

Dobson, C. M. (2002). Protein-misfolding diseases: Getting out of shape. *Nature, 418*(6899), 729–730.

Dobson, C. M. (2003). Protein folding and misfolding. *Nature, 426*(6968), 884–890.

Dupuis, N. F., Wu, C., Shea, J.-E., & Bowers, M. T. (2009). Human islet amyloid polypeptide monomers form ordered β-hairpins: A possible direct amyloidogenic precursor. *Journal of the American Chemical Society, 131*(51), 18283–18292.

Dupuis, N. F., Wu, C., Shea, J.-E., & Bowers, M. T. (2011). The amyloid formation mechanism in human iapp: Dimers have β-strand monomer- monomer interfaces. *Journal of the American Chemical Society, 133*(19), 7240–7243.

Eigen, M. (1996). Prionics or the kinetic basis of prion diseases. *Biophysical Chemistry, 63*(1), A1–A18.

Englander, S. W. (2000). Protein folding intermediates and pathways studied by hydrogen exchange. *Annual Review of Biophysics and Biomolecular Structure, 29*(1), 213–238.

Esler, W. P., Stimson, E. R., Jennings, J. M., Vinters, H. V., Ghilardi, J. R., Lee, J. P., et al. (2000). Alzheimer's disease amyloid propagation by a template-dependent dock-lock mechanism. *Biochemistry, 39*(21), 6288–6295.

Fawzi, N. L., Phillips, A. H., Ruscio, J. Z., Doucleff, M., Wemmer, D. E., et al. (2008). Structure and dynamics of the aβ21–30 peptide from the interplay of nmr experiments and molecular simulations. *Journal of the American Chemical Society, 130*(19), 6145–6158.

Ferreiro, D. U., Hegler, J. A., Komives, E. A., & Wolynes, P. G. (2007). Localizing frustration in native proteins and protein assemblies. *Proceedings of the National Academy of Sciences, 104*(50), 19819–19824.

Flöck, D., Colacino, S., Colombo, G., & Di Nola, A. (2006). Misfolding of the amyloid β-protein: A molecular dynamics study. *Proteins: Structure, Function, and Bioinformatics, 62*(1), 183–192.

Frauenfelder, H., Sligar, S. G., & Wolynes, P. G. (1991). The energy landscapes and motions of proteins. *Science, 254*(5038), 1598–1603.

Friedman, R., & Caflisch, A. (2011). Surfactant effects on amyloid aggregation kinetics. *Journal of Molecular Biology, 414*(2), 303–312.

Govaerts, C., Wille, H., Prusiner, S. B., & Cohen, F. E. (2004). Evidence for assembly of prions with left-handed β-helices into trimers. *Proceedings of the National Academy of Sciences, 101*(22), 8342–8347.

Harper, J. D., & Lansbury, P. T., Jr. (1997). Models of amyloid seeding in alzheimer's disease and scrapie: Mechanistic truths and physiological consequences of the time-dependent solubility of amyloid proteins. *Annual Review of Biochemistry, 66*(1), 385–407.
Herczenik, E., & Gebbink, M. F. (2008). Molecular and cellular aspects of protein misfolding and disease. *The FASEB Journal, 22*(7), 2115–2133.
Hipp, M. S., Park, S.-H., & Hartl, F. U. (2014). Proteostasis impairment in protein-misfolding and-aggregation diseases. *Trends in Cell Biology, 24*(9), 506–514.
Ingólfsson, H. I., Lopez, C. A., Uusitalo, J. J., de Jong, D. H., Gopal, S. M., Periole, X., et al. (2014). The power of coarse graining in biomolecular simulations. *Wiley Interdisciplinary Reviews: Computational Molecular Science, 4*(3), 225–248.
Irbäck, A., Linnemann, N., Linse, B., Wallin, S., et al. (2013). Aggregate geometry in amyloid fibril nucleation. *Physical Review Letters, 110*(5), 058101.
Jackson, M. P., & Hewitt, E. W. (2016). Cellular proteostasis: Degradation of misfolded proteins by lysosomes. *Essays in Biochemistry, 60*(2), 173–180.
Karanicolas, J., & Brooks, C. L., III (2002). The origins of asymmetry in the folding transition states of protein l and protein g. *Protein Science, 11*(10), 2351–2361.
Kashchiev, D., & Auer, S. (2010). Nucleation of amyloid fibrils. *The Journal of Chemical Physics, 132*(21), 215101.
Khandogin, J., & Brooks, C. L. (2007). Linking folding with aggregation in alzheimer's β-amyloid peptides. *Proceedings of the National Academy of Sciences, 104*(43), 16880–16885.
Koike, M., Fukushi, J., Ichinohe, Y., Higashimae, N., Fujishiro, M., Sasaki, C., et al. (2010). Valosin-containing protein (vcp) in novel feedback machinery between abnormal protein accumulation and transcriptional suppression. *Journal of Biological Chemistry, 285*(28), 21736–21749.
Kumar, A., Baruah, A., & Biswas, P. (2017). Role of local and nonlocal interactions in folding and misfolding of globular proteins. *The Journal of Chemical Physics, 146*(6), 065102.
Kumar, A., & Biswas, P. (2019a). Effect of correlated pair mutations in protein misfolding. *The Journal of Physical Chemistry B, 123*(24), 5069–5078.
Kumar, A., & Biswas, P. (2019b). Effect of site-directed point mutations on protein misfolding: A simulation study. *Proteins: Structure, Function, and Bioinformatics, 87*, 760–773.
Labbadia, J., & Morimoto, R. I. (2015). The biology of proteostasis in aging and disease. *Annual Review of Biochemistry, 84*, 435–464.
Laio, A., & Parrinello, M. (2002). Escaping free-energy minima. *Proceedings of the National Academy of Sciences, 99*(20), 12562–12566.
Lam, A., Teplow, D., Stanley, H., & Urbanc, B. (2008). Effects of the arctic (e22 g) mutation on amyloid β-protein folding: Discrete molecular dynamics study. *Journal of the American Chemical Society, 130*(51), 17413–17422.
Lee, C., & Ham, S. (2011). Characterizing amyloid-beta protein misfolding from molecular dynamics simulations with explicit water. *Journal of Computational Chemistry, 32*(2), 349–355.
Lee, O.-S., Cho, V., & Schatz, G. C. (2012). Modeling the self-assembly of peptide amphiphiles into fibers using coarse-grained molecular dynamics. *Nano Letters, 12*(9), 4907–4913.
Li, H., Helling, R., Tang, C., & Wingreen, N. (1996). Emergence of preferred structures in a simple model of protein folding. *Science, 273*(5275), 666–669.
Li, J., Uversky, V. N., & Fink, A. L. (2001). Effect of familial Parkinson's disease point mutations a30p and a53t on the structural properties, aggregation, and fibrillation of human α-synuclein. *Biochemistry, 40*(38), 11604–11613.
Li, M. S., Klimov, D., Straub, J., & Thirumalai, D. (2008). Probing the mechanisms of fibril formation using lattice models. *The Journal of Chemical Physics, 129*(17), 175101.
Lindorff-Larsen, K., Piana, S., Dror, R. O., & Shaw, D. E. (2011). How fast-folding proteins fold. *Science, 334*(6055), 517–520.

Lipfert, J., Franklin, J., Wu, F., & Doniach, S. (2005). Protein misfolding and amyloid formation for the peptide gnnqqny from yeast prion protein sup35: Simulation by reaction path annealing. *Journal of Molecular Biology, 349*(3), 648–658.

Locker, C. R., & Hernandez, R. (2001). A minimalist model protein with multiple folding funnels. *Proceedings of the National Academy of Sciences, 98*(16), 9074–9079.

Locker, C. R., & Hernandez, R. (2004). Folding behavior of model proteins with weak energetic frustration. *The Journal of Chemical Physics, 120*(23), 11292–11303.

Lu, Y., Wei, G., & Derreumaux, P. (2012). Structural, thermodynamical, and dynamical properties of oligomers formed by the amyloid nnqq peptide: Insights from coarse-grained simulations. *The Journal of Chemical Physics, 137*(2), 025101.

Luca, S., Yau, W.-M., Leapman, R., & Tycko, R. (2007). Peptide conformation and supramolecular organization in amylin fibrils: Constraints from solid-state nmr. *Biochemistry, 46*(47), 13505–13522.

Ma, B., & Nussinov, R. (2002). Stabilities and conformations of alzheimer's β-amyloid peptide oligomers (aβ16–22, aβ16–35, and aβ10–35): Sequence effects. *Proceedings of the National Academy of Sciences, 99*(22), 14126–14131.

Machuqueiro, M., & Baptista, A. M. (2006). Constant-ph molecular dynamics with ionic strength effects: Protonation- conformation coupling in decalysine. *The Journal of Physical Chemistry B, 110*(6), 2927–2933.

Machuqueiro, M., & Baptista, A. M. (2009). Molecular dynamics at constant ph and reduction potential: Application to cytochrome c 3. *Journal of the American Chemical Society, 131*(35), 12586–12594.

Magno, A., Pellarin, R., & Caflisch, A. (2012). Mechanisms and kinetics of amyloid aggregation investigated by a phenomenological coarse-grained model. In *Computational modeling of biological systems* (pp. 191–214). Springer.

Massi, F., & Straub, J. E. (2001). Energy landscape theory for alzheimer's amyloid β-peptide fibril elongation. *Proteins: Structure, Function, and Bioinformatics, 42*(2), 217–229.

Matthes, D., Gapsys, V., & de Groot, B. L. (2012). Driving forces and structural determinants of steric zipper peptide oligomer formation elucidated by atomistic simulations. *Journal of Molecular Biology, 421*(2–3), 390–416.

Matysiak, S., & Clementi, C. (2006). Minimalist protein model as a diagnostic tool for misfolding and aggregation. *Journal of Molecular Biology, 363*(1), 297–308.

Mello, C. C., & Barrick, D. (2004). An experimentally determined protein folding energy landscape. *Proceedings of the National Academy of Sciences, 101*(39), 14102–14107.

Miller, C., Zerze, G. H., & Mittal, J. (2013). Molecular simulations indicate marked differences in the structure of amylin mutants, correlated with known aggregation propensity. *The Journal of Physical Chemistry B, 117*(50), 16066–16075.

Monticelli, L., Kandasamy, S. K., Periole, X., Larson, R. G., Tieleman, D. P., & Marrink, S.-J. (2008). The martini coarse-grained force field: Extension to proteins. *Journal of Chemical Theory and Computation, 4*(5), 819–834.

Morriss-Andrews, A., & Shea, J.-E. (2014). Simulations of protein aggregation: Insights from atomistic and coarse-grained models. *The Journal of Physical Chemistry Letters, 5*(11), 1899–1908.

Morriss-Andrews, A., & Shea, J.-E. (2015). Computational studies of protein aggregation: Methods and applications. *Annual Review of Physical Chemistry, 66*, 643–666.

Naganathan, A. N., & Muñoz, V. (2005). Scaling of folding times with protein size. *Journal of the American Chemical Society, 127*(2), 480–481.

Nasica-Labouze, J., Meli, M., Derreumaux, P., Colombo, G., & Mousseau, N. (2011). A multiscale approach to characterize the early aggregation steps of the amyloid-forming peptide gnnqqny from the yeast prion sup-35. *PLoS Computational Biology, 7*(5). e1002051.

Nasica-Labouze, J., Nguyen, P. H., Sterpone, F., Berthoumieu, O., Buchete, N.-V., Cote, S., et al. (2015). Amyloid β protein and alzheimers disease: When computer simulations complement experimental studies. *Chemical Reviews, 115*(9), 3518–3563.

Nguyen, H., Maier, J., Huang, H., Perrone, V., & Simmerling, C. (2014). Folding simulations for proteins with diverse topologies are accessible in days with a physics-based force field and implicit solvent. *Journal of the American Chemical Society, 136*(40), 13959–13962.

Nguyen, H. D., & Hall, C. K. (2004). Molecular dynamics simulations of spontaneous fibril formation by random-coil peptides. *Proceedings of the National Academy of Sciences, 101*(46), 16180–16185.

Nguyen, H. D., & Hall, C. K. (2006). Spontaneous fibril formation by polyalanines; discontinuous molecular dynamics simulations. *Journal of the American Chemical Society, 128*(6), 1890–1901.

Nguyen, P. H., Li, M. S., Stock, G., Straub, J. E., & Thirumalai, D. (2007). Monomer adds to preformed structured oligomers of aβ-peptides by a two-stage dock–lock mechanism. *Proceedings of the National Academy of Sciences, 104*(1), 111–116.

Nilsson, M. R., Driscoll, M., & Raleigh, D. P. (2002). Low levels of asparagine deamidation can have a dramatic effect on aggregation of amyloidogenic peptides: Implications for the study of amyloid formation. *Protein Science, 11*(2), 342–349.

Nishino, M., Sugita, Y., Yoda, T., & Okamoto, Y. (2005). Structures of a peptide fragment of β2-microglobulin studied by replica-exchange molecular dynamics simulations—towards the understanding of the mechanism of amyloid formation. *FEBS Letters, 579*(24), 5425–5429.

Otzen, D. E., Miron, S., Akke, M., & Oliveberg, M. (2004). Transient aggregation and stable dimerization induced by introducing an alzheimer sequence into a water-soluble protein. *Biochemistry, 43*(41), 12964–12978.

Otzen, D. E., & Oliveberg, M. (1999). Salt-induced detour through compact regions of the protein folding landscape. *Proceedings of the National Academy of Sciences, 96*(21), 11746–11751.

Pellarin, R., Guarnera, E., & Caflisch, A. (2007). Pathways and intermediates of amyloid fibril formation. *Journal of Molecular Biology, 374*(4), 917–924.

Phelps, E. M., & Hall, C. K. (2012). Structural transitions and oligomerization along polyalanine fibril formation pathways from computer simulations. *Proteins: Structure, Function, and Bioinformatics, 80*(6), 1582–1597.

Qiao, Q., Bowman, G. R., & Huang, X. (2013). Dynamics of an intrinsically disordered protein reveal metastable conformations that potentially seed aggregation. *Journal of the American Chemical Society, 135*(43), 16092–16101.

Reddy, A. S., Wang, L., Singh, S., Ling, Y. L., Buchanan, L., Zanni, M. T., et al. (2010). Stable and metastable states of human amylin in solution. *Biophysical Journal, 99*(7), 2208–2216.

Rosenman, D. J., Connors, C. R., Chen, W., Wang, C., & García, A. E. (2013). Aβ monomers transiently sample oligomer and fibril-like configurations: Ensemble characterization using a combined md/nmr approach. *Journal of Molecular Biology, 425*(18), 3338–3359.

Santini, S., Mousseau, N., & Derreumaux, P. (2004). In silico assembly of alzheimer's aβ16-22 peptide into β-sheets. *Journal of the American Chemical Society, 126*(37), 11509–11516.

Scherzinger, E., Sittler, A., Schweiger, K., Heiser, V., Lurz, R., Hasenbank, R., et al. (1999). Self-assembly of polyglutamine-containing huntingtin fragments into amyloid-like fibrils: Implications for huntingtons disease pathology. *Proceedings of the National Academy of Sciences, 96*(8), 4604–4609.

Schor, M., Vreede, J., & Bolhuis, P. G. (2012). Elucidating the locking mechanism of peptides onto growing amyloid fibrils through transition path sampling. *Biophysical Journal, 103*(6), 1296–1304.

Selkoe, D. J. (1999). Translating cell biology into therapeutic advances in alzheimer's disease. *Nature, 399*(6738), A23–A31.

Serio, T. R., Cashikar, A. G., Kowal, A. S., Sawicki, G. J., Moslehi, J. J., Serpell, L., et al. (2000). Nucleated conformational conversion and the replication of conformational information by a prion determinant. *Science, 289*(5483), 1317–1321.

Sgourakis, N. G., Yan, Y., McCallum, S. A., Wang, C., & Garcia, A. E. (2007). The alzheimers peptides aβ40 and 42 adopt distinct conformations in water: A combined md/nmr study. *Journal of Molecular Biology, 368*(5), 1448–1457.

Shakhnovich, E., Farztdinov, G., Gutin, A., & Karplus, M. (1991). Protein folding bottlenecks: A lattice Monte Carlo simulation. *Physical Review Letters, 67*(12), 1665–1668.

Shakhnovich, E., & Gutin, A. (1990). Implications of thermodynamics of protein folding for evolution of primary sequences. *Nature, 346*(6286), 773–775.

Shaw, D. E., Deneroff, M. M., Dror, R. O., Kuskin, J. S., Larson, R. H., Salmon, J. K., et al. (2007). Anton, a special-purpose machine for molecular dynamics simulation. *ACM SIGARCH - Computer Architecture News, 35*(2), 1–12.

Shaw, D. E., Maragakis, P., Lindorff-Larsen, K., Piana, S., Dror, R. O., Eastwood, M. P., et al. (2010). Atomic-level characterization of the structural dynamics of proteins. *Science, 330*(6002), 341–346.

Sørensen, J., Periole, X., Skeby, K. K., Marrink, S.-J., & Schiøtt, B. (2011). Protofibrillar assembly toward the formation of amyloid fibrils. *The Journal of Physical Chemistry Letters, 2*(19), 2385–2390.

Soto, C. (2012). Transmissible proteins: Expanding the prion heresy. *Cell, 149*(5), 968–977.

Soto, C., Estrada, L., & Castilla, J. (2006). Amyloids, prions and the inherent infectious nature of misfolded protein aggregates. *Trends in Biochemical Sciences, 31*(3), 150–155.

Sterpone, F., Melchionna, S., Tuffery, P., Pasquali, S., Mousseau, N., Cragnolini, T., et al. (2014). The opep protein model: From single molecules, amyloid formation, crowding and hydrodynamics to dna/rna systems. *Chemical Society Reviews, 43*(13), 4871–4893.

Stoycheva, A. D., Brooks, C. L., III, & Onuchic, J. N. (2004). Gatekeepers in the ribosomal protein s6: Thermodynamics, kinetics, and folding pathways revealed by a minimalist protein model. *Journal of Molecular Biology, 340*(3), 571–585.

Straub, J. E., & Thirumalai, D. (2011). Toward a molecular theory of early and late events in monomer to amyloid fibril formation. *Annual Review of Physical Chemistry, 62*, 437–463.

Sugita, Y., & Okamoto, Y. (1999). Replica-exchange molecular dynamics method for protein folding. *Chemical Physics Letters, 314*(1–2), 141–151.

Tang, K. E., & Dill, K. A. (1999). How experiments see fluctuations of native proteins: Perspective from an exact model. *International Journal of Quantum Chemistry, 75*(3), 147–164.

Thirumalai, D., Reddy, G., & Straub, J. E. (2011). Role of water in protein aggregation and amyloid polymorphism. *Accounts of Chemical Research, 45*(1), 83–92.

Tozzini, V. (2005). Coarse-grained models for proteins. *Current Opinion in Structural Biology, 15*(2), 144–150.

Tran, L., & Ha-Duong, T. (2015). Exploring the alzheimer amyloid-β peptide conformational ensemble: A review of molecular dynamics approaches. *Peptides, 69*, 86–91.

Tsai, H.-H. G., Reches, M., Tsai, C.-J., Gunasekaran, K., Gazit, E., & Nussinov, R. (2005). Energy landscape of amyloidogenic peptide oligomerization by parallel-tempering molecular dynamics simulation: Significant role of asn ladder. *Proceedings of the National Academy of Sciences, 102*(23), 8174–8179.

Vácha, R., & Frenkel, D. (2011). Relation between molecular shape and the morphology of self-assembling aggregates: A simulation study. *Biophysical Journal, 101*(6), 1432–1439.

van der Kamp, M. W., & Daggett, V. (2011). Molecular dynamics as an approach to study prion protein misfolding and the effect of pathogenic mutations. In *Prion proteins* (pp. 169–197). Springer.

Vila-Vicosa, D., Campos, S. R., Baptista, A. M., & Machuqueiro, M. (2012). Reversibility of prion misfolding: Insights from constant-ph molecular dynamics simulations. *The Journal of Physical Chemistry B, 116*(30), 8812–8821.

Voelz, V. A., Bowman, G. R., Beauchamp, K., & Pande, V. S. (2010). Molecular simulation of ab initio protein folding for a millisecond folder ntl9 (1- 39). *Journal of the American Chemical Society, 132*(5), 1526–1528.

Volkman, B. F., Lipson, D., Wemmer, D. E., & Kern, D. (2001). Two-state allosteric behavior in a single-domain signaling protein. *Science, 291*(5512), 2429–2433.

Wagoner, V. A., Cheon, M., Chang, I., & Hall, C. K. (2011). Computer simulation study of amyloid fibril formation by palindromic sequences in prion peptides. *Proteins: Structure, Function, and Bioinformatics, 79*(7), 2132–2145.

Wagoner, V. A., Cheon, M., Chang, I., & Hall, C. K. (2012). Fibrillization propensity for short designed hexapeptides predicted by computer simulation. *Journal of Molecular Biology, 416*(4), 598–609.

Wang, Y., & Voth, G. A. (2010). Molecular dynamics simulations of polyglutamine aggregation using solvent-free multiscale coarse-grained models. *The Journal of Physical Chemistry B, 114*(26), 8735–8743.

Wei, L., Jiang, P., Xu, W., Li, H., Zhang, H., Yan, L., et al. (2011). The molecular basis of distinct aggregation pathways of islet amyloid polypeptide. *Journal of Biological Chemistry, 286*(8), 6291–6300.

Wille, H., Bian, W., McDonald, M., Kendall, A., Colby, D. W., Bloch, L., et al. (2009). Natural and synthetic prion structure from x-ray fiber diffraction. *Proceedings of the National Academy of Sciences, 106*(40), 16990–16995.

Wille, H., Michelitsch, M. D., Guénebaut, V., Supattapone, S., Serban, A., Cohen, F. E., et al. (2002). Structural studies of the scrapie prion protein by electron crystallography. *Proceedings of the National Academy of Sciences, 99*(6), 3563–3568.

Williams, D. M., & Pukala, T. L. (2013). Novel insights into protein misfolding diseases revealed by ion mobility-mass spectrometry. *Mass Spectrometry Reviews, 32*(3), 169–187.

Wolynes, P. G., Onuchic, J. N., Thirumalai, D., et al. (1995). Navigating the folding routes. *Science*, 1619–1620.

Xi, W., Li, W., & Wang, W. (2012). Template induced conformational change of amyloid-β monomer. *The Journal of Physical Chemistry B, 116*(25), 7398–7405.

Yang, D., & Kay, L. E. (1996). Contributions to conformational entropy arising from bond vector fluctuations measured from nmr-derived order parameters: Application to protein folding. *Journal of Molecular Biology, 263*(2), 369–382.

Yoon, J., Park, J., Jang, S., Lee, K., & Shin, S. (2008). Conformational characteristics of unstructured peptides: A-synuclein. *Journal of Biomolecular Structure and Dynamics, 25*(5), 505–515.

Yue, K., Fiebig, K. M., Thomas, P. D., Chan, H. S., Shakhnovich, E. I., & Dill, K. A. (1995). A test of lattice protein folding algorithms. *Proceedings of the National Academy of Sciences, 92*(1), 325–329.

Zhang, J., & Muthukumar, M. (2009). Simulations of nucleation and elongation of amyloid fibrils. *The Journal of Chemical Physics, 130*(3), 035102.

CHAPTER TWO

Combining molecular dynamics simulations and experimental analyses in protein misfolding

Holger Wille[a,b,c,*], Lyudmyla Dorosh[d], Sara Amidian[a,b], Gerold Schmitt-Ulms[e] and Maria Stepanova[d,*]
[a]Department of Biochemistry, University of Alberta, Edmonton, Canada
[b]Centre for Prions and Protein Folding Diseases, University of Alberta, Edmonton, Canada
[c]Neuroscience and Mental Health Institute, University of Alberta, Edmonton, Canada
[d]Department of Electrical and Computer Engineering, University of Alberta, Edmonton, Canada
[e]Laboratory Medicine and Pathobiology, University of Toronto, Toronto, Canada
*Corresponding author: E-mails: wille@ualberta.ca, ms1@ualberta.ca

Contents

Advances in Protein Chemistry and Structural Biology, Volume 118
ISSN 1876-1623
https://doi.org/10.1016/bs.apcsb.2019.10.001
© 2020 Elsevier Inc.
All rights reserved.

Abstract

The fold of a protein determines its function and its misfolding can result in loss-of-function defects. In addition, for certain proteins their misfolding can lead to gain-of-function toxicities resulting in protein misfolding diseases such as Alzheimer's, Parkinson's, or the prion diseases. In all of these diseases one or more proteins misfold and aggregate into disease-specific assemblies, often in the form of fibrillar amyloid deposits. Most, if not all, protein misfolding diseases share a fundamental molecular mechanism that governs the misfolding and subsequent aggregation. A wide variety of experimental methods have contributed to our knowledge about misfolded protein aggregates, some of which are briefly described in this review. The misfolding mechanism itself is difficult to investigate, as the necessary timescale and resolution of the misfolding events often lie outside of the observable parameter space. Molecular dynamics simulations fill this gap by virtue of their intrinsic, molecular perspective and the step-by-step iterative process that forms the basis of the simulations. This review focuses on molecular dynamics simulations and how they combine with experimental analyses to provide detailed insights into protein misfolding and the ensuing diseases.

1. Introduction

The diverse biological functions of proteins depend on their unique ability to adopt specific three-dimensional configurations (conformations). The folding of most proteins into a well-defined three-dimensional native structure is deemed a fundamental attribute of living systems (Dill & MacCallum, 2012; Dobson, 2003; Kendrew et al., 1958). The native conformations of proteins determine their functions, biological activity, stability, and interactions with other molecules. However under conditions that are not yet fully understood, protein chains may misfold into abnormal conformations. Such misfolded constructs are prone to aggregate into large amyloid fibrils dominated by cross-β structure, which may accumulate in the body (Chiti & Dobson, 2006; Dobson, 2004; Jucker & Walker, 2013; Tanaka & Komi, 2015). The realization of this process has resulted in the discovery of a previously unknown mechanism of disease in humans and animals (Prusiner, 1998; Stahl & Prusiner, 1991). Protein folding diseases were initially described as neurodegenerative disorders associated with misfolding and aggregation of the prion protein (PrP) in mammals; however, subsequently this concept has evolved into a significantly broader vision (Knowles, Vendruscolo, & Dobson, 2014, 2015; Aguzzi & Lakkaraju, 2016; Jucker & Walker, 2013; Prusiner, 2012; Walker & Jucker, 2015). To date, approximately 50 disorders have been associated with the misfolding of various proteins, and this number continues to increase (Chiti & Dobson, 2006, 2017; Iadanza, Jackson, Hewitt, Ranson, & Radford, 2018; Invernizzi, Papaleo, Sabate, & Ventura, 2012). Out of these, only a few diseases are neurodegenerative. Perhaps the best-known examples of such include the transmissible spongiform encephalopathies (TSEs) associated with misfolded PrPs (also known as prion diseases); Alzheimer's disease (AD) with amyloid-β peptide (Aβ) and tau protein; and Parkinson's disease (PD) with α-synuclein. The respective fibrillar aggregates are found predominantly in the central nervous system. However, the many other confirmed misfolding diseases involve conditions other than neurodegeneration, with the corresponding deposits accumulating in various tissues and organs including heart, liver, and kidney (Chiti & Dobson, 2017; Khan, Zakariya, & Khan, 2018). For instance, misfolding of islet amyloid polypeptide (IAPP) is implicated in type II diabetes, with fibrillar deposits accumulating in the pancreas.

2. Basic concepts of protein misfolding and aggregation

The general attributes of misfolding diseases vary broadly. However, despite the variety of their pathogeneses, most, if not all, misfolding diseases share a fundamental molecular mechanism. The resulting misfolding products are being studied using a wide variety of experimental methods, some of which are briefly introduced in this review. The misfolding mechanism itself is difficult to study with most experimental approaches, as the necessary timescale and resolution of the misfolding events often lie outside of the (easily) observable parameter space. Molecular dynamics simulations can fill this gap by virtue of their intrinsic, molecular perspective and the step-by-step iterative process that forms the basis of the simulations.

The initial spontaneous misfolding, or introduction of infectious proteinaceous particles commonly termed prions, leads to a cascade of self-replicating misfolding events often producing β-sheet-rich amyloid deposits (Prusiner, 2012; Hartl, 2017; Jucker & Walker, 2013; Knowles, Vendruscolo, & Dobson, 2015; Riek & Eisenberg, 2016; Walker & Jucker, 2015; Soto & Pritzkow, 2018). The self-replicating misfolding process has been investigated intensely over the last decade. It has been found that β-sheet-rich amyloid species, once they have nucleated and started growing, provide molecular templates for further misfolding. Thereby catalyzing the formation of additional copies through a positive-feedback reaction loop,

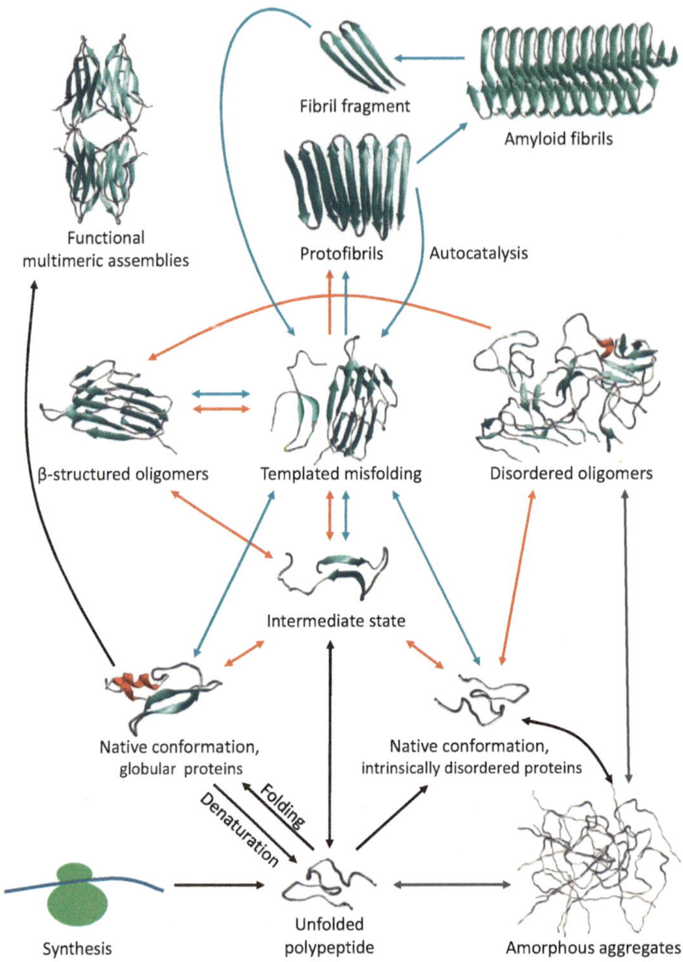

Fig. 1 Misfolding and aggregation pathways in globular and intrinsically disordered proteins. *Red arrows* mark pathways of homogeneous primary nucleation, which initiates the amyloidogenic process (Michaels et al., 2018). For globular proteins, primary nucleation is believed to involve adoption of an aggregation-competent intermediate state of unknown structure, followed by a formation of β-sheet-rich oligomeric aggregates. Intrinsically disordered proteins tend to aggregate into largely disordered oligomers with some transient secondary structure, which subsequently undergo a conversion into more ordered, β-sheet-rich entities (Breydo & Uversky, 2015; Cremades & Dobson, 2018). β-sheet-rich oligomers of either origin possess the ability of recruiting natively folded proteins or intermediate conformers into the process of templated misfolding (Hartl, 2017; Riek & Eisenberg, 2016; Soto & Pritzkow, 2018) resulting in the growth of amyloid-like proto-fibrils. Upon reaching a critical concentration, the accumulating proto-fibrils and fibril fragments give rise to an autocatalytic self-replication process known as secondary nucleation (Michaels et al., 2018; Törnquist et al., 2018),

also known as the secondary nucleation mechanism (Michaels et al., 2018; Törnquist et al., 2018; Ke et al., 2017; Nagel-Steger, Owen, & Strodel, 2016; Knowles et al., 2014; Riek & Eisenberg, 2016), see also Fig. 1. Experiments indicate that the initial multimeric species (oligomers) that arise at early stages of the spontaneous aggregation, tend to adopt a largely unstructured morphology, at least for intrinsically disordered proteins (IDPs) such as Aβ peptide, IAPP, or α-synuclein (Breydo & Uversky, 2015; Cremades & Dobson, 2018; Luo, Wärmländer, Gräslund, & Abrahams, 2014). Oligomeric species with pronounced β structure seem to nucleate out of the less structured oligomers (Breydo & Uversky, 2015; Cremades & Dobson, 2018). For globular proteins, or those containing a globular domain (Breydo & Uversky, 2015; Bemporad & Chiti, 2012; Chiti & Dobson, 2009; Requena & Wille, 2014; Flores-Fernández, Rathod, & Wille, 2018; Uversky & Fink, 2004), a prevailing expectation is that the misfolding is initiated by the proteins adopting an intermediate, partially folded conformation. However it is not entirely clear yet, what the specific structure of such intermediate state is, and what molecular events the intermediate state provokes. Advancing our understanding of this initial conformational change is particularly important, since the toxicity is attributed primarily to the relatively small, prefibrillar oligomers rather than mature fibrils (Cremades & Dobson, 2018; Soto & Pritzkow, 2018). The toxicity of the oligomers is hypothetically linked to the exposure of hydrophobic groups and unpaired β-strands at their surfaces (Balchin, Hayer-Hartl, & Hartl, 2016; Bemporad & Chiti, 2012; Breydo & Uversky, 2015; Cremades & Dobson, 2018; Eisele et al., 2015). However, the specific structure and morphology of the oligomers and other species arising in the process of misfolding has yet to be determined. Fig. 1 summarizes our current understanding of the principal misfolding and aggregation of globular and intrinsically disordered proteins.

Molecular-level factors determining the propensity of proteins to misfold have been investigated extensively over the recent years. Although no universal one-for-all amyloidogenic amino acid sequence or structure motif of the native folds could be determined, per-residue indicators of

as shown by *blue arrows*. This includes nucleation of β-sheet rich oligomers from natively folded monomers, mediated by existing fibrillar species. Double-ended arrows reflect the dynamic nature of early oligomeric species, and the reversibility of most conversion events (Törnquist et al., 2018). *Black arrows* show examples of off-pathway conversions.

"misfoldability" have been identified (Belli, Ramazzotti, & Chiti, 2011; Chiti & Dobson, 2017; Meric, Robinson, & Roberts, 2017; Monsellier, Ramazzotti, Taddei, & Chiti, 2008; Trainor, Broom, & Meiering, 2017). High propensity to misfold and aggregate formation have been linked, in the first place, to low net charge, high hydrophobicity, and strong β-propensity of residues. Predicted scores of the misfolding/aggregation propensity agree well with available experiments for disease-related proteins (Chiti & Dobson, 2017). However, extensive screening of the proteome for aggregation scores has indicated that most proteins, irrespective of their implication in a disease, contain at least one aggregation-prone region, and many contain several ones (Goldschmidt, Teng, Riek, & Eisenberg, 2010; Monsellier et al., 2008; Rousseau, Serrano, & Schymkowitz, 2006; Trainor et al., 2017). Furthermore, evidence has emerged that many if not all proteins, with or without known link to a disease, are able to misfold and aggregate into β-sheet rich fibrils, simply as a result of their backbone's propensity to form hydrogen bonds (Braselmann, Chaney, & Clark, 2013; Knowles et al., 2014, 2015). The vision of protein folding as a diffusional search on a rough free energy landscape has led to a novel paradigm of the amyloid state as a generic conformation, which might be even more stable than functional native folds (Gershenson, Gierasch, Pastore, & Radford, 2014; Hartl, 2017; Knowles et al., 2014). From this standpoint, the native fold of a protein emerges as one of several alternate conformations (Braselmann et al., 2013; Gershenson et al., 2014) maintained by a subtle balance of intra- and inter-molecular interactions (hydrogen bonds, electrostatic interactions, van der Waals forces, hydrophobic interactions) (Dill & MacCallum, 2012; Pace, Scholtz, & Grimsley, 2014; Zhou & Pang, 2018), kinetic factors (Ciryam, Kundra, Morimoto, Dobson, & Vendruscolo, 2015; Gsponer & Babu, 2012; Tartaglia, Pechmann, Dobson, & Vendruscolo, 2007), and the cellular proteostasis machinery (Aguzzi & Lakkaraju, 2016; Balch, Morimoto, Dillin, & Kelly, 2008; Balchin et al., 2016). This understanding is consistent with the tendency of misfolding diseases to emerge at advanced age, when the efficiency of the proteostasis machinery declines (Klaips, Jayaraj, & Hartl, 2017; Labbadia & Morimoto, 2015; Stroo, Koopman, Nollen, & Mata-Cabana, 2017).

This paradigm offers a general framework for anti-misfolding strategies by inducing subtle changes in a protein's structure (via mutations) or in the proteostatic systems (through ligands, chaperones, etc.) (Aguzzi & Lakkaraju, 2016; Balchin et al., 2016; Calamini, Silva, Madoux, Hutt, & et. al, 2011; Garcia-Seisdedos, Empereur-Mot, Elad, & Levy, 2017;

Horwich, 2014; Labbadia & Morimoto, 2015; Monsellier et al., 2008). Anti-misfolding therapies, presently under intense development, strive to discover factors that could stabilize native conformations of proteins, de-toxify products of misfolding, or stimulate cellular defenses (Ankarcrona et al., 2016; Cremades & Dobson, 2018; Eisele et al., 2015; Iadanza, Jackson et al., 2018; Sweeney et al., 2017). However, in most cases a disease-modifying therapy still has to be discovered. For this to be accomplished, a detailed, molecular-level understanding of the nature, structure, and dynamics of the misfolding process needs to be achieved in the first place. Clearly, for therapies to rationally target the very origin of misfolding diseases, pathways of protein misfolding, as well as the structure of toxic prefibrillar oligomers and amyloid fibrils require a detailed exploration.

3. Biophysical analyses of protein misfolding

Due to advances in experimental methods, researchers have gained access to powerful tools for characterizing the structures and dynamics of proteins in atomic details (Bedem & Fraser, 2015; Dill & MacCallum, 2012). X-ray diffraction (XRD) crystallography and solution-state nuclear magnetic resonance (NMR) spectroscopy in particular have been instrumental in determining native structures of globular proteins such as recombinant PrP (Moore, Taubner, & Priola, S.A., 2009; Riesner, 2003). Unlike native globular folds, products of misfolding have been difficult to study by classic structural biology methods. The insolubility of amyloid fibrils and the scarcity of well-ordered 3D crystalline forms has hindered high-resolution imaging of fibrils until the recent years, when the advent of solid-state NMR spectroscopy (ssNMR) and cryo-electron microscopy cryo-(EM) has made structural analyses of amyloid fibrils with near-atomic resolution possible (Eisenberg & Sawaya, 2017; Fernández, Rathod, & Wille, 2018; Iadanza, Jackson et al., 2018; Riek & Eisenberg, 2016). Small oligomers that arise at early stages of misfolding are even more complex to investigate. Although putatively soluble, the oligomers by their nature are a heterogeneous ensemble of short-lived species, often present in low quantities in comparison with the initial (native) and final (fibrillar) compounds. High-resolution characterization would require each oligomeric species to be selectively enriched and stabilized against further transformations (Cremades & Dobson, 2018; Nagel-Steger et al., 2016). Acknowledging this as an ultimate but demanding objective, most studies combine multiple complementary techniques such as dynamic light scattering (DLS), circular dichroism (CD) spectroscopy, Fourier transform infrared (FTIR) spectroscopy, single-molecule Förster resonance energy transfer (FRET), atomic force microscopy (AFM), and X-ray, NMR, and EM based methods, to obtain accessible structural information. Here we outline the applications of selected biophysical methods, focusing in particular on structural characterization of misfolded intermediates, oligomers, and amyloid fibrils.

3.1 X-ray diffraction and small-angle X-ray scattering

XRD crystallography has been widely used for determining atomic structures of proteins for decades (Dill & MacCallum, 2012; Shi, 2014). Many globular proteins form crystalline structures, which cause a diffraction of incident X-rays into specific directions. The resulting diffraction patterns can be used to recover the electron-density distribution in the sample. The method has been applied extensively to identify high-resolution tertiary structures of globular amyloidogenic proteins such as PrP (Moore, Taubner, & Priola, S.A., 2009; Riesner, 2003), and is deemed the most accurate technique for this purpose. However, high quality 3D crystals are required to realize these advantages, limiting the analyses of native PrP to recombinantly produced protein variants. In application to amyloid fibrils, the required crystalline material could be obtained only for small peptides, which crystallize into well-ordered nano- or micro-crystals (Balbirnie, Grothe, & Eisenberg, 2001; Eisenberg & Sawaya, 2017; Sawaya et al., 2007). More than 100 high-resolution amyloidogenic structures have been obtained from XRD of such samples (Eisenberg & Sawaya, 2017). These structures commonly encompass a motif known as steric zipper, which contains tightly packed β-sheets stabilized by hydrogen bonds (Eisenberg & Sawaya, 2017; Riek & Eisenberg, 2016). These results have played an important role in elucidating the structural origins of amyloid stability; however, because only short fragments of proteins have been crystalized, relevance to actual full-length misfolded species is not immediately guaranteed.

These limitations have prompted X-ray fiber diffraction analyses, where amyloid fibrils themselves are used to obtain XRD patterns (Eisenberg & Sawaya, 2017; Fernández, Rathod, & Wille, 2018; Iadanza, Jackson et al., 2018; Riek & Eisenberg, 2016). The diffraction patterns contain strong meridional signals corresponding to a distance of 4.7–4.8 Å, which is interpreted as the spacing between β-strands oriented perpendicularly to the fiber axis (Fitzpatrick et al., 2017, 2013; Gremer et al., 2017; Sunde et al., 1997; Wan et al., 2015; Wille et al., 2009). Such structural motif, commonly referred to as the cross-β structure, has been observed in all known amyloid fibrils (Iadanza, Jackson et al., 2018). The fiber diffraction analyses are limited by the availability of well-aligned fibril specimens.

Complementary to X-ray diffraction crystallography, small-angle X-ray scattering (SAXS) provides a low-resolution measure of sizes and shapes in a wide range of nanoscale dimensions (1−100 nm) without the need for sample orientation. In a typical SAXS experiment, an X-ray beam undergoes scattering in a solution, powder, or other isotropic material, and the intensity of scattered X-rays is recorded as a function of the scattering angle (Bizien et al., 2016; Pillon & Guarné, 2017). Using dedicated software packages, structural information such as the radius of gyration or surface-to-volume ratio can be deduced from these data. Reconstruction of hypothetical coarse-grained structures of molecules or supramolecular aggregates is also possible. In amyloidogenic conversion studies, SAXS is used primarily to evaluate the size and shape of oligomers in solution (Lorenzen et al. 2014; Amenitsch, Benetti, Ramos, Legname, & Requena, 2013; Redecke et al., 2007; Rezaei et al., 2005).

3.2 Nuclear magnetic resonance analyses

Nuclear magnetic resonance (NMR) methods are unique in the diversity of site-specific descriptors for molecular structure and dynamics that they can provide (Habenstein & Loquet, 2016; Kleckner & Foster, 2011; Ziarek, Baptista, & Wagner, 2018). In brief, these methods exploit perturbations of a nucleus' magnetic moment (spin) by electromagnetic waves in radiofrequency (RF) regimes. The perturbations of the spins are accompanied by changes in macroscopic magnetization of the sample, which are recorded in a form of an alternating current (known as the free induction decay) induced in a receiver coil (Kleckner & Foster, 2011). Such NMR signal can be observed only for nuclei that possess a net spin. In biological samples, the most abundant naturally occurring isotope is hydrogen ^{1}H with a spin of ½. It is also possible to label the samples with deuterium ^{2}H (spin 1), or isotopes ^{13}C or ^{15}N (spin ½). For an excitation of the nucleus' spin to occur, the nucleus needs to be placed in a magnetic field, and an RF wave with a resonance frequency is required that matches the nucleus' Larmor precessional frequency. The latter depends on the applied magnetic field, magnetic properties of the nucleus, status of the electron shell around it, and the chemical environment. In NMR experiments, the resonance is achieved either by varying the strength of the applied magnetic field while keeping the RF frequency constant, or by varying the frequency of RF pulses while keeping the applied magnetic field constant. Once a resonance excitation of a certain population of nuclei has occurred, the resulting NMR signal is acquired and Fourier-transformed into a spectral line with a specific position (commonly referred to as the chemical shift), intensity, and linewidth. Because these quantities depend on the physico-chemical environment of the corresponding nuclei, the NMR signal's Fourier spectra carry a rich pool of site-specific information regarding chemical bonding, secondary structure, exposure to solvents and co-solvents, and other characteristics. In addition to the various structural data, information about molecular motions (dynamics) also can be obtained including relaxation behaviors, correlation times, and chemical exchange rates (Kleckner & Foster, 2011). Using dedicated software, molecular models can be generated that satisfy the entire pool of collected NMR data, commonly denoted as constraints. Acquisition and processing of NMR data are more straightforward when the molecules are in solution and tumble isotropically (Ziarek et al., 2018); however, solid-state NMR techniques are also available (Habenstein & Loquet, 2016). This makes NMR capable of probing all protein folds from native conformations and early intermediate states to oligomers, protofibrillar species, and fibrils.

Numerous studies have employed solution NMR to probe native folds of globular proteins, complementing XRD studies. As an example, Fig. 2A shows a 3D structure of monomeric native PrP derived from solution NMR data (Zahn et al., 2000). The structure contains a globular domain composed of three α-helices H1, H2, and H3, an antiparallel β-sheet S1–S2, and an intrinsically disordered N-terminal tail. Importantly, solution NMR also allows characterizing early misfolding intermediates and soluble oligomers, most of which are not amenable to X-ray crystallography (Ahmed & Melacini, 2018; Bemporad & Chiti, 2012; Mercer et al., 2018; Redfield, 2004; Sengupta & Udgaonkar, 2018). NMR characterization of a PrP solution with increasing concentrations of urea has indicated that the β-sheet S1–S2 might be the most vulnerable to unfolding (Julien, Chatterjee, Thiessen, Graether, & Sykes, 2009). Recently, extensive changes in the dynamics of early PrP oligomers in comparison to monomeric recombinant mouse PrP have been observed from solution NMR experiments (Glaves, Ladner-Keay, Bjorndahl, Wishart, & Sykes, 2018). According to this study, the N-terminal region of moPrP between residues 90 and 114 remains mobile in the oligomeric state. In contrast, the mobility is pronouncedly decreased between residues ~127 and 225, with the exception of several localized spots of high mobility, suggesting that these regions might be involved in early oligomeric interactions (Glaves et al., 2018).

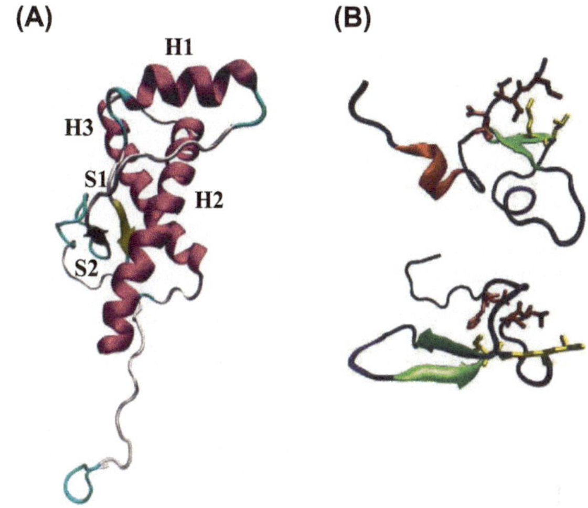

Fig. 2 (A) 3D structure of human $PrP_{88\text{-}230}$ determined from solution NMR data (Zahn et al., 2000). The structure contains a globular domain composed of three α-helices H1, H2, and H3, an antiparallel β-sheet S1−S2, and an intrinsically disordered N-terminal tail. These structural motifs are highly conserved across species and mutated variants of PrP. (B) representative structures of $A\beta_{1\text{-}42}$ monomeric chains that comply with solution NMR data (Ball et al., 2013) with locations of main transient elements of secondary structure indicated. *(A) is reproduced with permission from Blinov et al. (2009), © 2009 American Chemical Society. (B) is reproduced from Ball et al. (2013), © 2013 Elsevier under the STM permission.*

For monomeric, intrinsically disordered proteins such as $A\beta_{1\text{-}40/42}$ peptides or α-synuclein in solution, NMR experiments have revealed an ensemble of conformations containing predominantly random coils and occasionally other secondary structures (Roche, Shen, Lee, Ying, & Bax, 2016; Lokappa et al., 2014; Ball, Phillips, Wemmer, & Head-Gordon, 2013; Alderson & Markley, 2013). Fig. 2B presents examples of $A\beta_{1\text{-}42}$ monomeric chains in solution, with locations of the most persistent elements of secondary structure marked. Interestingly, NMR fingerprints of monomeric $A\beta_{1\text{-}40}$ and $A\beta_{1\text{-}42}$ have been reported to bear a close resemblance (Riek, Dobeli, Wipf, & Wuthrich, 2001; Roche et al., 2016), whereas those of $A\beta_{1\text{-}40/42}$ monomers adsorbed on a fibril surface have been found to differ substantially (Fawzi, Ying, Ghirlando, Torchia, & Clore, 2011). For Aβ oligomers, coexistence of disordered oligomers and β-sheet rich species (Kotler et al., 2015; Luo et al., 2014) has been reported. Overall, the structures and dynamics of small soluble oligomers appear to be highly diverse and sensitive to the specific conditions of the experiment (Ahmed & Melacini, 2018; Bemporad & Chiti, 2012; Nagel-Steger et al., 2016).

The advent of high-resolution solid-state NMR (ssNMR) spectroscopy made it possible to characterize the final product of misfolding, large insoluble amyloid fibrils (Eisenberg & Sawaya, 2017; Loquet et al., 2018; van der Wel, 2017) to near-atomic resolution. In this method, fibrils from isotopically labeled recombinant proteins are used. In order to study the structure of brain-derived fibrils, isotopically-labeled, recombinant proteins are seeded by pathogenic fibrils (Lu et al., 2013). It is presumed that the resulting fibrils will have the same structure as the brain-derived fibrils. The resulting set of ssMNR constraints then contains structural information regarding the identities of residues, recognition of parallel versus antiparallel β-sheets, hydrogen bonds, and other inter-residue contacts (Eisenberg & Sawaya, 2017). In distinction from other high-resolution methods, the lack of well-defined crystalline periodicity in fibrils does not prevent a reliable characterization by ssNMR. Structural deciphering of fibrils and other products of misfolding is progressing quickly (Iadanza, Jackson et al., 2018; Loquet et al., 2018; van der Wel, 2017). High-resolution 3D structural models have been developed for fibrils of $A\beta_{1\text{-}40/42}$ peptide (Colvin et al., 2016; Gremer et al., 2017; Schütz et al., 2015; Sgourakis, Yau, & Qiang, 2015; Wälti et al., 2016; Xiao et al., 2015), α-synuclein (Tuttle et al., 2016), HET-s (Wasmer et al., 2008), and several other proteins and peptides (Loquet et al., 2018; van der Wel, 2017). The greatest accuracy is achieved when residue-specific constraints form ssNMR are used to refine high-resolution 3D structures obtained from cryo-EM techniques (Glaves et al., 2018; Iadanza, Silvers et al., 2018; Gremer et al., 2017; Fitzpatrick et al., 2013).

3.3 Transmission electron microscopy and cryomicroscopy

In transmission electron microscopy (EM), a sample is exposed to a beam of electrons and transmitted electrons are recorded, producing an image of the sample. EM has a rich history of biological applications (Orlova & Saibil, 2011) due to a promise of sub-nanometric resolution; possibility of 3D image reconstruction; and tolerance to small sample sizes. Challenges have included the requirement to conduct exposures in a vacuum chamber; damage of biological samples in the process of this; and low contrast of imaging. A widely used technique, negative staining EM, resolves these by embedding biological samples into a thin layer of solid material containing a heavy metal salt (De Carloa & Harris, 2011). This simple technique dramatically increases the image contrast, and somewhat reduces a sample's damage during the exposure. Negative staining EM images have been used widely to measure the size of oligomers or fibrils in two dimensions (Wan et al., 2015; Younan & Viles, 2015; Lorenzen et al., 2014; Bjorndahl et al., 2011; Ahmed et al., 2010; Wegmann et al., 2010; Redecke et al., 2007; Arimon et al., 2005; Kuwata et al., 2003). An example of EM imaging of $A\beta_{42}$ oligomers, protofibrils, and mature fibrils is given in Fig. 3. Such imaging is often referred to as low-resolution EM. Three-dimensional reconstructions, or inferences into internal structure, are possible only in special cases (Arimon et al., 2005; Sachse et al., 2006; Tattum et al., 2006; Terry et al., 2016). Usually such details are difficult to obtain from negative staining EM because most electron beam scattering occurs in the surrounding stain layer.

A pivotal invention has been freezing of biological samples to cryogenic temperatures in a thin layer of non-crystalline liquid (vitreous ice) in order to protect their structural integrity against degradation during EM exposures (Adrian, Dubochet, Lepault, & McDowall, 1984; Dubochet, Lepault, Freeman, Berriman, & Homo, 1982; Taylor & Glaeser, 1976). This gave rise to cryo-EM, a method described as a revolution in structural biology in the recent years (Callaway, 2015; Shen, 2018). As the contrast of cryo-EM imaging is typically low, it took nearly three decades after its invention until the progress of direct electron detectors in tandem with advanced image processing procedures allowed achieving an unprecedented structural resolution (Bai, McMullan, & Scheres, 2015; Flores-Fernández et al., 2018; Nogales, 2016; Nogales & Scheres, 2015).

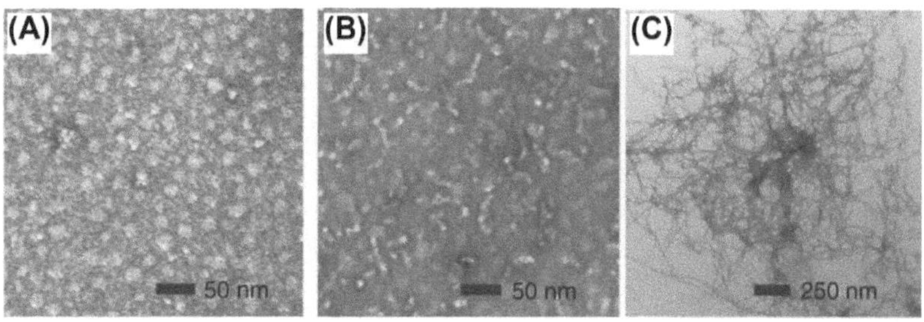

Fig. 3 Negative staining EM characterization of $A\beta_{1\text{-}42}$ oligomers (A), protofibrils (B) and mature fibrils (C), at different stages of misfolding (Ahmed et al., 2010). Oligomers are seen as round particles with average widths of ~10–15 nm. Subsequently elongated protofibrils are formed, which eventually develop dense networks of mature fibrils. Complementary ssNMR studies (Ahmed et al., 2010) have revealed parallel, in-register β-sheets in the resulting fibrils, but not in the oligomers. *Reproduced by permission from Springer Nature Ahmed et al. (2010), © 2010 Springer Nature.*

Combining cryo-EM images from amyloidal fibrils with a regular helical twist allowed to generate 3D reconstructions with atomic detail as has been reported for fibrils of several proteins, such as A$\beta_{1\text{-}42}$ (Gremer et al., 2017), α-synuclein (Guerrero-Ferreira et al., 2018; Li et al., 2018), tau (Fitzpatrick et al., 2017), and β_2-microglobulin (Iadanza, Silvers et al., 2018). Most fibrils were found to consist of two intertwined protofilaments, each containing protein chains stacked on top of each other and forming ladder-like β-sheets in an architecture known as parallel, in-register cross-β-structure. In most cases a slight tilt was observed in each layer, resulting in a helical twist of the two protofilaments around the fibril axis.

Cryo-EM reconstructions of prion protein fibrils (Vazquez-Fernandez et al., 2016) also revealed two intertwined protofilaments (Fig. 4). However, the data obtained suggest the key structural motif of the prion protein fibrils

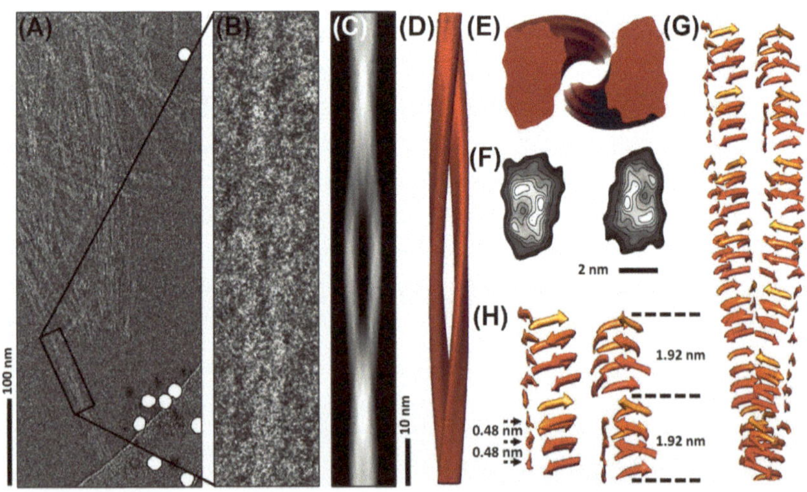

Fig. 4 Cryo-EM analysis of infectious prion protein amyloid fibrils. (A) Section of a cryo electron micrograph showing prion fibrils lacking the glycosylphosphatidylinositol (GPI) anchor. A single isolated and twisted fibril used for the 3D reconstruction is enclosed by a black box. (B) Close-up view of the isolated prion fibril. (C) Reprojected image of the 3D fibril map for comparison with the unprocessed image (B). (D) 3D reconstruction of the GPI-anchorless prion fibril. (E) Cross section of the reconstructed fibril showing two distinct protofilaments. (F) Contoured density maps of the cross section with lines contoured at increasing levels of 0.125 σ. (G) Cartoon depicting the proposed configuration of the polypeptide chains in the prion fibril. (H) Close-up view of the possible β-sheet stacking in a four-rung β-solenoid architecture. Different colors represent different β-solenoid rungs (Vazquez-Fernandez et al., 2016; Zweckstetter, Requena, & Wille, 2017). *Reproduced from Zweckstetter et al. (2017), under the Creative Commons License.*

to be a 4-rung β-solenoid, which comprises parallel but not in-register cross-β-strands (Flores-Fernández et al., 2018; Vazquez-Fernandez et al., 2016). Full atomic-level mapping of PrP fibrils has not been achieved yet due to the structural heterogeneity of the samples, which is a fundamental limitation of EM-based 3D reconstruction techniques. The most accurate and powerful approach combines ssNMR constraints and X-ray fiber diffraction data to validate augment cryo-EM based 3D reconstructions of amyloid fibrils (Fitzpatrick et al., 2013; Gremer et al., 2017; Iadanza, Silvers et al., 2018). A recent example of the $A\beta_{1-42}$ fibril structure determined by combining cryo-EM and ssNMR data (Gremer et al., 2017) is shown in Fig. 5. Each $A\beta_{1-42}$ molecule has been found to adopt an LS-shaped structure, in which the N terminus is L-shaped and the C terminus is S-shaped. The molecules form a staggered architecture with screw symmetry relative the fibril axis, resulting in distinct surfaces on the two ends of each filament (Fig. 5). Hydrophobic interactions and salt bridges have been found to play an important role in the stabilization of the fibril.

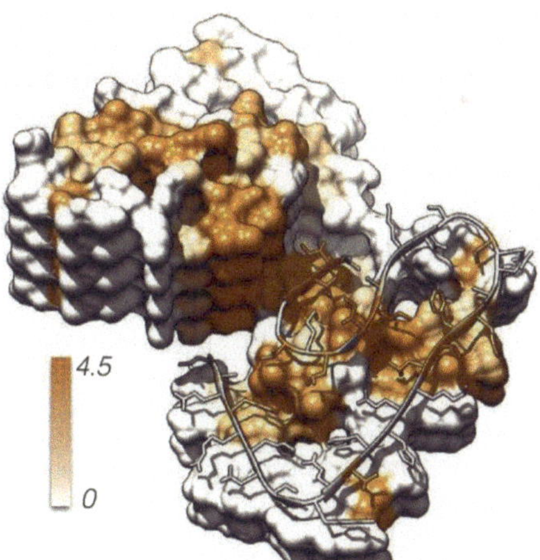

Fig. 5 Fragment of an atomic model for $A\beta_{1-42}$ fibrils determined by cryo-EM at 4.0 Å resolution combined with ssNMR characterization (Gremer et al., 2017). The colors represent surface-mapped hydrophobicity. $A\beta_{1-42}$ chains in the fibril adopt a screw symmetry with a rise of ~4.7Å from the N to C terminus. The interface between two protofilaments is formed by the hydrophobic C termini. The polar side chains are facing the outside of the protofilaments, protecting the hydrophobic C termini from the solvent. *Reproduced from Gremer et al. (2017), © 2010 AAAS with permission from American Association for the Advancement of Science.*

3.4 Atomic force microscopy

In AFM, a mechanical probe (sharp tip) attached to a flexible micro-cantilever is scanned over the object of interest, which is adsorbed onto a flat surface (Binnig, Quate, & Gerber, 1986). In modern instruments the positions of the tip, as well as van-der-Waals and electrostatic forces between the tip and the surface, are monitored ultra-precisely using a laser beam focused on the cantilever and reflected into a position-sensitive photodiode (Bruggink, Müller, Kuiperij, & Verbeek, 2012; Hinterdorfer & Dufrêne, 2006). Piezoelectric scanning of the surface allows collecting topography maps, which visualize the surface with a vertical resolution in sub-nanometric regimes (Alsteens et al., 2017; Dufrêne et al., 2017). Common applications of AFM include measurements of the horizontal dimensions and the height of supramolecular aggregates and individual molecules, either in buffer solution or at ambient conditions. The main limitations of the method are the need to adsorb objects of interest on a flat horizontal surface and the relatively slow data collection rate due to the need to scan each image line-by-line. Various stages of amyloid conversion, from monomers to oligomers and fibrils, have been visualized by AFM (Morel, Carrasco, Jurado, Marco, & Conejero-Lara, 2018; Chen et al., 2015; Luo et al., 2014; Fitzpatrick et al., 2013; Ahmed et al., 2010; Wegmann et al., 2010). The results are in remarkable agreement with the transmission EM imaging data (Arimon et al., 2005; Chen et al., 2015; Cremades & Dobson, 2018; Wegmann et al., 2010). Moreover, modern high-speed AFM instruments offer an added capability of monitoring the real-time evolution of monomers, oligomers, and fibrils *in situ* (Banerjee, Sun, Hayden, Teplow, & Lyubchenko, 2017; Watanabe-Nakayama et al., 2016; Zhang et al., 2013, 2018).

3.5 Optical characterization

Interaction of visible, infrared, or ultraviolet light with molecules forms the basis for a large and diverse group of characterization techniques. Some of these methods are among the most convenient and accessible for tackling misfolding and aggregation processes.

The dynamic light scattering (DLS) method allows determining the diffusion coefficient and the hydrodynamic radius of molecules or supramolecular aggregates in solution (Li, Lubchenko, & Vekilov, 2011; Stetefeld, McKenna, & Patel, 2016). In DLS, the intensity of laser light scattered by a solution is measured as a function of time. The Brownian motion of molecules or oligomeric aggregates in the solution causes fluctuations of the intensity of the scattered light. These fluctuating intensities are subjected to signal processing analyses allowing physical interpretations regarding the oligomer/aggregate size, for which specialized software is available. Metrics, such as the diffusion coefficient, the solution viscosity, or the distribution of hydrodynamic radii can be obtained, and DLS is routinely used to evaluate the size of oligomers or protofibrils in solution (Kotler et al., 2015; Morel et al., 2018; Redecke et al., 2007; Sabareesan & Udgaonkar, 2016). However, accurate DLS measurements are subject to conditions and limitations, such as exact knowledge of the temperature, or single-scattering of light in the sample. The method is applicable to molecules or aggregates with sizes from approximately 1 nm, and prone to biases in heterogeneous samples (Bruggink et al., 2012; Kundel et al., 2018). For these reasons, most studies employ DLS as a complementary method in combination with other techniques. Successful geometrical characterizations by DLS would be consistent with the SAXS data (Redecke et al., 2007; Rezaei et al., 2005), as well as low-resolution EM and AFM analyses (Chen et al., 2015; Morel et al., 2018; Redecke et al., 2007).

Circular dichroism (CD) spectroscopy is one of the most straightforward techniques to characterize the structure of proteins. The method exploits differences in absorption of right- and left-circularly-polarized light by optically active atomic groups (chromophores) that possess a structural asymmetry (chirality). In most CD instruments, differences in absorbance of the right- and left-circularly polarized components, expressed as ellipticity in degrees (Kelly, Jess, & Price, 2005), is measured as a function of the wavelength of the light. Since CD spectra are sensitive to the geometry of dihedral bonds in chromophores, they allow identifying the prevalent elements of secondary and tertiary structure. In the far UV region (below 250 nm) α-helices, β-structures, random coils, and other elements can be distinguished (Bruggink et al., 2012; Kelly et al., 2005). CD spectroscopy is extremely useful to characterize misfolding intermediates and oligomers in the course of amyloidogenic conversion. For PrP subjected to conversion-promoting conditions, far UV CD analyses consistently indicate a loss in α-helical content and a build-up of β-sheet secondary structure (Glaves et al., 2018; Sabareesan & Udgaonkar, 2016; Bjorndahl et al., 2011; Redecke et al., 2007). Aggregation of intrinsically disordered proteins is accompanied by a replacement of initial random coil conformation with β-structured content (Kotler et al., 2015; Luo et al., 2014; Nath, Meuvis, Hendrix, Carl, & Engelborghs, 2010).

Another powerful conformation-sensitive technique, Fourier transform infrared (FTIR) spectroscopy, analyzes the absorption of radiation with wavenumbers in a mid-infrared (IR) region. Many molecular vibrations fall into the IR frequency regimes, where they give rise to absorption bands when the molecule is exposed to the radiation. The spectral position of FTIR adsorption bands depends on the strength and geometry of the bonds involving the vibrating atomic groups. In proteins, these are determined by the secondary structure and dihedral torsional angles (Barth & Zscherp, 2002; Kong & Yu, 2007; Zandomeneghi, Krebs, McCammon, & Fandrich, 2004). Based on extensive experimental and theoretical analyses, a consensus has emerged for associating α-helices, β-sheets, turns, random coils, and other structures with certain FTIR absorption bands (Barth & Zscherp, 2002; Kong & Yu, 2007; Shivu et al., 2013). Moreover, an important milestone has been establishing differences between FTIR fingerprints from parallel cross-β-sheet motifs found in mature amyloidal fibrils, and those from native proteins or oligomers (Sarroukh, Goormaghtigh, Ruysschaert, & Raussens, 2013; Shivu et al., 2013; Zandomeneghi et al., 2004). β-strands from native proteins have been found to produce different FTIR bands than cross-β-sheets from fibrils, suggesting that the formation of the fibrils involves a substantial structural reorganization even for proteins containing many β-strands in their native form (Zandomeneghi et al., 2004). In turn, oligomeric species formed in the process of aggregation often exhibit FTIR fingerprints characteristic of anti-parallel β-sheets (Lorenzen et al. 2014; Sarroukh et al., 2013), which are not observed in most mature amyloid fibrils with a few exceptions (Breydo & Uversky, 2015). This sensitivity to secondary structure conformations, complemented by fast acquisition times, make FTIR spectroscopy very convenient for tracking down interconversions of conformers in the course of misfolding and aggregation (Morel et al., 2018; Sabareesan & Udgaonkar, 2016; Ahmed et al., 2010; Rezaei et al., 2005). Integration with AFM mapping, currently under active development, promises imaging and analysis capabilities with spatial precision below the diffraction limit (Xiao & Schultz, 2017).

Absorption of electromagnetic radiation by certain atomic groups (fluorophores) can lead to emission of light of lower energy, a process known as fluorescence. In fluorescence assays, the intensity and/or spectral distribution of emitted light are measured (Bruggink et al., 2012; Kundel et al., 2018; Munishkina & Fink, 2007). In proteins, tryptophan (W), tyrosine (Y), and phenylalanine (F) are natural fluorophores, who's emission spectra (shape and intensity) are dependent on the protonation state or hydrophobicity of the local environment. Numerous studies have used intrinsic fluorescence to tackle the amyloidogenic conversion (Munishkina & Fink, 2007). However, the most common technique employs labeling by fluorescent probes (extrinsic fluorophores) such as thioflavin-T (ThT) (Groenning, 2010; Hawe, Sutter, & Jiskoot, 2008; Younan & Viles, 2015), whose fluorescent quantum yield increases dramatically upon binding to β-sheet rich structures. ThT assays are widely used to track β-conversion, often accompanied by CD characterization of the secondary structure. Alternatively, it is possible to label hydrophobic groups using 1-anilinonaphthalene-8-sulfonate (ANS) or 4,4′-dianilino-1,1′-binaphthyl-5,5′-disulfonate (bis-ANS) probes, which exhibit an increase in fluorescence in hydrophobic environments (Hawe et al., 2008; Younan & Viles, 2015). Due to their simplicity, fluorescence assays are used extensively to detect and monitor misfolded, oligomeric, or fibrillar species (Morel et al., 2018; Chen et al., 2015; Kotler et al., 2015; Lorenzen et al. 2014; Luo et al., 2014; Lokappa et al., 2014). However, such assays provide information about all fluorescent species, with limited capabilities of distinguishing specific components in heterogeneous ensembles.

This limitation is largely overcome in the Förster resonance energy transfer (FRET) technique, which exploits the dipole-dipole coupling between two fluorophores, a donor and an acceptor, with matching emission/adsorption spectra. In a traditional form of FRET, a donor-acceptor pair is introduced in a single molecule. Absorption of electromagnetic radiation by the donor is followed by a non-radiative transfer of energy to the acceptor, which relaxes through fluorescence (Kundel et al., 2018; Munishkina & Fink, 2007; Orte, Clarke, & Klenerman, 2011; Teunissen, Pérez-Medina, Meijerink, & Mulder, 2018). Since the coupling is strongly distance-dependent, and also sensitive to mutual orientations of the donor-acceptor pair, FRET fluorescence analyses provide information about the size and conformation of the labeled molecule. For example, FRET studies indicate that the conversion of α-synuclein *in vitro* produces a population of transient early oligomers (Nath et al., 2010), which may undergo a change of conformation (Cremades et al., 2012; Nath et al., 2010) or dissociate (Horrocks et al., 2015) prior the formation of β-sheet rich fibrillar species.

3.6 Mass-spectrometry

Mass spectrometry measures the masses of charged analytes. Because there is more than one way to accomplish this goal, mass spectrometers come in a large variety of formats exploiting diverse physical principles (Domon &

Aebersold, 2006). Although mass spectrometry has been around for more than a century, protein mass spectrometry only took off once methods for the ionization of relatively large analytes became available (Karas, Bachmann, & Hillenkamp, 1985; Yamashita & Fenn, 1984). Because small analytes continue to exhibit more favorable ionization characteristics, the vast majority of protein mass spectrometry studies to date make use of a bottom-up workflow that infers the identity of proteins from measurements of their peptides. Insights into the fold of proteins typically gets lost in these applications unless the experimental design is specifically tailored to probe for this information (see below).

Consequently, when mass spectrometry is applied to protein folding research, it often is to merely identify a protein involved in protein folding diseases or to study its post-translational modifications (PTMs), rather than to investigate the protein misfolding events themselves. Ground-breaking studies undertaken in the 1990s established, for instance, that the normal cellular prion protein cannot be distinguished from its disease-associated conformer by the presence or absence of specific post-translational modifications (Rudd et al., 1999; Stahl, Borchelt, Hsiao, & Prusiner, 1987; Stahl & Prusiner, 1991). More recently, several groups have published in-depths characterizations of misfolded proteins. For the Tau protein, these studies confirmed previous antibody-based characterizations of its hyperphosphorylation and the presence of several additional PTMs (Morris et al., 2015). Powered by the advent of effective methods for the relative quantitation of peptides, mass spectrometry not only provides binary information regarding the presence or absence of phosphorylation events but can reveal insights into the relative occupancy of PTM acceptor sites (Liu, Song, Nisbet, & Gotz, 2016).

The discovery of most proteins underlying protein misfolding diseases preceded the most rapid phase of technological innovation in mass spectrometry that we have witnessed in the past 25 years. A notable exception to this statement represents TDP-43, one of the most studied neurodegenerative disease proteins to date. TDP-43 was identified by first isolating antibodies specific to insoluble material present in the brains of individuals with amyotrophic lateral sclerosis and frontotemporal lobar degeneration and subsequently employing immunoprecipitation mass spectrometry to identify the protein these antibodies were directed to (Neumann et al., 2006).

Rapid gains in speed, sensitivity and resolution of mass spectrometers have allowed investigators to turn toward studying the molecular environment of proteins that are central to these diseases, including the amyloid precursor protein (Bai et al., 2008), the Tau protein (Gunawardana et al., 2015) and TDP-43 (Chou et al., 2018). For the prion protein, a direct comparison of its molecular environment in several cell models was recently reported (Ghodrati et al., 2018). With these kinds of studies, the inherent power of mass spectrometry as a platform for discovery research can at times reveal unexpected insights. An interactome study of the prion protein has for example led to the discovery of the evolutionary origins of the cellular prion protein, owed to the fact that this protein inherited from its ancestral ZIP zinc transporter the ability to interact with contemporary ZIP family members (Schmitt-Ulms, Ehsani, Watts, Westaway, & Wille, 2009). A separate investigation into interactions of oligomeric Aβ revealed a direct interaction with somatostatin, a small cyclic endoneuropeptide (Wang et al., 2017). This interaction has recently been validated in molecular dynamics simulations and is intriguing because SST represents one of a small number of natural amyloids. Thus, it is conceivable that there can be crosstalk between diffusible SST amyloids that are released into the synaptic cleft from dense core granules and endoproteolytically generated Aβ peptides.

If the objective is to apply mass spectrometry to interrogate the fold of a protein using a bottom up strategy, investigators have the choice to use chemical crosslinkers that stabilize intramolecular interactions prior to digestion. Although this approach continues to grapple with the challenge that crosslinked peptides are underrepresented in protein digests relative to unmodified peptides, recent advances have paved the way for future applications of crosslinking mass spectrometry in the study of dynamic protein folding events (Chen & Rappsilber, 2018). Insights into the fold can also be obtained by exploiting differences in the accessibility of amino acids to chemical reactions. In one elegant approach, referred to as protein 'painting', proteins are exposed to chemicals that readily interact with exposed protein surfaces, thereby blocking trypsin cleavage sites. In addition to providing insights into the fold of protein, this strategy can inform about protein-protein interfaces, and was initially shown to reveal contact regions in a three-way interaction between the interleukin 1β (IL1b) ligand, its receptor IL1RI and the accessory protein IL1RAcP (Luchini, Espina, & Liotta, 2014).

A highly popular alternative methodology relies on the exchange of hydrogen atoms for deuterium. The latter has been shown to occur rapidly when hydrogens are exposed at the cell surface, but proceeds slowly when they are buried within protein folds or trapped in hydrogen bonds. The technique, commonly referred to as hydrogen-deuterium exchange (HDX) requires protein preparations to be exposed to a deuterated solvent for a certain labeling time prior to digestion and mass spectrometry analysis (Kostyukevich et al., 2018). In addition to revealing information about the fold of proteins, it has become a mainstay for the study of protein-ligand interactions and to characterize the fold and dynamics of intrinsically disordered proteins or membrane proteins (Oganesyan, Lento, & Wilson, 2018). Leading-edge implementations of this technique can provide residue resolution and offer time-resolved insights into conformational changes that occur during the folding or misfolding of a protein. A milestone 2009 report showed that HDX can be used to characterize the early formation of two types of SH3 domain aggregates. Whereas one was amorphous, providing no HDX protection, the other one was characterized by partial HDX protection, yet both appeared to represent on pathway precursors to amyloid fibril formation (Carulla et al., 2009). More recently, kinetic HDX mass spectrometry experiments validated the existence of a core region within the N-terminus of PrP that harbors at its center a well-characterized hydrophobic stretch and appears to promote the propensity of mutant PrP variants to stabilize early oligomerization, i.e., ahead of more substantial conformational changes involving the C-terminal PrP domain (Sabareesan & Udgaonkar, 2016; Singh, Sabareesan, Mathew, & Udgaonkar, 2012).

In recent years, whole protein mass spectrometry analyses have increasingly garnered interest. In the protein folding/misfolding field, special attention is afforded to an auxiliary technique known as ion mobility spectrometry (IMS). IMS can separate proteins of alternative folds by exploiting differences in the speed with which they traverse a buffer gas-filled tube (Lanucara, Holman, Gray, & Eyers, 2014). Increasingly, IMS modules are placed on the front end of mass spectrometers for the separation of proteins. Because the mobility of a protein during IMS reflects its rotationally averaged cross-sectional area, this approach does, however, not just provide a convenient orthogonal modality for the separation of proteins but can also reveal valuable insights about the fold of a protein. Moreover, when combined with ligand binding studies or downstream analyses of PTMs, IMS can provide insights on how a given ligand interaction or PTM affects the fold of a protein. A leading-edge application of this technology provided unprecedented insights underlying the conformational conversion from random assembly to beta-sheet during amyloid fibril formation (Bleiholder, Dupuis, Wyttenbach, & Bowers, 2011).

4. Molecular simulations of misfolding proteins

The molecular dynamics (MD) simulations method was developed to study movements of atoms/molecules by numerical solution for equations of motion of a system of interacting particles. This inherently interdisciplinary method incorporates mathematics, physics, computer science, chemistry, and biology, and is applied broadly in chemical physics, computational chemistry, materials science, structural biology, and drug design. Since their invention, MD simulations have successfully complemented experimental methods, providing inputs that are difficult or impossible to obtain experimentally. Explicit prediction of the precise positions of each atom with a time resolution down to 10^{-15} s has allowed researchers to see the motion of proteins in such details that could not be achieved otherwise. The first application of MD in molecular biology that involved simulation of a protein in 1977 (McCammon, Gelin, & Karplus, 1977) changed the perception of proteins as rigid structures, acknowledging their dynamic nature. Subsequently, MD simulation studies have been employed extensively to provide insights into the nature of biomolecular motions and how they are related to molecule function; helped to understand mechanisms of conformational changes; or predict outcomes of such changes, to name just a few of many applications. The current vision of biomolecules as dynamic systems, in which motions play a fundamental role, and conformational changes are often necessary for proper functioning (Dill & MacCallum, 2012; Henzler-Wildman & Kern, 2007), stemmed to a significant extent from the results of MD simulations. Due to importance of such studies, the MD methods and approaches have been undergoing continuous development since their creation (Dror, Dirks, Grossman, Xu, & Shaw, 2012; Karplus & McCammon, 2002). Here we overview the aspects of MD simulations that might be useful for applications targeting protein misfolding and aggregation.

4.1 Atomistic modeling of proteins

Solving the equations of atomic motions underlying the MD simulation method starts with setting the initial hypothetical positions and velocities of all atoms in the system. Unless *de novo* prediction of protein folding is pursued, the starting configurations are normally obtained from the Protein Data Bank (PDB) (Rose et al., 2017). These starting structures are determined by X-ray crystallography, NMR techniques, or cryo-EM. Only high-resolution structures can be used, and often positions of missing atoms need to be assigned during the structure preparation (for example, parts of protein that are too flexible to be determined experimentally, or the hydrogen atoms that not observed in X-ray experiments, have to be added *in-silico*). Building such missing parts, or introducing mutations or amino acid insertions, can be done using several software packages such as Accelrys Discovery Studio Visualiser, VMD (Humphrey, Dalke, & Schulten, 1996) or Chimera (Pettersen et al., 2004). To set the desired pH level for the simulations, a protein is evaluated using the PROPKA graphical user interface (Rostkowski, Olsson, Søndergaard, & Jensen, 2011) and the protonation of pertinent residues is changed. Once the starting configuration is prepared, solvent molecules are added. The initial velocities for both protein and solvent atoms are assigned randomly using the Maxwell-Boltzmann distribution at the chosen temperature. To provide reliable results, several control simulations are usually run with different initial velocities of all atoms.

4.2 Force fields for protein and water models

The forces that are acting on each atom during the MD simulations are derived from a classical potential energy function, which is called the force field. A force field represents the dependence of the potential energy of the system on the position of each atom, taking into account details of chemical bonds such as the energy, length, and angles with respect to other bonds, as well as non-bonded interactions (Mackerell, 2004). In turn, the corresponding forces determine how each atom's coordinates and velocities change with time. The force fields are parameterized either empirically to match experimental observables, or from quantum-mechanical predictions, or both. Most of existing force fields have been designed to properly model folding patterns of globular proteins from experiment. For example, the AMBER99SB-ILDN (Lindorff-Larsen et al., 2010), CHARMM36m (Huang et al., 2017), and CHARMM22* (Piana, Lindorff-Larsen, & Shaw, 2011) force fields have been shown to accurately reproduce NMR-derived structures for ubiquitin, GB3, HEWL, and BPTI in their folded state (Lindorff-Larsen et al., 2012; Robustelli, Piana, & Shaw, 2018).

Most of current simulation studies of PrP adopt the Amber99SB-ILDN, CHARMM22, or GROMOS96 43A1 force fields. Simulations of PrP using a special polarizable force field with an on-the-fly charge update (Xu, Lazim, Mei, & Zhang, 2012) have resulted in an elongation of the β-strands during the simulations. For the intrinsically disordered Aβ peptide, extensive comparative analyses of existing force fields have been performed (Carballo-Pacheco & Strodel, 2017; Man, Nguyen, & Derreumaux,

2017b; Rauscher et al., 2015; Rosenman, Wang, & García, 2016; Smith, Rao, Segelken, & Cruz, 2015; Somavarapu & Kepp, 2015; Watts, Gregory, Frisbie, & Lovas, 2018). Most studies report differences in conformations, secondary structure content, and hydrogen bonding among simulations with different force fields. Although there is no consensus among the studies, a slightly better agreement with NMR and/or CD experiments on the Aβ peptide has been reported for the CHARMM22* (Carballo-Pacheco & Strodel, 2017; Rauscher et al., 2015; Rosenman et al., 2016; Somavarapu & Kepp, 2015; Watts et al., 2018), AMBER99SB-ILDN (Carballo-Pacheco & Strodel, 2017; Man et al., 2017b; Somavarapu & Kepp, 2015), and OPLS-AA (Carballo-Pacheco & Strodel, 2017; Man et al., 2017b; Smith et al., 2015) force fields. One of concerns is that classical non-polarizable force fields tend to predict overly structured or compact conformations for IDPs, which in their native state are more solvent exposed in comparison with globular proteins (Carballo-Pacheco & Strodel, 2017; Rauscher et al., 2015; Rosenman et al., 2016). Since the reason of this discrepancy is rooted in the tradition to parameterize force fields so that they predict stable globular structures rather than expanded conformations, a tangible path forward would involve a more balanced parameterization (Chong, Chatterjee, & Ham, 2017; Huang & MacKerell, 2018; Robustelli et al., 2018; Song, Luo, & Chen, 2017). One highly regarded route toward universal force fields is the inclusion of polarizability (enabling partial atomic charges) in MD simulations (Baker, 2015; Huang & MacKerell, 2018); however, this comes at a cost of increased computational requirements for the simulations. Extensive comparative analysis of aggregation of small peptides using AMBER99SB-ILDN, CHARMM36, and GROMOS9643A1 (Matthes, Gapsys, Brennecke, & de Groot, 2016) has concluded that the existing non-polarizable force fields are able to capture the global features of the peptide oligomerization process, and to distinguish between aggregation-prone and non-fibrillizing sequences.

Water is second major constituent of biomolecular systems. A comprehensive review of water models has been published recently (Onufriev & Izadi, 2017). Among the classical, explicit all-atom models, the TIP3P (Jorgensen, Chandrasekhar, Madura, Impey, & Klein, 1983), TIP4P-Ew (Horn et al., 2004), and SPC/E (Berendsen, Grigera, & Straatsma, 1987) models are most common in biomolecular applications. Although initially most force fields for proteins were co-optimized with specific water models (the AMBER and CHARMM force fields with the TIP3P and modified TIP3P models, respectively; the GROMOS force field with the SPC and

SPC/E models; and the OPLS AA force field with the TIP4P model), different combinations of protein force fields and water models are employed frequently. For example, the SPC/E model was found to outperform other non-polarizable 3-point models in predictions of radial distribution functions, self-diffusion coefficients, and dielectric properties of bulk water (Mark & Nilsson, 2001; van der Spoel, van Maaren, & Berendsen, 1998; Vega & Abascal, 2011), which has stimulated its use with various force fields.

The importance of a careful validation of the water models in MD simulations is well-recognized (Piana, Donchev, Robustelli, & Shaw, 2015; Rauscher et al., 2015; Simone, Dodson, Verma, Zagari, & Fraternali, 2005; Somavarapu & Kepp, 2015; Zuegg & Gready, 1999). For globular proteins, the TIP3P water model in combination with the CHARMMC22* force field was found to match experimental NMR ensembles very well (Robustelli et al., 2018). On the other hand, the performance of the TIP3P model was found to be inferior to the SPC/E and TIP4P-Ew models when combined with the AMBER14SB force field (Persson, Söderhjelm, & Halle, 2018). For intrinsically disordered proteins, combinations of either TIP3P, TIP4P, or SPC water models with the CHARMMC22* force field produced relatively good predictions (Robustelli et al., 2018; Somavarapu & Kepp, 2015). For $A\beta_{1-40}$ chains, the hydrophobic surface was found to increase in the order TIP3P > TIP4P > SPC irrespective of the protein's force field (Somavarapu & Kepp, 2015), suggesting that the SPC water model might be preferable for extended conformations. In contrast, the TIP3P model due to a higher polarity may hinder interactions with hydrophobic groups, resulting in collapsed conformations of the peptide. However, a study of $A\beta_{21-30}$ fragments in various force fields (Smith et al., 2015) concluded that the results of the simulations are largely independent of the choice of water models. A novel, improved TIP4P-D water model with stronger dispersion interactions (Piana et al., 2015) was recently shown to properly reproduce conformations of both globular and intrinsically disordered proteins when it is combined with the AMBER99SB-ILDN force field (Piana et al., 2015; Robustelli et al., 2018). Emerging polarizable water models are also highly regarded, with the expectation that a trade-off of their accuracy and complexity will be achieved in the future (Onufriev & Izadi, 2017).

4.3 The challenges of system size and simulation time, and alternative approaches

All-atom MD simulations provide a superb spatial and temporal resolution of the predictions. However, when doing such simulations for a typical protein in explicit solvent, the resulting system would contain many thousands of atoms. Aggregates consisting of several protein molecules in water would lead to even larger systems. The prediction of conformational changes as a function of time becomes challenging when the size of the system increases due to the computational requirements. With current computational powers, simulations up to the microseconds regimes are conceivable for systems of up to 10^5 atoms (Lane, Shukla, Beauchamp, & Pande, 2013). Microsecond-long MD simulations have been achieved (Piana, Klepeis, & Shaw, 2014), allowing for a more complete conformational sampling, and providing means for a more rigorous testing of force fields and other models. In applications for the misfolding of proteins, such timescales might be sufficient to capture early misfolding events, given that they occur close to local energy minima. However, some of misfolding events may involve crossing of certain barriers in the free energy landscape, limiting the likelihood of capturing such events in a short MD trajectory.

In an effort to overcome the challenges of computational time, various methods have been employed. A simple and widely used approach is to generate several independent MD trajectories simultaneously. Drawing upon the capabilities in parallel computing, this allows accumulating more exhaustive conformational sets, as well as improving statistical significance of the predictions. A different possibility is to overcome the free energy barriers by increasing the temperature of system. A method, which allows drawing upon this while generating physiologically relevant conformational ensembles, is known as the replica exchange MD (REMD) (Bernardi, Melo, & Schulten, 2015; Sugita & Okamoto, 1999; Wei, Xi, Nussinov, & Ma, 2016). In this method, several independent MD simulations (replicas) of the system are done in parallel at different temperatures, and instantaneous conformations are exchanged between replicas of equal potential energy with a probability calculated using the Metropolis criterion (Sugita & Okamoto, 1999). The REMD method is frequently used to model misfolding and aggregation phenomena (Nagel-Steger et al., 2016), as it allows reaching conformations that may only be accessible at elevated temperatures. The novel method of Markov state models (MSMs) (Chodera & Noé, 2014; Shukla, Hernández, Weber, & Pande, 2015; Sirur, De Sancho, & Best,

2016; Wei et al., 2016) strives to identify kinetically relevant conformational states and inter-state transitions through an application of the mathematical formalism known as the symbolic dynamics (Hao & Zheng, 1998). The MSM method postulates that evolution of a system can be represented as a Markov chain of memory-less transitions between discrete "microstates", which can be identified from relatively short MD or REMD trajectories through sophisticated adaptive sampling algorithms.

The accessibility of enhanced energy barrier crossing (REMD), or the potential of explicitly predicting kinetically relevant aspects of conformational evolution (MSM) is, of course, an advantage. However, these and related methods involve a departure from the dynamical nature of the initial MD trajectories, where each conformation is linked to the previous ones through dynamic equations of motion. An exchange of conformations between independent trajectories disrupts the dynamical process, whereas the adoption of a Markov sequence of transitions between "microstates" limits the applicability to a so-called Markovian partition (Hao & Zheng, 1998), a requirement whose fulfillment is extremely difficult to verify. In order to take memory effects of the sampling into account, the metadynamics method has been suggested (Barducci, Bonomi, & Parrinello, 2011; Barducci et al., 2006). In this method, adaptive, biased MD simulations are done using collective variables, allowing to selectively populate desired configurations away from the local free energy minima.

A much-discussed alternative is combining of implicit solvent models (Onufriev & Izadi, 2017) with coarse-grained representation of the protein (Kmiecik et al., 2016; Saunders & Voth, 2013). In the latter method groups of atoms, or even entire residues or peptide chains, are treated as one particle. This decreases the computational time, whereas higher resolution conformations can be recovered from the final results (Habibi, Rottler, & Plotkin, 2017a, 2017b; Rojas, Maisuradze, Kachlishvili, Scheraga, & Maisuradze, 2017). However, the decrease in computational demand is achieved at the expense of strong dependence on the detail of the coarse-graining and the choice of the corresponding force field. Despite their limitations in simulation time, presently all-atom classical MD simulations and REMD offer an optimal trade-off between reliability and efficiency to study protein misfolding.

4.4 Analysis of molecular dynamics trajectories

The positions and velocities of atoms as functions of time as retrieved from MD trajectories per se might be a source of valuable information. However, in most cases drawing conclusion about the nature of misfolding events requires post-simulation processing and interpretation of the data in terms of physical, chemical, and structural biology descriptors that could be compared with experiments, or used to interpret experimental findings. For these purposes, many techniques are available and used extensively.

Since amyloidogenic misfolding, by definition, results in a conversion of proteins from their native form into cross-β structured fibrils, changes in secondary structure (SS) as predicted by MD simulations are among the most commonly reported analyses. Fig. 6A shows an example of SS analysis for native PrP. Variations in secondary structure predicted in MD simulations are commonly interpreted as an indication of weak conformational integrity, local structural instability, and/or a sign of early stages for misfolding (Barducci, Chelli, Procacci, & Schettino, 2005; Blinov, Berjanskii, Wishart, & Stepanova, 2009; Chakroun et al., 2013; Cheng & Daggett, 2014; Hannaoui et al., 2017; Menon & Sengupta, 2017; Zhang, 2010). Notable examples of SS changes in PrP include displacement and/or loss of integrity of helix H1 (Chakroun et al., 2013; Tseng, Yu, & Lee, 2009), loss of contacts between H1 and loop LH2H3 (Cheng & Daggett, 2014; De Simone, Zagari, & Derreumaux, 2007; Meli, Gasset, & Colombo, 2011), or partial unfolding of the C-termini of helices H2 and H3 (Chandrasekaran & Rajasekaran, 2016; Hannaoui et al., 2017; Meli et al., 2011; Tomobe, Yamamoto, Akimoto, Yasui, & Yasuoka, 2016) as seen in experiments (Sengupta & Udgaonkar, 2018).

Most amyloidogenic proteins contain at least one intrinsically disordered domain, but some lack structured domains entirely and exist as completely intrinsically disordered proteins (IDPs) (Babu, 2016; Uversky, 2010). Such peptide chains lack a stable structure under physiological conditions, and adopt a continuum of conformations rich in random coils. In PrP, the first ~100 residues of the N-terminal region (23–126) are largely unstructured (Fig. 2A). According to a model of prion fibrils derived from NMR experiments, a region of the N-terminal tail 106–126 may adopt a β-hairpin structure in misfolded PrP (Abskharon et al., 2014; Groveman et al., 2014). Indeed, it is not unusual for MD simulations to predict the appearance of transient SS in the tail region, including β-strands (Cheng & Daggett, 2014; Cong et al., 2013; Hannaoui et al., 2017; Mercer et al., 2018). By engaging a templating mechanism and biased MD, it was possible to simulate the conversion of the folded domain of PrP into a right-handed β-solenoid structure, which is thought to approximate the structure of PrP^{Sc} (Spagnolli et al., 2019).

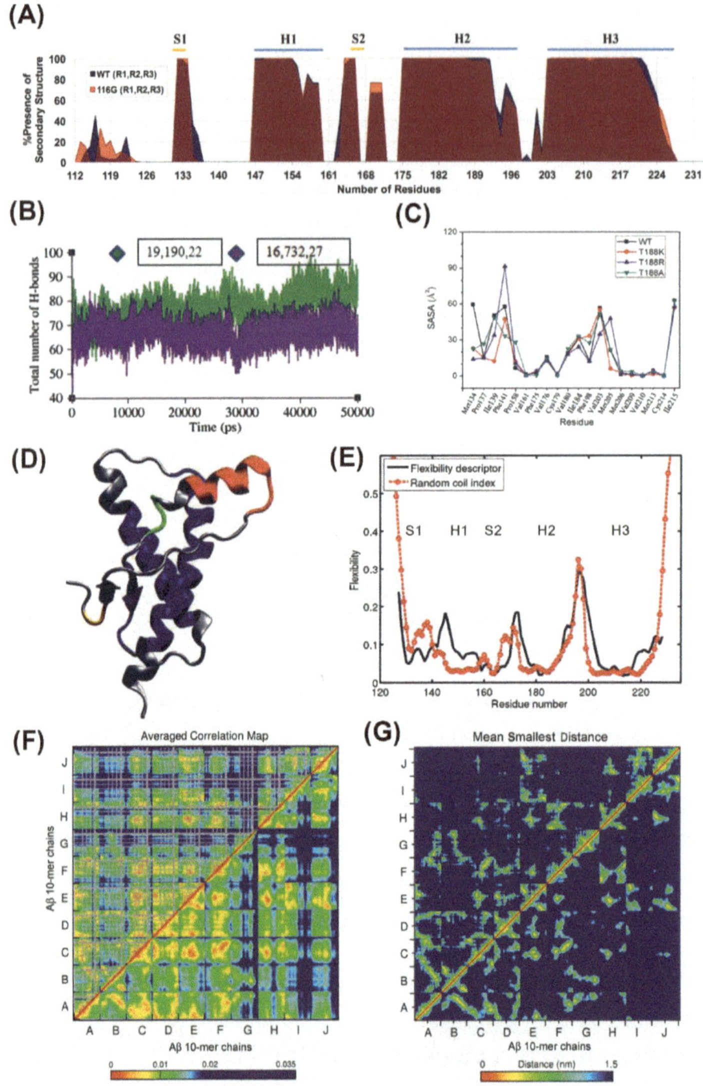

Fig. 6 Examples of MD data analyses for amyloidogenic proteins. (A) percentage of secondary structure in wild-type (WT) and mutant (A116G) cervid PrP (Hannaoui et al., 2017). (B) total number of hydrogen bonds as a function of MD simulation time, and the corresponding averages, for WT (green) and mutant V210I (indigo) human PrP (Chandrasekaran & Rajasekaran, 2016). (C) SASA of WT human PrP and mutants Y188K, T188R, and T188A (Guo et al., 2012). (D) ECD domains of correlated motion color-mapped onto the secondary structure of human PrP (Blinov et al., 2009). The blue color corresponds to the largest dynamics domain; it is followed by red, green, and yellow colors in order of decreasing domain size. Off-domain regions are shown in gray. (E) ECD backbone flexibility profile (black) compared to NMR-derived random coil index

Structurally, the Aβ peptide, α-synuclein, and tau protein in their entirety adopt disordered conformations in water (Fig. 2B), although $A\beta_{1-40/42}$ peptides are mainly hydrophobic, and do not strictly classify as extended IDP. MD studies of Aβ peptide and α-synuclein have been reviewed recently (Berhanu & Hansmann, 2014; Coskuner-Weber & Uversky, 2018; Nagel-Steger et al., 2016; Tran & Ha-Duong, 2015). In $A\beta_{1-42}$ and $A\beta_{1-40}$ monomers or their fragments, MD models predict transient β-content at many locations, especially in regions 2–6, 16–21 and 27–36, which are also relatively less prone to express turns/bends (Rosenman et al., 2016; Tran & Ha-Duong, 2015). In Aβ dimers, these regions show a propensity to develop stable inter-peptide β-sheets (Man et al., 2017b; Man, Nguyen, & Derreumaux, 2017a; Nagel-Steger et al., 2016), with apparent preference for anti-parallel orientation. Transient α-helices are observed occasionally in MD simulations of Aβ (Man et al., 2017a, 2017b; Tran & Ha-Duong, 2015).

Root mean square deviation (RMSD) of backbone atoms in conformations generated by MD from an initial structure is a second major structural characteristic in MD analyses. RMSD provides a simple quantitative assessment of general stability of the analyzed structure, and as such it is commonly used to verify convergence of MD simulations. It is natural to associate large RMSDs of globular domains in proteins with their structural vulnerability, as for example in case of pathogenic PrP mutants (Behmard, Abdolmaleki, & Asadabadi, 2012; Doss et al., 2013; Hannaoui et al., 2017; Menon & Sengupta, 2017; Zhang, 2010). A closely related characteristic, root mean-square fluctuations (RMSFs), are calculated at a per-residue

(red) of human PrP (Blinov et al., 2009). (F) ECD pair correlation map for a self-assembled $A\beta_{17-42}$ oligomer consisting of ten collapsed chains (Dorosh & Stepanova, 2017). In the map, strong side-chain ECD correlations (above the diagonal) and backbone correlations (below the diagonal) are colored red, orange, and yellow. Weaker correlations are colored blue, and gray indicates the absence of side-chains for glycines. (G) Mean smallest distance map for the $A\beta_{17-42}$ oligomer (Dorosh & Stepanova, 2017). Short mean distances between side-chains (above the diagonal) and main-chain (below the diagonal) are colored in red, orange, and yellow, and longer distances are colored in blue. *(A) is reproduced from Hannaoui et al. (2017), under the Creative Commons License; (B) and (F,G) are reproduced from Chandrasekaran and Rajasekaran (2016), and Dorosh & Stepanova (2017), © 2016 RSC respectively, with permission from the Royal Society of Chemistry; (C) is reproduced from Guo et al. (2012), © 2012 Elsevier, under the STM permission; (D,E) are reproduced with permission from Blinov et al. (2009), © 2009 American Chemical Society.*

basis, and provide an indication for local backbone mobility. In globular proteins, RMSF are usually high in regions of flexibile loops, and relatively low for stable SS elements. RMSF are frequently used to assess local stability of PrP (Behmard et al., 2012; Chamachi & Chakrabarty, 2017; Cheng & Daggett, 2014; Doss et al., 2013; Garrec, Tavernelli, & Rothlisberger, 2013; Hannaoui et al., 2017) and other globular proteins.

To characterize dimensions of proteins or their aggregates, the radius of gyration (Rg) is frequently used. Rg is determined as the root-mean-square distance of atoms or residues from the center of mass of the chain or aggregate. It is a simple but useful measure of effective sizes, which can be compared directly with DLS or SAXS measurements. As such, Rg is utilized broadly to assess the quality of MD models for Aβ peptide and α-synuclein (Coskuner-Weber & Uversky, 2018; Man et al., 2017b; Rosenman et al., 2016). A decrease in Rg is interpreted as a greater compactness of the system, resulting in more inter-atomic interactions and a lesser solvent exposure of the chains.

Electrostatic interactions between polar groups determine stability of major SS elements in a protein, and also play an important role in holding the tertiary structure together (Huggins, 2016; Pace et al., 2014; Zhou & Pang, 2018). Importantly, identification of hydrogen bonds (HBs) and salt bridges (SB) formed or lost provides crucial atomic-level detail behind the propensity of a protein to misfold (Caldarulo, Barducci, Wüthrich, & Parrinello, 2017; Chen, van der Kamp, & Daggett, 2014; Guest, Cashman, & Plotkin, 2010), or on the stability of an aggregate (Dorosh & Stepanova, 2017; Man et al., 2017b; Nguyen, Li, Stock, Straub, & Thirumalai, 2007). In PrP for example, the bonding network determined from MD simulations was found to depend on temperature (Chamachi & Chakrabarty, 2017), pH (Garrec et al., 2013; Giachin et al., 2015), presence of metal ions (Giachin et al., 2015), or certain mutations (Chandrasekaran & Rajasekaran, 2016; Jahandideh, Jamalan, & Faridounnia, 2015; Zhou, Shi, Liu, Liu, & Yao, 2016), see also Fig. 6B. It has been hypothesized that the onset of PrP misfolding involves a disruption of the bonding network between helix H2 and other structural elements, resulting in an exposure of hydrophobic region between helices H2, H3, and the β-sheet bundle S1−S2 (Chakroun et al., 2013; Chen et al., 2014; Cheng & Daggett, 2014; Menon & Sengupta, 2015). To characterize such changes, calculations of the solvent accessible surface area (SASA) are employed extensively (Chakroun et al., 2013; Chen et al., 2014; Menon & Sengupta, 2015), as illustrated in Fig. 6C. An increased exposure of hydrophobic residues is often interpreted as a sign of high structural vulnerability in an aqueous environment. For the Aβ peptide, hydrophobic SASA determines the dock-lock process of fibril elongation (Nguyen et al., 2007). In an even broader context, the toxicity of misfolded species is attributed to aberrant interactions of hydrophobic groups at their surfaces (Bemporad & Chiti, 2012; Breydo & Uversky, 2015; Chiti & Dobson, 2017; Hartl, 2017), making insights from SASA analyses unique for understanding the misfolding mechanisms. In MD simulations of aggregation, the SASA of individual chains is interpreted in terms of their propensity to engage in various interactions, whereas buildup of inter-molecular HB and SH is employed as a probe of oligomerization (Dorosh & Stepanova, 2017; Man et al., 2017b; Zhou et al., 2016).

One of strengths of MD simulations is that the resulting trajectories capture motions of all atoms in a protein. However, descriptors such as the secondary structure, or the number of HB and SB, or SASA, although tremendously important, are rather "static" in a sense that they capture one particular snapshot of a trajectory, or an average over a selection of snapshots. To understand molecular mechanisms of misfolding in greater depth, descriptors representing structural stability in the context of atomic motion (dynamics) are required. The novel essential collective dynamics (ECD) method relies upon a recently developed statistical-mechanical framework (Blinov et al., 2009; Issack, Berjanskii, Wishart, & Stepanova, 2012; Potapov & Stepanova, 2012; Stepanova, 2007), according to which a macromolecule can be described by generalized Langevin equations with a set of essential collective coordinates. The latter can be identified as principal eigenvectors of a covariance matrix, which is calculated by applying the principal component analysis on MD trajectories. Usually, 10–30 essential collective coordinates are sufficient to sample approximately 90% of the displacement for a typical MD trajectory of a protein. In the ECD method, an all-atom projected image of the protein is constructed in a multi-dimensional space of the essential collective coordinates, such that each atom is represented by a point (atom-image). It has been shown that such projected image represents the degree of dynamic correlation (coupling) between the protein's atoms: two atom-images located close to each other indicate that motions of the corresponding atoms are strongly correlated regardless of their proximity in secondary or tertiary structure of the protein, whereas more distant atom-images correspond to a relatively independent motions (Stepanova, 2007).

A suite of simple dynamical descriptors has been derived within the ECD framework. For example ECD domains of collective motion, which represent relatively rigid parts of the protein composed of atoms moving coherently (Fig. 6D), can be identified through a nearest-neighbor clustering of the protein's atom-images (Stepanova, 2007). Within the same framework, the ECD flexibility descriptor (Fig. 6E) represents the level of dynamic coupling of the motion of individual atoms or residues with the entire molecule (Blinov et al., 2009; Dorosh & Stepanova, 2017; Issack et al., 2012; Mane & Stepanova, 2016; Santo, Berjanskii, Wishart, & Stepanova, 2011). In globular proteins, high levels of the flexibility descriptor usually correspond to flexible loops, whereas minima indicate α-helices or β-sheets, reminiscent of RMSF profiles. Moreover, merely visualizing the distances between atom-images of a protein in a form of a map provides a simple but effective way to assess pair correlations of motion between individual atoms, or groups of atoms (Dorosh & Stepanova, 2017; Issack et al., 2012; Mane & Stepanova, 2016), see also Fig. 6F. As the figure illustrates, ECD pair correlations often tend to follow the corresponding inter-atomic distances (Fig. 6G); however unlike the distances, the pair correlations also capture mediated indirect interactions. It has been proven analytically (Potapov & Stepanova, 2012) that the ECD framework represents invariant (highly stable) correlations of atomic motions in a protein. As such, the predictions can be extrapolated beyond the timescale of MD trajectories employed, making exhaustive sampling of MD conformations unnecessary for accurate predictions. Indeed, ECD dynamic domains and backbone flexibility profiles derived from nanosecond-scale MD simulations agree very well with NMR-derived structural data that represent longer timescales (Blinov et al., 2009; Issack et al., 2012; Santo et al., 2011; Stepanova, 2007). The ECD method has been successfully employed to characterize dynamical stability of PrP monomers and dimers (Blinov et al., 2009; Issack et al., 2012; Santo et al., 2011), α-synuclein dimers and tetramers (Mane & Stepanova, 2016), and self-assembled Aβ oligomers (Dorosh & Stepanova, 2017).

5. Integrative approaches in understanding misfolding and aggregation

An overarching line of inquiry in the studies of protein misfolding is, what chains of molecular events lead to pathogenic transformations related to diseases, and what is the structure and morphology of the corresponding intermediate states, oligomers, and other misfolded species. In particular, it is important to understand specific mechanisms of both primary and secondary nucleation (Fig. 1). Primary nucleation is initiated by spontaneous aggregation of monomeric proteins. This is preceded or accompanied by proteins adopting un-natural aggregation-competent conformations. Secondary nucleation involves an autocatalytic aggregation of native proteins or their misfolded intermediate states in the presence of existing fibrillar species (Cremades & Dobson, 2018; Ke et al., 2017; Knowles et al., 2014; Törnquist et al., 2018), which enhances the amyloidogenic process through a positive-feedback reaction loop (Michaels et al., 2018). This mechanism also is believed to be responsible for transmissible types of misfolding diseases (Chiti & Dobson, 2017; Soto & Pritzkow, 2018). As such, both types of nucleation require a detailed investigation in molecular-level detail. Possible influences of various extrinsic factors also beg for a better understanding. Although no single experimental method or computational model would be capable of capturing all aspects of the amyloidogenic processes in their entirety, observations from diverse biophysical and biochemical assays complemented by insights from MD modeling are enabling an unprecedented balance of depth and breadth in this challenging research area. Below we present recent examples of integrative analyses combining multiple techniques to decipher mechanisms of misfolding and aggregation. By no means does this selection of examples pretend to be exhaustive. Due to the large number of excellent studies published over the recent years, complete coverage of all achievements would not be possible in a brief review like this. However, we hope that this non-limiting set of examples will stimulate both new integrative studies, and the usage of existing knowledge to advance research on protein misfolding.

5.1 Effects of mutations, truncations and insertions on protein structure and stability

Although most protein misfolding diseases are believed to occur sporadically, roughly 10% are hereditary, or familial — related to mutations in particular proteins (Chiti & Dobson, 2006, 2017; Cremades & Dobson, 2018). This raises a question, how mutations, or other variations of the amino acid sequence, influence misfolding of proteins at the molecular level.

In case of PrP, more than 50 mutations have been associated with familial types of SEs, especially the Gerstmann-Sträussler-Scheinker syndrome (GSS) and familial Creutzfeldt-Jakob disease (CJD), as well as fatal familial insomnia (FFI) (Minikel et al., 2016; Schmitz et al., 2017; Takada et al., 2017). Some mutations have not been linked to a disease, or their role is unknown. Influence of pathogenic mutations onto the native structure of recombinant PrP monomers has been addressed in many studies, primarily by comparing their XRD- or NMR-based high-resolution structures, as well as other NMR-derived structural descriptors, with those of WT PrP (Biljan, Ilc, & Plavec, 2017; Yu et al., 2016; Giachin, Biljan, Ilc, Plavec, & Legname, 2013; Lee et al., 2010; Bae et al., 2009; and references therein). Often, such analyses find that the global backbone structure of monomeric mutant PrP is remarkably similar to that of WT PrP (Bae et al., 2009; Biljan et al., 2017; Yu et al., 2016). However, several point mutations have been found to induce pronounced structural perturbations. In case of the mutation Q212P which is implicated in GSS, P212 disrupts helix H3, resulting in a change of the mutual orientations of helices H2 and H3 accompanied by a loss of contacts between loop LS2H2 and the C-terminus of H3, exposing hydrophobic regions to the solvent (Biljan et al., 2017; Giachin et al., 2013). The substitution D178 N (CJD and FFI) alters interactions of H2 with loop LS2H2, whereas the substitution F198S (GSS) influences mobility of the loop and adjacent areas, (Lee et al., 2010).

MD models by their nature are very well suited for probing the molecular-level impact of mutations, or other variations of the amino acid sequence. Replacements of individual residues, deletions, or insertions in the initial structure can be done easily in most cases, and the resulting alternations in conformational dynamics can be tracked down directly. A vast number of studies employed MD simulations to investigate the effect of point mutations on PrP (Rossetti & Carloni, 2017; Meli et al., 2011; Rossetti, Cong, Caliandro, Legname, & Carloni, 2011; Behmard et al., 2012; Guo, Ning, Ren, Liu, & Yao, 2012; Cong et al., 2013; Doss et al., 2013; Jahandideh et al., 2015; Chandrasekaran & Rajasekaran, 2016; van der Kamp & Daggett, 2010a; and references therein). Numerous MD studies indicate that the pathogenic mutations V180I, F198S, and V210I in the area of the hydrophobic core of PrP may destabilize the structure and dynamics of the bonding network across helices H2, H3, and the S1–S2 bundle (van der Kamp & Daggett, 2010a; Rossetti et al., 2011; Meli et al., 2011; Behmard et al., 2012; Doss et al., 2013; Jahandideh et al., 2015; Chandrasekaran & Rajasekaran, 2016). As a consequence, various characteristics such as the secondary structure (Fig. 6A), and the number of hydrogen bonds (Fig. 6B) of PrP are frequently found altered. For instance, MD studies of PrP monomers and dimers indicated that the T183A mutation may facilitate a conversion $\alpha \rightarrow \beta$ in helix H2, (Chebaro & Derreumaux, 2009). As expected, mutants such as the substitution Q212P exhibit significant differences of bonding between helix H3 and other regions in comparison to WT PrP

(Jahandideh et al., 2015; Rossetti et al., 2011). Other pathogenic substitutions in the globular domain, D178N, T188R/K/A and E200K, also have been found to alter the bonding network in PrP (Doss et al., 2013; Guo et al., 2012; Meli et al., 2011; Rossetti et al., 2011). Comparative analyses of N-terminal mutations Q52P, P101L, P104L/T/S, G113V, and A116V in moPrP (Cong et al., 2013) revealed substantial variations in main-chain flexibility of the N-terminal region across the mutants, with an increase in β-content for P101L (GSS). A recent MD study of a LGGLGGYV insertion between G125 and G126 of moPrP with a M128V-substitution, which caused GSS in the original carrier and in transgenic mouse models (Mercer et al., 2018), showed increased β-content around the insert, and also a substantial disruption of helix H3 (Fig. 7), which may explain the rapid misfolding and aggregation of this construct observed experimentally.

The comparison of numerous MD results for various PrP mutations is challenging due to the often different conditions used, such as the temperature, pH, PrP construct length, and force fields; as well as often subtle differences across the mutants. Only few studies have investigated large sets of mutations and compared their effects on the PrP structure within the same computational framework (Cong et al., 2013; Doss et al., 2013; Meli et al., 2011; Rosetti et al., 2011; van der Kamp & Daggett, 2010a). In an MD study of 40 point mutations (Rossetti et al., 2011) common structural traits were identified for many pathogenic mutations, such as a loss of salt bridges in the H2–H3 regions, along with a loss of π-stacking interactions of loop LS2H2 with helix H3, resulting in an exposure of residue Y169. Such exposures have been hypothesized to facilitate pathogenic misfolding of PrP (Rossetti & Carloni, 2017). However overall, both computational (Meli et al., 2011; Rossetti et al., 2011; van der Kamp & Daggett, 2010a) and experimental (Bae et al., 2009; Biljan et al., 2017; Yu et al., 2016) studies indicate that not all pathogenic mutations necessarily decrease the structural stability of monomeric PrP, and the identified variations in the monomer stability do not correlate unambiguously with the pathogenic conversion susceptibility (Yu et al., 2016; Meli et al., 2011; Sabareesan & Udgaonkar, 2016). It has been hypothesized that mutations may either induce an increase in the population of partially folded intermediate states of PrP (Apetri, Surewicz, & Surewicz, 2004), or accelerate oligomerization that precedes misfolding (Sabareesan & Udgaonkar, 2016). Generalizing, one may expect pathogenic mutations to favor the conversion process through a multitude of mechanisms (Chiti & Dobson, 2017), resulting in different kinetic pathways,

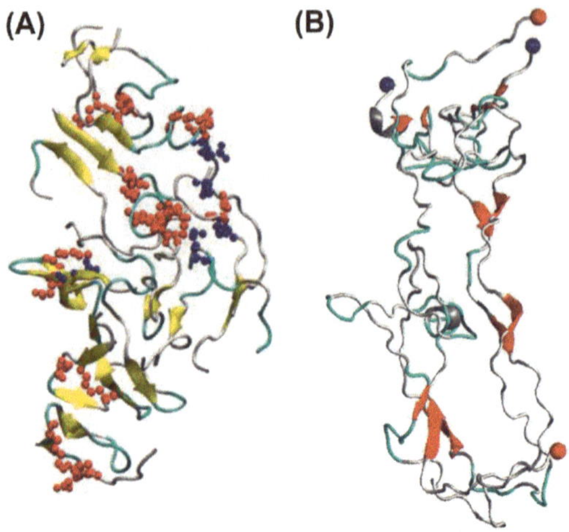

Fig. 7 Examples of unstructured oligomers of $A\beta_{17\text{-}42}$ (A) and α-synuclein (B), as predicted by MD simulations. Panel (A) depicts an $A\beta_{17\text{-}42}$ decamer formed by spontaneous aggregation of ten randomly positioned Aβ chains in water after 200 ns simulations (Dorosh & Stepanova, 2017). Residues participating in intra-chain and inter-chain salt bridges are shown by red and blue spheres, respectively, and β-sheet content is highlighted in yellow. Although both parallel and antiparallel β-sheets are observed, the quaternary structure is dominated by random coils and loops without a pronounced long-range alignment. Panel (B) shows a collapsed α-synuclein dimer after a 100 ns MD simulation of two α-synuclein chains, which initially were aligned head-to-head in an unfolded conformation (Mane & Stepanova, 2016). N− and C-termini are indicated by blue and red spheres, respectively, and β-sheet content is highlighted in red. Despite the collapsed quaternary structure with many random coils, the central parts of the two chains retain their alignment relative to each other (Mane & Stepanova, 2016). However, the C-termini unbuckled and adopted antiparallel positions with respect to adjacent regions, giving rise to several antiparallel β-sheets. *(A) is reproduced with permission from Dorosh & Stepanova (2017), © 2016 RSC; (B) is reproduced from Mane and Stepanova (2016) under the Creative Commons License.*

quaternary organizations, and toxicities (Meli et al., 2011). This suggests shifting the focus from mutation-driven perturbations of monomeric native PrP to the role of mutations in a broader context of the conversion/aggregation process (Hadži, Ondračka, Jerala, & Hafner-Bratkovič, 2015; Sabareesan & Udgaonkar, 2016; Yu et al., 2016).

The impacts of pathogenic mutations onto monomeric conformations and aggregation propensities of intrinsically disordered proteins have been examined extensively, both experimentally and by MD simulations (Breydo, Redington, & Uversky, 2017; Chiti & Dobson, 2017; Chong et al., 2017; Coskuner-Weber & Uversky, 2018). Familial types of

Parkinsonism are related to point mutations A30P, E46K, H50Q, G51D, A53E, and A53T in α-synuclein (Wong & Kranic, 2017). Solution-state NMR and CD analyses of mutant proteins (Ghosh et al., 2014; Ghosh et al., 2013; Bertoncini, Fernandez, Griesinger, Jovin, & Zweckstetter, 2005) show structural features characteristic of unfolded conformations similar to WT α-synuclein. *In-vitro* aggregation analyses combining fluorometry, CD and FTIR spectroscopies complemented by transmission EM or AFM imaging indicate that substitutions E46K, H50Q, and A53T may facilitate formation of amyloidal fibrils (Conway, Harper, & Lansbury, 1998; Ghosh et al., 2013; Ghosh et al., 2014; Greenbaum et al., 2005), whereas substitutions G51D and A53E rather decrease the aggregation rate (Ghosh et al., 2014; Fares et al., 2014). The impact of the A30P mutation appears more sensitive to experimental conditions than that of the other substitutions, with a propensity to promote formation of oligomers rather than fibrils (Ghosh et al., 2014; Conway et al., 2000; Conway et al., 1998). Recently, solution NMR experiments have captured pH-depended rearrangements of interatomic contacts between the mutation sites and the C-terminus in monomeric mutant proteins, which may explain differences in the corresponding *in-vitro* aggregation rates (Ranjan & Kumar, 2017). In agreement with these experiments, MD simulations of human α-synuclein mutants A30P (Wise-Scira, Aloglu, Dunn, Sakallioglu, & Coskuner, 2013) and A53T (Coskuner & Wise-Scira, 2013; Losasso, Pietropaolo, Zannoni, Gustincich, & Carloni, 2011) revealed significant local and long-distance impacts of the mutations. This included a loss or decrease of long-range contacts between both terminal domains and the central region known as the NAC domain (A53T); or between the C-terminal domain and the NAC domain (A30P) in comparison with WT α-synuclein. This has led to an overall decrease of compactness of the conformational ensembles in these mutants. In contrast, the mutant E46K (Wise-Scira, Dunn, Aloglu, Sakallioglu, & Coskuner, 2013b) exhibited a greater likelihood of long-range contacts between the C-terminal, the NAC domain, and the adjacent region in the N-terminal domain, as well as a greater overall compactness. These findings suggest that seeding behaviors of the E46K mutant in water may involve fundamentally different mechanisms than those of the A30P and A53T mutants, explaining different conformational landscapes of pre-fibrillar oligomeric states observed recently by NMR (Bhattacharyya et al., 2018). Interestingly, all pathogenic mutations are located in the N-terminal domain, which is intrinsically disordered in water; however, it adopts a stable α-helical structure upon binding to lipid membranes. The question to what extent insertion into a membrane might play a role, has been addressed in a recent MD study (Tsigelny et al., 2015). The study indicated that pathogenic mutants tend to adopt conformations that favor more frequent interactions of the N-terminal domain with the membrane, in comparison with WT α-synuclein. This in turn has led to a hypothesis, that insertion into a membrane environment may play a crucial role in the oligomerization of pathogenic α-synuclein mutants (Tsigelny et al., 2015). Fluorometric analyses complemented by AFM imaging (Flagmeier et al., 2016) confirmed a strong influence of the mutations onto the aggregation propensities of membrane-bound α-synuclein. However, no clear links between these trends and the pathogenic phenotypes have been established yet (Chiti & Dobson, 2017).

The Aβ peptide exists in multiple alloforms. The most abundant ones, $A\beta_{1\text{-}42}$ and $A\beta_{1\text{-}40}$, differ by two C-terminal residues, I41 and A42. Although $A\beta_{1\text{-}40}$ variant is more abundant than $A\beta_{1\text{-}42}$, the latter is believed to be more pathogenic (Murphy & LeVine III, 2010; Nasica-Labouze et al., 2015). Many *in-vitro* studies indicate that the aggregation rates are significantly higher for $A\beta_{1\text{-}42}$ than for $A\beta_{1\text{-}40}$ (Bitan et al., 2003; Jarrett, Berger, & Lansbury, 1993; Meisl et al., 2014). However, solution-state NMR (Riek et al., 2001; Roche et al., 2016) and single-molecule FRET (Meng et al., 2018) revealed close structural similarities of monomeric $A\beta_{1\text{-}40}$ and $A\beta_{1\text{-}42}$, suggesting that the differences in aggregation rates originate from different mechanisms, rather than the initial conformations of the monomers. Specific predictions based on MD simulations vary somewhat across different studies; however, most authors agree that the two alloforms of Aβ tend to exhibit different dynamic trends when modeled in a single framework. The presence of the two hydrophobic C-terminal residues may lead to a higher propensity to form β-sheets in the C-terminal and central regions (Côté, Derreumaux, & Mousseau, 2011; Lin, Bowman, Beauchamp, & Pande, 2012; Meng et al., 2018; Rosenman et al., 2016) and a greater hydrophobic collapse (Song, Wang, Colletier, Yang, & Xu, 2015) of $A\beta_{1\text{-}42}$. Because of this length dependence, caution is recommended when extrapolating results from $A\beta_{1\text{-}40}$ to $A\beta_{1\text{-}42}$ and *vice versa* (Coskuner-Weber & Uversky, 2018).

28 pathogenic mutations within the 42-residue Aβ sequence have been reported (Hunter & Brayne, 2018). Not all of these mutants have been studied in detail yet; however most of those which have been characterized, exhibit faster aggregation kinetics in comparison to their WT counterparts (Hatami, Monjazeb, Milton, & Glabe, 2017; Nasica-Labouze et al., 2015).

In-vitro studies by a combination of biophysical and biochemical methods indicate that N-terminal mutants H6R and D7N are prone to form both fibrils (Hatami et al., 2017) and oligomeric aggregates (Ono, Condron, & Teplow, 2010) at a higher rate than WT Aβ. Substitutions in central region A21G and E22K were reported to favor predominantly amorphous aggregates (Hatami et al., 2017); whereas substitutions E22Q/G and E23N (Betts et al., 2008; Hatami et al., 2017), as well as the E22Δ deletion (Ovchinnikova, Finder, Vodopivec, Nitsch, & Glockshuber, 2011), were found to be fibrillogenic. In most cases, the observed increase in aggregation propensity of both $A\beta_{1\text{-}40}$ and $A\beta_{1\text{-}42}$ mutants tends to correlate with a greater cytotoxicity (Murakami et al., 2002; Ono et al., 2010; Ovchinnikova et al., 2011). Mass-spectrometry (Gessel, Bernstein, Kemper, Teplow, & Bowers, 2012), CD (Yamamoto, Hasegawa, Matsuzaki, Naiki, & Yanagisawa, 2004) and FTIR (Perálvarez-Marín et al., 2009) analyses suggest that the various mutants populate different aggregation states and express different secondary structures at early stages of the process. MD simulations clarify possible origins of these differences (Coskuner-Weber & Uversky, 2018; Nasica-Labouze et al., 2015). Changes of the electrostatic profile due to the substitutions at position 22 were found to exert both local and distant alternations on transient secondary structure and conformations in $A\beta_{1\text{-}42}$ (Coskuner, Wise-Scira, Perry, & Kitahara, 2012; Lin et al., 2012; Lin & Pande, 2012) and $A\beta_{1\text{-}40}$ (Coskuner et al., 2012; Côté et al., 2011; Rosenman et al., 2016). The substitution D23N in $A\beta_{1\text{-}40}$ dimers has altered structural motifs in the C-terminal region (Côté, Laghaei, Derreumaux, & Mousseau, 2012), whereas the mutations H6R (Viet, Nguyen, Derreumaux, & Li, 2014) and D7N (Viet, Nguyen, Ngo, Li, & Derreumaux, 2013) were found to influence the dynamics of salt-bridging, and in the case of D7N, also alter the flexibility of residues 23–28 in the dimers. Overall, the MD results indicate that the various mutations influence the conformational dynamics of Aβ in distinct ways, resulting in different inter-chain contact preferences during aggregation.

Comparison of the mutants E22K/G/Δ and D23N in $A\beta_{1\text{-}40}$ monomers with several force fields and water models (Rosenman et al., 2016) also suggests that prevalent intra-molecular contacts may be influenced by the force field. For example, OPLS-AA simulations predicted a register shift of anti-parallel contacts in the central region of $A\beta_{1\text{-}40}$ E22K mutant, whereas AMBER99SB-ILDN simulations predicted merely an increase in transient β-content in comparison with WT $A\beta_{1\text{-}40}$. While these differences are relatively subtle, they may lead to different interpretations of fast aggregation rates observed in the mutant experimentally (Rosenman et al., 2016). Careful comparison of such predictions with experimental observables is required to draw ultimate conclusions regarding molecular mechanisms of Aβ seeding.

5.2 Intermediate states, oligomers, and molecular basis of aggregation

Relatively small, soluble pre-fibrillar species are hypothetically the most toxic products of amyloidogenic misfolding and aggregation (Cremades & Dobson, 2018; Soto & Pritzkow, 2018). At the same time, due to their transient nature and high variability, pre-fibrillar species are difficult to investigate both experimentally and computationally. It has been hypothesized that globular proteins may adopt an un-natural, aggregation-competent intermediate state prior or during oligomerization (Apetri et al., 2004; Sabareesan & Udgaonkar, 2016; Uversky & Fink, 2004). However, the specific structure of such intermediate states is not known. As hypothetical precursors of primary nucleation, dimers and larger multimeric assemblies of PrP have been studied by MD simulations (Chamachi & Chakrabarty, 2016; Chebaro & Derreumaux, 2009; Issack et al., 2012; Lou et al., 2015). In most cases, such simulations predict the globular domain of PrP to retain its integrity at physiological temperatures and pH levels. Recent solution-state NMR characterization complemented by CD spectroscopy (Glaves et al., 2018) has confirmed that recombinant moPrP oligomers (created by shaking *in-vitro*) may retain major structural elements of native monomers, yet with an increase in β-sheet content and alterations in mobility. In particular, a pronounced decrease of mobility of most residues between position ~127 and 225 was observed in the oligomers (Glaves et al., 2018). Early oligomers from other globular proteins such as insulin or superoxide dismutase 1 (SOD1) were also found to largely preserve their native folds at early stages of primary nucleation (Breydo & Uversky, 2015; Chiti & Dobson, 2017), although details of these structures have yet to be determined. In the absence of unambiguous models for oligomers, MD simulations of amyloidogenic PrP fragments have been reported, especially addressing the $PrP_{106-126}$ fragment, which is known to adopt a β-hairpin strand-loop-strand structure (Nagel-Steger et al., 2016). Pre-built $moPrP_{106-126}$ octamers composed of two four-stranded parallel β-sheets remained stable during MD simulations (Kuwata et al., 2003). However, REMD simulations of two, three, and four randomly positioned huPrP$_{106\text{-}126}$ chains produced disordered aggregates with many random coils (Ning et al. 2014).

Intrinsically disordered proteins have been reported to aggregate into oligomeric species of various morphologies (Morel et al., 2018; Luo et al., 2014; Lorenzen et al. 2014). Fig. 3A shows, as an example, a low-resolution EM image of A$\beta_{1\text{-}42}$ oligomers formed *in-vitro*. Characterization of A$\beta_{1\text{-}40/42}$ and α-synuclein oligomers by means of solution-state NMR (Ahmed et al., 2010; Luo et al., 2014; Kotler et al., 2015), FTIR (Morel et al., 2018; Lorenzen et al. 2014), CD (Luo et al., 2014; Kotler et al., 2015; Nath et al., 2010), FRET (Cremades et al., 2012; Nath et al., 2010) and fluorometry (Luo et al., 2014; Morel et al., 2018) indicate a tendency of adopting largely unstructured morphologies, initially without pronounced secondary structure. Unstructured morphologies without pronounced alignment of the chains were also obtained from recent MD and REMD simulations for dimers of A$\beta_{1\text{-}40}$ (Tarus et al., 2015) and A$\beta_{1\text{-}42}$ (Man et al., 2017b), as well as for self-assembled multimeric aggregates of A$\beta_{1\text{-}42}$ (Barz, Olubiyi, & Strodel, 2014) or A$\beta_{17\text{-}42}$ (Dorosh & Stepanova, 2017), in agreement with earlier kinetic Monte-Carlo modeling (Cheon et al., 2007). An example can be seen in Fig. 7A. Parallel and antiparallel dimers and tetramers of α-synuclein have also shown a propensity to collapse into random coil-rich structures in the course of MD simulations (Fig. 7B), yet retain some of the initial alignment of the chains relative to each other (Mane & Stepanova, 2016; Bloch and Miller, 2017). Extensive comparative analyses of aggregation of six small peptides using all-atom unbiased MD simulations with three different force fields (Matthes et al., 2016) has also indicated that the spontaneous oligomerization process initially yields a heterogeneous ensemble of conformations, with the majority of structures at least partially disordered. Hydrophobic collapse accompanied by expulsion of water from inter-chain space and inter-peptide hydrogen bonding appear to drive the formation of such disordered oligomeric aggregates (Cheon et al., 2007; Dorosh & Stepanova, 2017; Matthes et al., 2016; Nagel-Steger et al., 2016).

Characterization of Aβ and α-synuclein aggregates *in vitro* by means of solution-state NMR (Ahmed et al., 2010; Luo et al., 2014), FTIR (Morel et al., 2018), CD (Luo et al., 2014; Nath et al., 2010), fluorometry (Luo et al., 2014; Morel et al., 2018), and FRET (Nath et al., 2010; Cremades et al., 2012) suggests that initially unstructured oligomers may undergo a reorganization into β-sheet rich conformations (Ahmed et al., 2010; Nath et al., 2010; Cremades et al., 2012; Cremades & Dobson, 2018), or mediate formation of β-sheet rich species (Luo et al., 2014; Morel et al., 2018).

According to NMR (Morel et al., 2018; Yu et al., 2009) and FTIR (Chandra et al., 2017; Lorenzen et al. 2014) analyses, small β-sheet rich oligomers tend to express anti-parallel β-sheets, which are atypical for mature amyloid fibrils. Details how exactly the unstructured aggregates nucleate β-structured entities are not completely understood yet. Bis-ANS fluorescence assays (Morel et al., 2018; Younan & Viles, 2015) indicate that formation of β-sheet-rich species is accompanied by an increase in hydrophobicity, suggesting a different mechanism than an initial hydrophobic collapse. On theoretical grounds (Cheon et al., 2007) it has been suggested that the primary nucleation process is driven by a competition of hydrophobic coalescence producing unstructured conformations, and hydrogen bonding resulting in a build-up of β-sheets. According to all-atom MD simulations (Dorosh & Stepanova, 2017), rapid aggregation of $A\beta_{1\text{-}42}$ chains into an unstructured oligomer is followed by a long re-organization of bonding networks and building β-structure. Kinetic Monte-Carlo modeling (Cheon et al., 2007) and more recent all-atom MD simulations (Matthes et al., 2016) of a system of small peptides demonstrated a possibility of a two-step process, with hydrophobicity-driven assembly of peptide monomers into disordered aggregates, followed by a hydrogen bonding-driven formation of β-sheet rich elements (Fig. 8). In the MD simulation (Matthes et al., 2016), anti-parallel β-sheets appeared early in the oligomerization process, and initially disordered oligomeric states did not prevent the conversion into ordered β-sheet rich structures. In terms of collective dynamics, sub-aggregates with strong inter-chain dynamic correlations of motion, as revealed by an ECD analysis of self-assembled Aβ oligomers (Fig. 6D), might indicate an ongoing nucleation of β-sheet rich conformers (Dorosh & Stepanova, 2017).

Elongation of existing β-structured oligomers or proto-fibrillar structures (Fig. 9) is expected to occur in two steps, a fast and reversible "dock" step when a monomer is deposited onto an edge of a fibril, and a slower "lock" process when the deposited chain becomes associated with the fibril and adopts its structure (Esler et al., 2000). Simulations of the "dock-lock" process have been reported for the Aβ peptide (Bacci, Vymětal, Mihajlovic, Caflisch, & Vitalis, 2017; Nguyen et al., 2007; Rodriguez, Chen, Plascencia-Villa, & Perry, 2018; Schwierz, Frost, Geissler, & Zacharias, 2016) and recently PrP (Spagnolli et al., 2019). From these simulations, hydrophobic interactions and de-solvation of both the fibril edge and the deposited chain were identified as the main factors that control both the docking and locking steps (Bacci et al., 2017; Rodriguez et al.,

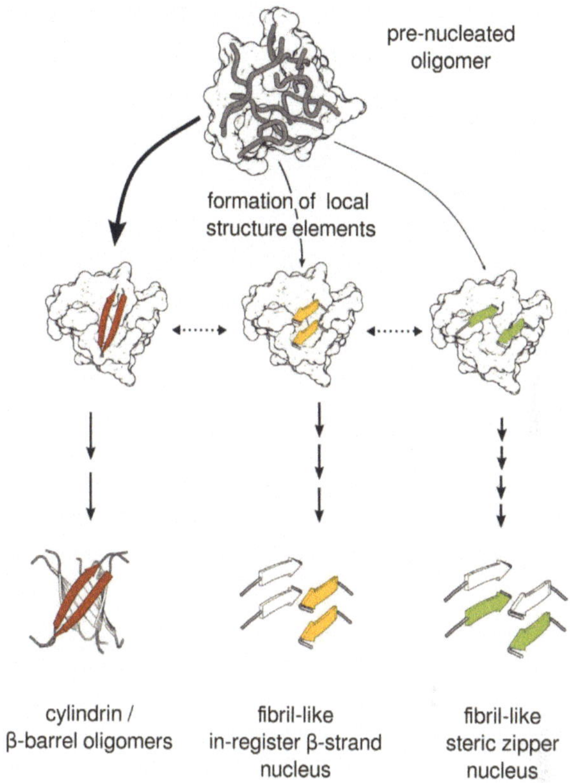

Fig. 8 Schematics of nucleation for β-structures from disordered oligomers (Matthes et al., 2016). Hydrophobicity-driven assembly of peptide monomers into disordered aggregates is followed by spontaneous hydrogen bonding-driven build-up of β-sheet rich fibrillar elements. *Reproduced from Matthes et al. (2016) under the Creative Commons License.*

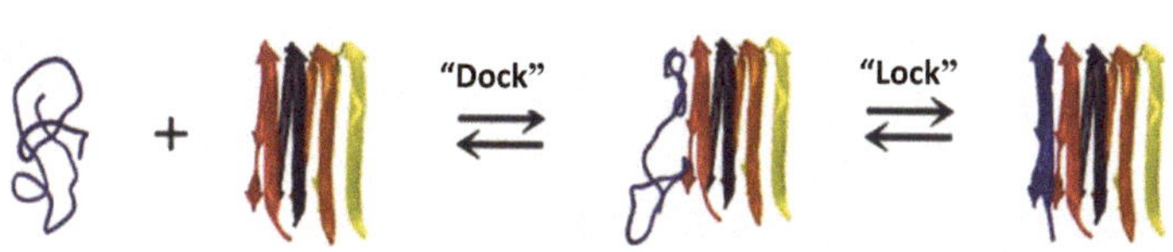

Fig. 9 Schematics of fibril elongation through a dock-lock mechanism (Bacci et al., 2017). The deposited chain is shown in blue. At the initial "dock" stage, a chain is deposited onto to the edge of the amyloid template, where it fluctuates in a reversible manner. Subsequently at the "lock" stage, the deposited chain becomes a part of stable β-sheet structure of the template. *Reproduced with permission from Bacci et al. (2017), © 2017 American Chemical Society.*

2018; Schwierz et al., 2016). Most changes in the conformation of the deposited Aβ monomer were found to occur during the fast "dock" process (Nguyen et al., 2007), which is mediated by transient hydrogen bonds (Schwierz et al., 2016). This is followed by more stable hydrogen bonding responsible for build-up of in-register β-structure during the much slower "lock" step (Nguyen et al., 2007; Schwierz et al., 2016), which is anticipated to be rate-limiting for fibril elongation. Unlike hydrogen bonding, hydrophobic side-chain interactions were found to promote disordered docking of Aβ monomers at fibril's surface rather than in-register alignment (Takeda & Klimov, 2009). The threading of incoming PrP molecules onto the β-solenoid fold of the PrP^{Sc} model also involved a similar and step-wize "dock" and "lock" process, only that the whole procedure was repeated four times for the four rungs of the putative PrP^{Sc} conformation (Spagnolli et al., 2019).

Distinct from primary nucleation events that occur spontaneously, secondary nucleation is catalyzed by the presence of fibrillar species, leading to exponential proliferation of fibrillar structure (Cremades & Dobson, 2018; Knowles et al., 2014; Riek & Eisenberg, 2016; Törnquist et al., 2018). This understanding emerged from experimentally observed kinetics of fibril growth, which cannot be explained by homogeneous, primary nucleation and fibril elongation alone (Cohen et al., 2013, 2018; Michaels et al., 2018; Törnquist et al., 2018). It is expected that secondary nucleation occurs at side surfaces of fibrils (Jeong, Ansaloni, Mezzenga, Lashuel, & Dietler, 2013; Törnquist et al., 2018), although breakage of filaments that increases the number of elongation sites may also play a role (Cohen et al., 2013; Michaels; Knowles, 2014). MD modeling of the deposition of Aβ monomers onto lateral surfaces of fibrillar filaments has revealed a significant contribution of hydrophobic collapse, resulting in disordered morphologies of deposited chains (Barz & Strodel, 2016; Schwierz, Frost, Geissler, & Zacharias, 2017). This appears to contradict the expectation that a fibril's side surface may stabilize monomers in nucleation-competent conformations (Törnquist et al., 2018). Future studies are required to clarify the specific molecular mechanisms of secondary nucleation.

5.3 Influence of extrinsic factors

The challenge of identifying misfolding-competent intermediate states that could initiate or accompany the process of amyloidogenic aggregation has prompted interest in the denaturing action of heat, acidic environment, and chemical agents. On the other hand, the ongoing search for potential therapeutic compounds that could target various steps of the amyloidogenic conversion also requires a thorough understanding of the action of small molecules and other external factors at the molecular level.

Recent MD and REMD simulations, for instance, have addressed the impact of heating onto the structure of PrP and other globular proteins (Chamachi & Chakrabarty, 2017; Steckmann, Bhandari, Chapagain, & Gerstman, 2017; Chen, Duan, Zhu, & Zhang, 2013; and references therein). In a REMD study, thermal unfolding of the small signaling protein ccβ was found to result in the self-assembly of unstructured aggregates, which quickly developed stable steric-zipper β-sheet content in certain temperature regimes (Steckmann et al., 2017).

Exposure of PrP to acidic conditions has been addressed thoroughly with the expectation of capturing misfolded conformations and the misfolding pathways that lead to them. According to solution-state NMR studies, the globular domain of native PrP retains its structural integrity at moderately acidic pH levels (as reviewed by Biljan et al., 2017). However, conformational changes have been reported at pH values below approximately 4.5 (Calzolai & Zahn, 2003; Prigent & Rezaei, 2011), especially with an addition of salts and denaturants such as urea (Bjorndahl et al., 2011; Moulick, Das, & Udgaonkar, 2015; Sengupta & Udgaonkar, 2018; Singh & Udgaonkar, 2016). At a pH of 4, the combined NMR and CD characterization of mouse $PrP_{23\text{-}231}$ (Moulick et al., 2015) revealed the existence of at least two partially unfolded conformers, one with a disordered C-terminal of H3, and another one exhibiting a disruption or detachment of the S1–S2 bundle from the H2–H3 core region. At a pH of 3, structural changes between β-strand S2 and helix H2 and adjacent residues in helix H3 were observed by NMR in Syrian hamster $PrP_{90\text{-}232}$ (Bjorndahl et al., 2011). The putative trigger of acid-induced changes in PrP, protonation of oxidation-susceptible (titratable) residues and the resulting changes in electrostatic bonding networks, have been addressed by numerous MD simulations (Gao, Zhu, Zhang, Zhang, & Mei, 2018; Chen et al., 2014; Cheng & Daggett, 2014; Chakroun et al., 2013; Garrec et al., 2013; van der Kamp & Daggett, 2010b; and references therein). The simulations have predicted an electrostatic repulsion between the protonated imidazole ring of H187 and the nearby side-chain R136 of mouse PrP, which displaces both residues from their native positions, also affecting the C-terminal part of H2 and the loop LH1S1 located nearby (Garrec et al., 2013). This destabilization mechanism was subsequently confirmed experimentally through site-directed mutagenesis of moPrP followed by mass spectrometry assays (Singh & Udgaonkar, 2016). These experimental observations (Singh &

Udgaonkar, 2016) have been explained by a detachment of the N-terminal part of helix H1 from the H2—H3 core exposing hydrophobic regions to solvent, as commonly observed in MD simulations of PrP at acidic pH (Chakroun et al., 2013; Chen et al., 2014; Gao et al., 2018; van der Kamp & Daggett, 2010b). The resulting destabilization of the hydrophobic core is expected to open several pathways of potential misfolding (Cheng & Daggett, 2014; Garrec et al., 2013; Sengupta & Udgaonkar, 2018) eventually resulting in oligomerization. Indeed, PrP is well-known to oligomerize at acidic pH *in-vitro* (Bjorndahl et al., 2011; Prigent & Rezaei, 2011; Rezaei et al., 2005; Sengupta, Bhate, Das, & Udgaonkar, 2017; Singh & Udgaonkar, 2016). CD, fluorometry, and FTIR assays indicate a significant increase of β-sheet content in such oligomers (Bjorndahl et al., 2011; Rezaei et al., 2005). The presence of salt is believed to facilitate oligomerization by shielding of electrostatic repulsion between monomers, as well as disrupting some of intra-molecular electrostatic bonds (Sengupta et al., 2017). The addition of denaturants such as urea facilitates aggregation, although the specific molecular action has not been fully clarified yet. The convertibility of the resulting PrP oligomers into fibrillar aggregates has been demonstrated by AFM and EM imaging (Bjorndahl et al., 2011; Singh et al., 2012). However, their relevance to the pathogenic conformation and misfolding pathways is questionable (Vazquez-Fernandez et al., 2016).

The application of several biophysical methods including hydrogen-deuterium exchange based mass spectrometry and NMR spectroscopy to study intermediate species during aggregation of a small globular protein, the SH3 domain, under acidic pH conditions (Carulla et al., 2009) revealed the existence of two distinct pathways leading to fibril formation. At pH 2.0 the dominance of amorphous aggregates was observed; whereas at pH 1.5 relatively disordered initial species evolved into ordered protofibrillar aggregates. In either case, well-defined amyloid fibrils of similar morphology developed ultimately (Carulla et al., 2009).

Transition metal ions and metal-organic compounds are ubiquitous regulatory components of cellular functions, and as such they are anticipated to play a role both in the pathogenesis of misfolding diseases, and as potential therapeutic agents. The influence of Cu^{2+}, Fe^{2+}/Fe^{3+}, Zn^{2+}, and other metal ions onto the structures and aggregation propensities of amyloidogenic proteins has been reviewed extensively (Breydo et al., 2017; Kepp, 2017; Kozlowski, Luczkowski, Remell, & Valensin, 2012; Miller, Ma, & Nussinov, 2012; Nasica-Labouze et al., 2015; Zhou & Xiao, 2013). Depending on the stoichiometry and experimental conditions, a variety of effects of the ions have been reported, from promoting the formation of amorphous or fibrillar species to stabilizing particular conformers of PrP (Kozlowski et al., 2012; Zhou & Xiao, 2013), Aβ peptides (Kepp, 2017; Miller et al., 2012; Nasica-Labouze et al., 2015), α-synuclein (Breydo, Wu, & Uversky, 2012; Kozlowski et al., 2012), and of other proteins (Breydo et al., 2017). The pathways that control this influence are expected to involve both molecular mechanisms and the overall aggregation kinetics (Breydo et al., 2017; Kozlowski et al., 2012; Nasica-Labouze et al., 2015). According to solution-state NMR analyses, the coordination of monomeric proteins by metal ions is mediated by specific residues, especially histidines (Miller et al., 2012). For oligomeric aggregates of $A\beta_{17\text{-}42}$ peptides that do not contain a histidine, MD simulations predict stable chelation of Cu^{2+} and Fe^{2+} ions by acidic side chains and C-terminal carboxyl groups, which also may influence the aggregation behavior (Dorosh & Stepanova, 2017). Continuing experimental and computational studies are required to improve our understanding of the impacts of metal ions and metal-organic compounds on protein misfolding and aggregation.

A tremendous effort has been invested over the last decade into the search of compounds that could inhibit or transform the various stages of the amyloidogenic process (Ankarcrona et al., 2016; Cremades & Dobson, 2018; Eisele et al., 2015; Iadanza, Jackson et al., 2018; Jucker & Walker, 2013; Sweeney et al., 2017). The rational design of drug candidates that could stabilize the native conformations of pathogenic proteins is seen as a promising anti-misfolding strategy for globular proteins (Chiti & Dobson, 2017). This strategy relies largely upon the ability to predict the specific interactions a compound is capable of forming with a particular target structure. Compounds that are scored high against selected criteria *in-silico*, are subjected to an array of *in-vitro* and *in-vivo* assays in order to measure their effects onto protein folding/misfolding, aggregation, and toxicity. In native PrP^C for instance, binding into the pocket between the loop LH1S2, the C-terminal part of helix H2, and loop LH2H3 has been suggested as a putative hallmark of potential anti-misfolding compounds (Kuwata et al., 2007). Virtual screening of compounds against this criterion followed by *in-vitro* cell culture assays and *in-vivo* neurotoxicity tests identified 2-pyrrolidin-1-yl-N-[4-[4-(2-pyrrolidin-1-yl-acetylamino)-benzyl]-phenyl]-acetamide, termed GN8, as a potential inhibitor of PrP misfolding (Kuwata et al., 2007). More recently (Hyeon et al., 2015), the PrP-GN8 complex has been employed as a model of desired molecular recognition (pharmacophore) for a next-generation virtual screening of ~700,000 compounds, which yielded 37 anti-misfolding drug candidates. Among these, a benoxazole derivative denoted as BMD42-29 has shown the strongest binding to PrP, along with a promising *in-vitro* performance (Hyeon et al., 2015). Broader *in-silico* searches for PrP docking regions (Ferreira et al., 2014; Ishibashi et al., 2016) have confirmed the importance of the pocket between loops LH1S2 and LH2H3, where various potential anti-misfolding compounds have shown a high binding affinity. The stabilizing effect has been attributed to a protection of the hydrophobic core and the C-terminal region of helix H2 against misfolding, although the specific molecular interactions may vary across different compounds (Ishibashi et al., 2016; Zhou, Liu, An, Yao, & Liu, 2017).

In case of intrinsically disordered proteins, the design of anti-amyloid inhibitors is more complicated because of the dynamic and transient nature of the target structures (Chiti & Dobson, 2017; Iadanza, Jackson et al., 2018). The large number of published *in-vitro* studies suggests that in general, charged compounds tend to promote the formation of amyloid fibrils; whereas soluble aromatic compounds such as epigallocatechin gallate (EGCG) or dopamine, tend to interact with hydrophobic residues or clusters promoting the formation of amorphous aggregates rather than fibrils (as reviewed by Refs. Breydo et al., 2017). However, inhibition of fibril elongation has been found ineffective as this may lead to an increase in toxicity in the long term (Linse, 2019; Habchi, Chia et al., 2016). Recent strategies for anti-amyloid drug design pursue selective inhibition of either primary nucleation to block the onset of aggregation (Habchi, Arosio et al., 2016; Aprile et al., 2016), or secondary nucleation to prevent the proliferation of β-sheet rich oligomers (Cohen et al., 2015; Aprile et al., 2016; Linse, 2019). The challenge is that traditional, structure-based drug discovery techniques require molecular-level details of the target entities, which in case of IDPs are not sufficiently clear, despite intense efforts (Heller, Bonomi, & Vendruscolo, 2018). Various strategies have been proposed, such as antibody screening against hypothetical amyloidogenic epitopes (Aprile et al., 2016); small molecule docking against known targets found in interactome studies (Habchi, Chia et al., 2016; Hassan, Raza, Abbasi, Moustafa, & Seo, 2019); or virtual screening of compounds reported in the literature to influence the aggregation of IDPs (Joshi et al., 2016; Habchi, Arosio et al., 2016). Potential inhibitors that resulted from the various types of virtual screening have been subjected to extensive *in-vitro* testing with an emphasis on their influence onto primary and secondary nucleation pathways, as well as cytotoxicity

(Habchi, Arosio et al., 2016; Habchi, Chia et al., 2016; Aprile et al., 2016; Chia et al., 2018). A promising success story has been the identification of the Brichos chaperone domain as a selective inhibitor of secondary nucleation for $A\beta_{1\text{-}42}$ (Cohen et al., 2015). More recently, the Brichos domain was also found to inhibit the aggregation and toxicity of IAPP (Oskarsson et al., 2018).

Confinement on various biological surfaces, especially those of phospholipid membranes, is an integral part of *in-vivo* proteostasis. Membranes influence the lifecycle of amyloidogenic proteins in various ways, such as membrane-anchoring of precursors (in case of Aβ) or native conformers (PrP); or a conformational switch from unstructured to an α-helical form upon membrane binding (α-synuclein and IAPP). It is known that membrane interactions may modulate amyloidogenic aggregation, and that at least a part of the resulting toxic effects involve membrane damage (Cecchi & Stefani, 2013). Over the recent years the role of membrane interactions in protein misfolding and aggregation has been investigated intensely through a multitude of biophysical methods (Sciacca, Tempra, Scollo, Milardi, & La Rosa, 2018). Most studies use artificial phospholipid vesicles as *in-vitro* models of membranes. It has been demonstrated that surfaces of vesicles may catalyze the nucleation of fibrillar species of α-synuclein (Galvagnion et al., 2015), Aβ peptides (Kinoshita et al., 2017; Korshavn et al., 2017), and IAPP (Divakara et al., 2019). The presence of common membrane constituents such as cholesterol was reported to enhance the catalytic effect (Sciacca et al., 2016; Habchi et al., 2018). However the occurrence of nucleation events, as well as the action of additives, depends on the phospholipid composition and the inclusion of other ingredients (Korshavn et al., 2017; Sciacca et al., 2018). Aggregation of IDPs at the surface of membranes is often accompanied by a disruption of the membrane integrity (Korshavn et al., 2017; Sciacca et al., 2016, 2018; Breydo et al., 2017). Hypothetical models of membrane damage include destabilizing interactions of amyloid oligomers with membrane's surfaces; penetration of amyloid oligomers into phospholipid bilayers creating leakage channels; and extraction of lipids from membranes through hydrophobic interactions with the oligomers (Sciacca et al., 2018; Press-Sandler & Miller, 2018), as illustrated in Fig. 10. Recent MD simulations have addressed important details of the interaction of α-synuclein, Aβ, and IAPP aggregates with phospholipid membranes (Brown & Bevan, 2016; Press-Sandler & Miller, 2018; Sciacca et al., 2018; Tsigelny et al., 2015; Zhang et al., 2017). The challenge,

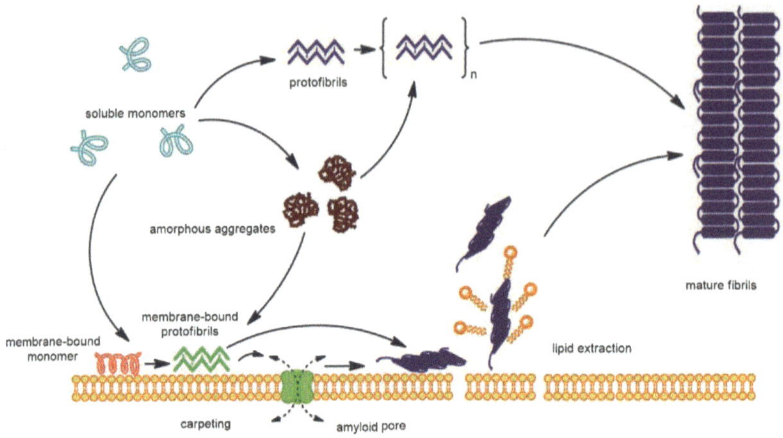

Fig. 10 Schematics of membrane disruption by amyloidogenic aggregation of IDPs (Sciacca et al., 2018). The membrane surface catalyzes primary nucleation of β-sheet rich fibrillar oligomers from adsorbed monomers. The fibrillar oligomers may disturb the membrane by merely remaining at its surface (termed the "carpeting" effect). Alternatively, some of the oligomers may assemble inside the membrane or embed into it creating an amyloid pore. Furthermore, hydrophobic interactions with the fibrillar oligomers may result in a "detergent-like" removal of lipids from the membrane. *Reproduced from Sciacca et al. (2018), © 2018 Elsevier, under the STM permission.*

however, is that the exact structures of the toxic aggregates that are involved in such interactions are not known in sufficient detail as required to fully decipher the mechanisms of the membrane disruption.

6. Open questions and challenges

In recent years, cryo electron microscopy and solid-state NMR spectroscopy have provided an ever increasing number of high-resolution structures of amyloid fibrils and misfolded protein aggregates including structures of fibrillized Aβ, α-synuclein, the tau protein, and many others (reviewed in Fitzpatrick & Saibil, 2019). These structures provide a wealth of new insights into the structural changes that occur during the protein misfolding that is thought to cause the underlying protein misfolding diseases such as Alzheimer's disease, Parkinson's disease, to name just a couple. Nevertheless, the detailed mechanisms that cause the misfolding and the molecular rearrangements that result in the altered structures are still poorly understood. MD simulations and the wealth of analytical approaches that depend on their output will be able to contribute to our understanding of the molecular causes of the diseases. In turn these insights may get translated into disease modifying or prophylactic treatment options, which are urgently needed.

Acknowledgments

HW, GSM, and MS acknowledge support from the Alberta Prion Research Institute (awards 201600028, 201600029, and 201700016). SA is supported by a Faculty of Medicine & Dentistry 75th Anniversary Award (University of Alberta). HW and MS acknowledge support and resources from the Compute Canada's Resources for Research Groups facilities.

References

Abskharon, R. N. N., Giachin, G., Wohlkonig, A., Soror, S. H., Pardon, E., Legname, G., et al. (2014). Probing the N-terminal β-sheet conversion in the crystal structure of the human prion protein bound to a nanobody. *Journal of the American Chemical Society, 136*(3), 937–944. http://dx.doi.org/10.1021/ja407527p.

Accelrys Discovery Studio Visualiser. (n.d.). (Version 4.0). BIOVIA. Retrieved from http://accelrys.com/.

Adrian, M., Dubochet, J., Lepault, J., & McDowall, A. W. (1984). Cryo-electron microscopy of viruses. *Nature, 308*(1), 32–36. http://dx.doi.org/10.1038/308032a0.

Aguzzi, A., & Lakkaraju, A. K. K. (2016). Cell biology of prions and prionoids: A status report. *Trends in Cell Biology, 26*(1), 40–51. http://dx.doi.org/10.1016/j.tcb.2015.08.007.

Ahmed, M., Davis, J., Aucoin, D., Sato, T., Ahuja, S., Aimoto, S., Elliott, J. I., Van Nostrand, V. E., et al. (2010). Structural conversion of neurotoxic amyloid-β_{1-42} oligomers to fibrils. *Nature Structural and Molecular Biology, 17*(5), 561–577. http://dx.doi.org/10.1038/nsmb.1799.

Ahmed, R., & Melacini, G. (2018). A solution NMR toolset to probe the molecular mechanisms of amyloid inhibitors. *Chemical Communications, 54*(37), 4644–4652. http://dx.doi.org/10.1039/c8cc01380b.

Alderson, T. R., & Markley, J. L. (2013). Biophysical characterization of α-synuclein and its controversial structure. *Intrinsically Disordered Proteins, 1*(1), 18–39. http://dx.doi.org/10.4161/idp.26255.

Alsteens, D., Gaub, H. E., Newton, R., Pfreundschuh, M., Gerber, C., & Müller, D. J. (2017). Atomic force microscopy-based characterization and design of biointerfaces. *Nature Reviews, 2*(5), 17008. http://dx.doi.org/10.1038/natrevmats.2017.8.

Amenitsch, H., Benetti, F., Ramos, A., Legname, G., & Requena, J. R. (2013). SAXS. structural studies of PrP^{Sc} reveals ~11 diameter of basic double intertwined fibrils. *Prion, 7*(6), 496–500. http://dx.doi.org/10.4161/pri.27190.

Ankarcrona, M., Winblad, B., Monteiro, C., Fearns, C., Powers, E. T., Johansson, J., et al. (2016). Current and future treatment of amyloid diseases. *Journal of Internal Medicine, 280*(2), 177–202. http://dx.doi.org/10.1111/joim.12506.

Apetri, A., Surewicz, K., & Surewicz, W. K. (2004). The effect of disease-associated mutations on the folding pathway of human prion protein. *The Journal of Biological Chemistry, 279*(17), 18008–18014. http://dx.doi.org/10.1074/jbc.M313581200.

Aprile, F. A., Sormanni, P., Perni, M., Arosio, P., Linse, S., Knowles, T. P. J., et al. (2016). Selective targeting of primary and secondary nucleation pathways in Aβ42 aggregation using a rational antibody scanning method. *Science Advances, 3*(6), e1700488. http://dx.doi.org/10.1126/sciadv.1700488.

Arimon, M., Díez-Pérez, I., Kogan, M. J., Durany, N., Giralt, E., Sanz, F., et al. (2005). Fine structure study of Aβ1-42 fibrillogenesis with atomic force microscopy. *The FASEB Journal, 19*(10), 1344–1346. http://dx.doi.org/10.1096/fj.04-3137fje.

Babu, M. M. (2016). The contribution of intrinsically disordered regions to protein function, cellular complexity, and human disease. *Biochemical Society Transactions, 44*(5), 1185–1200. http://dx.doi.org/10.1042/BST20160172.

Bacci, M., Vymětal, J., Mihajlovic, M., Caflisch, A., & Vitalis, A. (2017). Amyloid β fibril elongation by monomers involves disorder at the tip. *Journal of Chemical Theory and Computation, 13*(10), 5117–5130. http://dx.doi.org/10.1021/acs.jctc.7b00662.

Bae, S. H., Legname, G., Serban, A., Prusiner, S. B., Wright, P. E., & Dyson, H. J. (2009). Prion proteins with pathogenic and protective mutations show similar structure and dynamics. *Biochemistry, 48*(34), 8120–8128. http://dx.doi.org/10.1021/bi900923b.

Bai, X., McMullan, G., & Scheres, S. H. W. (2015). How cryo-EM is revolutionizing structural biology. *Trends in Biochemical Sciences, 40*(1), 49–57. http://dx.doi.org/10.1016/j.tibs.2014.10.005.

Bai, Y., Markham, K., Chen, F., Weerasekera, R., Watts, J., Horne, P., et al. (2008). The in vivo brain interactome of the amyloid precursor protein. *Molecular & Cellular Proteomics,* 7(1), 15–34. http://dx.doi.org/10.1074/mcp.M700077-MCP200.

Baker, C. M. (2015). Polarizable force fields for molecular dynamics simulations of biomolecules. *Wiley Interdisciplinary Reviews: Computational Molecular Science, 5*(2), 241–254. http://dx.doi.org/10.1002/wcms.1215.

Balbirnie, M., Grothe, R., & Eisenberg, D. S. (2001). An amyloid-forming peptide from the yeast prion Sup35 reveals a dehydrated -sheet structure for amyloid. *Proceedings of the National Academy of Sciences of the United States of America, 98*(5), 2375–2380. http://dx.doi.org/10.1073/pnas.041617698.

Balchin, D., Hayer-Hartl, M., & Hartl, F. U. (2016). In vivo aspects of protein folding and quality control. *Science, 353*(6294), aac4354. http://dx.doi.org/10.1126/science.aac4354.

Balch, W. E., Morimoto, R. I., Dillin, A., & Kelly, J. W. (2008). Adapting proteostasis for disease intervention. *Science, 319*(5865), 916–919. http://dx.doi.org/10.1126/science.1141448.

Ball, K. A., Phillips, A. H., Wemmer, D. E., & Head-Gordon, T. (2013). Differences in β-strand populations of monomeric Aβ40 and Aβ42. *Biophysical Journal, 104*(12), 2714–2724. http://dx.doi.org/10.1016/j.bpj.2013.04.056.

Banerjee, S., Sun, Z., Hayden, E. Y., Teplow, D. B., & Lyubchenko, Y. L. (2017). Nanoscale dynamics of amyloid β-42 oligomers as revealed by high-speed atomic force microscopy. *ACS Nano, 11*(12), 12202–12209. http://dx.doi.org/10.1021/acsnano.7b05434.

Barducci, A., Bonomi, M., & Parrinello, M. (2011). Metadynamics: Metadynamics. *Wiley Interdisciplinary Reviews: Computational Molecular Science, 1*(5), 826–843. http://dx.doi.org/10.1002/wcms.31.

Barducci, A., Chelli, R., Procacci, P., & Schettino, V. (2005). Misfolding pathways of the prion protein probed by molecular dynamics simulations. *Biophysical Journal, 88*(2), 1334–1343. http://dx.doi.org/10.1529/biophysj.104.049882.

Barducci, A., Chelli, R., Procacci, P., Schettino, V., Gervasio, F. L., & Parrinello, M. (2006). Metadynamics simulation of prion protein: β-Structure stability and the early stages of misfolding. *Journal of the American Chemical Society, 128*(8), 2705–2710. http://dx.doi.org/10.1021/ja057076l.

Barth, A., & Zscherp, C. (2002). What vibrations tell us about proteins. *Quarterly Reviews of Biophysics, 35*(4), 369–430. http://dx.doi.org/10.1017/S0033583502003815.

Barz, B., Olubiyi, O. O., & Strodel, B. (2014). Early amyloid β-protein aggregation precedes conformational change. *Chemical Communications, 50*(40), 5373–5375. http://dx.doi.org/10.1039/c3cc48704k.

Barz, B., & Strodel, B. (2016). Understanding amyloid-β oligomerization at the molecular level: The role of the fibril surface. *Chemistry: A European Journal, 22*(26), 8768–8772. http://dx.doi.org/10.1002/chem.201601701.

van den Bedem, H., & Fraser, J. S. (2015). Integrative, dynamic structural biology at atomic resolution—it's about time. *Nature Methods, 12*(4), 307–318. http://dx.doi.org/10.1038/nmeth.3324.

Behmard, E., Abdolmaleki, P., & Asadabadi, E. B. (2012). Mutation in a valine residue induces drastic changes in 3D structure of human prion protein. *Frontiers in Life Science, 6*(1–2), 47–51. http://dx.doi.org/10.1080/21553769.2013.775078.

Belli, M., Ramazzotti, M., & Chiti, F. (2011). Prediction of amyloid aggregation in vivo. *EMBO Reports, 12*(7), 657–663. http://dx.doi.org/10.1038/embor.2011.116.

Bemporad, F., & Chiti, F. (2012). Protein misfolded oligomers: Experimental approaches, mechanism of formation, and structure-toxicity relationships. *Chemistry & Biology, 19*(3), 315–327. http://dx.doi.org/10.1016/j.chembiol.2012.02.003.

Berendsen, H. J. C., Grigera, J. R., & Straatsma, T. P. (1987). The missing term in effective pair potentials. *The Journal of Physical Chemistry, 91*(24), 6269–6271. http://dx.doi.org/10.1021/j100308a038.

Berhanu, W. M., & Hansmann, U. H. E. (2014). Stability of amyloid oligomers. *Advances in Protein Chemistry and Structural Biology, 96*, 113–141. http://dx.doi.org/10.1016/bs.apcsb.2014.06.006.

Bernardi, R. C., Melo, M. C. R., & Schulten, K. (2015). Enhanced sampling techniques in molecular dynamics simulations of biological systems. *Biochimica et Biophysica Acta (BBA) - General Subjects, 1850*(5), 872–877. http://dx.doi.org/10.1016/j.bbagen.2014.10.019.

Bertoncini, C. W., Fernandez, C. O., Griesinger, C., Jovin, T. M., & Zweckstetter, M. (2005). Familial mutants of alpha-synuclein with increased neurotoxicity have a destabilized conformation. *The Journal of Biological Chemistry, 280*(35), 30649–30652. http://dx.doi.org/10.1074/jbc.C500288200.

Betts, V., Leissring, M. A., Dolios, G., Wang, R., Selkoe, D. J., & Walsh, D. M. (2008). Aggregation and catabolism of disease-associated intra-Aβ mutations: Reduced proteolysis of AβA21G by neprilysin. *Neurobiology of Disease, 31*(3), 442–450. http://dx.doi.org/10.1016/j.nbd.2008.06.001.

Bhattacharyya, D., Kumar, R., Mehra, S., Ghosh, A., Maji, S. K., & Bhunia, A. (2018). Multitude NMR studies of α-synuclein familial mutants: Probing their differential aggregation propensities. *Chemical Communications, 54*(29), 3605–3608. http://dx.doi.org/10.1039/C7CC09597J.

Biljan, I., Ilc, G., & Plavec, J. (2017). Understanding the effect of disease-related mutations on human prion protein structure: Insights from NMR spectroscopy. *Progress in Molecular Biology and Translational Science, 150*, 83–103. http://dx.doi.org/10.1016/bs.pmbts.2017.06.006.

Binnig, G., Quate, C. F., & Gerber, C. (1986). Atomic force microscope. *Physical Review Letters, 56*(9), 931–933. http://dx.doi.org/10.1103/PhysRevLett.56.930.

Bitan, G., Kirkitadze, M. D., Lomakin, A., Vollers, S. S., Benedek, G. B., & Teplow, D. B. (2003). Amyloid β-protein (Aβ) assembly: Aβ40 and Aβ42 oligomerize through distinct pathways. *Proceedings of the National Academy of Sciences of the United States of America, 100*(1), 330–335. http://dx.doi.org/10.1073/pnas.222681699.

Bizien, T., urand, D., Roblin, P., Tureau, A., Vachette, P., & Pérez, J. (2016). A brief survey of state-of-the-art bioSAXS. *Protein & Peptide Letters, 23*(3), 217–231. http://dx.doi.org/10.2174/0929866523666160106153655.

Bjorndahl, T. C., Zhou, G.-P., Liu, X., Perez-Pineiro, R., Semenchenko, V., Saleem, F., et al. (2011). Detailed biophysical characterization of the acid-induced PrPc to PrPβ conversion process. *Biochemistry, 50*(7), 1162–1173. http://dx.doi.org/10.1021/bi101435c.

Bleiholder, C., Dupuis, N. F., Wyttenbach, T., & Bowers, M. T. (2011). Ion mobility-mass spectrometry reveals a conformational conversion from random assembly to β-sheet in amyloid fibril formation. *Nature Chemistry, 3*(2), 172–177. http://dx.doi.org/10.1038/nchem.945.

Blinov, N., Berjanskii, M., Wishart, D. S., & Stepanova, M. (2009). Structural domains and main-chain flexibility in prion proteins. *Biochemistry, 48*(7), 1488–1497. http://dx.doi.org/10.1021/bi802043h.

Bloch, D. N., & Miller, Y. (2017). Study of molecular mechanisms of α-synuclein assembly: Insight into a cross-β structure in the N-termini of new α-synuclein fibrils. *ACS Omega, 2*(7), 3363–3370. http://dx.doi.org/10.1021/acsomega.7b00459.

Braselmann, E., Chaney, J. L., & Clark, P. L. (2013). Folding the proteome. *Trends in Biochemical Sciences, 38*(7), 337–344. http://dx.doi.org/10.1016/j.tibs.2013.05.001.

Breydo, L., Redington, J. M., & Uversky, V. N. (2017). Effects of intrinsic and extrinsic factors on aggregation of physiologically important intrinsically disordered proteins. *International Review of Cell and Molecular Biology, 329*, 145–185. http://dx.doi.org/10.1016/bs.ircmb.2016.08.011.

Breydo, L., & Uversky, V. N. (2015). Structural, morphological, and functional diversity of amyloid oligomers. *FEBS Letters, 589*(19A), 2640–2648. http://dx.doi.org/10.1016/j.febslet.2015.07.013.

Breydo, L., Wu, J. W., & Uversky, V. N. (2012). α-synuclein misfolding and Parkinson's disease. *Biochimica et Biophysica Acta, 1822*(2), 261–285. http://dx.doi.org/10.1016/j.bbadis.2011.10.002.

Brown, A. M., & Bevan, D. R. (2016). Molecular dynamics simulations of amyloid β-peptide (1-42): Tetramer formation and membrane interactions. *Biophysical Journal, 111*(5), 937–949. http://dx.doi.org/10.1016/j.bpj.2016.08.001.

Bruggink, K. A., Müller, M., Kuiperij, H. B., & Verbeek, M. M. (2012). Methods for analysis of amyloid-beta aggregates. *Journal of Alzheimer's Disease, 28*(4), 735–758. http://dx.doi.org/10.3233/JAD-2011-111421.

Calamini, B., Silva, M. C., Madoux, F., Hutt, D. M., et al. (2011). Small-molecule proteostasis regulators for protein conformational diseases. *Nature Chemical Biology, 8*(2), 185–196. http://dx.doi.org/10.1038/nchembio.763.

Caldarulo, E., Barducci, A., Wüthrich, K., & Parrinello, M. (2017). Prion protein β2–α2 loop conformational landscape. *Proceedings of the National Academy of Sciences of the United States of America, 114*(36), 9617–9622. http://dx.doi.org/10.1073/pnas.1712155114.

Callaway, E. (2015). The revolution will not be crystallized. *Nature, 525*(7568), 172–174. http://dx.doi.org/10.1038/525172a.

Calzolai, L., & Zahn, R. (2003). Influence of pH on NMR structure and stability of the human prion protein globular domain. *The Journal of Biological Chemistry, 278*(37), 35592–35596. http://dx.doi.org/10.1074/jbc.M303005200.

Carballo-Pacheco, M., & Strodel, B. (2017). Comparison of force fields for Alzheimer's A : A case study for intrinsically disordered proteins. *Protein Science, 26*(2), 174–185. http://dx.doi.org/10.1002/pro.3064.

Carulla, N., Zhou, M., Arimon, M., Gairí, M., Giralt, E., Robinson, C. V., et al. (2009). Experimental characterization of disordered and ordered aggregates populated during the process of amyloid fibril formation. *Proceedings of the National Academy of Sciences of the United States of America, 106*(19), 7828–7833. http://dx.doi.org/10.1073/pnas.0812227106.

Cecchi, C., & Stefani, M. (2013). The amyloid-cell membrane system. The interplay between the biophysical features of oligomers/fibrils and cell membrane defines amyloid toxicity. *Biophysical Chemistry, 182*, 30–43. http://dx.doi.org/10.1016/j.bpc.2013.06.003.

Chakroun, N., Fornili, A., Prigent, S., Kleinjung, J., Dreiss, C. A., Rezaei, H., et al. (2013). Decrypting prion protein conversion into a β-rich conformer by molecular dynamics. *Journal of Chemical Theory and Computation, 9*(5), 2455–2465. http://dx.doi.org/10.1021/ct301118j.

Chamachi, N. G., & Chakrabarty, S. (2016). Replica exchange molecular dynamics study of dimerization in prion protein: Multiple modes of interaction and stabilization. *The Journal of Physical Chemistry, 56*(6), 833–844. http://dx.doi.org/10.1021/acs.jpcb.6b03690.

Chamachi, N. G., & Chakrabarty, S. (2017). Temperature-induced misfolding in prion protein: Evidence of multiple partially disordered states stabilized by non-native hydrogen bonds. *Biochemistry, 56*(6), 833–844. http://dx.doi.org/10.1021/acs.biochem.6b01042.

Chandra, B., Bhowmik, D., Maity, B. K., Mote, K. R., Dhara, D., Venkatramani, R., et al. (2017). Major reaction coordinates linking transient amyloid-β oligomers to fibrils measured at atomic level. *Biophysical Journal, 113*(4), 805–816. http://dx.doi.org/10.1016/j.bpj.2017.06.068.

Chandrasekaran, P., & Rajasekaran, R. (2016). Detailed computational analysis revealed mutation V210I on PrP induced conformational conversion on β2-α2 loop and α2-α3. *Molecular BioSystems, 12*(10), 3223–3233. http://dx.doi.org/10.1039/c6mb00342g.

Chebaro, Y., & Derreumaux, P. (2009). The conversion of helix H2 to β-sheet is accelerated in the monomer and dimer of the prion protein upon T183A mutation. *The Journal of Physical Chemistry. B, 113*(19), 6942–6948. http://dx.doi.org/10.1021/jp900334s.

Chen, S. W., Drakulic, S., Deas, E., Ouberai, M., Aprile, F. A., Arranz, R., et al. (2015). Structural characterization of toxic oligomers that are kinetically trapped during α-synuclein fibril formation. *Proceedings of the National Academy of Sciences of the United States of America, 112*(16), E1994–E2003. http://dx.doi.org/10.1073/pnas.1421204112.

Chen, X., Duan, D., Zhu, S., & Zhang, J. (2013). Molecular dynamics simulation of temperature induced unfolding of animal prion protein. *Journal of Molecular Modeling, 9*(10), 4433–4441. http://dx.doi.org/10.1007/s00894-013-1955-0.

Chen, Z. A., & Rappsilber, J. (2018). Protein dynamics in solution by quantitative crosslinking/mass spectrometry. *Trends in Biochemical Sciences, 43*(11), 908–920. http://dx.doi.org/10.1016/j.tibs.2018.09.003.

Chen, W., van der Kamp, M. W., & Daggett, V. (2014). Structural and dynamic properties of the human prion protein. *Biophysical Journal, 106*(5), 1152–1163. http://dx.doi.org/10.1016/j.bpj.2013.12.053.

Cheng, C. J., & Daggett, V. (2014). Molecular dynamics simulations capture the misfolding of the bovine prion protein at acidic pH. *Biomolecules, 4*(1), 181–201. http://dx.doi.org/10.3390/biom4010181.

Cheon, M., Chang, I., Mohanty, S., Luheshi, L. M., Dobson, C. M., Vendruscolo, M., et al. (2007). Structural reorganisation and potential toxicity of oligomeric species formed during the assembly of amyloid fibrils. *PLoS Computational Biology, 3*(9), 1727–1738. http://dx.doi.org/10.1371/journal.pcbi.0030173.

Chia, S., Habchi, J., Michaels, T. C. T., Cohen, S. I. A., Linse, S., Dobson, C. M., et al. (2018). SAR by kinetics for drug discovery in protein misfolding diseases. *Proceedings of the National Academy of Sciences of the United States of America, 115*(41), 10245–10250. http://dx.doi.org/10.1073/pnas.1807884115.

Chiti, F., & Dobson, C. M. (2006). Protein misfolding, functional amyloid, and human disease. *Annual Review of Biochemistry, 75*(1), 333–366. http://dx.doi.org/10.1146/annurev.biochem.75.101304.123901.

Chiti, F., & Dobson, C. M. (2009). Amyloid formation by globular proteins under native conditions. *Nature Chemical Biology, 5*(1), 15–22. http://dx.doi.org/10.1038/nchembio.131.

Chiti, F., & Dobson, C. M. (2017). Protein misfolding, amyloid formation, and human disease: A summary of progress over the last decade. *Annual Review of Biochemistry, 86*(1), 27–68. http://dx.doi.org/10.1146/annurev-biochem-061516-045115.

Chodera, J. D., & Noé, F. (2014). Markov state models of biomolecular conformational dynamics. *Current Opinion in Structural Biology, 25*, 135–144. http://dx.doi.org/10.1016/j.sbi.2014.04.002.

Chong, S.-H., Chatterjee, P., & Ham, S. (2017). Computer simulations of intrinsically disordered proteins. *Annual Review of Physical Chemistry, 68*(1), 117–134. http://dx.doi.org/10.1146/annurev-physchem-052516-050843.

Chou, C. C., Zhang, Y., Umoh, M. E., Vaughan, S. W., Lorenzini, I., Liu, F., et al. (2018). TDP-43 pathology disrupts nuclear pore complexes and nucleocytoplasmic transport in ALS/FTD. *Nature Neuroscience, 21*(2), 228–239. http://dx.doi.org/10.1038/s41593-017-0047-3.

Ciryam, P., Kundra, R., Morimoto, R. I., Dobson, C. M., & Vendruscolo, M. (2015). Supersaturation is a major driving force for protein aggregation in neurodegenerative diseases. *Trends in Pharmacological Sciences, 36*(2), 72–77. http://dx.doi.org/10.1016/j.tips.2014.12.004.

Cohen, S. I. A., Arosio, P., Presto, J., Kurudenkandy, F. R., et al. (2015). A molecular chaperone breaks the catalytic cycle that generates toxic Aβ oligomers. *Nature Structural & Molecular Bioligy, 22*(3), 207–213. http://dx.doi.org/10.1038/nsmb.2971.

Cohen, S. I. A., Cukalevski, R., Michaels, T. C. T., Šarić, A., Törnquist, M., Vendruscolo, M., et al. (2018). Distinct thermodynamic signatures of oligomer generation in the aggregation of the amyloid-β peptide. *Nature Chemistry, 10*(5), 523–531. http://dx.doi.org/10.1038/s41557-018-0023-x.

Cohen, S. I. A., Linse, S., Luheshi, L. M., Hellstrand, E., White, D. A., Rajah, L., et al. (2013). Proliferation of amyloid-β42 aggregates occurs through a secondary nucleation mechanism. *Proceedings of the National Academy of Sciences of the United States of America, 110*(24), 9758–9763. http://dx.doi.org/10.1073/pnas.1218402110.

Colvin, M. T., Silvers, R., Ni, Q. Z., Can, T. V., Sergeyev, I., Rosay, M., et al. (2016). Atomic resolution structure of monomorphic Aβ42 amyloid fibrils. *Journal of American Chemical Society, 138*(30), 9663–9674. http://dx.doi.org/10.1021/jacs.6b05129.

Cong, X., Casiraghi, N., Rossetti, G., Mohanty, S., Giachin, G., Legname, G., et al. (2013). Role of prion disease-linked mutations in the intrinsically disordered N-terminal domain of the prion protein. *Journal of Chemical Theory and Computation, 9*(11), 5158–5167. http://dx.doi.org/10.1021/ct400534k.

Conway, K. A., Harper, J. D., & Lansbury, P. T. (1998). Accelerated in vitro fibril formation by a mutant α-synuclein linked to early-onset Parkinson disease. *Nature Medicine, 4*(11), 1318–1320. http://dx.doi.org/10.1038/3311.

Conway, K. A., Lee, S.-J., Rochet, J.-C., Ding, T. T., Williamson, R. E., & Lansbury, P. T. (2000). Acceleration of oligomerization, not fibrillization, is a shared property of both α-synuclein mutations linked to early-onset Parkinson's disease: Implications for pathogenesis and therapy. *Proceedings of the National Academy of Sciences of the United States of America, 97*(2), 571–576. http://dx.doi.org/10.1073/pnas.97.2.571.

Coskuner-Weber, O., & Uversky, V. N. (2018). Insights into the molecular mechanisms of Alzheimer's and Parkinson's diseases with molecular simulations: Understanding the roles of artificial and pathological missense mutations in intrinsically disordered proteins related to pathology. *International Journal of Molecular Sciences, 19*(2). http://dx.doi.org/10.3390/ijms19020336.

Coskuner, O., & Wise-Scira, O. (2013). Structures and free energy landscapes of the A53T mutant-type α-synuclein protein and impact of A53T mutation on the structures of the wild-type α-synuclein protein with dynamics. *ACS Chemical Neuroscience, 4*(7), 1101–1113. http://dx.doi.org/10.1021/cn400041j.

Coskuner, O., Wise-Scira, O., Perry, G., & Kitahara, T. (2012). The structures of the E22Δ mutant-type amyloid-β alloforms and the impact of E22Δ mutation on the structures of the wild-type amyloid-β alloforms. *ACS Chemical Neuroscience, 4*(2), 310–320. http://dx.doi.org/10.1021/cn300149j.

Côté, S., Derreumaux, P., & Mousseau, N. (2011). Distinct morphologies for amyloid beta protein monomer: $A\beta_{1-40}$, $A\beta_{1-42}$, and $A\beta_{1-40}$(D23N). *Journal of Chemical Theory and Computation, 7*(8), 2584–2592. http://dx.doi.org/10.1021/ct1006967.

Côté, S., Laghaei, R., Derreumaux, P., & Mousseau, N. (2012). Distinct dimerization for various alloforms of the amyloid-beta protein: $A\beta_{1-40}$, $A\beta_{1-42}$, and $A\beta_{1-40}$(D23N). *The Journal of Physical Chemistry B, 116*(13), 4043–4055. http://dx.doi.org/10.1021/ct1006967.

Cremades, N., Cohen, S. I. A., Deas, E., Abramov, A. Y., Chen, A. Y., Orte, A., et al. (2012). Direct observation of the interconversion of normal and toxic forms of α-synuclein. *Cell, 149*(5), 1048–1059. http://dx.doi.org/10.1016/j.cell.2012.03.037.

Cremades, N., & Dobson, C. M. (2018). The contribution of biophysical and structural studies of protein self-assembly to the design of therapeutic strategies for amyloid diseases. *Neurobiology of Disease, 109*, 178–190. http://dx.doi.org/10.1016/j.nbd.2017.07.009.

De Carloa, S., & Harris, J. R. (2011). Negative staining and cryo-negative staining of macromolecules and viruses for TEM. *Micron, 42*(2), 117–131. http://dx.doi.org/10.1016/j.micron.2010.06.003.

De Simone, A., Zagari, A., & Derreumaux, P. (2007). Structural and hydration properties of the partially unfolded states of the prion protein. *Biophysical Journal, 93*(4), 1284–1292. http://dx.doi.org/10.1529/biophysj.107.108613.

Dill, K. A., & MacCallum, J. L. (2012). The protein-folding problem, 50 years on. *Science, 338*(6110), 1042–1046. http://dx.doi.org/10.1126/science.1219021.

Divakara, M. B., Martinez, D., Ravi, A., Bhavana, V., Ramana, V., Habenstein, B., et al. (2019). Molecular mechanisms for the destabilization of model membranes by islet amyloid polypeptide. *Biophysical Chemistry, 245*, 34–40. http://dx.doi.org/10.1016/j.bpc.2018.12.002.

Dobson, C. M. (2003). Protein folding and misfolding. *Nature, 426*(6968), 884–890. http://dx.doi.org/10.1038/nature02261.

Dobson, C. M. (2004). Principles of protein folding, misfolding and aggregation. *Seminars in Cell & Developmental Biology, 15*(1), 3–16. http://dx.doi.org/10.1016/j.semcdb.2003.12.008.

Domon, B., & Aebersold, R. (2006). Mass spectrometry and protein analysis. *Science, 312*(5771), 212–217. http://dx.doi.org/10.1126/science.1124619.

Dorosh, L., & Stepanova, M. (2017). Probing oligomerization of amyloid β peptide in silico. *Molecular BioSystems, 13*(1), 165–182. http://dx.doi.org/10.1039/C6MB00441E.

Doss, C. G. P., Rajith, B., Rajasekaran, R., Srajan, J., Nagasundaram, N., & Debajyoti, C. (2013). In silico analysis of prion protein mutants: A comparative study by molecular dynamics approach. *Cell Biochemistry and Biophysics, 67*(3), 1307–1318. http://dx.doi.org/10.1007/s12013-013-9663-z.

Dror, R. O., Dirks, R. M., Grossman, J. P., Xu, H., & Shaw, D. E. (2012). Biomolecular simulation: A computational microscope for molecular biology. *Annual Review of Biophysics, 41*, 429–452. http://dx.doi.org/10.1146/annurev-biophys-042910-155245.

Dubochet, J., Lepault, J., Freeman, R., Berriman, J. A., & Homo, J. C. (1982). Electron microscopy of frozen water and aqueous solutions. *Journal of Microscopy, 128*(3), 219–237. http://dx.doi.org/10.1111/j.1365-2818.1982.tb04625.x.

Dufrêne, I. F., Ando, T., Garcia, R., Alsteens, D., Martinez-Martin, D., Engel, A., et al. (2017). Imaging modes of atomic force microscopy for application in molecular and cell biology. *Nature Nanotechnology, 12*(4), 295–307. http://dx.doi.org/10.1038/nnano.2017.45.

Eisele, Y. S., Monteiro, C., Fearns, C., Encalada, S. E., Wiseman, R. L., Powers, E. T., et al. (2015). Targeting protein aggregation for the treatment of degenerative diseases. *Nature Reviews Drug Discovery, 14*(11), 759–780. http://dx.doi.org/10.1038/nrd4593.

Eisenberg, D. S., & Sawaya, M. R. (2017). Structural studies of amyloid proteins at the molecular level. *Annual Review of Biochemistry, 86*(1), 69–95. http://dx.doi.org/10.1146/annurev-biochem-061516-045104.

Esler, W. P., Stimson, E. R., Jennings, J. M., Vinters, H. V., Ghilardi, J. R., Lee, J. P., et al. (2000). Alzheimer's disease amyloid propagation by a template-dependent dock-lock mechanism. *Biochemistry, 39*(21), 6288–6295. http://dx.doi.org/10.1021/bi992933h.

Fares, M.-B., Ait-Bouziad, N., Dikiy, I., Mbefo, M. K., Jovičić, A., Kiely, A., et al. (2014). The novel Parkinson's disease linked mutation G51D attenuates in vitro aggregation and membrane binding of α-synuclein, and enhances its secretion and nuclear localization in cells. *Human Molecular Genetics, 23*(17), 4491–4509. http://dx.doi.org/10.1093/hmg/ddu165.

Fawzi, N. L., Ying, J., Ghirlando, R., Torchia, D. A., & Clore, G. M. (2011). Atomic-resolution dynamics on the surface of amyloid-b protofibrils probed by solution NMR. *Nature, 480*(7376), 268–272. http://dx.doi.org/10.1038/nature10577.

Ferreira, N. C., Marques, I. A., Conceição, W. A., Macedo, B., et al. (2014). Anti-prion activity of a panel of aromatic chemical compounds: In vitro and in silico approaches. *PLOS One, 9*(1), e84531. http://dx.doi.org/10.1371/journal.pone.0084531.

Fitzpatrick, A. W. P., Debelouchina, G. T., Bayro, M. J., Clare, D. K., et al. (2013). Atomic structure and hierarchical assembly of a cross-beta amyloid fibril. *Proceedings of the National Academy of Sciences of the United States of America, 110*(14), 5468–5473. http://dx.doi.org/10.1073/pnas.1219476110.

Fitzpatrick, A. W. P., Falcon, B., He, S., Murzin, A. G., Murshudov, G., Garringer, H. J., et al. (2017). Cryo-EM structures of tau filaments from Alzheimer's disease. *Nature, 547*(7662), 185–190. http://dx.doi.org/10.1038/nature23002.

Fitzpatrick, A. W., & Saibil, H. R. (2019). Cryo-EM of amyloid fibrils and cellular aggregates. *Current Opinion in Structural Biology, 58*, 34–42. http://dx.doi.org/10.1016/j.sbi.2019.05.003.

Flagmeier, P., Meisl, G., Vendruscolo, M., Knowles, T. P., Dobson, C. M., Buell, A. K., et al. (2016). Mutations associated with familial Parkinson's disease alter the initiation and amplification steps of α-synuclein aggregation. *Proceedings of the National Academy of Sciences of the United States of America, 113*(37), 10328–10333. http://dx.doi.org/10.1073/pnas.1604645113.

Flores-Fernández, J. M., Rathod, V., & Wille, H. (2018). Comparing the folds of prions and other pathogenic amyloids. *Pathogens,* 7(2), 50. http://dx.doi.org/10.3390/pathogens7020050.

Galvagnion, C., Buell, A. K., Meisl, G., Michaels, T. C., Vendruscolo, M., Knowles, T. P., et al. (2015). Lipid vesicles trigger α-synuclein aggregation by stimulating primary nucleation. *Nature Chemical Biology, 11*(3), 229–234. http://dx.doi.org/10.1038/nchembio.1750.

Gao, Y., Zhu, T., Zhang, C., Zhang, J. Z. H., & Mei, Y. (2018). Comparison of the unfolding and oligomerization of human prion protein under acidic and neutral environments by molecular dynamics simulations. *Chemical Physics Letters, 706*, 594–600. http://dx.doi.org/10.1016/j.cplett.2018.07.014.

Garcia-Seisdedos, H., Empereur-Mot, C., Elad, N., & Levy, E. D. (2017). Proteins evolve on the edge of supramolecular self-assembly. *Nature, 548*(7666), 244–247. http://dx.doi.org/10.1038/nature23320.

Garrec, J., Tavernelli, I., & Rothlisberger, U. (2013). Two misfolding routes for the prion protein around pH 4.5. *PLOS Computational Biology, 9*(5), e1003057. http://dx.doi.org/10.1371/journal.pcbi.1003057.

Gershenson, A., Gierasch, L. M., Pastore, A., & Radford, S. E. (2014). Energy landscapes of functional proteins are inherently risky. *Nature Chemical Biology, 10*(11), 884–891. http://dx.doi.org/10.1038/nchembio.1670.

Gessel, M. M., Bernstein, S., Kemper, M., Teplow, D. B., & Bowers, M. T. (2012). Familial Alzheimer's disease mutations differentially alter amyloid β-protein oligomerization. *ACS Chemical Neuroscience, 3*(11), 909–918. http://dx.doi.org/10.1021/cn300050d.

Ghodrati, F., Mehrabian, M., Williams, D., Halgas, O., Bourkas, M. E. C., Watts, J. C., et al. (2018). The prion protein is embedded in a molecular environment that modulates transforming growth factor β and integrin signaling. *Scientific Reports, 8*(1), 8654. http://dx.doi.org/10.1038/s41598-018-26685-x.

Ghosh, D., Mondal, M., Mohite, G. M., Singh, P. K., Ranjan, P., Anoop, A., et al. (2013). The Parkinson's disease-associated H50Q mutation accelerates α-synuclein aggregation *in vitro*. *Biochemistry, 52*(40), 6925–6927. http://dx.doi.org/10.1021/bi400999d.

Ghosh, D., Sahay, S., Ranjan, P., Salot, S., Mohite, G. M., et al. (2014). The newly discovered Parkinson's disease associated Finnish mutation (A53E) attenuates α-synuclein aggregation andmembrane binding. *Biochemistry, 53*(41), 6419–6421. http://dx.doi.org/10.1021/bi5010365.

Giachin, G., Biljan, I., Ilc, G., Plavec, J., & Legname, G. (2013). Probing early misfolding events in prion protein mutants by NMR spectroscopy. *Molecules, 18*(8), 9451–9476. https://www.mdpi.com/1420-3049/18/8/9451.

Giachin, G., Mai, P. T., Tran, T. H., Salzano, G., Benetti, F., Migliorati, V., et al. (2015). The non-octarepeat copper binding site of the prion protein is a key regulator of prion conversion. *Scientific Reports, 5*, 15253. http://dx.doi.org/10.1038/srep15253.

Glaves, J. P., Ladner-Keay, C. L., Bjorndahl, T. C., Wishart, D. S., & Sykes, B. D. (2018). Residue-specific mobility changes in soluble oligomers of the prion protein define regions involved in aggregation. *BBA -Proteins and Proteomics, 1866*(9), 982–988. http://dx.doi.org/10.1016/j.bbapap.2018.06.005.

Goldschmidt, L., Teng, P. K., Riek, R., & Eisenberg, D. (2010). Identifying the amylome, proteins capable of forming amyloid-like fibrils. *Proceedings of the National Academy of Sciences of the United States of America, 107*(8), 3487–3492. http://dx.doi.org/10.1073/pnas.0915166107.

Greenbaum, E. A., Graves, C. L., Mishizen-Eberz, A. J., Lupoli, M. A., Lynch, D. R., Englander, S. W., et al. (2005). The E46K mutation in α-synuclein increases amyloid fibril formation. *Journal of Biological Chemistry, 280*(9), 7800–7807. http://dx.doi.org/10.1074/jbc.M411638200.

Gremer, L., Schölzel, D., Schenk, C., Reinartz, E., Labahn, J., Ravelli, R. B. G., et al. (2017). Fibril structure of amyloid-β(1–42) by cryo–electron microscopy. *Science, 358*(6359), 114–119. http://dx.doi.org/10.1126/science.aao2825.

Groenning, M. (2010). Binding mode of Thioflavin T and other molecular probes in the context of amyloid fibrils—current status. *Journal of Chemical Biology, 3*(1), 1–18. http://dx.doi.org/10.1007/s12154-009-0027-5.

Groveman, B. R., Dolan, M. A., Taubner, L. M., Kraus, A., Wickner, R. B., & Caughey, B. (2014). Parallel in-register intermolecular β-sheet architectures for prion-seeded prion protein (PrP) amyloids. *The Journal of Biological Chemistry, 289*(35), 24129–24142. http://dx.doi.org/10.1074/jbc.M114.578344.

Gsponer, J., & Babu, M. M. (2012). Cellular strategies for regulating functional and nonfunctional protein aggregation. *Cell Reports, 2*(5), 1425–1437. http://dx.doi.org/10.1016/j.celrep.2012.09.036.

Guerrero-Ferreira, R., Taylor, N. M. I., Mona, D., Ringler, P., Lauer, M. E., Riek, R., et al. (2018). Cryo-EM structure of alpha-synuclein fibrils. *eLIFE, 7*, e36402. http://dx.doi.org/10.7554/eLife.36402.009.

Guest, W. C., Cashman, N. R., & Plotkin, S. S. (2010). Electrostatics in the stability and misfolding of the prion protein: Salt bridges, self energy, and solvation. *Biochemistry and Cell Biology, 88*(2), 371–381. http://dx.doi.org/10.1139/o09-180.

Gunawardana, C. G., Mehrabian, M., Wang, X., Mueller, I., Lubambo, I. B., Jonkman, J. E., et al. (2015). The human tau interactome: Binding to the ribonucleoproteome, and impaired binding of the proline-to-leucine mutant at position 301 (P301L) to chaperones and the proteasome. *Molecular & Cellular Proteomics, 14*(11), 3000–3014. http://dx.doi.org/10.1074/mcp.M115.050724.

Guo, J., Ning, L., Ren, H., Liu, H., & Yao, X. (2012). Influence of the pathogenic mutations T188K/R/A on the structural stability and misfolding of human prion protein: Insight from molecular dynamics simulations. *Biochimica et Biophysica Acta, 1820*(2), 116–123. http://dx.doi.org/10.1016/j.bbagen.2011.11.013.

Habchi, J., Arosio, P., Perni, M., Costa, A. R., et al. (2016). An anticancer drug suppresses the primary nucleation reaction that initiates the production of the toxic Aβ42 aggregates linked with Alzheimer's disease. *Science Advances, 2*(2), e1501244. http://dx.doi.org/10.1126/sciadv.1501244.

Habchi, J., Chia, S., Galvagnion, C., Michaels, T. C. T., et al. (2018). Cholesterol catalyses Aβ42 aggregation through a heterogeneous nucleation pathway in the presence of lipid membranes. *Nature Chemistry, 10*(6), 673–683. http://dx.doi.org/10.1038/s41557-018-0031-x.

Habchi, J., Chia, S., Limbocker, R., Mannini, B., et al. (2016). Systematic development of small molecules to inhibit specific microscopic steps of Aβ42 aggregation in Alzheimer's disease. *Proceedings of the National Academy of Sciences of the United States of America, 114*(2), E200–E208. http://dx.doi.org/10.1073/pnas.1615613114.

Habenstein, B., & Loquet, A. (2016). Solid-state NMR: An emerging technique in structural biology of self-assemblies. *Biophysical Chemistry, 210*(1), 14–26. http://dx.doi.org/10.1016/j.bpc.2015.07.003.

Habibi, M., Rottler, J., & Plotkin, S. S. (2017a). As simple as possible but not simpler: On the reliability of protein coarse-grained models. *Biophysical Journal, 112*(3), 176a. http://dx.doi.org/10.1016/j.bpj.2016.11.974.

Habibi, M., Rottler, J., & Plotkin, S. S. (2017b). The unfolding mechanism of monomeric mutant SOD1 by simulated force spectroscopy. *Biochimica et Biophysica Acta (BBA) - Proteins and Proteomics, 1865*(11, Part B), 1631–1642. http://dx.doi.org/10.1016/j.bbapap.2017.06.009.

Hadži, S., Ondračka, A., Jerala, R., & Hafner-Bratkovič, I. (2015). Pathological mutations H187R and E196K facilitate subdomain separation and prion protein conversion by destabilization of the native structure. *The FASEB Journal, 29*(3), 882–893. http://dx.doi.org/10.1096/fj.14-255646.

Hannaoui, S., Amidian, S., Cheng, Y. C., Velásquez, C. D., Dorosh, L., Law, S., et al. (2017). Destabilizing polymorphism in cervid prion protein hydrophobic core determines prion conformation and conversion efficiency. *PLOS Pathogens, 13*(8), e1006553. http://dx.doi.org/10.1371/journal.ppat.1006553.

Hao, B., & Zheng, W. (1998). *Applied symbolic dynamics and chaos*. World Scientific.

Hartl, F. U. (2017). Protein misfolding diseases. *Annual Review of Biochemistry, 86*(1), 21–26. http://dx.doi.org/10.1146/annurev-biochem-061516-044518.

Hassan, M., Raza, H., Abbasi, M. A., Moustafa, A. A., & Seo, S. Y. (2019). The exploration of novel Alzheimer's therapeutic agents from the pool of FDA approved medicines using drug repositioning, enzyme inhibition and kinetic mechanism approaches. *Biomedicine & Pharmacotherapy, 109*, 2513–2526. http://dx.doi.org/10.1016/j.biopha.2018.11.115.

Hatami, A., Monjazeb, S., Milton, S., & Glabe, C. G. (2017). Familial alzheimer's disease mutations within the amyloid precursor protein alter the aggregation and conformation of the amyloid-β peptide. *Journal of Biological Chemistry, 292*(8), 3172–3185. http://dx.doi.org/10.1074/jbc.M116.755264.

Hawe, A., Sutter, M., & Jiskoot, W. (2008). Extrinsic fluorescent dyes as tools for protein characterization. *Pharmaceutical Research, 25*(7), 1487–1499. http://dx.doi.org/10.1007/s11095-007-9516-9.

Heller, G. T., Bonomi, M., & Vendruscolo, M. (2018). Structural ensemble modulation upon small-molecule binding to disordered proteins. *Journal of Molecular Biology, 430*(16), 2288–2292. http://dx.doi.org/10.1016/j.jmb.2018.03.015.

Henzler-Wildman, K., & Kern, D. (2007). Dynamic personalities of proteins. *Nature, 450*(7172), 964–972. http://dx.doi.org/10.1038/nature06522.

Hinterdorfer, P., & Dufrêne, Y. F. (2006). Detection and localization of single molecular recognition events using atomic force microscopy. *Nature Methods, 3*(5), 347–355. http://dx.doi.org/10.1038/NMETH871.

Horn, H. W., Swope, W. C., Pitera, J. W., Madura, J. D., Dick, T. J., Hura, G. L., et al. (2004). Development of an improved four-site water model for biomolecular simulations: TIP4P-Ew. *The Journal of Chemical Physics, 120*(20), 9665–9678. http://dx.doi.org/10.1063/1.1683075.

Horrocks, M. H., Tosatto, L., Dear, A. J., Garcia, G. A., Iljina, M., Cremades, N., et al. (2015). Fast flow microfluidics and single-molecule fluorescence for the rapid haracterization of α-synuclein oligomers. *Analytical Chemistry, 87*(17), 8818–8826. http://dx.doi.org/10.1021/acs.analchem.5b01811.

Horwich, A. L. (2014). Molecular chaperones in cellular protein folding: The birth of a field. *Cell, 157*(2), 285–288. http://dx.doi.org/10.1016/j.cell.2014.03.029.

Huang, J., & MacKerell, A. D. (2018). Force field development and simulations of intrinsically disordered proteins. *Current Opinion in Structural Biology, 48*, 40–48. http://dx.doi.org/10.1016/j.sbi.2017.10.008.

Huang, J., Rauscher, S., Nawrocki, G., Ran, T., Feig, M., de Groot, B. L., et al. (2017). CHARMM36m: An improved force field for folded and intrinsically disordered proteins. *Nature Methods, 14*(1), 71–73. http://dx.doi.org/10.1038/nmeth.4067.

Huggins, D. J. (2016). Studying the role of cooperative hydration in stabilizing folded protein states. *Journal of Structural Biology, 196*(3), 394–406. http://dx.doi.org/10.1016/j.jsb.2016.09.003.

Humphrey, W., Dalke, A., & Schulten, K. (1996). VMD: Visual molecular dynamics. *Journal of Molecular Graphics, 14*(1), 33–38. http://dx.doi.org/10.1016/0263-7855(96)00018-5.

Hunter, S., & Brayne, C. (2018). Understanding the roles of mutations in the amyloid precursor protein in Alzheimer disease. *Molecular Psychiatry, 23*(1), 81–93. http://dx.doi.org/10.1038/mp.2017.218.

Hyeon, J. W., Choi, J., Kim, S. Y., Govindaraj, R. G., Hwang, K. J., Lee, Y. S., et al. (2015). Discovery of novel anti-prion compounds using in silico and in vitro approaches. *Scientific Reports, 5*, 14944. http://dx.doi.org/10.1038/srep14944.

Iadanza, M. G., Jackson, M. P., Hewitt, E. W., Ranson, N. A., & Radford, S. E. (2018a). A new era for understanding amyloid structures and disease. *Nature Reviews Molecular Cell Biology, 19*(12), 755–773. http://dx.doi.org/10.1038/s41580-018-0060-8.

Iadanza, M. G., Silvers, R., Boardman, J., Smith, H. I., et al. (2018b). The structure of a β_2-microglobulin fibril suggests a molecular basis for its amyloid polymorphism. *Nature Communications, 9*(1), 4517. http://dx.doi.org/10.1038/s41467-018-06761-6.

Invernizzi, G., Papaleo, E., Sabate, R., & Ventura, S. (2012). Protein aggregation: Mechanisms and functional consequences. *The International Journal of Biochemistry & Cell Biology, 44*(9), 1541–1554. http://dx.doi.org/10.1016/j.biocel.2012.05.023.

Ishibashi, D., Nakagaki, T., Ishikawa, T., Atarashi, R., Watanabe, K., Cruz, F. A., et al. (2016). Structure-based drug discovery for prion disease using a novel binding simulation. *EBioMedicine, 9*, 238–249. http://dx.doi.org/10.1016/j.ebiom.2016.06.010.

Issack, B. B., Berjanskii, M., Wishart, D. S., & Stepanova, M. (2012). Exploring the essential collective dynamics of interacting proteins: Application to prion protein dimers. *Proteins: Structure, Function, and Bioinformatics, 80*(7), 1847–1865. http://dx.doi.org/10.1002/prot.24082.

Jahandideh, S., Jamalan, M., & Faridounnia, M. (2015). Molecular dynamics study of the dominant-negative E219K polymorphism in human prion protein. *Journal of Biomolecular Structure & Dynamics, 33*(6), 1315–1325. http://dx.doi.org/10.1080/07391102.2014.945486.

Jarrett, J. T., Berger, E. P., & Lansbury, P. T. (1993). The carboxy terminus of the β-amyloid protein is critical for the seeding of amyloid formation—implications for the pathogenesis of Alzheimer's disease. *Biochemistry, 32*, 4693–4697. http://dx.doi.org/10.1021/bi00069a001.

Jeong, J. S., Ansaloni, A., Mezzenga, R., Lashuel, H. A., & Dietler, G. (2013). Novel mechanistic insight into the molecular basis of amyloid polymorphism and secondary nucleation during amyloid formation. *Journal of Molecular Biology, 425*(10), 1765–1781. http://dx.doi.org/10.1016/j.jmb.2013.02.005.

Jorgensen, W. L., Chandrasekhar, J., Madura, J. D., Impey, R. W., & Klein, M. L. (1983). Comparison of simple potential functions for simulating liquid water. *The Journal of Chemical Physics, 79*(2), 926–935. http://dx.doi.org/10.1063/1.445869.

Joshi, P., Chia, S., Habchi, J., Knowles, T. P., Dobson, C. M., & Vendruscolo, M. (2016). A fragment-based method of creating small-molecule libraries to target the aggregation of intrinsically disordered proteins. *ACS Combinatorial Science, 18*(3), 144–153. http://dx.doi.org/10.1021/acscombsci.5b00129.

Jucker, M., & Walker, L. C. (2013). Self-propagation of pathogenic protein aggregates in neurodegenerative diseases. *Nature, 501*(7465), 45–51. http://dx.doi.org/10.1038/nature12481.

Julien, O., Chatterjee, S., Thiessen, A., Graether, S. P., & Sykes, B. D. (2009). Differential stability of the bovine prion protein upon urea unfolding. *Protein Science, 18*(10), 2172–2182. http://dx.doi.org/10.1002/pro.231.

van der Kamp, M. W., & Daggett, V. (2010a). Pathogenic mutations in the hydrophobic core of the human prion protein can promote structural instability and misfolding. *Journal of Molecular Biology, 404*(4), 732–748. http://dx.doi.org/10.1016/j.jmb.2010.09.060.

van der Kamp, M. W., & Daggett, V. (2010b). Influence of pH on the human prion protein: Insights into the early steps of misfolding. *Biophysical Journal, 99*(7), 2289–2298. http://dx.doi.org/10.1016/j.bpj.2010.07.063.

Karas, M., Bachmann, D., & Hillenkamp, F. (1985). Influence of the wavelength in high-irradiance ultraviolet laser desorption mass spectrometry of organic molecules. *Analytical Chemistry, 57*(14), 2935–2939. http://dx.doi.org/10.1021/ac00291a042.

Karplus, M., & McCammon, J. A. (2002). Molecular dynamics simulations of biomolecules. *Nature Structural & Molecular Biology, 9*(9), 646–652. http://dx.doi.org/10.1038/nsb0902-646.

Ke, P. C., Sani, M.-A., Ding, F., Kakinen, A., Javed, I., Separovic, F., et al. (2017). Implications of peptide assemblies in amyloid diseases. *Chemical Society Reviews, 46*(21), 6492–6531. http://dx.doi.org/10.1039/C7CS00372B.

Kelly, S. M., Jess, T. J., & Price, N. C. (2005). How to study proteins by circular dichroism. *Biochimica et Biophysica Acta, 1751*(2), 119–139. http://dx.doi.org/10.1016/j.bbapap.2005.06.005.

Kendrew, J. C., Bodo, G., Dintzis, H. M., Parrish, R. G., Wyckoff, H., & Phillips, D. C. (1958). A three-dimensional model of the myoglobin molecule obtained by X-ray analysis. *Nature, 181*(4610), 662–666. http://dx.doi.org/10.1038/181662a0.

Kepp, K. P. (2017). Alzheimer's disease: How metal ions define β-amyloid function. *Coordination Chemistry Reviews, 351*, 127–159. http://dx.doi.org/10.1016/j.ccr.2017.05.007.

Khan, M. V., Zakariya, S. M., & Khan, R. H. (2018). Protein folding, misfolding and aggregation: A tale of constructive to destructive assembly. *International Journal of Biological Macromolecules, 112*, 217–229. http://dx.doi.org/10.1016/j.ijbiomac.2018.01.099.

Kinoshita, M., Kakimoto, E., Terakawa, M. S., Lin, Y., Ikenoue, T., So, M., et al. (2017). Model membrane size-dependent amyloidogenesis of Alzheimer's amyloid-β peptides. *Physical Chemistry Chemical Physics, 19*(24), 16257–16266. http://dx.doi.org/10.1039/c6cp07774a.

Klaips, C. L., Jayaraj, G. G., & Hartl, F. U. (2017). Pathways of cellular proteostasis in aging and disease. *Journal of Cell Biology, 217*(1), 51. http://dx.doi.org/10.1083/jcb.201709072.

Kleckner, I. R., & Foster, M. P. (2011). An introduction to NMR-based approaches for measuring protein dynamics. *Biochimica at Biophysica Acta, 1814*(8), 9942–9968. http://dx.doi.org/10.1016/j.bbapap.2010.10.012.

Kmiecik, S., Gront, D., Kolinski, M., Wieteska, L., Dawid, A. E., & Kolinski, A. (2016). Coarse-grained protein models and their applications. *Chemical Reviews, 116*(14), 7898–7936. http://dx.doi.org/10.1021/acs.chemrev.6b00163.

Knowles, T. P. J., Vendruscolo, M., & Dobson, C. M. (2014). The amyloid state and its association with protein misfolding diseases. *Nature Reviews Molecular Cell Biology, 15*(6), 384–396. http://dx.doi.org/10.1038/nrm3810.

Knowles, T. P. J., Vendruscolo, M., & Dobson, C. M. (2015). The physical basis of protein misfolding disorders. *Physics Today, 68*(3), 36–41. http://dx.doi.org/10.1063/PT.3.2719.

Kong, J., & Yu, S. (2007). Fourier transform infrared spectroscopic analysis of protein secondary structures. *Acta Biochimica et Biophysica Sinica, 39*(8), 549–559. http://dx.doi.org/10.1111/j.1745-7270.2007.00320.x.

Korshavn, K. J., Satriano, C., Lin, Y., Zhang, R., et al. (2017). Reduced lipid bilayer thickness pegulates the aggregation and cytotoxicity of amyloid-β. *Journal of Biological Chemistry, 292*(11), 4638–4650. http://dx.doi.org/10.1074/jbc.M116.764092.

Kostyukevich, Y., Acter, T., Zherebker, A., Ahmed, A., Kim, S., & Nikolaev, E. (2018). Hydrogen/deuterium exchange in mass spectrometry. *Mass Spectrometry Reviews, 37*(6), 811–853. http://dx.doi.org/10.1002/mas.21565.

Kotler, S. A., Brender, J. R., Vivekanandan, S., Suzuki, Y., Yamamoto, K., Monette, M., et al. (2015). High-resolution NMR characterization of low abundance oligomers of amyloid-β without purification. *Scientific Reports, 5*, 11811. http://dx.doi.org/10.1038/srep11811.

Kozlowski, H., Luczkowski, M., Remell, M., & Valensin, D. (2012). Copper, zinc and iron in neurodegenerative diseases (Alzheimer's, Parkinson's and prion diseases). *Coordination Chemistry Reviews, 256*(19–20), 2129–2141. http://dx.doi.org/10.1016/j.ccr.2012.03.013.

Kundel, F., Tosatto, L., Whiten, D. R., Wirthensohn, D. C., Horrocks, M. H., & Klenerman, D. (2018). Shedding light on aberrant interactions – a review of modern tools for studying protein aggregates. *The FEBS Journal, 285*(19), 3604–3630. http://dx.doi.org/10.1111/febs.14409.

Kuwata, K., Matumoto, T., Cheng, H., Nagayama, K., James, T. L., & Roder, H. (2003). NMR-detected hydrogen exchange and molecular dynamics simulations provide structural insight into fibril formation of prion protein fragment 106–126. *Proceedings of the National Academy of Sciences of the United States of America, 100*, 14790–14795. http://dx.doi.org/10.1073/pnas.2433563100.

Kuwata, K., Nishida, N., Matsumoto, T., Kamatari, Y. O., Hosokawa-Muto, J., Kodama, K., et al. (2007). Hot spots in prion protein for pathogenic conversion. *Proceedings of the National Academy of Sciences of the United States of America, 104*(29), 11921–11926. http://dx.doi.org/10.1073/pnas.0702671104.

Labbadia, J., & Morimoto, R. I. (2015). The biology of proteostasis in aging and disease. *Annual Review of Biochemistry, 84*, 435–464. http://dx.doi.org/10.1146/annurev-biochem-060614-033955.

Lane, T. J., Shukla, D., Beauchamp, K. A., & Pande, V. S. (2013). To milliseconds and beyond: Challenges in the simulation of protein folding. *Current Opinion in Structural Biology, 23*(1), 58–65. http://dx.doi.org/10.1016/j.sbi.2012.11.002.

Lanucara, F., Holman, S. W., Gray, C. J., & Eyers, C. E. (2014). The power of ion mobility-mass spectrometry for structural characterization and the study of conformational dynamics. *Nature Chemistry, 6*(4), 281–294. http://dx.doi.org/10.1038/nchem.1889.

Lee, S., Antony, L., Hartmann, R., Knaus, K. J., Surewicz, K., Surewicz, W. K., et al. (2010). Conformational diversity in prion protein variants influences intermolecular b-sheet formation. *The EMBO Journal, 29*(1), 251–262. http://dx.doi.org/10.1038/emboj.2009.333.

Li, B. S., Ge, P., Murray, K. A., Sheth, P., Zhang, M., Nair, G., et al. (2018). Cryo-EM of full-length alpha-synuclein reveals fibril polymorphs with a common structural kernel. *Nature Communications, 9*(1), 3609. http://dx.doi.org/10.1038/s41467-018-05971-2.

Li, Y., Lubchenko, V., & Vekilov, P. G. (2011). The use of dynamic light scattering and Brownian microscopy to characterize protein aggregation. *Review of Scientific Instruments, 82*(5), 053106. http://dx.doi.org/10.1063/1.3592581.

Lin, Y.-S., Bowman, G. R., Beauchamp, K. A., & Pande, V. S. (2012). Investigating how peptide length and a pathogenic mutation modify the structural ensemble of amyloid beta monomer. *Biophysical Journal, 102*(2), 315–324. http://dx.doi.org/10.1016/j.bpj.2011.12.002.

Lin, Y.-S., & Pande, V. S. (2012). Effects of familial mutations on the monomer structure of $A\beta_{42}$. *Biophysical Journal, 103*(12), L47–L49. http://dx.doi.org/10.1016/j.bpj.2012.11.009.

Lindorff-Larsen, K., Maragakis, P., Piana, S., Eastwood, M. P., Dror, R. O., & Shaw, D. E. (2012). Systematic validation of protein force fields against experimental data. *PLoS ONE*, 7(2), e32131. http://dx.doi.org/10.1371/journal.pone.0032131.

Lindorff-Larsen, K., Piana, S., Palmo, K., Maragakis, P., Klepeis, J. L., Dror, R. O., et al. (2010). Improved side-chain torsion potentials for the Amber ff99SB protein force field. *Proteins, 78*(8), 1950–1958. http://dx.doi.org/10.1002/prot.22711.

Linse, S. (2019). Mechanism of amyloid protein aggregation and the role of inhibitors. *Pure and Applied Chemistry, 91*(2), 211–229. http://dx.doi.org/10.1515/pac-2018-1017.

Liu, C., Song, X., Nisbet, R., & Gotz, J. (2016). Co-immunoprecipitation with tau isoform-specific antibodies reveals distinct protein interactions and highlights a putative role for 2N tau in disease. *Journal of Biological Chemistry, 291*(15), 8173–8188. http://dx.doi.org/10.1074/jbc.M115.641902.

Lokappa, S. B., Suk, J. E., Balasubramanian, A., Samanta, S., Situ, A. J., & Ulmer, T. S. (2014). Sequence and membrane determinants of the random coil–helix transition of α-synuclein. *Journal of Molecular Biology, 426*(10), 2130–2144. http://dx.doi.org/10.1016/j.jmb.2014.02.024.

Loquet, A., El Mammeri, N., Stanek, J., Berbon, M., Bardiaux, B., Pintacuda, G., et al. (2018). 3D structure determination of amyloid fibrils using solid-state NMR spectroscopy. *Methods, 138–139*, 26–38. http://dx.doi.org/10.1016/j.ymeth.2018.03.014.

Lorenzen, N., et al. (2014). The role of stable α-synuclein oligomers in the molecular events underlying amyloid formation. *Journal of the American Chemical Society, 136*(10), 3859–3868. http://dx.doi.org/10.1021/ja411577t.

Losasso, V., Pietropaolo, A., Zannoni, C., Gustincich, S., & Carloni, P. (2011). Structural role of compensatory amino acid replacements in the α-synuclein protein. *Biochemistry, 50*(32), 6994–7001. http://dx.doi.org/10.1021/bi2007564.

Lou, Z., Wang, B., Guo, C., Wang, K., Zhang, H., & Xu, B. (2015). Molecular-level insights of early-stage prion protein aggregation on mica and gold surface determined by AFM imaging and molecular simulation. *Colloids and Surfaces B: Biointerfaces, 135*, 371–378. http://dx.doi.org/10.1016/j.colsurfb.2015.07.053.

Lu, J.-X., Qiang, W., Yau, W.-M., Schwieters, C. D., Meredith, S. C., & Tycko, R. (2013). Molecular structure of β-amyloid fibrils in Alzheimer's disease brain tissue. *Cell, 154*(6), 1257–1268. http://dx.doi.org/10.1016/j.cell.2013.08.035.

Luchini, A., Espina, V., & Liotta, L. A. (2014). Protein painting reveals solvent-excluded drug targets hidden within native protein–protein interfaces. *Nature Communications, 5*, 4413. http://dx.doi.org/10.1038/ncomms5413.

Luo, J., Wärmländer, S. K. T. S., Gräslund, A., & Abrahams, J. P. (2014). Alzheimer peptides aggregate into transient nanoglobules that nucleate fibrils. *Biochemistry, 53*(40), 6302–6308. http://dx.doi.org/10.1021/bi5003579.

Mackerell, A. D. (2004). Empirical force fields for biological macromolecules: Overview and issues. *Journal of Computational Chemistry, 25*(13), 1584–1604. http://dx.doi.org/10.1002/jcc.20082.

Man, V. H., Nguyen, P. H., & Derreumaux, P. (2017a). Conformational ensembles of the wild-type and S8C Aβ1-42 dimers. *The Journal of Physical Chemistry. B, 121*(11), 2434–2442. http://dx.doi.org/10.1021/acs.jpcb.7b00267.

Man, V. H., Nguyen, P. H., & Derreumaux, P. (2017b). High-resolution structures of the amyloid-β 1–42 dimers from the comparison of four atomistic force fields. *The Journal of Physical Chemistry B, 121*(24), 5977–5987. http://dx.doi.org/10.1021/acs.jpcb.7b04689.

Mane, J. Y., & Stepanova, M. (2016). Understanding the dynamics of monomeric, dimeric, and tetrameric α-synuclein structures in water. *FEBS Open Bio, 6*(7), 666–686. http://dx.doi.org/10.1002/2211-5463.12069.

Mark, P., & Nilsson, L. (2001). Structure and dynamics of the TIP3P, SPC, and SPC/E water models at 298 K. *The Journal of Physical Chemistry A, 105*(43), 9954–9960. http://dx.doi.org/10.1021/jp003020w.

Matthes, D., Gapsys, V., Brennecke, J. T., & de Groot, B. L. (2016). An atomistic view of amyloidogenic self-assembly: Structure and dynamics of heterogeneous conformational states in the pre-nucleation phase. *Scientific Reports, 6*, 33156. http://dx.doi.org/10.1038/srep33156.

McCammon, J. A., Gelin, B. R., & Karplus, M. (1977). Dynamics of folded proteins. *Nature, 267*(5612), 585–590. http://dx.doi.org/10.1038/267585a0.

Meisl, G., Yang, X., Hellstrand, E., Frohm, B., Kirkegaard, J. B., Cohen, S. I., et al. (2014). Differences in nucleation behavior underlie the contrasting aggregation kinetics of the Aβ40 and Aβ42 peptides. *Proceedings of the National Academy of Sciences of the United States of America, 111*(26), 9384–9389. http://dx.doi.org/10.1073/pnas.1401564111.

Meli, M., Gasset, M., & Colombo, G. (2011). Dynamic diagnosis of familial prion diseases supports the β2-α2 loop as a universal interference target. *PLoS ONE, 6*(4), e19093. http://dx.doi.org/10.1371/journal.pone.0019093.

Meng, F., Bellaiche, M. M. J., Kim, J. Y., Zerze, G. H., Best, R. B., & Chung, H. S. (2018). Highly disordered amyloid-β monomer probed by single-molecule FRET and MD simulation. *Biophysical Journal, 114*(4), 870–884. http://dx.doi.org/10.1016/j.bpj.2017.12.025.

Menon, S., & Sengupta, N. (2015). Perturbations in inter-domain associations may trigger the onset of pathogenic transformations in PrP^C: Insights from atomistic simulations. *Molecular BioSystems, 11*(5), 1443–1453. http://dx.doi.org/10.1039/C4MB00689E.

Menon, S., & Sengupta, N. (2017). Influence of hyperglycemic conditions on self-association of the Alzheimer's amyloid β (Aβ1−42) peptide. *ACS Omega, 2*(5), 2134−2147. http://dx.doi.org/10.1021/acsomega.7b00018.

Mercer, R. C. C., Daude, N., Dorosh, L., Fu, Z.-L., Mays, C. E., Gapeshina, H., et al. (2018). A novel Gerstmann-Sträussler-Scheinker disease mutation defines a precursor for amyloidogenic 8 kDa PrP fragments and reveals N-terminal structural changes shared by other GSS alleles. *PLoS Pathogens, 14*(1), e1006826. http://dx.doi.org/10.1371/journal.ppat.1006826.

Meric, G., Robinson, A. S., & Roberts, C. J. (2017). Driving forces for nonnative protein aggregation and approaches to predict aggregation-prone regions. *Annual Review of Chemical and Biomolecular Engineering, 8*, 139−159. http://dx.doi.org/10.1146/annurev-chembioeng-060816-101404.

Michaels, T. C. T., & Knowles, T. P. J. (2014). Role of filament annealing in the kinetics and thermodynamics of nucleated polymerization. *The Journal of Chemical Physics, 140*(21), 214904. http://dx.doi.org/10.1063/1.4880121.

Michaels, T. C. T., Šarić, A., Habchi, J., Chia, S., Meisl, G., Vendruscolo, M., et al. (2018). Chemical kinetics for bridging molecular mechanisms and macroscopic measurements of amyloid fibril formation. *Annual Review of Physical Chemistry, 69*, 273−298. http://dx.doi.org/10.1146/annurev-physchem-050317-021322.

Miller, Y., Ma, B. Y., & Nussinov, R. (2012). Metal binding sites in amyloid oligomers: Complexes and mechanisms. *Coordination Chemistry Reviews, 256*(19−20), 2245−2252. http://dx.doi.org/10.1016/j.ccr.2011.12.022.

Minikel, E. V., Vallabh, S. M., Lek, M., Estrada, K., et al. (2016). Quantifying prion disease penetrance using large population control cohorts. *Science Translational Medicine, 8*(322), 322ra9. http://dx.doi.org/10.1126/scitranslmed.aad5169.

Monsellier, E., Ramazzotti, M., Taddei, N., & Chiti, F. (2008). Aggregation propensity of the human proteome. *PLoS Computational Biology, 4*(10), e1000199. http://dx.doi.org/10.1371/journal.pcbi.1000199.

Moore, R. A., Taubner, L. M., & Priola, S. A. (2009). Prion protein misfolding and disease. *Current Opinion in Structural Biology, 19*(1), 14−22. http://dx.doi.org/10.1016/j.sbi.2008.12.007.

Morel, B., Carrasco, M. P., Jurado, S., Marco, C., & Conejero-Lara, F. (2018). Dynamic micellar oligomers of amyloid beta peptides play a crucial role in their aggregation mechanisms. *Physical Chemistry Chemical Physics, 20*(31), 20597−20614. http://dx.doi.org/10.1039/c8cp02685h.

Morris, M., Knudsen, G. M., Maeda, S., Trinidad, J. C., Ioanoviciu, A., Burlingame, A. L., et al. (2015). Tau post-translational modifications in wild-type and human amyloid precursor protein transgenic mice. *Nature Neuroscience, 18*(8), 1183−1189. http://dx.doi.org/10.1038/nn.4067.

Moulick, R., Das, R., & Udgaonkar, J. B. (2015). Partially unfolded forms of the prion protein populated under misfolding-promoting conditions. *Journal of Biological Chemistry, 290*(42), 25227−25240. http://dx.doi.org/10.1074/jbc.M115.677575.

Munishkina, L. A., & Fink, A. L. (2007). Fluorescence as a method to reveal structures and membrane-interactions of amyloidogenic proteins. *Biochimica et Biophysica Acta, 1768*(8), 1862−1885. http://dx.doi.org/10.1016/j.bbamem.2007.03.015.

Murakami, K., Irie, K., Morimoto, A., Ohigashi, H., Shindo, M., Nagao, M., et al. (2002). Synthesis, aggregation, neurotoxicity, and secondary structure of various Aβ1-42 mutants of familial Alzheimer's disease at positions 21-23. *Biochemical and Biophysical Research Communications, 294*(1), 5−10. http://dx.doi.org/10.1016/S0006-291X(02)00430-8.

Murphy, M. P., & LeVine, H., III (2010). Alzheimer's disease and the β-amyloid peptide. *Journal of Alzheimer's Disease, 19*(1), 311. http://dx.doi.org/10.3233/JAD-2010-1221.

Nagel-Steger, L., Owen, M. C., & Strodel, B. (2016). An account of amyloid oligomers: Facts and figures obtained from experiments and simulations. *ChemBioChem, 17*(8), 657–676. http://dx.doi.org/10.1002/cbic.201500623.

Nasica-Labouze, J., Nguyen, P. H., Sterpone, F., Berthoumieu, O., Buchete, N.-V., Coté, S., et al. (2015). Amyloid β protein and Alzheimer's disease: When computer simulations complement experimental studies. *Chemical Reviews, 115*(9), 3518–3563. http://dx.doi.org/10.1021/cr500638n.

Nath, S., Meuvis, J., Hendrix, J., Carl, S. A., & Engelborghs, Y. (2010). Early aggregation steps in α-synuclein as measured by FCS and FRET: Evidence for a contagious conformational change. *Biophysical Journal, 98*(7), 1302–1311. http://dx.doi.org/10.1016/j.bpj.2009.12.4290.

Neumann, M., Sampathu, D. M., Kwong, L. K., Truax, A. C., Micsenyi, M. C., Chou, T. T., et al. (2006). Ubiquitinated TDP-43 in frontotemporal lobar degeneration and amyotrophic lateral sclerosis. *Science, 314*(5796), 130–133. http://dx.doi.org/10.1126/science.1134108.

Nguyen, P. H., Li, M. S., Stock, G., Straub, J. E., & Thirumalai, D. (2007). Monomer adds to preformed structured oligomers of Aβ-peptides by a two-stage dock–lock mechanism. *Proceedings of the National Academy of Sciences of the United States of America, 104*(1), 111–116. http://dx.doi.org/10.1073/pnas.0607440104.

Ning, L., Guo, J., Bai, Q., Jin, N., Liu, H., & Yao, X. (2014). Structural diversity and initial oligomerization of PrP106–126 studied by replica-exchange and conventional molecular dynamics simulations. *PLoS ONE, 9*, e87266. http://dx.doi.org/10.1371/journal.pone.0087266.

Nogales, E. (2016). The development of cryo-EM into a mainstream structural biology technique. *Nature Methods, 13*(1), 24–27. http://dx.doi.org/10.1038/nmeth.3694.

Nogales, E., & Scheres, S. H. W. (2015). Cryo-em: A unique tool for the visualization of macromolecular complexity. *Molecular Cell, 58*(4), 677–689. http://dx.doi.org/10.1016/j.molcel.2015.02.019.

Oganesyan, I., Lento, C., & Wilson, D. J. (2018). Contemporary hydrogen deuterium exchange mass spectrometry. *Methods, 144*, 27–42. http://dx.doi.org/10.1016/j.ymeth.2018.04.023.

Ono, K., Condron, M. M., & Teplow, D. B. (2010). Effects of the English (H6R) and Tottori (D7N) familial Alzheimer disease mutations on amyloid β-protein assembly and toxicity. *Journal of Biological Chemistry, 285*(30), 23186–23197. http://dx.doi.org/10.1074/jbc.M109.086496.

Onufriev, A. V., & Izadi, S. (2017). Water models for biomolecular simulations. *Wiley Interdisciplinary Reviews: Computational Molecular Science, 8*(2), e1347. http://dx.doi.org/10.1002/wcms.1347.

Orlova, E. V., & Saibil, H. R. (2011). Structural analysis of macromolecular assemblies by electron microscopy. *Chemical Reviews, 111*(12), 7710–7748. http://dx.doi.org/10.1021/cr100353t.

Orte, A., Clarke, R. W., & Klenerman, D. (2011). Single-molecule fluorescence coincidence spectroscopy and its application to resonance energy transfer. *ChemPhysChem, 12*(3), 491–499. http://dx.doi.org/10.1002/cphc.201000636.

Oskarsson, M. E., Hermansson, E., Wang, Y., Welsh, N., Presto, J., Johansson, J., et al. (2018). BRICHOS domain of Bri2 inhibits islet amyloid polypeptide (IAPP) fibril formation and toxicity in human beta cells. *Proceedings of the National Academy of Sciences of the United States of America, 115*(12), E2752–E2761. http://dx.doi.org/10.1073/pnas.1715951115.

Ovchinnikova, O. Y., Finder, V. H., Vodopivec, I., Nitsch, R. M., & Glockshuber, R. (2011). He Osaka FAD mutation E22Δ leads to the formation of a previously unknown type of amyloid β fibrils and modulates Aβ neurotoxicity. *Journal of Molecular Biology, 408*(4), 780–791. http://dx.doi.org/10.1016/j.jmb.2011.02.049.

Pace, C. N., Scholtz, J. M., & Grimsley, G. R. (2014). Forces stabilizing proteins. *FEBS Letters, 588*(14), 2177–2184. http://dx.doi.org/10.1016/j.febslet.2014.05.006.

Perálvarez-Marín, A., Mateos, L., Zhang, C., Singh, S., Cedazo-Mínguez, A., Visa, N., et al. (2009). Influence of residue 22 on the folding, aggregation profile, and toxicity of the Alzheimer's amyloid β peptide. *Biophysical Journal, 97*(1), 277–285. http://dx.doi.org/10.1016/j.bpj.2009.04.017.

Persson, F., Söderhjelm, P., & Halle, B. (2018). The spatial range of protein hydration. *The Journal of Chemical Physics, 148*(21), 215104. http://dx.doi.org/10.1063/1.5031005.

Pettersen, E. F., Goddard, T. D., Huang, C. C., Couch, G. S., Greenblatt, D. M., Meng, E. C., et al. (2004). UCSF Chimera–a visualization system for exploratory research and analysis. *Journal of Computational Chemistry, 25*(13), 1605–1612. http://dx.doi.org/10.1002/jcc.20084.

Piana, S., Donchev, A. G., Robustelli, P., & Shaw, D. E. (2015). Water dispersion interactions strongly influence simulated structural properties of disordered protein states. *The Journal of Physical Chemistry B, 119*(16), 5113–5123. http://dx.doi.org/10.1021/jp508971m.

Piana, S., Klepeis, J. L., & Shaw, D. E. (2014). Assessing the accuracy of physical models used in protein-folding simulations: Quantitative evidence from long molecular dynamics simulations. *Current Opinion in Structural Biology, 24*, 98–105. http://dx.doi.org/10.1016/j.sbi.2013.12.006.

Piana, S., Lindorff-Larsen, K., & Shaw, D. E. (2011). How robust are protein folding simulations with respect to force field parameterization? *Biophysical Journal, 100*(9), L47–L49. http://dx.doi.org/10.1016/j.bpj.2011.03.051.

Pillon, M. C., & Guarné, A. (2017). Complementary uses of small angle X-ray scattering and X-ray crystallography. *Biochimica et Biophysica Acta – Proteins and Proteomics, 1865*(11), 1623–1630. http://dx.doi.org/10.1016/j.bbapap.2017.07.013.

Potapov, A., & Stepanova, M. (2012). Conformational modes in biomolecules: Dynamics and approximate invariance. *Physical Review E, 85*(2), 020901. http://dx.doi.org/10.1103/PhysRevE.85.020901.

Press-Sandler, O., & Miller, Y. (2018). Molecular mechanisms of membrane-associated amyloid aggregation: Computational perspective and challenges. Biochimica et Biophysica Acta – Biomembranes, *1860*, 1889–1905. http://dx.doi.org/10.1016/j.bbamem.2018.03.014.

Prigent, S., & Rezaei, H. (2011). PrP assemblies: Spotting the responsible regions in prion propagation. *Prion, 5*(2), 69–75. http://dx.doi.org/10.4161/pri.5.2.16383.

Prusiner, S. B. (1998). Prions. *Proceedings of the National Academy of Sciences of the United States of America, 95*(23), 13363–13383. http://dx.doi.org/10.1073/pnas.95.23.13363.

Prusiner, S. B. (2012). A unifying role for prions in neurodegenerative diseases. *Science, 336*(6088), 1511–1513. http://dx.doi.org/10.1126/science.1222951.

Ranjan, P., & Kumar, A. (2017). Perturbation in long-range contacts modulates the kinetics of amyloid formation in α-synuclein familial mutants. *ACS Chemical Neuroscience, 8*(10), 2235–2246. http://dx.doi.org/10.1021/acschemneuro.7b00149.

Rauscher, S., Gapsys, V., Gajda, M. J., Zweckstetter, M., de Groot, B. L., & Grubmüller, H. (2015). Structural ensembles of intrinsically disordered proteins depend strongly on force field: A comparison to experiment. *Journal of Chemical Theory and Computation, 11*(11), 5513–5524. http://dx.doi.org/10.1021/acs.jctc.5b00736.

Redecke, L., vonBergen, M., Clos, J., Konarev, P. V., Svergun, D. I., Fittschen, U. E. A., et al. (2007). Structural characterization of β-sheeted oligomers formed on the pathway of oxidative prion protein aggregation in vitro. *Journal of Structural Biology, 157*(2), 308–320. http://dx.doi.org/10.1016/j.jsb.2006.06.013.

Redfield, C. (2004). Using nuclear magnetic resonance spectroscopy to study molen globule states of proteins. *Methods, 34*(1), 121–132. http://dx.doi.org/10.1016/j.ymeth.2004.03.009.

Requena, J. R., & Wille, H. (2014). The structure of the infectious prion protein. *Prion, 8*(1), 60–66. http://dx.doi.org/10.4161/pri.28368.

Rezaei, H., Eghiaian, F., Perez, J., Doublet, B., Choiset, Y., Haertle, T., et al. (2005). Sequential generation of two structurally distinct ovine prion protein soluble oligomers displaying different biochemical reactivities. *Journal of Molecular Biology, 347*(3), 665–679. http://dx.doi.org/10.1016/j.jmb.2005.01.043.

Riek, R., & Eisenberg, D. S. (2016). The activities of amyloids from a structural perspective. *Nature, 539*(7628), 227–235. http://dx.doi.org/10.1038/nature20416.

Riek, R. G., Dobeli, P. H., Wipf, B., & Wuthrich, K. (2001). NMR studies in aqueous solution fail to identify significant conformational differences between the monomeric forms of two Alzheimer peptides with widely different plaque-competence $Ab(1-40)^{ox}$ and $Ab(1-42)^{ox}$. *European Journal of Biochemistry, 268*(22), 5930–5936. http://dx.doi.org/10.1046/j.0014-2956.2001.02537.x.

Riesner, D. (2003). Biochemistry and structure of PrP^{C} and PrP^{Sc}. *British Medical Bulletin, 66*(1), 21–33.

Robustelli, P., Piana, S., & Shaw, D. E. (2018). Developing a molecular dynamics force field for both folded and disordered protein states. *Proceedings of the National Academy of Sciences of the United States of America.* http://dx.doi.org/10.1073/pnas.1800690115.

Roche, J., Shen, Y., Lee, J. H., Ying, J., & Bax, A. (2016). Monomeric $A\beta^{1-40}$ and $A\beta^{1-42}$ peptides in solution adopt very similar Ramachandran map distributions that closely resemble random coil. *Biochemistry, 55*(5), 762–775. http://dx.doi.org/10.1021/acs.biochem.5b01259.

Rodriguez, R. A., Chen, L. Y., Plascencia-Villa, G., & Perry, G. (2018). Thermodynamics of amyloid-β fibril elongation: Atomistic details of the transition state. *ACS Chemical Neuroscience, 9*(4), 783–789. http://dx.doi.org/10.1021/acschemneuro.7b00409.

Rojas, A., Maisuradze, N., Kachlishvili, K., Scheraga, H. A., & Maisuradze, G. G. (2017). Elucidating important sites and the mechanism for amyloid fibril formation by coarse-grained molecular dynamics. *ACS Chemical Neuroscience, 8*(1), 201–209. http://dx.doi.org/10.1021/acschemneuro.6b00331.

Rosenman, D. J., Wang, C., & García, A. E. (2016). Characterization of Aβ monomers through the convergence of ensemble properties among simulations with multiple force fields. *The Journal of Physical Chemistry B, 120*(2), 259–277. http://dx.doi.org/10.1021/acs.jpcb.5b09379.

Rose, P. W., Prlić, A., Altunkaya, A., Bi, C., Bradley, A. R., Christie, C. H., et al. (2017). The RCSB protein Data Bank: Integrative view of protein, gene and 3D structural information. *Nucleic Acids Research, 45*(Database issue), D271–D281. http://dx.doi.org/10.1093/nar/gkw1000.

Rossetti, G., & Carloni, P. (2017). Structural modeling of human prion protein's point mutations. *Progress in Molecular Biology and Translational Science, 150*, 105–122. http://dx.doi.org/10.1016/bs.pmbts.2017.07.001.

Rossetti, G., Cong, X., Caliandro, R., Legname, G., & Carloni, P. (2011). Common structural traits across pathogenic mutants of the human prion protein and their implications for familial prion diseases. *Journal of Molecular Biology, 411*(3), 700–712. http://dx.doi.org/10.1016/j.jmb.2011.06.008.

Rostkowski, M., Olsson, M. H., Søndergaard, C. R., & Jensen, J. H. (2011). Graphical analysis of pH-dependent properties of proteins predicted using PROPKA. *BMC Structural Biology, 11*, 6. http://dx.doi.org/10.1186/1472-6807-11-6.

Rousseau, F., Serrano, L., & Schymkowitz, J. W. H. (2006). How evolutionary pressure against protein aggregation shaped chaperone specificity. *Journal of Molecular Biology, 355*(5), 1037–1047. http://dx.doi.org/10.1016/j.jmb.2005.11.035.

Rudd, P. M., Endo, T., Colominas, C., Groth, D., Wheeler, S. F., Harvey, D. J., et al. (1999). Glycosylation differences between the normal and pathogenic prion protein isoforms. *Proceedings of the National Academy of Sciences of the United States of America, 96*(23), 13044–13049. http://dx.doi.org/10.1073/pnas.96.23.13044.

Sabareesan, A. T., & Udgaonkar, J. B. (2016). Pathogenic mutations within the disordered palindromic region of the prion protein induce structure therein and accelerate the formation of misfolded oligomers. *Journal of Molecular Biology, 428*(20), 3935–3947. http://dx.doi.org/10.1016/j.jmb.2016.08.015.

Sachse, C., Xu, C., Wieligmann, K., Diekmann, S., Grigorieff, N., & Fandrich, M. (2006). Quaternary structure of a mature amyloid fibril from Alzheimer's a beta(1-40) peptide. *Journal of Molecular Biology, 362*(2), 347–354. http://dx.doi.org/10.1016/j.jmb.2006.07.011.

Santo, K. P., Berjanskii, M., Wishart, D. S., & Stepanova, M. (2011). Comparative analysis of essential collective dynamics and NMR-derived flexibility profiles in evolutionarily diverse prion proteins. *Prion, 5*(3), 188–200. http://dx.doi.org/10.4161/pri.5.3.16097.

Sarroukh, R., Goormaghtigh, E., Ruysschaert, J. M., & Raussens, V. (2013). Atr-ftir: A "rejuvenated" tool to investigate amyloid proteins. *Biochimica et Biophysica Acta, 1828*(10), 2328–2338. http://dx.doi.org/10.1016/j.bbamem.2013.04.012.

Saunders, M. G., & Voth, G. A. (2013). Coarse-graining methods for computational biology. *Annual Review of Biophysics, 42*, 73–93. http://dx.doi.org/10.1146/annurev-biophys-083012-130348.

Sawaya, M. R., Sambashivan, S., Nelson, R., Ivanova, M. I., Sievers, S. A., Apostol, M. I., et al. (2007). Atomic structures of amyloid cross-β spines reveal varied steric zippers. *Nature, 447*(7143), 453–457. http://dx.doi.org/10.1038/nature05695.

Schmitt-Ulms, G., Ehsani, S., Watts, J. C., Westaway, D., & Wille, H. (2009). Evolutionary descent of prion genes from the ZIP family of metal ion transporters. *PLoS One, 4*(9), e7208. http://dx.doi.org/10.1371/journal.pone.0007208.

Schmitz, M., Dittmar, K., Llorens, F., Gelpi, E., Ferrer, I., Schulz-Schaeffer, W. J., et al. (2017). Hereditary human prion diseases: An update. *Molecular Neurobiology, 54*(6), 4138–4149. http://dx.doi.org/10.1007/s12035-016-9918-y.

Schütz, A. K., Vagt, T., Huber, M., Ovchinnikova, O. Y., Cadalbert, R., Wall, J., et al. (2015). Atomic-resolution three-dimensional structure of amyloid β fibrils bearing the osaka mutation. *Angewandte Chemie International Edition, 54*(1), 331–335. http://dx.doi.org/10.1002/anie.201408598.

Schwierz, N., Frost, C. V., Geissler, P. L., & Zacharias, M. (2016). Dynamics of seeded $A\beta_{40}$-fibril growth from atomistic molecular dynamics simulations: Kinetic trapping and reduced water mobility in the locking step. *Journal of the American Chemical Society, 38*, 527–539. http://dx.doi.org/10.1021/jacs.5b08717.

Schwierz, N., Frost, C. V., Geissler, P. L., & Zacharias, M. (2017). From Aβ filament to fibril: Molecular mechanism of surface-activated secondary nucleation from all-atom MD simulations. *The Journal of Physical Chemistry B, 121*, 671–682. http://dx.doi.org/10.1021/acs.jpcb.6b10189.

Sciacca, M. F., Lolicato, F., Di Mauro, G., Milardi, D., D'Urso, L., Satriano, C., et al. (2016). The role of cholesterol in driving IAPP-membrane interactions. *Biophysical Journal, 111*(1), 140–151. http://dx.doi.org/10.1016/j.bpj.2016.05.050.

Sciacca, M. F. M., Tempra, C., Scollo, F., Milardi, D., & La Rosa, C. (2018). Amyloid growth and membrane damage: Current themes and emerging perspectives from theory and experiments on Aβ and hIAPP. *Biochimica et Biophysica Acta – Biomembranes, 1860*(9), 1625–1638. http://dx.doi.org/10.1016/j.bbamem.2018.02.022.

Sengupta, I., Bhate, S. H., Das, R., & Udgaonkar, J. B. (2017). Salt-mediated oligomerization of the mouse prion protein monitored by real-time NMR. *Journal of Molecular Biology, 429*(12), 1852–1872. http://dx.doi.org/10.1016/j.jmb.2017.05.006.

Sengupta, I., & Udgaonkar, J. B. (2018). Structural mechanisms of oligomer and amyloid fibril formation by the prion protein. *Chemical Communications, 54*(49), 6230–6242. http://dx.doi.org/10.1039/c8cc03053g.

Sgourakis, N. G., Yau, W.-M., & Qiang, W. (2015). Modeling an in-register, parallel "Iowa" Aβ fibril structure using solid-state NMR data from labeled samples with Rosetta. *Structure, 23*(1), 216–227. http://dx.doi.org/10.1016/j.str.2014.10.022.

Shen, P. S. (2018). The 2017 Nobel prize in chemistry: Cryo-em comes of age. *Analytical and Bioanalytical Chemistry, 410*(8), 2053–2057. http://dx.doi.org/10.1007/s00216-018-0899-8.

Shi, Y. A. (2014). Glimpse of structural biology through X-ray crystallography. *Cell, 159*(5), 995–1014.

Shivu, B., Seshadri, S., Li, J., Oberg, K. A., Uversky, V. N., & Fink, A. L. (2013). Distinct β-sheet structure in protein aggregates determined by ATR–FTIR spectroscopy. *Biochemistry, 52*(31), 5176–5183. http://dx.doi.org/10.1021/bi400625v.

Shukla, D., Hernández, C. X., Weber, J. K., & Pande, V. S. (2015). Markov state models provide insights into dynamic modulation of protein function. *Accounts of Chemical Research, 48*(2), 414–422. http://dx.doi.org/10.1021/ar5002999.

Simone, A. D., Dodson, G. G., Verma, C. S., Zagari, A., & Fraternali, F. (2005). Prion and water: Tight and dynamical hydration sites have a key role in structural stability. *Proceedings of the National Academy of Sciences of the United States of America, 102*(21), 7535–7540. http://dx.doi.org/10.1073/pnas.0501748102.

Singh, J., Sabareesan, A. T., Mathew, M. K., & Udgaonkar, J. B. (2012). Development of the structural core and of conformational heterogeneity during the conversion of oligomers of the mouse prion protein to worm-like amyloid fibrils. *Journal of Molecular Biology, 423*(2), 217–231. http://dx.doi.org/10.1016/j.jmb.2012.06.040.

Singh, J., & Udgaonkar, J. B. (2016). Unraveling the molecular mechanism of pH-induced misfolding and oligomerization of the prion protein. *Journal of Molecular Biology, 428*(6), 1345–1355. http://dx.doi.org/10.1016/j.jmb.2016.01.030.

Sirur, A., De Sancho, D., & Best, R. B. (2016). Markov state models of protein misfolding. *The Journal of Chemical Physics, 144*(7), 075101. http://dx.doi.org/10.1063/1.4941579.

Smith, M. D., Rao, J. S., Segelken, E., & Cruz, L. (2015). Force-field induced bias in the structure of Aβ21–30: A comparison of OPLS, AMBER, CHARMM, and GROMOS force fields. *Journal of Chemical Information and Modeling, 55*(12), 2587–2595. http://dx.doi.org/10.1021/acs.jcim.5b00308.

Somavarapu, A. K., & Kepp, K. P. (2015). The dependence of amyloid-β dynamics on protein force fields and water models. *ChemPhysChem, 16*(15), 3278–3289. http://dx.doi.org/10.1002/cphc.201500415.

Song, D., Luo, R., & Chen, H.-F. (2017). The IDP-specific force field ff14IDPSFF improves the conformer sampling of intrinsically disordered proteins. *Journal of Chemical Information and Modeling, 57*(5), 1166–1178. http://dx.doi.org/10.1021/acs.jcim.7b00135.

Song, W., Wang, Y., Colletier, J.-P., Yang, H., & Xu, Y. (2015). Varied probability of staying collapsed/extended at the conformational equilibrium of monomeric $A\beta_{40}$ and $A\beta_{42}$. *Scientific Reports, 5*, 035101. http://dx.doi.org/10.1038/srep11024.

Soto, C., & Pritzkow, S. (2018). Protein misfolding, aggregation, and conformational strains in neurodegenerative diseases. *Nature Neuroscience, 21*(10), 1332–1340. http://dx.doi.org/10.1038/s41593-018-0235-9.

Spagnolli, G., Rigoli, M., Orioli, S., Sevillano, A. M., Faccioli, P., Wille, H., et al. (2019). Full atomistic model of prion structure and conversion. *PLoS Pathogens, 15*(7), e1007864. http://dx.doi.org/10.1371/journal.ppat.1007864.

van der Spoel, D., van Maaren, P. J., & Berendsen, H. J. C. (1998). A systematic study of water models for molecular simulation: Derivation of water models optimized for use with a reaction field. *The Journal of Chemical Physics, 108*(24), 10220–10230. http://dx.doi.org/10.1063/1.476482.

Stahl, N., Borchelt, D. R., Hsiao, K., & Prusiner, S. B. (1987). Scrapie prion protein contains a phosphatidylinositol glycolipid. *Cell, 51*(2), 229–240. http://dx.doi.org/10.1016/0092-8674(87)90150-4.

Stahl, N., & Prusiner, S. B. (1991). Prions and prion proteins. *FASEB Journal, 5*(13), 2799–2807. http://dx.doi.org/10.1096/fasebj.5.13.1916104.

Steckmann, T., Bhandari, Y. R., Chapagain, P. P., & Gerstman, B. S. (2017). Cooperative structural transitions in amyloid-like aggregation. *The Journal of Chemical Physics, 146*(13), 135103. http://dx.doi.org/10.1063/1.4979516.

Stepanova, M. (2007). Dynamics of essential collective motions in proteins: Theory. *Physical Review. E, Statistical, Nonlinear, and Soft Matter Physics, 76*(5 Pt 1), 051918. http://dx.doi.org/10.1103/PhysRevE.76.051918.

Stetefeld, J., McKenna, S. A., & Patel, T. R. (2016). Dynamic light scattering: A practical guide and applications in biomedical sciences. *Biophysical Reviews, 8*(4), 409–427. http://dx.doi.org/10.1007/s12551-016-0218-6.

Stroo, E., Koopman, M., Nollen, E. A. A., & Mata-Cabana, A. (2017). Cellular regulation of amyloid formation in aging and disease. *Frontiers in Neuroscience, 11*. http://dx.doi.org/10.3389/fnins.2017.00064.

Sugita, Y., & Okamoto, Y. (1999). Replica-exchange molecular dynamics method for protein folding. *Chemical Physics Letters, 314*(1), 141–151. http://dx.doi.org/10.1016/S0009-2614(99)01123-9.

Sunde, M., Serpell, L. C., Bartlam, M., Fraser, P. E., Pepys, M. B., & Blake, C. C. (1997). Common core structure of amyloid fibrils by synchrotron X-ray diffraction. *Journal of Molecular Biology, 273*(3), 729–739. http://dx.doi.org/10.1006/jmbi.1997.1348.

Sweeney, P., Park, H., Baumann, M., Dunlop, J., Frydman, J., Kopito, R., et al. (2017). Protein misfolding in neurodegenerative diseases: Implications and strategies. *Translational Neurodegeneration, 6*. http://dx.doi.org/10.1186/s40035-017-0077-5.

Takada, L. T., Kim, M.-O., Cleveland, R. W., Wong, K., Forner, S. A., Gala, I. I., et al. (2017). Genetic prion disease: Experience of a rapidly progressive dementia center in the United States and a review of the literature. *American Journal of Medical Genetics B, 174*(1), 36–69. http://dx.doi.org/10.1002/ajmg.b.32505.

Takeda, T., & Klimov, D. K. (2009). Probing energetics of Aβ fibril elongation by molecular dynamics simulations. *Biophysical Journal, 96*(11), 4428–4437. http://dx.doi.org/10.1016/j.bpj.2009.03.015.

Tanaka, M., & Komi, Y. (2015). Layers of structure and function in protein aggregation. *Nature Chemical Biology, 11*(6), 373–377. http://dx.doi.org/10.1038/nchembio.1818.

Tartaglia, G. G., Pechmann, S., Dobson, C. M., & Vendruscolo, M. (2007). Life on the edge: A link between gene expression levels and aggregation rates of human proteins. *Trends in Biochemical Sciences, 32*(5), 204–206. http://dx.doi.org/10.1016/j.tibs.2007.03.005.

Tarus, B., Tran, T. T., Nasica-Labouze, J., Sterpone, F., Nguyen, P. H., & Derreumaux, P. (2015). Structures of the Alzheimer's wild-type A$\beta_{1\text{-}40}$ dimer from atomistic simulations. *The Journal of Physical Chemistry B, 119*(33), 10478–10487. http://dx.doi.org/10.1021/acs.jpcb.5b05593.

Tattum, M. H., Cohen-Krausz, S., Thumanu, K., Wharton, C. W., Khalili-Shirazi, A., Jackson, G. S., et al. (2006). Elongated oligomers assemble into mammalian PrP amyloid fibrils. *Journal of Molecular Biology, 357*(3), 975–985. http://dx.doi.org/10.1016/j.jmb.2006.01.052.

Taylor, K. A., & Glaeser, R. M. (1976). Electron microscopy of frozen hydrated biological specimens. *Journal of Ultrastructure Research, 55*(3), 448–456. http://dx.doi.org/10.1016/S0022-5320(76)80099-8.

Terry, C., Wenborn, A., Gros, N., Sells, J., Joiner, S., et al. (2016). Ex vivo mammalian prions are formed of paired double helical prion protein fibrils. *Open Biology, 6*(5), 160035. http://dx.doi.org/10.1098/rsob.160035.

Teunissen, A. J. P., Pérez-Medina, C., Meijerink, A., & Mulder, W. J. M. (2018). Investigating supramolecular systems using Förster resonance energy transfer. *Chemical Society Reviews, 47*(18), 7027–7044. http://dx.doi.org/10.1039/c8cs00278a.

Törnquist, M., Michaels, T. C. T., Sanagavarapu, K., Yang, X., Meisl, G., Cohen, S. I. A., et al. (2018). Secondary nucleation in amyloid formation. *ChemComm, 54*(63), 8667–8684. http://dx.doi.org/10.1039/c8cc02204f.

Trainor, K., Broom, A., & Meiering, E. M. (2017). Exploring the relationships between protein sequence, structure and solubility. *Current Opinion in Structural Biology, 42*, 136–146. http://dx.doi.org/10.1016/j.sbi.2017.01.004.

Tran, L., & Ha-Duong, T. (2015). Exploring the alzheimer amyloid-β peptide conformational ensemble: A review of molecular dynamics approaches. *Peptides, 69*, 86–91. http://dx.doi.org/10.1016/j.peptides.2015.04.009.

Tseng, C.-Y., Yu, C.-P., & Lee, H. C. (2009). Integrity of H1 helix in prion protein revealed by molecular dynamic simulations to be especially vulnerable to changes in the relative orientation of H1 and its S1 flank. *European Biophysics Journal, 38*(5), 601–611. http://dx.doi.org/10.1007/s00249-009-0414-4.

Tsigelny, I. F., Sharikov, Y., Kouznetsova, V. L., Greenberg, G. P., Wrasidlo, W., Overk, C., et al. (2015). Molecular determinants of α-synuclein mutants' oligomerization and membrane interactions. *ACS Chemical Neuroscience, 6*(3), 403–416. http://dx.doi.org/10.1021/cn500332w.

Tuttle, M. D., Comellas, G., Nieuwkoop, A. J., Covell, D. J., Berthold, D. A., Kloepper, K. D., et al. (2016). Solid-state NMR structure of a pathogenic fibril of full-length human α-synuclein. *Nature Structural & Molecular Biology, 23*(5), 409–415. http://dx.doi.org/10.1038/nsmb.3194.

Uversky, V. N. (2010). Targeting intrinsically disordered proteins in neurodegenerative and protein dysfunction diseases: Another illustration of the D2 concept. *Expert Review of Proteomics,* 7(4), 543–564. http://dx.doi.org/10.1586/epr.10.36.

Uversky, V. N., & Fink, A. L. (2004). Conformational constraints for amyloid fibrillation: The importance of being unfolded. *Biochimica et Biophysica Acta - Proteins and Proteomics, 1698*(2), 131–153. http://dx.doi.org/10.1016/j.bbapap.2003.12.008.

Vazquez-Fernandez, E., Vos, M. R., Afanasyev, P., Cebey, L., Sevillano, A. M., Vidal, E., et al. (2016). The structural architecture of an infectious mammalian prion using electron cryomicroscopy. *PLoS Pathogens, 12*(9), e1005835. http://dx.doi.org/10.1371/journal.ppat.1005835.

Vega, C., & Abascal, J. L. F. (2011). Simulating water with rigid non-polarizable models: A general perspective. *Physical Chemistry Chemical Physics, 13*(44), 19663–19688. http://dx.doi.org/10.1039/C1CP22168J.

Viet, M. H., Nguyen, P. H., Derreumaux, P., & Li, M. S. (2014). Effect of the English familial disease mutation (H6R) on the monomers and dimers of Aβ40 and Aβ42. *ACS Chemical Neuroscience, 5*(8), 646–657. http://dx.doi.org/10.1021/cn500007j.

Viet, M. H., Nguyen, P. H., Ngo, S. T., Li, M. S., & Derreumaux, P. (2013). Effect of the Tottori familial disease mutation (D7N) on the monomers and dimers of $A\beta_{40}$ and $A\beta_{42}$. *ACS Chemical Neuroscience, 4*(11), 1446–1457. http://dx.doi.org/10.1021/cn400110d.

Walker, L. C., & Jucker, M. (2015). Neurodegenerative diseases: Expanding the prion concept. *Annual Review of Neuroscience, 38*(1), 87–103. http://dx.doi.org/10.1146/annurev-neuro-071714-033828.

Wälti, M. A., Ravotti, F., Arai, H., Glabe, C. G., Wall, J. S., Böckmann, A., et al. (2016). Atomic-resolution structure of a disease-relevant Aβ(1−42) amyloid fibril. *Proceedings of the National Academy of Sciences of the United States of America, 113*(34), E4976−E4984. http://dx.doi.org/10.1073/pnas.1600749113.

Wan, W., Wille, H., Stöhr, J., Kendall, A., Bian, W., McDonald, M., et al. (2015). Structural studies of truncated forms of the prion protein PrP. *Biophysical Journal, 108*(6), 1548−1554. http://dx.doi.org/10.1016/j.bpj.2015.01.008.

Wang, H., Muiznieks, L. D., Ghosh, P., Williams, D., Solarski, M., Fang, A., et al. (2017). Somatostatin binds to the human amyloid beta peptide and favors the formation of distinct oligomers. *eLife, 6*, e28401. http://dx.doi.org/10.7554/eLife.28401.

Wasmer, C., Lange, A., Van Melckebeke, H., Siemer, A. B., Riek, R., & Meier, B. H. (2008). Amyloid fibrils of the HET-s(218-289) prion form a beta solenoid with a triangular hydrophobic core. *Science, 319*(5869), 1523−1526. http://dx.doi.org/10.1126/science.1151839.

Watanabe-Nakayama, T., Ono, K., Itami, M., Takahashi, R., Teplow, D. B., & Yamada, M. (2016). High-speed atomic force microscopy reveals structural dynamics of amyloid $\beta_{1\text{-}42}$ aggregates. *Proceedings of the National Academy of Sciences of the United States of America, 113*(21), 5835−5840. http://dx.doi.org/10.1073/pnas.1524807113.

Watts, C. R., Gregory, A., Frisbie, C., & Lovas, S. (2018). Effects of force fields on the conformational and dynamic properties of amyloid β(1-40) dimer explored by replica exchange molecular dynamics simulations. *Proteins, 86*(3), 279−300. http://dx.doi.org/10.1002/prot.25439.

Wegmann, S., Jung, Y. J., Chinnathambi, S., Mandelkow, E. M., Mandelkow, E., & Muller, D. J. (2010). Human tau isoforms assemble into ribbon-like fibrils that display polymorphic structure and stability. *The Journal of Biological Chemistry, 285*(35), 27302−27313. http://dx.doi.org/10.1074/jbc.M110.145318.

Wei, G., Xi, W., Nussinov, R., & Ma, B. (2016). Protein ensembles: How does nature harness thermodynamic fluctuations for life? The diverse functional roles of conformational ensembles in the cell. *Chemical Reviews, 116*(11), 6516−6551. http://dx.doi.org/10.1021/acs.chemrev.5b00562.

van der Wel, P. C. A. (2017). Insights into protein misfolding and aggregation enabled by solid-state NMR spectroscopy. *Solid State Nuclear Magnetic Resonance, 88*(1), 1−14. http://dx.doi.org/10.1016/j.ssnmr.2017.10.001.

Wille, H., Bian, W., McDonald, M., Kendall, A., Colby, D. W., Bloch, L., et al. (2009). Natural and synthetic prion structure from X-ray fiber diffraction. *Proceedings of the National Academy of Sciences of the United States of America, 106*(40), 16990−16995. http://dx.doi.org/10.1073/pnas.0909006106.

Wise-Scira, O., Aloglu, A. K., Dunn, A., Sakallioglu, I. T., & Coskuner, O. (2013a). Structures and free energy fandscapes of the fild-type and A30P mutant-type α-synuclein proteins with dynamics. *ACS Chemical Neuroscience, 4*(3), 486−497. http://dx.doi.org/10.1021/cn300198q.

Wise-Scira, O., Dunn, A., Aloglu, A. K., Sakallioglu, I. T., & Coskuner, O. (2013b). Structures of the E46K mutant-type α-synuclein protein and impact of E46K mutation on the structures of the wild-type α-synuclein protein. *ACS Chemical Neuroscience, 4*(3), 498−508. http://dx.doi.org/10.1021/cn3002027.

Wong, Y. C., & Kranic, D. (2017). α-Synuclein toxicity in neurodegeneration: Mechanism and therapeutic strategies. *Nature Medicine, 23*(2), 1−13. http://dx.doi.org/10.1038/nm.4269.

Xiao, L., & Schultz, Z. D. (2017). Spectroscopic imaging at the nanoscale: Technologies and recent applications. *Analytical Chemistry, 90*(1), 440−458. http://dx.doi.org/10.1021/acs.analchem.7b04151.

Xiao, Y., Ma, B., McElheny, D., Parthasarathy, S., Long, F., Hoshi, M., et al. (2015). Aβ(1–42) fibril structure illuminates self-recognition and replication of amyloid in Alzheimer's disease. *Nature Structural & Molecular Biology, 22*(6), 499–505. http://dx.doi.org/10.1038/nsmb.2991.

Xu, Z., Lazim, R., Mei, Y., & Zhang, D. (2012). Stability of the β-structure in prion protein: A molecular dynamics study based on polarized force field. *Chemical Physics Letters, 539*(540), 239–244. http://dx.doi.org/10.1016/j.cplett.2012.05.025.

Yamamoto, N., Hasegawa, K., Matsuzaki, K., Naiki, H., & Yanagisawa, K. (2004). Environment- and mutation-dependent aggregation behavior of Alzheimer amyloid beta-protein. *Journal of Neurochemistry, 90*(1), 62–69. http://dx.doi.org/10.1111/j.1471-4159.2004.02459.x.

Yamashita, M., & Fenn, J. B. (1984). Electrospray ion source. Another variation on the free-jet theme. *The Journal of Physical Chemistry, 88*(20), 4451–4459. http://dx.doi.org/10.1021/j150664a002.

Younan, N. D., & Viles, J. H. (2015). A comparison of three fluorophores for the detection of amyloid fibers and prefibrillar oligomeric assemblies. ThT (thioflavin T); ANS (1-anilinonaphthalene-8-sulfonic acid); and bisANS (4,4′-dianilino-1,1′-binaphthyl-5,5′-disulfonic acid). *Biochemistry, 54*(28), 4297–4306. http://dx.doi.org/10.1021/acs.biochem.5b00309.

Yu, L., Edalji, R., Harlan, J. E., Holzman, T. F., Lopez, A. P., et al. (2009). Structural characterization of a soluble amyloid beta-peptide oligomer. *Biochemistry, 48*(9), 1870–1877. http://dx.doi.org/10.1021/bi802046n.

Yu, Y., Yu, Z., Zheng, Z., Wang, H., Wu, X., Guo, C., et al. (2016). Distinct effects of mutations on biophysical properties of human prion protein monomers and oligomers. *Acta Biochimica et Biophysica Sinica, 48*(11), 1016–1025. http://dx.doi.org/10.1093/abbs/gmw094.

Zahn, R., Liu, A., Luhrs, T., Riek, R., Von Schroetter, C., Lopez Garcia, F., et al. (2000). NMR solution structure of the human prion protein. *Proceedings of the National Academy of Sciences of the United States of America, 97*(1), 145–150. http://dx.doi.org/10.1073/pnas.97.1.145.

Zandomeneghi, G., Krebs, M. R., McCammon, M. G., & Fandrich, M. (2004). FTIR reveals structural differences between native β-sheet proteins and amyloid fibrils. *Protein Science, 13*(12), 3314–3321. http://dx.doi.org/10.1110/ps.041024904.

Zhang, J. (2010). Studies on the structural stability of rabbit prion probed by molecular dynamics simulations of its wild-type and mutants. *Journal of Theoretical Biology, 264*(1), 119–122. http://dx.doi.org/10.1016/j.jtbi.2010.01.024.

Zhang, M., Ren, B., Liu, Y., Liang, G., Sun, Y., Xu, L., et al. (2017). Membrane interactions of hIAPP monomer and oligomer with lipid membranes by molecular dynamics simulations. *ACS Chemical Neuroscience, 8*(8), 1789–1800. http://dx.doi.org/10.1021/acschemneuro.7b00160.

Zhang, S., et al. (2013). Coexistence of ribbon and helical fibrils originating from hIAPP(20-29) revealed by quantitative nanomechanical atomic force microscopy. *Proceedings of the National Academy of Sciences of the United States of America, 110*(8), 2798–2803. http://dx.doi.org/10.1073/pnas.1209955110.

Zhang, Y., Hashemi, M., Lv, Z., Williams, B., Popov, K. I., Dokholyan, N. V., et al. (2018). High-speed atomic force microscopy reveals structural dynamics of alpha-synuclein monomers and dimers. *Journal of Chemical Physics, 148*(12), 123322. http://dx.doi.org/10.1063/1.5008874.

Zhou, H.-X., & Pang, X. (2018). Electrostatic interactions in protein structure, folding, binding, and condensation. *Chemical Reviews, 118*(4), 1691–1741. http://dx.doi.org/10.1021/acs.chemrev.7b00305.

Zhou, S., Liu, X., An, X., Yao, X., & Liu, H. (2017). Molecular dynamics simulation study on the binding and stabilization mechanism of antiprion compounds to the "hot spot" region of PrP^C. *ACS Chemical Neuroscience, 8*(11), 2446–2456. http://dx.doi.org/10.1021/acschemneuro.7b00214.

Zhou, S., Shi, D., Liu, X., Liu, H., & Yao, X. (2016). Protective V127 prion variant prevents prion disease by interrupting the formation of dimer and fibril from molecular dynamics simulations. *Scientific Reports, 6*, 21804. http://dx.doi.org/10.1038/srep21804.

Zhou, Z., & Xiao, G. (2013). Conformational conversion of prion protein in prion diseases. *Acta Biochimica Et Biophysica Sinica, 45*(6), 465–476. http://dx.doi.org/10.1093/abbs/gmt027.

Ziarek, J. J., Baptista, D., & Wagner, G. (2018). Recent developments in solution nuclear magnetic resonance (NMR)-based molecular biology. *Journal of Molecular Medicine, 96*(1), 1–8. http://dx.doi.org/10.1007/s00109-017-1560-2.

Zuegg, J., & Gready, J. E. (1999). Molecular dynamics simulations of human prion protein: Importance of correct treatment of electrostatic interactions. *Biochemistry, 38*(42), 13862–13876. http://dx.doi.org/10.1021/bi991469d.

Zweckstetter, M., Requena, J. R., & Wille, H. (2017). Elucidating the structure of an infectious protein. *PLOS Pathogens, 13*(4), e1006229. http://dx.doi.org/10.1371/journal.ppat.1006229.

CHAPTER THREE

Mass spectrometric approaches for profiling protein folding and stability

Joseph C. Genereux*
Department of Chemistry, University of California, Riverside, CA, United States
*Corresponding author: Email: josephg@ucr.edu

Contents

Abstract

Protein stability reports on protein homeostasis, function, and binding interactions, such as to other proteins, metabolites and drugs. As such, there is a pressing need for technologies that can report on protein stability. The ideal technique could be

Advances in Protein Chemistry and Structural Biology, Volume 118
ISSN 1876-1623
https://doi.org/10.1016/bs.apcsb.2019.09.006
© 2020 Elsevier Inc.
All rights reserved.

applied in vitro or in vivo systems, proteome-wide, independently of matrix, under native conditions, with residue-level resolution, and on protein at endogenous levels. Mass spectrometry has rapidly become a preferred technology for identifying and quantifying proteins. As such, it has been increasingly incorporated into methodologies for interrogating protein stability and folding. Although no single technology can satisfy all desired applications, several emerging approaches have shown outstanding success at providing biological insight into the stability of the proteome. This chapter outlines some of these recent emerging technologies.

1. Introduction

Protein structure determines protein function. Localization of an individual residue in three-dimensional space can contribute electron density to enable a reaction, modulate a charge field to maintain a protein-protein interaction, or block otherwise complementary regions to prevent unproductive interactions. At the molecular level, the relative orientation of two domains in the same protein can determine the function of the entire complex. Dynamics increase the heterogeneity of protein function, as conformational excursions enable a protein to switch between binding modes, to partition between cellular phases, or allow off-equilibrium intermediates that permit nucleation and irreversible aggregation.

Characterizing protein conformations provides insight into protein function, and lays the groundwork for designing strategies to modulate it. Bulk measurements of a protein's stability reveal insight into its function, half-life, and protein homeostasis. Because binding interactions, both with proteins and ligands, influence stability, protein stability can serve as a proxy for specific interactions. Recognition of this relationship has led to approaches for identifying drug targets and interaction networks. With the chemogenomics revolution rapidly providing us biologically active molecules of unknown mechanisms, approaches to characterize chemical targets inside living cells have already become essential to modern drug discovery and chemical biology (McFedries, Schwaid, & Saghatelian, 2013).

Recent improvements in fractionation, resolution, sensitivity, and particularly quantification have enlarged the mass spectrometry toolbox for proteome-wide measuring protein stability. In this chapter, key terminology for mass spectrometry-based proteomics is briefly introduced, followed by a summary of major quantitative methods in use today. Then, a series of established and emerging techniques for applying mass spectrometry to characterize protein stability are described.

2. Terminology

Detailed introductions to mass spectrometry (Gross, 2011) and proteomics (Y. Zhang, Fonslow, Shan, Baek, & Yates, 2013) are available. Mass spectrometry measures the **mass-to-charge ratio (m/z; unit of Th)** of ions in the gas phase. All **analytes** must be **ionized** and promoted to the gas phase, which usually occurs in an **ion source**. The number of charges on a species is the **charge state**, which for peptides is typically between 1 and 5 and for whole proteins about 1 per 2 kDa. Intensities of are detected in the **mass analyzer**. The mass spectrum, without fragmentation, can be referred to as the **MS1** scan. Alternatively, ions in a small m/z range can be isolated for **fragmentation**. Peptides and proteins are typically fragmented by Collision Induced Dissociation (**CID**), Higher-energy C-trap (Collisional) Dissociation (**HCD**; only available on Orbitrap instrumentation) Electron Transfer Dissociation (**ETD**), or Electron Capture Dissociation (**ECD**). These fragment ions are scanned to generate an **MS2** spectrum; n rounds of isolation/fragmentation generate an **MSn** spectrum.

Analytes are initially present in a **matrix**, containing all material that is not part of the analyte. Important matrix effects in mass spectrometry include **interference**, when an ion with the same m/z as the analyte artificially increases the signal, and **ion suppression**, where a matrix element affects the ionization or transmission ratio of the analyte. Methods to simplify the matrix include **fractionation**; for proteins and peptides this is most commonly accomplished through liquid chromatography (**LC**).

Peptides in particular are most often prepared as **tryptic** digests. Trypsin specifically cleaves C-terminal to lysine and arginine, generating peptides with charge states ≥ 2. In **shotgun proteomics** of complex mixtures, the instrument uses data-dependent analysis (**DDA**) to identify signals that are likely to be peptides. These potential peptides are isolated, fragmented, and the MS2 matched against a **spectral library** of all possible tryptic peptides from the proteome under study (Eng, McCormack, & Yates, 1994; Tyanova, Temu, & Cox, 2016). MS2 that match a peptide sequence from the library are denoted as **Peptide Spectral Matches** (**PSMs**). Alternatively, m/z windows that together encompass the entire spectral range can be fragmented in every cycle, thus achieving data-independent analysis (**DIA**) (Ludwig et al., 2018). At sufficiently high resolution, these MS2 can be deconvoluted to identify the peptide precursors observed in the MS1.

3. Quantitation in mass spectrometry

Mass spectrometry is inherently semi-quantitative. The number of ion counts observed for an individual peptide or protein changes in response to the analyte concentration, but is also impacted by the peptide's particular ionizability, charge state, potential suppression from the matrix, detector performance (and type of mass analyzer employed), and factors specific to the individual instrument. While qualitative measurements are adequate to identify the nature of a species based on its m/z, standardization is required for quantification.

3.1 Spectral counts

A straightforward method to quantify protein levels is by **spectral counts**, the number of observed spectra that are matched to peptides assigned to a given protein (Lundgren, Hwang, Wu, & Han, 2010) (Fig. 1A). The number of spectral counts accurately represents the levels of an individual protein between samples (Zybailov, Coleman, Florens, & Washburn, 2005). However, the error in spectral counting is on order of the square root of the count (Gregori, Villarreal, Sánchez, Baselga, & Villanueva, 2013). Most proteins are assigned fewer than 5 spectral counts in a typical shotgun proteomics run, such that many replicates are necessary to distinguish small changes in protein levels by spectral counting.

3.2 Label free quantification

On low resolution instruments, and with complex samples, extracted ion chromatograms often have too much interference for reliable quantification. With the advent of rapid high-resolution mass spectrometers, the area under the curve for an MS1 peak can be used to confidently quantify a peptide peak without being overwhelmed by interference (Cox et al., 2014) (Fig. 1B). In this case, retention time information or a clear MS2 scan in the elution window are necessary to identify the precursor. Elution standards allow the retention window to be normalized with high confidence (Escher et al., 2012), further increasing confidence in the assignment of a given peak to a specific peptide, and serve as a reference for quantification.

3.3 Targeted proteomics

During DDA, the instrument is constantly attempting to identify, isolate, fragment, and scan additional peptides. If only a limited number of peptides

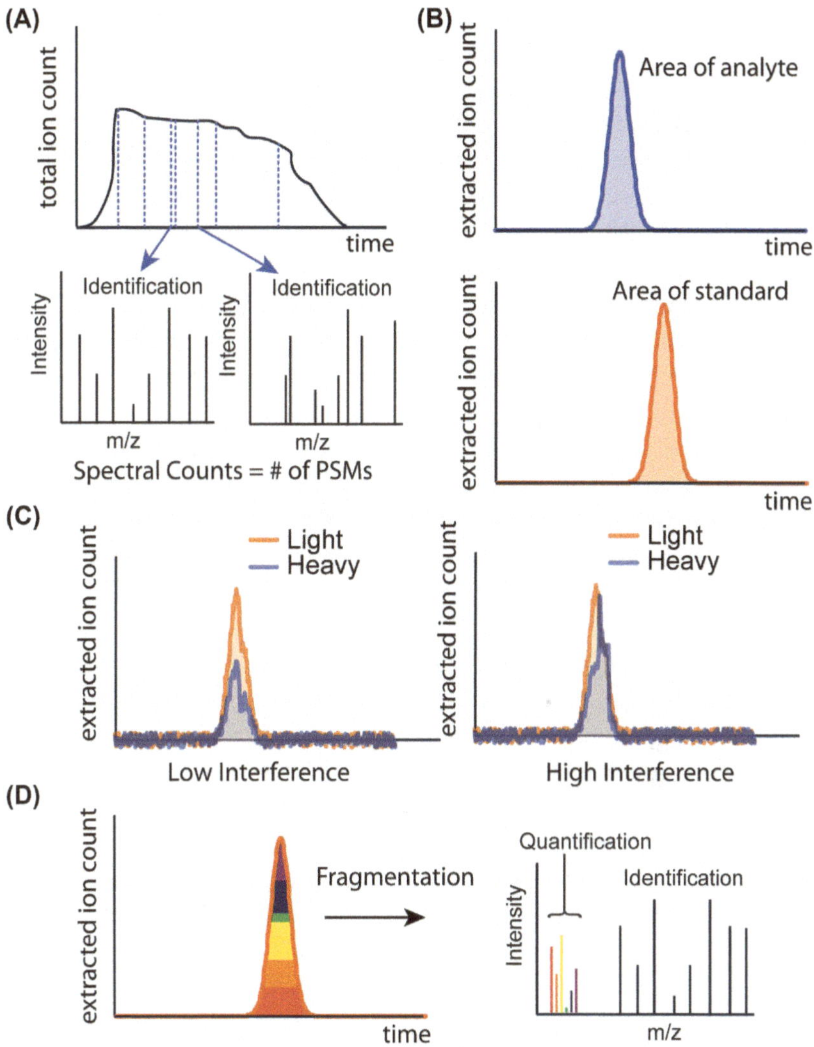

Fig. 1 Protein quantification methods in mass spectrometry. (A) The sum total of Peptide Spectral Matches that correspond to peptides in a given protein is a semiquantitative measure of the amount of that protein in a sample. If each dotted blue line corresponds to a single PSM for the same protein, then in this example the protein would be assigned seven spectral counts. (B) If a clean, interference-free precursor peak appears in the extracted ion chromatogram, then the ratio of Areas Under the Curve between two runs is the same as the ratio of that peptide between the two samples. (C) If the samples are differentially isotopically labeled, then they can be run together and the peptide ratio similarly calculated from the peak integrations. Furthermore, since they are chemically identical, the peak high correlation along the peak also reflects the level of interference. (D) Pooled samples labeled with different isobaric tags yield reporter ions in the MS2 spectra. The reporter ion ratios in turn reflect the relative level of that peptide between the different samples.

are of interest, this instrument time is wasted. In Selected Ion Monitoring (**SIM**), the instrument monitors a small window near the analyte *m*/*z*, thus enhancing the sampling frequency and improving quantification (Picotti & Aebersold, 2012; Schilling et al., 2015). Selected Reaction Monitoring (**SRM**) achieves higher specificity by measuring a precursor-fragment pair, or **transition**, that is specific to the analyte. In Multiple Reaction Monitoring (**MRM**), the analyzer serially scans through several transitions, increasing both sensitivity and specificity, the latter by requiring that relative transition heights for an analyte match a known standard. In Parallel Reaction Monitoring (**PRM**), the entire fragmentation spectrum is acquired, increasing confidence in both specificity and quantification (Peterson, Russell, Bailey, Westphall, & Coon, 2012).

A technical limitation for targeted methods is **cycle time**; every transition requires time to accumulate the signal. Monitoring 20 transitions for 300 ms each will lead to 6 s between sampling of each individual ion. If an ion elutes over 30 s, it will at best be sampled only five times, which is inadequate for robust quantification by peak area integration. To decrease cycle time, transitions can be **scheduled** for where the analyte is expected to elute (Picotti, Bodenmiller, Mueller, Domon, & Aebersold, 2009). This elution window can be determined from previous runs, publicly available library databases (Zauber, Kirchner, & Selbach, 2018), or prediction based on peptide hydrophobicity (Krokhin et al., 2004).

3.4 Stable isotopes

For absolute quantification (AQUA), and to remove run-to-run variability, isotopically labeled peptides can be used as the reference standard (Kettenbach, Rush, & Gerber, 2011). This stable isotope dilution method offers two benefits beyond standardization. First, these peptide standards can be used to trigger monitoring of the target peptide, enabling smaller windows and lower cycle times (Gallien, Kim, & Domon, 2015). Second, the standard and target peptides are chemically identical, such that their elution and ionization properties are identical. Hence, the peak shape correlation in the MS1 establishes the extent of interference (Fig. 1C).

Use of an isotopically labeled proteome offers standards against all peptides in a sample. To compare two conditions directly, one sample can be prepared from isotopically "heavy"-labeled cells, while the other is prepared from isotopically "light"-labeled cells. If handling metazoan cells, the label would usually be introduced via isotopically labeled amino acids (SILAC; Stable Isotopic Labeling of Amino acids in Cell culture). In unicellular organisms, nitrogen sources in the media can be labeled with ^{15}N. Samples are mixed and prepared for mass spectrometry. If ran using DDA, either the light or heavy peptide is adequate for identification. In the MS1, the ratio between the co-eluting heavy and light peaks provides the relative concentration of the two peptides, while the correlation between the two provides a measure of confidence in the integration. Although isotopically labeled proteomes do not provide absolute quantification, they provide excellent quantification of fold changes between the samples. Furthermore, a heavy-labeled reference sample can be used as a spike-in standard to then compare a series of unlabeled experiments, alleviating the requirement to label experimental samples. This approach has worked well in ex vivo studies, wherein SILAM (Stable Isotopic Labeling of Amino acids in Mammals) extracts, while laborious to generate, have served as standards for dozens of experiments (McClatchy & Yates, 2014).

3.5 Isobaric tags

Several methods for determining protein stability require peptide levels to be measured as a response to treatments that modulate stability (e.g. temperature, chaotrope, etc.). These approaches are enabled by isobaric multiplexing, a technology that provides head-to-head comparison of up to 11 conditions (and at time of publication, 16) in every reporter ion scan (Fig. 1D) (Braun et al., 2015; Frost, Greer, & Li, 2015; Pappireddi, Martin, & Wühr, 2019; Rauniyar & Yates, 2014; ; Thompson et al., 2003; Unwin, 2010). Members of a set of isobaric labels are chemically identical, and have three parts. The first part is the reporter ion region, which will differ in mass across the set. For isotopomeric reagents, this difference will be as small as a few mTh. Next, a balance region is isotopically labeled to ensure that the total molecular weight of the labels is constant. Finally, a reactive group allows peptide modification. Isobaric labeling is typically performed after peptide digestion, employing isobaric tags that harbor an NHS ester and label at lysine and the peptide N-terminus.

After quenching, the reaction is pooled and analyzed by shotgun proteomics. After a peptide is isolated based on observation in an MS1 scan, fragmentation releases the reporter ions, which are optimized to cleave between the reporter and balance region. Because multiple precursor ions are typically present in the isolation region, and reporter ions cannot be differentiated after cleavage, the quantification of isobaric tags can deviate strongly from the actual values. Since most proteins under most conditions do not change, this leads to moderate to severe underestimation (as much as

80%). To overcome this ratio compression, CID fragmentation to generate MS2, followed by MS3 on b-ions (or y-ions with C terminal lysine) to produce the reporter ions, is often used (Ting, Rad, Gygi, & Haas, 2011). Alternatively, methods for excluding ions based on the precursor purity and exploiting the TMT balance region (which is still attached to the b-ions) have been described (Savitski et al., 2013; Sonnett, Yeung, & Wühr, 2018; Wühr et al., 2012).

4. Fractionation methods for characterizing the destabilized proteome

Protein identification by mass spectrometry is an effective technique in a mature field. Hence, any method that can physically sequester proteins that are conformationally unique can take advantage of the power of mass spectrometry to determine which proteins are partitioned into which fraction.

4.1 Thermal destabilization

All proteins can misfold (Vendruscolo, Knowles, & Dobson, 2011; Walther et al., 2015). Upon destabilization, proteins will exposure hydrophobic surfaces that enable non-native interactions. Oligomerization beyond a critical nucleus size leads to irreversible downhill polymerization to insoluble aggregates. The classic example is cooking an egg; the ovalbumin protein denatures and aggregates into a firm, opaque gel (Varela, England, & Cavagnero, 2019).

Although melting temperature is distinct from $\Delta G_{unfolding}$ (Becktel & Schellman, 1987), they are highly correlated (Plata & Vitkup, 2018). The melting temperature of a protein can be assessed by heating aliquots to different temperatures, centrifuging, and measuring the remaining soluble fraction. Ligand binding generally stabilizes a protein (but not exclusively, as covalent linkages can *destabilize* (Nnadi et al., 2018)); if there was no favorable enthalpy the binding interaction would be unlikely to occur. Hence, a molecule is likely a ligand if it changes the melting temperature of a protein. Thermal shift assays have been around for many years, with end analysis by spectroscopy, mass spectrometry, or any technology that can characterize protein concentration. However, only recently have the groups of Nordlund and Savitski integrated thermal aggregation with quantitative multiplexed proteomics to enable robust proteome-wide profiling (Franken et al., 2015; Molina et al., 2013); the use of isobaric tags ensures that all

~10,000 proteins identified in a single run are quantified across all temperatures. This approach is called CEllular Thermal Shift Assay (CETSA) or Thermal Protein Profiling (TPP).

As an example of the type of insight that CETSA can provide, Nordlund and colleagues recently characterized protein destabilization as a function of cellular oxidation (Sun et al., 2019). The outline of the experiment is provided in Fig. 2A. The reduced:oxidized glutathione ratio was varied in the HepG2 cellular lysates and the solubility of the proteome at 52 °C assessed. Clear sigmoidal relationships were observed between protein stability and redox state of the cell (Fig. 2B). They expanded the study to live cells, varying exogenous peroxide to control the redox gradient. Fig. 2C shows a plot of proteins delineated by sigmoidal quality (abscissa) and the magnitude of the peroxide response (ordinate), with stabilized proteins (largely those maintaining redox homeostasis) exhibiting a positive and destabilized proteins exhibiting a negative area under the curve. Interestingly, ATP-binding proteins were largely destabilized, but only following cellular treatment, consistent with peroxide-induced ATP depletion. This highlights the ability of the technique to identify thermal destabilization that is mediated by

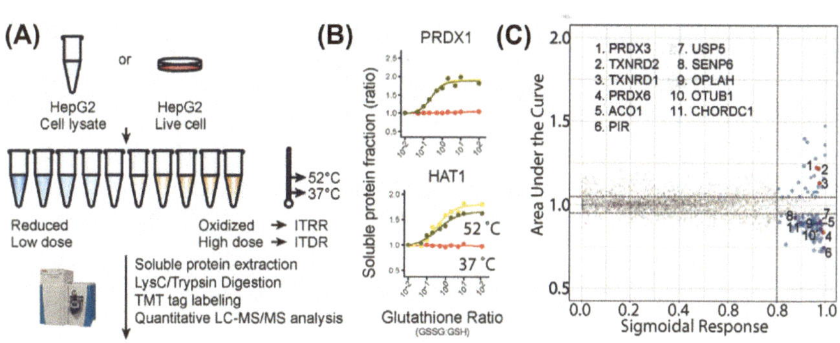

Fig. 2 In CETSA, insoluble protein is separated from soluble protein by heat induced denaturation and aggregation. (A) In this experiment, a gradient of redox states was applied to living HepG2 cells or lysate, and soluble protein isolated after treatment at 37 or 52 °C. (B) For many lysate proteins, smooth sigmoidal relationships were found between the relative amount of protein soluble at high temperature and the redox state of the sample. In this case, redox state was modulated by the reduced/oxidized glutathione ratio. (C) For this plot, the redox state in living cells was modulated by adding peroxide. The abscissa represents the fit to a sigmoidal curve, and the ordinate the relative stabilization (positive) or destabilization (negative) of proteins. Colored circles represent proteins exceeding ratio filters. *From Sun, W., Dai, L., Yu, H., Puspita, B., Zhao, T., Li, F., et al. (2019). Monitoring structural modulation of redox-sensitive proteins in cells with MS-CETSA.* Redox Biology, 24, *101168. https://doi.org/10.1016/j.redox.2019.101168.*

effects downstream of the particular stress. This study makes clear the ease and robustness of the technique.

Another pair of studies characterized how protein stability is modulated by the cell cycle (Becher et al., 2018; Dai et al., 2018). Strikingly, the results of these studies were qualitatively and quantitatively similar, demonstrating the robustness this approach can achieve. For example, both studies found CDK1 was substantially stabilized during S and G2 phases. Both authors were able to assign putative complexes under the hypothesis that proteins whose stabilities share identical temperature dependences across the cell cycle are likely to be in the same complexes, and these assignments were consistent with but expanded the extant literature.

Despite its success, CETSA carries several caveats. First, it requires absolute reproducibility in a stochastic process (heat-induced aggregation) across many complex samples handled in parallel. Second, aggregation is inherently a far off-equilibrium process. Aggregation temperatures do not always coincide to the melting of the *folded* population, nor to other biophysical parameters.

4.2 Isolation of aggresomes

The cell defends against toxic misfolded protein accumulation by segregating aggregates in aggresomes (Farrawell et al., 2015). Proteins in these environments can then be extracted for refolding or degradation. Several studies have purified interactors of aggresome or other aggregate components to identify potential substrates (Li et al., 2010; O'Connell et al., 2014; Wang et al., 2009; Zaarur et al., 2015). Unfortunately, none of these has clearly delineated proteins involved in maintaining aggresome proteostasis from substrates. This is due in part to the still emerging understanding of the functional roles that these bodies play, and the potential that factors besides stability could influence partitioning.

4.3 Conformational antibodies

The paratrope-epitope interaction surface constitutes around 750 Å^2 (Davies, Padlan, & Sheriff, 1990). This interaction takes place in two-dimensional space, and hence is frequently sensitive to the folding of the protein target. For example, antibodies against folded and misfolded p53 can be used to identify the relative stability of the protein population (Trinidad et al., 2013). Other antibodies, such as A11, are selective for the common amyloid fold found in soluble oligomers (Kayed et al., 2003, 2007; Lambert et al., 2007). Palhano and colleagues used the LOC antibody that recognizes a universal amyloid epitope to purify amyloid from *C. elegans* extracts (Greiner, Kelly, & Palhano, 2014), but could not confidently assign Type I and Type II errors. A similar strategy using the A11 antibody in cells revealed binding sites on a variety of chaperones, but did not lead to the discovery of novel aggregating proteins (Yoshiike et al., 2008). Despite the initial disappointments, future development in conformational antibodies for destabilized proteins could still hold promise for characterizing the misfolded proteome.

5. Enzymatic methods for identifying protein stability

Enzymes have evolved to be exquisitely sensitive to the physical and chemical properties of their substrates. Hence, enzymes can serve to transduce a conformational state into a signal, usually an irreversible chemical modification, that is distinguishable by mass spectrometry.

5.1 Limited proteolysis

Buried protein domains cannot enter the protease active site and are more resistant to degradation than extended regions. This property has long been used as a proxy for protein conformation, for example to differentiate the pre- and post-fusion states of viral hemagglutinin (White et al., 2015). The introduction of pulse proteolysis by the Marqusee lab standardized proteolytic sensitivity as a method to interrogate the proteome (Park & Marqusee, 2005). In this technique, varying concentrations of protease (typically thermolysin, subtilisin, protease K, or pronase) are added to a protein sample for a brief (minutes) time to partially digest the sample. Any protein that receives even a single cut will migrate faster on a denaturing gel. The concentration of protease necessary to decrease the intact population of a protein target is a proxy for the relative stability of that protein. Any perturbation that stabilizes (or destabilizes) the protein, such as through mutation, post-translational modification, or ligand binding, will lead to an increase (or decrease) in the concentration of protease necessary to decrease the intact protein population. As a recent example, Schroder and coworkers validated that the small molecule inflammasome inhibitor MCC950 directly binds the NOD-like receptor NLRP3 (Coll et al., 2019). By assaying protection against proteolysis by pronase for a series of NLRP3 variants and truncations, they further identified the NLRP3 domain that binds MCC950.

This approach was extended to characterize unknown targets of metabolites or drugs in complex mixtures by the development of Drug Affinity Responsive Target Stability (DARTS) in the Huang lab (Lomenick et al., 2009). In this technique, a ligand is added either to cells or directly to lysate and lysates analyzed by pulse proteolysis. Next, treated and untreated samples are compared by denaturing gel electrophoresis followed by Coomassie. Bands that are more intense in the treated as opposed to untreated samples correspond to proteins that were stabilized by the ligand, and protein identifications are then determined by mass spectrometry. This approach (using 2-D gel separation) was successfully applied in mouse neural lysate to identify CRMP2 as the target of naringenin, a component of Drynaria Rhyzome and an inducer of axonal growth (Yang, Kuboyama, & Tohda, 2017).

However, development as a proteome-wide profiling technique lagged due to the challenge of identifying conformotypic peptides. Higher resolution mass analyzers, and particularly DIA which allows all potential cleavage products to be quantified, has enabled proteome-wide pulse proteolysis, termed limited proteolysis (LiP), to be realized. Once a conformotypic peptide is identified, targeted experiments such as SRM can be used to further characterize the effects of perturbations on the stability of a specific protein (Feng et al., 2014).

To date, the major proteome-wide applications of limited proteolysis (LiP) have come from the Picotti group. In an important study, Picotti and colleagues determined the variation in stability across the bacterial proteome with the presence of metabolites such as ATP and phenylalanine (Piazza et al., 2018) (Fig. 3A and B). 231 proteins in *E coli* demonstrated increased stability in the presence of exogenous ATP, of which 104 were previously annotated ATP binders. 20 other metabolites were evaluated in *E. coli*, demonstrating the throughput of the technique. Interestingly, conformotypic peptides tend to be near the small molecule binding sites, demonstrating that LiP provides information regarding *where* a protein is destabilized, rather than just total stability of the protein (Fig. 3C). This is in contrast with CETSA, which only reports on global protein stability, and allowed functional insight into mechanism. Specifically, the localization of metabolite-sensitive conformotypic peptides to near the active sites of enzymes demonstrated that metabolites were selectively stabilizing these enzymes through direct active site binding.

Limited proteolysis can also successfully characterize protein stability modulation independent of ligand. In an expansive study that circumvented the challenge of identifying conformotypic peptide products, Picotti and

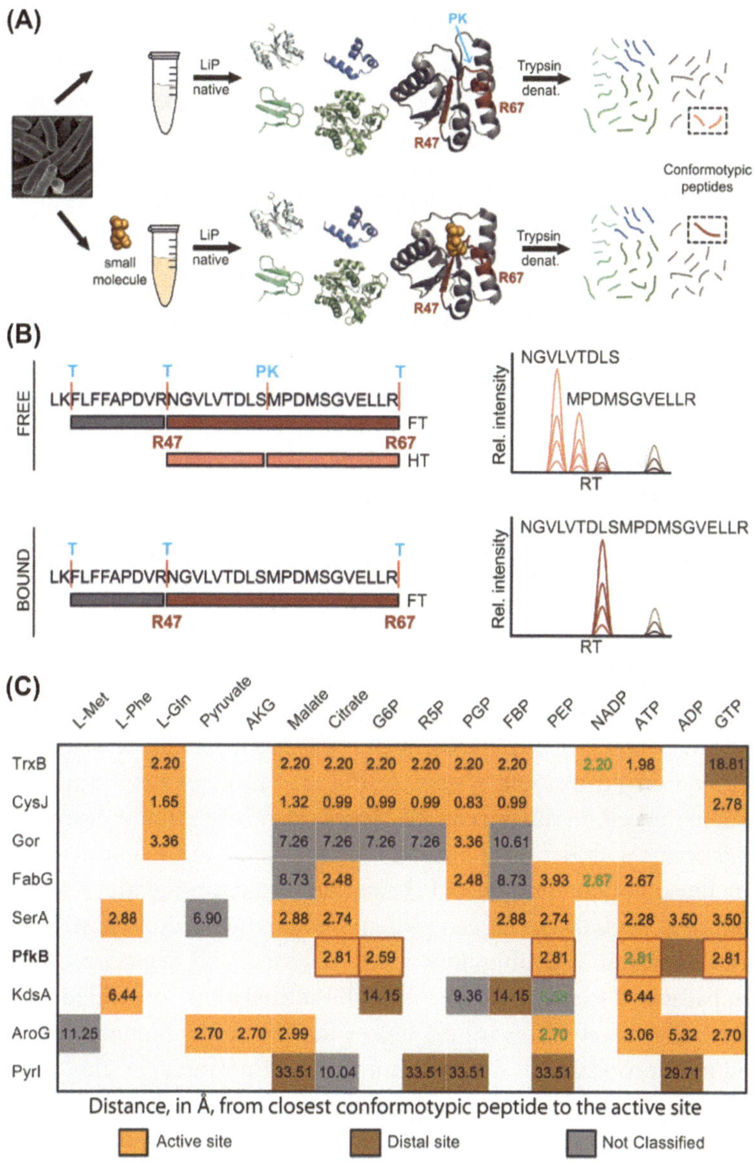

Fig. 3 Use of Limited Proteolysis (LiP) to obtain a proteome-wide map of metabolite binding sites on the *E. coli* proteome. (A) Illustration of how metabolite binding can protect a proteolytic site on a protein. This does not require direct binding of the metabolite to the conformotypic site. (B) Modulated proteolysis of the protein leads to a transition between intact and cleaved peptides near the conformotypic site. (C) Map of the distances between the active sites of the most promiscuous *E. coli* enzymes and the site of the closest conformotypic peptide for several studied metabolites. Most conformotypic peptides are near the active site, consistent with promiscuous enzymes being stabilized by binding of different metabolites at the active site. *From Piazza, I., Kochanowski, K., Cappelletti, V., Fuhrer, T., Noor, E., Sauer, U., et al. (2018). A map of protein-metabolite interactions reveals principles of chemical communication.* Cell, *172(1–2), 358–372.e23. https://doi.org/10.1016/j.cell.2017.12.006.*

coworkers compared the heat sensitivity of proteomes across bacteria, yeast, and human cells (Leuenberger et al., 2017). In this case, rather than seeking specific conformotypic peptides, the response of all tryptic peptides to heat was characterized. An impressive 25% of all identified peptides were found to follow a sigmoidal relationship with temperature, enabling reliable assignment of a melting temperature. Note that many proteins will not be subject to heat denaturation between 37 °C and 70 °C due to being heat-stable even at high temperature, or being natively unstructured and hence proteolytically sensitive at all temperatures. Extensive validation supported the quantitative accuracy of the derived melting temperatures.

This is an emerging technology and the full scale of applications has not yet been demonstrated. Several examples have successfully identified proteins stabilized by ligands in complex proteomes. The challenge in identifying conformotypic peptides translates into a high Type II error rate, consistent with what has been observed using the other cutting-edge technologies discussed in this review such as SPROX.

5.2 Ubiquitinylation

The cell has a rich diversity of pathways that identify and handle misfolded protein to prevent its toxic accumulation (Klaips, Jayaraj, & Hartl, 2018). The ubiquitin-proteasome system (UPS) is central to the elimination of protein in the cell. In this system, E3 ligases catalyze the modification of substrate proteins with the small protein ubiquitin. The most common modification, linear polyubiquitinylated chains connected through the K48 residue of ubiquitin, leads to the tagged protein being recruited to the proteasome, where the protein is deubiquitinylated, unfolded, and degraded. Although many pathways can lead to protein ubiquitinylation, misfolded proteins are generally directed toward E3 ligases for K48 polyubiquitinylation and subsequent proteasomal degradation. Ubiquitinylation in this pathway thus serves as a proxy for protein destabilization in two related senses: destabilized proteins tend to get ubiquitinylated, and ubiquitinylated proteins tend to have a shorter half-life.

Because ubiquitinylation stoichiometry can be somewhat low, enrichment is necessary for shotgun proteomics analysis. Intact ubiquitinylated proteins can be immunoprecipitated by anti-ubiquitin antibodies, or by introduction of affinity tagged ubiquitin and subsequent affinity purification (Danielsen et al., 2011; Peng et al., 2003). Alternatively, ubiquitinylated peptides can be enriched following digestion by antibodies against the tryptically exposed K-ε-GG modification at ubiquitinylated lysines (Kim et al.,

2011; Udeshi et al., 2013). Quantitation can be achieved by the use of SILAC or isobaric tags (Rose et al., 2016). Combining metabolic labeling with mass-offsetting N-terminal tags enables isotopomeric quantification (Cao et al., 2017). By comparing the ubiquitinylated proteome in the presence and absence of an E3 ligase inhibitor, K-ε-GG immunoprecipitation coupled with mass spectrometry allowed the substrate specificity of that E3 ligase class to be determined (Emanuele et al., 2011). The identified proteome correlated well with that found from a fluorescent assay of protein stability. Note, however, that this approach will identify both pro-degradation K48 polyubiquitinylation and non-UPS ubiquitinylation (i.e. monoubiquitination, or polyubiquitin linkages through K6, K11, and K63). An antibody that was specific toward K48 polyubiquitin linkages has been reported (Newton et al., 2008), but has not been widely used for AP-MS. Alternatively, the pro-degradation versus neutral ubiquitinylome can be determined by filtering for proteins that are sensitive to proteasomal inhibitor treatment. Bennett and co-workers exploited this approach to determine the subcellular localization of ubiquitinylated proteins that are proteasomally targeted (Gendron et al., 2016). They found that most destabilized ubiquitinylated proteins are cytosolic, and proposed that these are mostly nascent mistranslated proteins.

Several other groups have used changes in ubiquitinylation to identify modulated protein stability (Braten et al., 2016; ; Thompson et al., 2014). Theurillat and coworkers characterized the differential ubiquitinylome in the presence of cancer associated mutations to the E3 ligase SPOP. In this study, they immunoprecipitated K-ε-GG peptides from SILAC-labeled cells overexpressing the mutants or the wild-type protein. They identified two proteins, DEK and TRIM24, that were differentially ubiquitinylated, and hence demonstrated differential stability, leading to the oncogenic phenotypes.

6. Exploiting chemical reactivity to characterize protein stability

Changes in conformation are associated with changes in function; this is indeed the rationale underlying the interest in describing protein structure and conformation in the first place. The reactivity of any site on a protein will be governed by two factors: reactivity and accessibility. A buried site that is not accessible will not react. A variety of reactive species have been used for footprinting, of which we discuss a subset here (Fig. 4A–D),

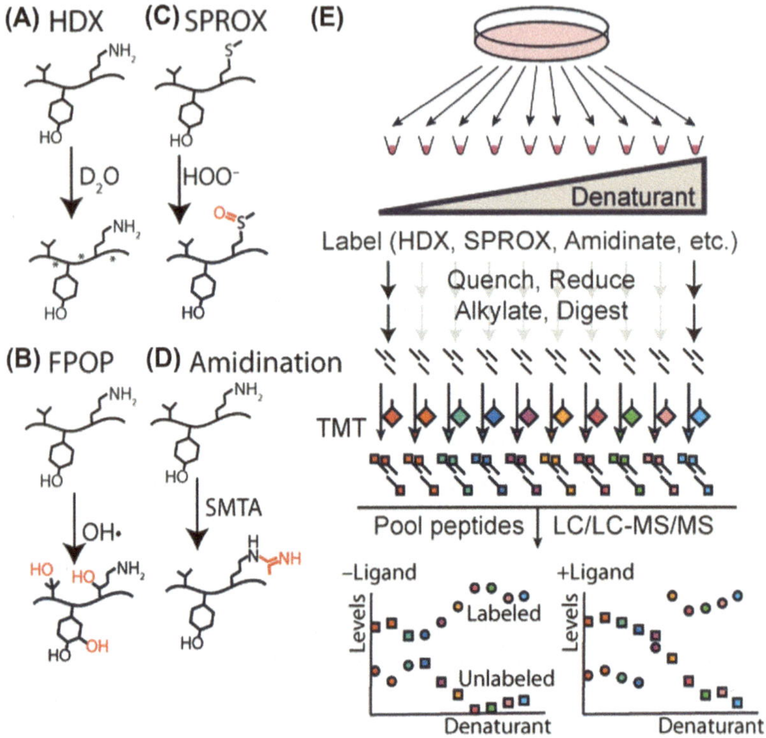

Fig. 4 Illustration of footprinting technologies, with chemical modifications marked in red. (A) Schematic of HDX labeling at the peptide backbone. (B) Schematic of FPOP hydroxylation at amino acid side chains. Carbonylation modifications are also common. (C) Schematic of selective SPROX sulfoxidation at methionine. (D) Schematic of amidination at lysine. Note that basicity at the amidinated lysine is preserved. (E) A generalized scheme for isobaric tagging for SUPREX style analysis in the presence and absence of a ligand. This approach is amenable with a variety of footprinting chemistries.

though other promising methods have also been reported (Luchini, Espina, & Liotta, 2014; Manzi et al., 2016; Moio et al., 2011).

6.1 Hydrogen-deuterium exchange

A widely used method to characterize protein folding and stability in vitro is hydrogen deuterium exchange (termed HX, HDX or HXMS) (Masson et al., 2019) (Fig. 4A). Differences in the local environment of the backbone, the presence of hydrogen bonds, and solvent accessibility modulate amide proton exchange kinetics from hundreds of milliseconds for completely exposed sites to days for highly buried sites. Benefits of using MS versus NMR to characterize HDX include the ability to handle large proteins and complexes, and amenability to complex matrices, as long as instrument resolution is adequate to preclude interference. Furthermore, measuring HDX following equilibration with varying concentrations of denaturant allows $\Delta G_{unfolding}$ to be inferred. This is the essence of Stability of Unpurified Proteins from Rates of H/D EXchange (SUPREX) (Ghaemmaghami, Fitzgerald, & Oas, 2000), and similar strategies are used for the oxidative and electrophilic profiling strategies described below (Fig. 4E).

Pulse deuterium labeling with an appropriate rapid dilution setup allows measurement of protein folding with millisecond resolution (Englander & Mayne, 2014; Walters, Mayne, Hinshaw, Sosnick, & Englander, 2013). Labeling polydispersity over time reveals conversions between discrete populations, allowing pathways for individual domains to be deconvoluted. Hartl and coworkers used time-resolved HDX to characterize the effect of the Hsp60 chaperone GroEL on the folding of substrate protein DapA (Fig. 5A and B) (Georgescauld et al., 2014). This work demonstrated that GroEL allows segmented folding of different domains of DapA, as opposed to slow holistic folding for the unassisted protein, suggesting that Hsp60 function can be profolding rather than solely anti-aggregation.

The power of HDX to provide comprehensive structural characterization is exemplified by an investigation into the mechanism of DNA-damage responsive PARP-1 activation (Dawicki-McKenna et al., 2015). PARP-1 coordinates the cellular response to DNA single strand breaks through its poly ADP-ribosylation activity (Tao, Gao, & Liu, 2009). HDX allowed the conformation of full length PARP-1 to be characterized both in the absence and presence of DNA harboring a single strand break (Fig. 5C), revealing that the PARP-1 HD domain unfolds upon the recognition event, freeing the ADP-ribosylation active site (Fig. 5D). Note that overlapping peptides were quantified over much of the sequence, increasing resolution of the inferred protein dynamics.

In another example, Walensky and collaborators measured the stability of prospective stapled peptide inhibitors of BAX, a critical protein in the apoptotic cascade (Barclay et al., 2015). Although stapling peptides changes distribution, structure, stability, and binding as compared to the linear sequences, how these properties change is hard to predict (Okamoto et al., 2014). The stability of a series of peptides, with residue specific resolution, was determined by HDX and consequent mass spectrometry. HDX-MS further determined each peptide's binding affinity to BAX, the binding site on BAX, the conformational change in BAX, and the overall effect on BAX stability associated with peptide binding. These changes were

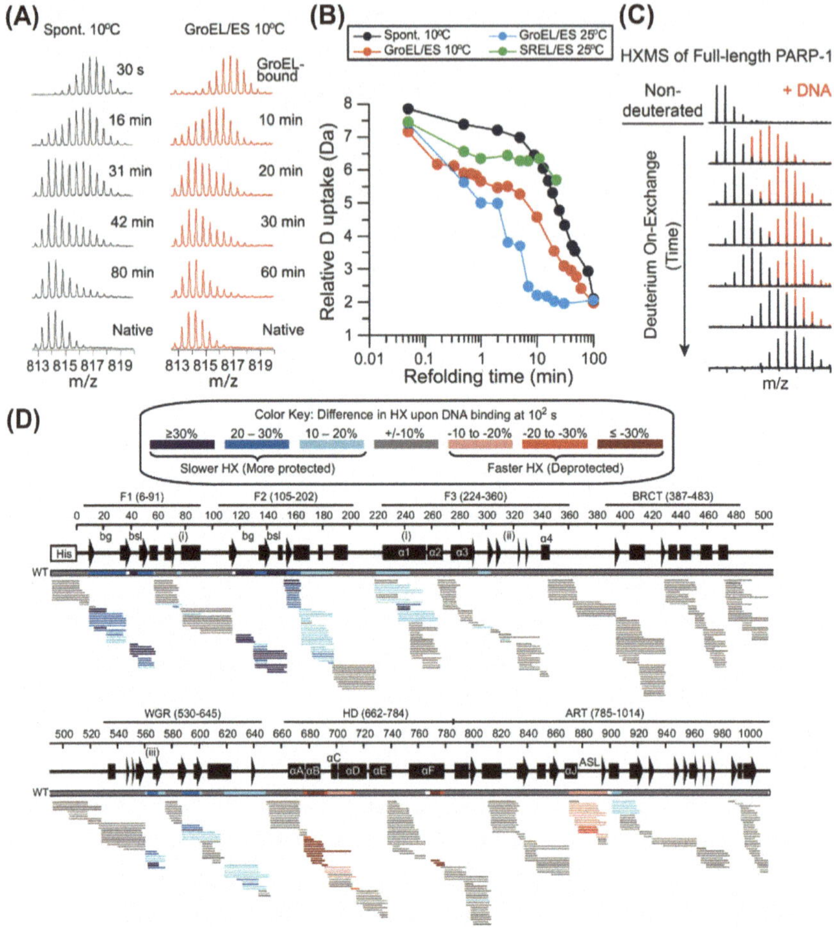

Fig. 5 Illustrations of HDX to characterize protein folding. (A) The time resolved HDX exchange (following 12 s D_2O incubation) for an interior peptide of DapA in the absence (left) and presence (right) of the GroEL/ES Hsp60 chaperone assembly. (B) Quantification of (A) (C) Time resolved HDX exchange of an HD domain peptide of PARP-1 following incubation with dumbbell DNA harboring a nucleotide gap that mimics a single strand break. (D) Relative HDX exchange after 100 s D_2O incubation across all of PARP-1 in the presence and absence of DNA. *Figures from Dawicki-McKenna, J. M., Langelier, M.-F., DeNizio, J. E., Riccio, A. A., Cao, C. D., Karch, K. R., et al. (2015). PARP-1 activation requires local unfolding of an autoinhibitory domain.* Molecular Cell, *60(5), 755–768. https://doi.org/10.1016/j.molcel.2015.10.013; Georgescauld, F., Popova, K., Gupta, A. J., Bracher, A., Engen, J. R., Hayer-Hartl, M., et al (2014). GroEL/ES chaperonin modulates the mechanism and accelerates the rate of TIM-barrel domain folding.* Cell, 157*(4), 922–934. https://doi.org/10.1016/j.cell.2014.03.038.*

confirmed by photocrosslinking mass spectrometry, NMR, and traditional binding assays. This study demonstrates how HDX can provide detailed information covering all aspects of protein structure and dynamics.

The technical challenges of applying HDX are substantial. To avoid biased exchange rates, hydrogen/deuterium scrambling must be prevented until analysis is complete. Toward this end, samples are immediately acidified to pH 2–3 (Tüchsen & Woodward, 1985), digested with an acid-stable protease (usually pepsin), maintained at 4 °C, and analyzed rapidly. Even with these precautions, back-exchange is hard to avoid; a few minutes between quenching and analysis can be sufficient for 50% exchange (Laine, Streaker, Nabavi, Fenselau, & Beckett, 2008). Proton exchange can occur during electrospray ionization (Buijs et al., 2001) and ion transmission (Guttman et al., 2016). CID can completely scramble labeling for some sequences (Rand & Jørgensen, 2007), hence ETD and ECD are often used instead to localize exchange to specific backbone sites (Kan, Walters, Mayne, & Englander, 2013). The background level of H-D exchange should be determined for all setups (Hoofnagle, Resing, & Ahn, 2004).

Use of pepsin increases sample complexity as opposed to trypsin, necessitating excellent fractionation and resolution to differentiate between peptides and limiting the use of HDX on endogenous proteins in complex samples. As instrument resolution and fractionation improve, this hurdle will eventually be surmounted. In the meantime, residue-selective and irreversible reactions have been investigated to allow measurements of protein folding, stability and conformations inside biological environments.

6.2 FPOP

The challenge of deuterium back-exchange during HDX would be relieved by a technology that generates an irreversible reaction product. Fast Photochemical Oxidation of Proteins (FPOPs), wherein protein footprinting is accomplished by employing hydroxyl radicals, provides that technology (Johnson, Di Stefano, & Jones, 2019; Li, Shi, & Gross, 2018). Hydroxyl radicals react with a wide range of backbone sites to generate known products (Xu & Chance, 2007), providing site-specific resolution approaching that obtained from HDX (Fig. 4B). FPOP employs a flow cell that enables rapid mixing of H_2O_2 ($\leq$20 mM) with the sample in a few μsec, followed by irradiation with a ~100 ms pulse near 250 nm to generate hydroxyl radicals. A radical scavenger limits the reaction time to between 100 ns and 10 μs, depending on the quencher. This set-up ensures that labeling is completed before the protein structure can respond to the modifications (Gau, Sharp,

Rempel, & Gross, 2009). The actual yield of radicals produced can be measured and used for normalization with a standard reporter (Niu et al., 2017; Xie & Sharp, 2015). Sample analysis can be accomplished using standard shotgun or targeted proteomics. Because it generates many chemically distinct peptide modifications, FPOP substantially increases the complexity of the matrix, necessitating high resolution mass analysis to avoid interference and making peptide identification challenging. A recent iterative multi-level platform allows completely automated FPOP peptide identifications and quantifications (Rinas, Espino, & Jones, 2016).

Though not as widely applied as HDX, FPOP has been widely used to characterize in vitro protein structure (Gau, Garai, Frieden, & Gross, 2011; Vahidi, Stocks, Liaghati-Mobarhan, & Konermann, 2012), folding (Chen, Rempel, Gau, & Gross, 2012), dynamics (Li, Rempel, & Gross, 2016), and binding (Zhang, Gau, Jones, Vidavsky, & Gross, 2011). A particularly exciting application worth highlighting is measurements of protein stability in living cells. Jones and coworkers optimized the flow setup of FPOP to allow irradiation of intact cells, while maintaining the stringent "single-hit" pulse system (Rinas, Mali, Espino, & Jones, 2016). They further extended this approach to *C. elegans*, finding similar reactivity to that predicted by solvent accessibility (Espino & Jones, 2019). The rapidity with which FPOP has been translated into highly complex samples is a testament to the robustness of the underlying approach. As protein coverage in complex samples increases, and commercial systems become available (Misra, Orlando, Weinberger, & Sharp, 2019), FPOP is likely to become a central technique for assaying protein stability in complex samples.

6.3 SPROX

In contrast to the high reactive hydroxyl radicals used in FPOP, hydrogen peroxide on its own primarily converts methionine to methionine sulfoxide (Fig. 4C) (Toennies, 1938). Cysteine oxidation products are readily reversed by reductants used in normal mass spectrometry workflows; irreversible sulfinic and sulfonic acids are produced at low stoichiometry (Walker, Bettinger, Welle, Hryhorenko, & Ghaemmaghami, 2019). Monitoring methionine oxidation is the essence of Stability of Proteins from Rates of OXidation (SPROX), originally developed by the Fitzgerald lab at Duke, wherein the reactivity of methionine toward hydrogen peroxide is used as a proxy for the solvent accessibility of the methionine residues (Kaur et al., 2018; Ogburn, Jin, Meng, & Fitzgerald, 2017; West, Tang, &

Fitzgerald, 2008). In combination with a gradient of chaotrope, stability isotherms in lyzed samples can be generated and unfolding thermodynamics inferred.

Limiting modification to a single product at a single residue greatly simplifies peptide identification as only a single modification needs to be considered while seeking spectral identifications: +15.9949 Da to methionine. This allows measurements to be readily performed in complex biological samples. Background oxidation of methionines should be minimized both before and after the labeling period, and controls are necessary to establish the background level of oxidation. The relative yield of methionine oxidation can be determined for both the modified and unmodified methionines by integrated ion intensities in the precursor scan, SILAC isotope ratios, targeted reaction monitoring, or reporter ion ratios following isobaric tagging (Geer & Fitzgerald, 2016; Geer Wallace et al., 2016; Ogburn et al., 2017; Tran, Adhikari, & Fitzgerald, 2014; Xu, Wallace, & Fitzgerald, 2016). Data can be fit according to a standard two state model for equilibrium between the unfolded and folded states. Methionine is the second rarest residue among vertebrates (Lin et al., 2017; Lodish, 2000), and surface methionines are not conformotypic as they will be oxidized for all conformations. Hence, SPROX cannot access the complete proteome.

In addition to many studies identifying ligand targets (Geer & Fitzgerald, 2016; Geer Wallace et al., 2016; Ogburn et al., 2017), SPROX was successfully applied to identify proteins with differential stabilities between cancer lines (Adhikari, West, & Fitzgerald, 2015). Once proteins are identified, targeted MS can be applied to follow the oxidation rate of specific proteins as a function of drug treatment (Jin, Wang, Gooden, Ball, & Fitzgerald, 2016). In a particularly impressive application of SPROX, Ghaemmigammi and colleagues recently applied SPROX to measure the $\Delta G_{unfolding}$ across peptides from over 1000 proteins in cellular lysate with exquisite precision (Walker et al., 2019). A key advance that enabled this coverage was the use of TMT multiplexing, which is compatible with a range of SUPREX-style experiments (Fig. 4E). This allows many denaturant conditions to be directly compared, greatly decreasing error. A variety of internal consistencies strongly validated the data from this study, and the authors also found that the chemical chaperone TMAO can stabilize the proteome *in general*. Although many investigations over the years had found specific proteins to be stabilized by TMAO, this was the first strong evidence for a *general* function of protein stabilization.

In summary, the ease of SPROX, both in the technical application and the data analysis, and the inherently quantitative physical parameters that it provides, places it as a first-line method for characterizing protein stability in complex biological samples.

6.4 Electrophilic profiling

The reactive nucleophiles cysteine and lysine are more common than methionine, but similarly generate homogenous products upon electrophilic modification. Because cysteine is subject to a wide diversity of post-translational modifications, changes in cysteine reactivity can indicate subtle changes in cellular function. On the other hand, the same sensitivity toward PTMs can convolute the use of cysteine for structural changes, unless it is naturally or deliberately buried (Isom, Castaneda, Cannon, & Garcia-Moreno E., 2011; Jha & Udgaonkar, 2007; Shen et al., 2018). Lysine is less reactive at neutral pH, but can be targeted independently of cysteine by amidination and amidification reactions, as the corresponding thioesters and thioamides rapidly hydrolyze. For SUPREX-style labeling experiments, where equilibration in denaturant is assumed, the concentration of label should be carefully optimized to ensure that labeling is not competitive with protein refolding kinetics (Isom, Castaneda, Cannon, & Garcia-Moreno, 2011). A benefit of electrophilic labels is that they can incorporate isotopic or isobaric tags, enabling direct quantitative comparison between two or more conditions in a single run.

NHS esters are the prototypical electrophile for bioconjugation at surfaceexposed lysines (Boutureira & Bernardes, 2015), though still sensitive to local reactivity (Ward, Kleinman, & Nomura, 2017). The popular isobaric labeling reagents TMT and iTRAQ are commercially available as NHS esters, providing reagents for electrophilic reactivity profiling that are already isobarically modified (Zhou & Vachet, 2013). This approach has been successfully applied to footprint the binding site between Receptor-Associated Protein and the LDL Receptor (Bloem et al., 2015).

Thioimidate reagents, largely developed and characterized by the Reilly group, are notable in that they are cell-penetrable, selectively label an abundant amino acid (lysine), provide a homogenous reaction product allowing straightforward spectral assignment, promote ionization during electrospray, and retain basicity at lysine (Beardsley & Reilly, 2003; Chang, Lauber, Running, Reilly, & Giedroc, 2011) (Fig. 4D). An establishing study compared ribosomal modifications in treated *E. coli* cells as opposed to cellular lysates (Jaffee, Lauber, Running, & Reilly, 2012). Nearly quantitative labeling at surface exposed lysine residues was observed, with minimal labeling at buried sites. Furthermore, the labeling in lysates and in cells was identical. Our group successfully applied S-methylthioacetamide (SMTA) labeling in conjunction with PRM to identify the fusion-associated conformational change in influenza virus hemagglutinin (Nguyen, Thurmond, Hai, & Genereux, 2019). Brodbelt and collaborators used S-ethylthioacetamide to label a domain of PARP-1 and horse heart myoglobin with and without acetonitrile denaturation. They then identified conformational changed by top-down MS in conjunction with UV photodissociation fragmentation (Cammarata, Lin, Pruet, Liu, & Brodbelt, 2014).

In a study that directly compared thioacetamide labeling to similar experiments performed using SPROX, Fitzgerald and collaborators subjected yeast lysates to a range of denaturant concentrations prior to SMTA labeling and quantification by iTRAQ and LC-MS (West et al., 2010; Xu, Falk, Hallen, & Fitzgerald, 2011). They found that although conformotypic peptides were found from only about 100 proteins, the approaches are highly complementary and together can increase the coverage of untargeted protein stability profiling. Both techniques, similar to any denaturation-based stability measurement, can only characterize proteins with two-state unfolding behavior.

6.5 Crosslinking and mass spectrometry (XL-MS)

Bifunctional crosslinkers place greater constraints on reactivity profiling by requiring a maximal distance between reactive sites on the protein, and provide the additional information as to *which* domains of the protein are in close association (Yu et al., 2016). As such, they are a direct measure of protein structure. In principle, any of the electrophilic labels above can be incorporated into a crosslinker, and crosslinker length is readily modulated. The major challenge is that crosslinked peptides exponentially convert the potential sequence search space from one to two dimensions. This both increases the computational demands of identification, as well as requiring more stringent statistical filters to maintain confidence in identifications.

The identification problem can be solved with MS-cleavable linkers (Paramelle, Miralles, Subra, & Martinez, 2013), incorporating labile groups such as RINK (Tang, Munske, Siems, & Bruce, 2005) or sulfoxides (Burke et al., 2015; Gutierrez et al., 2018). These sites are cleaved by CID, separating the crosslinked peptides in the gas phase. In the case of the Protein Interaction Reporters, a reporter ion is released with CID that triggers data-dependent isolation of the released peptides for MS3 fragmentation to generate spectra for identification. These technologies are further compatible with any of the modern quantification methods discussed above (Chavez et al., 2016; Yu et al., 2016), or isotopic tags can be introduced into the crosslinker itself. Higher yield of crosslinked peptides can be achieved by incorporating an affinity handle into the crosslinker as well.

Photoinduced crosslinkers have been successfully used to interrogate protein stability as well. In once recent example, pH-dependent conformational changes in human serum albumin and cytochrome C were quantified by measuring crosslinked products associated with each conformation (Muller, Graziadei, & Rappsilber, 2019). In this case, the proteins were pre-loaded with the photoactivated crosslinker sulfosuccinimidyl 4,4′ azidopentanoate at neutral pH (as succinimide reactivity is pH- and conformation-dependent). Upon pH modulation, photoactivation of the crosslinker allowed the formation of crosslinks that are dependent on structure but *not* electrophilic reactivity. Digestion and DIA allow quantification of a wide range of crosslinking products and consequent inference of structure.

7. Summary

The recent explosion in quantitative proteomics techniques has in turn driven new approaches to characterize protein stability and conformation. These methods offer biochemists a diverse toolbox for characterizing stability on a range of scales, from isolated proteins in vitro to endogenous proteins in the living cell. Maturing techniques such as CETSA, SPROX, and LiP exploit fundamentally different aspects of protein stability. As ever broadening subsets of the proteome are profiled by these complementary methods under similar stresses, new insight into how protein homeostasis is maintained will be achieved.

References

Adhikari, J., West, G. M., & Fitzgerald, M. C. (2015). Global analysis of protein folding thermodynamics for disease state characterization. *Journal of Proteome Research, 14*(5), 2287–2297. http://dx.doi.org/10.1021/acs.jproteome.5b00057.

Barclay, L. A., Wales, T. E., Garner, T. P., Wachter, F., Lee, S., Guerra, R. M., et al. (2015). Inhibition of pro-apoptotic BAX by a noncanonical interaction mechanism. *Molecular Cell, 57*(5), 873–886. http://dx.doi.org/10.1016/j.molcel.2015.01.014.

Beardsley, R. L., & Reilly, J. P. (2003). Quantitation using enhanced signal tags: A technique for comparative proteomics. *Journal of Proteome Research, 2*(1), 15–21. http://dx.doi.org/10.1021/pr0255506.

Becher, I., Andrés-Pons, A., Romanov, N., Stein, F., Schramm, M., Baudin, F., et al. (2018). Pervasive protein thermal stability variation during the cell cycle. *Cell, 173*(6), 1495–1507.e18. http://dx.doi.org/10.1016/j.cell.2018.03.053.

Becktel, W. J., & Schellman, J. A. (1987). Protein stability curves. *Biopolymers, 26*(11), 1859–1877. http://dx.doi.org/10.1002/bip.360261104.

Bloem, E., Ebberink, E. H. T. M., van den Biggelaar, M., van der Zwaan, C., Mertens, K., & Meijer, A. B. (2015). A novel chemical footprinting approach identifies critical lysine residues involved in the binding of receptor-associated protein to cluster II of LDL receptor-related protein. *Biochemical Journal, 468*(1), 65–72. http://dx.doi.org/10.1042/BJ20140977.

Boutureira, O., & Bernardes, G. J. L. (2015). Advances in chemical protein modification. *Chemical Reviews, 115*(5), 2174–2195. http://dx.doi.org/10.1021/cr500399p.

Braten, O., Livneh, I., Ziv, T., Admon, A., Kehat, I., Caspi, L. H., et al. (2016). Numerous proteins with unique characteristics are degraded by the 26S proteasome following monoubiquitination. *Proceedings of the National Academy of Sciences, 113*(32), E4639–E4647. http://dx.doi.org/10.1073/pnas.1608644113.

Braun, C. R., Bird, G. H., Wühr, M., Erickson, B. K., Rad, R., Walensky, L. D., et al. (2015). Generation of multiple reporter ions from a single isobaric reagent increases multiplexing capacity for quantitative proteomics. *Analytical Chemistry, 87*(19), 9855–9863. http://dx.doi.org/10.1021/acs.analchem.5b02307.

Buijs, J., Hagman, C., Håkansson, K., Richter, J. H., Håkansson, P., & Oscarsson, S. (2001). Inter- and intra-molecular migration of peptide amide hydrogens during electrospray ionization. *Journal of the American Society for Mass Spectrometry, 12*(4), 410–419. http://dx.doi.org/10.1016/S1044- 0305(01)00204-5.

Burke, A. M., Kandur, W., Novitsky, E. J., Kaake, R. M., Yu, C., Kao, A., et al. (2015). Synthesis of two new enrichable and MS-cleavable cross-linkers to define protein-protein interactions by mass spectrometry. *Organic and Biomolecular Chemistry, 13*(17), 5030–5037. http://dx.doi.org/10.1039/c5ob00488h.

Cammarata, M., Lin, K.-Y., Pruet, J., Liu, H., & Brodbelt, J. (2014). Probing the unfolding of myoglobin and domain C of PARP-1 with covalent labeling and top-down ultraviolet photodissociation mass spectrometry. *Analytical Chemistry, 86*(5), 2534–2542. http://dx.doi.org/10.1021/ac4036235.

Cao, T., Zhang, L., Zhang, Y., Yan, G., Fang, C., Bao, H., et al. (2017). Site-specific quantification of protein ubiquitination on MS2 fragment ion level via isobaric peptide labeling. *Analytical Chemistry, 89*(21), 11468–11475. http://dx.doi.org/10.1021/acs.analchem.7b02654.

Chang, F.-M. J., Lauber, M. A., Running, W. E., Reilly, J. P., & Giedroc, D. P. (2011). Ratiometric pulse–chase amidination mass spectrometry as a probe of biomolecular complex formation. *Analytical Chemistry, 83*(23), 9092–9099. http://dx.doi.org/10.1021/ac202154r.

Chavez, J. D., Eng, J. K., Schweppe, D. K., Cilia, M., Rivera, K., Zhong, X., et al. (2016). A general method for targeted quantitative cross-linking mass spectrometry. *PLoS One, 11*(12), e0167547. http://dx.doi.org/10.1371/journal.pone.0167547.

Chen, J., Rempel, D. L., Gau, B. C., & Gross, M. L. (2012). Fast photochemical oxidation of proteins and mass spectrometry follow submillisecond protein folding at the amino-acid level. *Journal of the American Chemical Society, 134*(45), 18724–18731. http://dx.doi.org/10.1021/ja307606f.

Coll, R. C., Hill, J. R., Day, C. J., Zamoshnikova, A., Boucher, D., Massey, N. L., et al. (2019). MCC950 directly targets the NLRP3 ATP-hydrolysis motif for inflammasome inhibition. *Nature Chemical Biology, 15*(6), 556–559. http://dx.doi.org/10.1038/s41589-019-0277-7.

Cox, J., Hein, M. Y., Luber, C. A., Paron, I., Nagaraj, N., & Mann, M. (2014). Accurate proteome-wide label-free quantification by delayed normalization and maximal peptide ratio extraction, termed MaxLFQ. *Molecular & Cellular Proteomics, 13*(9), 2513–2526. http://dx.doi.org/10.1074/mcp.M113.031591.

Dai, L., Zhao, T., Bisteau, X., Sun, W., Prabhu, N., Lim, Y. T., et al. (2018). Modulation of protein-interaction states through the cell cycle. *Cell, 173*(6). http://dx.doi.org/10.1016/j.cell.2018.03.065, 1481–1494.e13.

Danielsen, J. M. R., Sylvestersen, K. B., Bekker-Jensen, S., Szklarczyk, D., Poulsen, J. W., Horn, H., et al. (2011). Mass spectrometric analysis of lysine ubiquitylation reveals promiscuity at site level. *Molecular & Cellular Proteomics, 10*(3). http://dx.doi.org/10.1074/mcp.M110.003590.

Davies, D. R., Padlan, E. A., & Sheriff, S. (1990). Antibody-antigen complexes. *Annual Review of Biochemistry, 59*(1), 439–473. http://dx.doi.org/10.1146/annurev.bi.59.070190.002255.

Dawicki-McKenna, J. M., Langelier, M.-F., DeNizio, J. E., Riccio, A. A., Cao, C. D., Karch, K. R., et al. (2015). PARP-1 activation requires local unfolding of an autoinhibitory domain. *Molecular Cell, 60*(5), 755–768. http://dx.doi.org/10.1016/j.molcel.2015.10.013.

Emanuele, M. J., Elia, A. E. H., Xu, Q., Thoma, C. R., Izhar, L., Leng, Y., et al. (2011). Global identification of modular Cullin-RING ligase substrates. *Cell, 147*(2), 459–474. http://dx.doi.org/10.1016/j.cell.2011.09.019.

Englander, S. W., & Mayne, L. (2014). The nature of protein folding pathways. *Proceedings of the National Academy of Sciences, 111*(45), 15873–15880. http://dx.doi.org/10.1073/pnas.1411798111.

Eng, J. K., McCormack, A. L., & Yates, J. R. (1994). An approach to correlate tandem mass spectral data of peptides with amino acid sequences in a protein database. *Journal of the American Society for Mass Spectrometry, 5*(11), 976–989.

Escher, C., Reiter, L., MacLean, B., Ossola, R., Herzog, F., Chilton, J., et al. (2012). Using iRT, a normalized retention time for more targeted measurement of peptides. *Proteomics, 12*(8), 1111–1121. http://dx.doi.org/10.1002/pmic.201100463.

Espino, J. A., & Jones, L. M. (2019). Illuminating biological interactions with in vivo protein footprinting. *Analytical Chemistry*. http://dx.doi.org/10.1021/acs.analchem.9b00244.

Farrawell, N. E., Lambert-Smith, I. A., Warraich, S. T., Blair, I. P., Saunders, D. N., Hatters, D. M., et al. (2015). Distinct partitioning of ALS associated TDP-43, FUS and SOD1 mutants into cellular inclusions. *Scientific Reports, 5*(1), 13416. http://dx.doi.org/10.1038/srep13416.

Feng, Y., De Franceschi, G., Kahraman, A., Soste, M., Melnik, A., Boersema, P. J., et al. (2014). Global analysis of protein structural changes in complex proteomes. *Nature Biotechnology, 32*(10), 1036–1044. http://dx.doi.org/10.1038/nbt.2999.

Franken, H., Mathieson, T., Childs, D., Sweetman, G. M. A., Werner, T., Tögel, I., et al. (2015). Thermal proteome profiling for unbiased identification of direct and indirect drug targets using multiplexed quantitative mass spectrometry. *Nature Protocols, 10*(10), 1567–1593. http://dx.doi.org/10.1038/nprot.2015.101.

Frost, D. C., Greer, T., & Li, L. (2015). High-resolution enabled 12-plex DiLeu isobaric tags for quantitative proteomics. *Analytical Chemistry, 87*(3), 1646–1654. http://dx.doi.org/10.1021/ac503276z.

Gallien, S., Kim, S. Y., & Domon, B. (2015). Large-scale targeted proteomics using internal standard triggered-parallel reaction monitoring (IS-PRM). *Molecular & Cellular Proteomics, 14*(6), 1630–1644. http://dx.doi.org/10.1074/mcp.O114.043968.

Gau, B., Garai, K., Frieden, C., & Gross, M. L. (2011). Mass spectrometry-based protein footprinting characterizes the structures of oligomeric apolipoprotein E2, E3, and E4. *Biochemistry, 50*(38), 8117–8126. http://dx.doi.org/10.1021/bi200911c.

Gau, B. C., Sharp, J. S., Rempel, D. L., & Gross, M. L. (2009). Fast photochemical oxidation of protein footprints faster than protein unfolding. *Analytical Chemistry, 81*(16), 6563–6571. http://dx.doi.org/10.1021/ac901054w.

Geer Wallace, M. A., Kwon, D.-Y., Weitzel, D. H., Lee, C.-T., Stephenson, T. N., Chi, J.-T., et al. (2016). Discovery of manassantin a protein targets using large-scale protein folding and stability measurements. *Journal of Proteome Research, 15*(8), 2688–2696. http://dx.doi.org/10.1021/acs.jproteome.6b00237.

Geer, M. A., & Fitzgerald, M. C. (2016). Characterization of the *Saccharomyces cerevisiae* ATP-Interactome using the iTRAQ-SPROX Technique. *Journal of the American Society for Mass Spectrometry, 27*(2), 233–243. http://dx.doi.org/10.1007/s13361-015-1290-z.

Gendron, J. M., Webb, K., Yang, B., Rising, L., Zuzow, N., & Bennett, E. J. (2016). Using the ubiquitin-modified proteome to monitor distinct and spatially restricted protein homeostasis dysfunction. *Molecular & Cellular Proteomics, 15*(8), 2576–2593. http://dx.doi.org/10.1074/mcp.M116.058420.

Georgescauld, F., Popova, K., Gupta, A. J., Bracher, A., Engen, J. R., Hayer-Hartl, M., et al. (2014). GroEL/ES chaperonin modulates the mechanism and accelerates the rate of TIM-barrel domain folding. *Cell, 157*(4), 922–934. http://dx.doi.org/10.1016/j.cell.2014.03.038.

Ghaemmaghami, S., Fitzgerald, M. C., & Oas, T. G. (2000). A quantitative, highthroughput screen for protein stability. *Proceedings of the National Academy of Sciences of the United States of America, 97*(15), 8296–8301. http://dx.doi.org/10.1073/pnas.140111397.

Gregori, J., Villarreal, L., Sánchez, A., Baselga, J., & Villanueva, J. (2013). An effect size filter improves the reproducibility in spectral counting-based comparative proteomics. *Journal of Proteomics, 95*, 55–65. http://dx.doi.org/10.1016/j.jprot.2013.05.030.

Greiner, E. R., Kelly, J. W., & Palhano, F. L. (2014). Immunoprecipitation of amyloid fibrils by the use of an antibody that recognizes a generic epitope common to amyloid fibrils. *PLoS One, 9*(8), e105433. http://dx.doi.org/10.1371/journal.pone.0105433.

Gross, J. H. (2011). *Mass Spectrometry*. http://dx.doi.org/10.1007/978-3-642-10711-5.

Gutierrez, C. B., Block, S. A., Yu, C., Soohoo, S. M., Huszagh, A. S., Rychnovsky, S. D., et al. (2018). Development of a novel sulfoxide-containing MS-cleavable homobifunctional cysteine-reactive cross-linker for studying protein- protein interactions. *Analytical Chemistry, 90*(12), 7600–7607. http://dx.doi.org/10.1021/acs.analchem.8b01287.

Guttman, M., Wales, T. E., Whittington, D., Engen, J. R., Brown, J. M., & Lee, K. K. (2016). Tuning a high transmission ion guide to prevent gas-phase proton exchange during H/D exchange MS analysis. *Journal of the American Society for Mass Spectrometry, 27*(4), 662–668. http://dx.doi.org/10.1007/s13361- 015-1330-8.

Hoofnagle, A. N., Resing, K. A., & Ahn, N. G. (2004). Practical methods for deuterium exchange/mass spectrometry. In R. Seger (Ed.), *MAP kinase signaling protocols* (Vol. 0, pp. 283–298). http://dx.doi.org/10.1385/1-59259-671-1:283.

Isom, D. G., Castaneda, C. A., Cannon, B. R., & Garcia-Moreno, E.,B. (2011). Large shifts in pKa values of lysine residues buried inside a protein. *Proceedings of the National Academy of Sciences, 108*(13), 5260–5265. http://dx.doi.org/10.1073/pnas.1010750108.

Jaffee, E. G., Lauber, M. A., Running, W. E., & Reilly, J. P. (2012). In vitro and in vivo chemical labeling of ribosomal proteins: A quantitative comparison. *Analytical Chemistry, 84*(21), 9355–9361. http://dx.doi.org/10.1021/ac302115m.

Jha, S. K., & Udgaonkar, J. B. (2007). Exploring the cooperativity of the fast folding reaction of a small protein using pulsed thiol labeling and mass spectrometry. *Journal of Biological Chemistry, 282*(52), 37479–37491. http://dx.doi.org/10.1074/jbc.M706714200.

Jin, L., Wang, D., Gooden, D. M., Ball, C. H., & Fitzgerald, M. C. (2016). Targeted mass spectrometry-based approach for protein–ligand binding analyses in complex biological mixtures using a phenacyl bromide modification strategy. *Analytical Chemistry, 88*(22), 10987–10993. http://dx.doi.org/10.1021/acs.analchem.6b02658.

Johnson, D. T., Di Stefano, L. H., & Jones, L. M. (2019). Fast photochemical oxidation of proteins (FPOP): A powerful mass spectrometry-based structural proteomics tool. *The Journal of Biological Chemistry, 294*(32), 11969–11979. http://dx.doi.org/10.1074/jbc.REV119.006218.

Kan, Z.-Y., Walters, B. T., Mayne, L., & Englander, S. W. (2013). Protein hydrogen exchange at residue resolution by proteolytic fragmentation mass spectrometry analysis. *Proceedings of the National Academy of Sciences of the United States of America, 110*(41), 16438–16443. http://dx.doi.org/10.1073/pnas.1315532110.

Kaur, U., Meng, H., Lui, F., Ma, R., Ogburn, R. N., Johnson, J. H. R., et al. (2018). Proteome-wide structural biology: An emerging field for the structural analysis of proteins on the proteomic scale. *Journal of Proteome Research, 17*(11), 3614–3627.

Kayed, R., Head, E., Sarsoza, F., Saing, T., Cotman, C. W., Necula, M., et al. (2007). Fibril specific, conformation dependent antibodies recognize a generic epitope common to amyloid fibrils and fibrillar oligomers that is absent in prefibrillar oligomers. *Molecular Neurodegeneration, 2*(1), 18. http://dx.doi.org/10.1186/1750-1326-2-18.

Kayed, R., Head, E., Thompson, J. L., McIntire, T. M., Milton, S. C., Cotman, C. W., et al. (2003). Common structure of soluble amyloid oligomers implies common mechanism of pathogenesis. *Science (New York, N.Y.), 300*(5618), 486–489. http://dx.doi.org/10.1126/science.1079469.

Kettenbach, A. N., Rush, J., & Gerber, S. A. (2011). Absolute quantification of protein and post-translational modification abundance with stable isotope–labeled synthetic peptides. *Nature Protocols, 6*(2), 175–186. http://dx.doi.org/10.1038/nprot.2010.196.

Kim, W., Bennett, E. J., Huttlin, E. L., Guo, A., Li, J., Possemato, A., et al. (2011). Systematic and quantitative assessment of the ubiquitin-modified proteome. *Molecular Cell, 44*(2), 325–340. http://dx.doi.org/10.1016/j.molcel.2011.08.025.

Klaips, C. L., Jayaraj, G. G., & Hartl, F. U. (2018). Pathways of cellular proteostasis in aging and disease. *The Journal of Cell Biology, 217*(1), 51–63. http://dx.doi.org/10.1083/jcb.201709072.

Krokhin, O. V., Craig, R., Spicer, V., Ens, W., Standing, K. G., Beavis, R. C., et al. (2004). An improved model for prediction of retention times of tryptic peptides in ion pair reversed-phase HPLC: Its application to protein peptide mapping by off-line HPLC-MALDI MS. *Molecular & Cellular Proteomics, 3*(9), 908–919. http://dx.doi.org/10.1074/mcp.M400031-MCP200.

Laine, O., Streaker, E. D., Nabavi, M., Fenselau, C. C., & Beckett, D. (2008). Allosteric signaling in the biotin repressor occurs via local folding coupled to global dampening of protein dynamics. *Journal of Molecular Biology, 381*(1), 89–101. http://dx.doi.org/10.1016/j.jmb.2008.05.018.

Lambert, M. P., Velasco, P. T., Chang, L., Viola, K. L., Fernandez, S., Lacor, P. N., et al. (2007). Monoclonal antibodies that target pathological assemblies of Aβ. *Journal of Neurochemistry, 100*(1), 23–35. http://dx.doi.org/10.1111/j.1471-4159.2006.04157.x.

Leuenberger, P., Ganscha, S., Kahraman, A., Cappelletti, V., Boersema, P. J., von Mering, C., et al. (2017). Cell-wide analysis of protein thermal unfolding reveals determinants of thermostability. *Science, 355*(6327), eaai7825. http://dx.doi.org/10.1126/science.aai7825.

Li, K. S., Rempel, D. L., & Gross, M. L. (2016). Conformational-sensitive fast photochemical oxidation of proteins and mass spectrometry characterize amyloid beta 1–42 aggregation. *Journal of the American Chemical Society, 138*(37), 12090–12098. http://dx.doi.org/10.1021/jacs.6b07543.

Li, K. S., Shi, L., & Gross, M. L. (2018). Mass spectrometry-based fast photochemical oxidation of proteins (FPOP) for higher order structure characterization. *Accounts of Chemical Research, 51*(3), 736–744. http://dx.doi.org/10.1021/acs.accounts.7b00593.

Li, X., Zhang, Y., Xie, P., Piao, J., Hu, Y., Chang, M., et al. (2010). Proteomic characterization of an isolated fraction of synthetic proteasome inhibitor (PSI)-induced inclusions in PC12 cells might offer clues to aggresomes as a cellular defensive response against proteasome inhibition by PSI. *BMC Neuroscience, 11*(1), 95. http://dx.doi.org/10.1186/1471-2202-11-95.

Lin, S., Yang, X., Jia, S., Weeks, A. M., Hornsby, M., Lee, P. S., et al. (2017). Redox-based reagents for chemoselective methionine bioconjugation. *Science, 355*(6325), 597–602. http://dx.doi.org/10.1126/science.aal3316.

Lodish, H. F. (Ed.). (2000). *Molecular cell biology* (4th ed.). New York: W.H. Freeman.

Lomenick, B., Hao, R., Jonai, N., Chin, R. M., Aghajan, M., Warburton, S., et al. (2009). Target identification using drug affinity responsive target stability (DARTS). *Proceedings of the National Academy of Sciences, 106*(51), 21984–21989. http://dx.doi.org/10.1073/pnas.0910040106.

Luchini, A., Espina, V., & Liotta, L. A. (2014). Protein painting reveals solventexcluded drug targets hidden within native protein–protein interfaces. *Nature Communications, 5*(1), 4413. http://dx.doi.org/10.1038/ncomms5413.

Ludwig, C., Gillet, L., Rosenberger, G., Amon, S., Collins, B. C., & Aebersold, R. (2018). Data-independent acquisition-based SWATH-MS for quantitative proteomics: A tutorial. *Molecular Systems Biology, 14*(8), e8126. http://dx.doi.org/10.15252/msb.20178126.

Lundgren, D. H., Hwang, S.-I., Wu, L., & Han, D. K. (2010). Role of spectral counting in quantitative proteomics. *Expert Review of Proteomics,* 7(1), 39–53. http://dx.doi.org/10.1586/epr.09.69.

Manzi, L., Barrow, A. S., Scott, D., Layfield, R., Wright, T. G., Moses, J. E., et al. (2016). Carbene footprinting accurately maps binding sites in protein– ligand and protein–protein interactions. *Nature Communications,* 7(1), 13288. http://dx.doi.org/10.1038/ncomms13288.

Masson, G. R., Burke, J. E., Ahn, N. G., Anand, G. S., Borchers, C., Brier, S., et al. (2019). Recommendations for performing, interpreting and reporting hydrogen deuterium exchange mass spectrometry (HDX-MS) experiments. *Nature Methods, 16*(7), 595–602. http://dx.doi.org/10.1038/s41592-019- 0459-y.

McClatchy, D. B., & Yates, J. R. (2014). Stable isotope labeling in Mammals (SILAM). In D. Martins-de-Souza (Ed.), *Shotgun proteomics* (Vol. 1156, pp. 133–146). http://dx.doi.org/10.1007/978-1-4939-0685-7_8.

McFedries, A., Schwaid, A., & Saghatelian, A. (2013). Methods for the elucidation of protein-small molecule interactions. *Chemistry & Biology, 20*(5), 667–673. http://dx.doi.org/10.1016/j.chembiol.2013.04.008.

Misra, S. K., Orlando, R., Weinberger, S. R., & Sharp, J. S. (2019). Compensated hydroxyl radical protein footprinting measures buffer and excipient effects on conformation and aggregation in an adalimumab biosimilar. *The AAPS Journal, 21*(5), 87. http://dx.doi.org/10.1208/s12248-019-0358-2.

Moio, P., Kulyyassov, A., Vertut, D., Camoin, L., Ramankulov, E., Lipinski, M., et al. (2011). Exploring the use of dimethylsulfate for in vivo proteome footprinting. *Proteomics, 11*(2), 249–260. http://dx.doi.org/10.1002/pmic.200900832.

Molina, D. M., Jafari, R., Ignatushchenko, M., Seki, T., Larsson, E. A., Dan, C., et al. (2013). Monitoring drug target engagement in cells and tissues using the cellular thermal shift assay. *Science, 341*(6141), 84–87. http://dx.doi.org/10.1126/science.1233606.

Müller, F., Graziadei, A., & Rappsilber, J. (2019). Quantitative photo-crosslinking mass spectrometry revealing protein structure response to environmental changes. *Analytical Chemistry, 91*(14), 9041–9048. http://dx.doi.org/10.1021/acs.analchem.9b01339.

Newton, K., Matsumoto, M. L., Wertz, I. E., Kirkpatrick, D. S., Lill, J. R., Tan, J., et al. (2008). Ubiquitin chain editing revealed by polyubiquitin linkage- specific antibodies. *Cell, 134*(4), 668–678. http://dx.doi.org/10.1016/j.cell.2008.07.039.

Nguyen, K. K., Thurmond, S., Hai, R., & Genereux, J. C. (2019). Protein profiling and pseudo-parallel reaction monitoring to monitor a fusion-associated conformational change in hemagglutinin. *Analytical and Bioanalytical Chemistry, 411*(19), 4987–4998. http://dx.doi.org/10.1007/s00216-019-01921-z.

Niu, B., Mackness, B. C., Rempel, D. L., Zhang, H., Cui, W., Matthews, C. R., et al. (2017). Incorporation of a reporter peptide in FPOP compensates for adventitious scavengers and permits time-dependent measurements. *Journal of the American Society for Mass Spectrometry, 28*(2), 389–392. http://dx.doi.org/10.1007/s13361-016-1552-4.

Nnadi, C. I., Jenkins, M. L., Gentile, D. R., Bateman, L. A., Zaidman, D., Balius, T. E., et al. (2018). Novel K-ras G12C switch-II covalent binders destabilize ras and accelerate nucleotide exchange. *Journal of Chemical Information and Modeling, 58*(2), 464–471. http://dx.doi.org/10.1021/acs.jcim.7b00399.

Ogburn, R. N., Jin, L., Meng, H., & Fitzgerald, M. C. (2017). Discovery of tamoxifen and *N* -desmethyl tamoxifen protein targets in MCF-7 cells using large- scale protein folding and stability measurements. *Journal of Proteome Research, 16*(11), 4073–4085. http://dx.doi.org/10.1021/acs.jproteome.7b00442.

Okamoto, T., Segal, D., Zobel, K., Fedorova, A., Yang, H., Fairbrother, W. J., et al. (2014). Further insights into the effects of pre-organizing the BimBH3 helix. *ACS Chemical Biology, 9*(3), 838–839. http://dx.doi.org/10.1021/cb400638p.

O'Connell, J. D., Tsechansky, M., Royall, A., Boutz, D. R., Ellington, A. D., & Marcotte, E. M. (2014). A proteomic survey of widespread protein aggregation in yeast. *Molecular BioSystems, 10*(4), 851. http://dx.doi.org/10.1039/c3mb70508k.

Pappireddi, N., Martin, L., & Wühr, M. (2019). A review on quantitative multiplexed proteomics. *ChemBioChem, 20*(10), 1210–1224. http://dx.doi.org/10.1002/cbic.201800650.

Paramelle, D., Miralles, G., Subra, G., & Martinez, J. (2013). Chemical cross-linkers for protein structure studies by mass spectrometry. *Proteomics, 13*(3–4), 438–456. http://dx.doi.org/10.1002/pmic.201200305.

Park, C., & Marqusee, S. (2005). Pulse proteolysis: A simple method for quantitative determination of protein stability and ligand binding. *Nature Methods, 2*(3), 207–212. http://dx.doi.org/10.1038/nmeth740.

Peng, J., Schwartz, D., Elias, J. E., Thoreen, C. C., Cheng, D., Marsischky, G., et al. (2003). A proteomics approach to understanding protein ubiquitination. *Nature Biotechnology, 21*(8), 921–926. http://dx.doi.org/10.1038/nbt849.

Peterson, A. C., Russell, J. D., Bailey, D. J., Westphall, M. S., & Coon, J. J. (2012). Parallel reaction monitoring for high resolution and high mass accuracy quantitative, targeted proteomics. *Molecular & Cellular Proteomics, 11*(11), 1475–1488. http://dx.doi.org/10.1074/mcp.O112.020131.

Piazza, I., Kochanowski, K., Cappelletti, V., Fuhrer, T., Noor, E., Sauer, U., et al. (2018). A map of protein-metabolite interactions reveals principles of chemical communication. *Cell, 172*(1–2). http://dx.doi.org/10.1016/j.cell.2017.12.006, 358–372.e23.

Picotti, P., & Aebersold, R. (2012). Selected reaction monitoring–based proteomics: Workflows, potential, pitfalls and future directions. *Nature Methods, 9*(6), 555–566. http://dx.doi.org/10.1038/nmeth.2015.

Picotti, P., Bodenmiller, B., Mueller, L. N., Domon, B., & Aebersold, R. (2009). Full dynamic range proteome analysis of *S. cerevisiae* by targeted proteomics. *Cell, 138*(4), 795–806. http://dx.doi.org/10.1016/j.cell.2009.05.051.

Plata, G., & Vitkup, D. (2018). Protein stability and avoidance of toxic misfolding do not explain the sequence constraints of highly expressed proteins. *Molecular Biology and Evolution, 35*(3), 700–703. http://dx.doi.org/10.1093/molbev/msx323.

Rand, K. D., & Jørgensen, T. J. D. (2007). Development of a peptide probe for the occurrence of hydrogen ($^1H/^2H$) scrambling upon gas-phase fragmentation. *Analytical Chemistry, 79*(22), 8686–8693. http://dx.doi.org/10.1021/ac0710782.

Rauniyar, N., & Yates, J. R. (2014). Isobaric labeling-based relative quantification in shotgun proteomics. *Journal of Proteome Research, 13*(12), 5293–5309. http://dx.doi.org/10.1021/pr500880b.

Rinas, A., Espino, J. A., & Jones, L. M. (2016). An efficient quantitation strategy for hydroxyl radical-mediated protein footprinting using Proteome Discoverer. *Analytical and Bioanalytical Chemistry, 408*(11), 3021–3031. http://dx.doi.org/10.1007/s00216-016-9369-3.

Rinas, A., Mali, V. S., Espino, J. A., & Jones, L. M. (2016). Development of a microflow system for in-cell footprinting coupled with mass spectrometry. *Analytical Chemistry, 88*(20), 10052–10058. http://dx.doi.org/10.1021/acs.analchem.6b02357.

Rose, C. M., Isasa, M., Ordureau, A., Prado, M. A., Beausoleil, S. A., Jedrychowski, M. P., et al. (2016). Highly multiplexed quantitative mass spectrometry analysis of ubiquitylomes. *Cell Systems, 3*(4). http://dx.doi.org/10.1016/j.cels.2016.08.009, 395–403.e4.

Savitski, M. M., Mathieson, T., Zinn, N., Sweetman, G., Doce, C., Becher, I., et al. (2013). Measuring and managing ratio compression for accurate iTRAQ/TMT quantification. *Journal of Proteome Research, 12*(8), 3586–3598. http://dx.doi.org/10.1021/pr400098r.

Schilling, B., MacLean, B., Held, J. M., Sahu, A. K., Rardin, M. J., Sorensen, D. J., et al. (2015). Multiplexed, scheduled, high-resolution parallel reaction monitoring on a full scan QqTOF instrument with integrated data- dependent and targeted mass spectrometric workflows. *Analytical Chemistry, 87*(20), 10222–10229. http://dx.doi.org/10.1021/acs.analchem.5b02983.

Shen, G., Li, S., Cui, W., Liu, S., Yang, Y., Gross, M., et al. (2018). Membrane protein structure in live cells: Methodology for studying drug interaction by mass spectrometry-based footprinting. *Biochemistry, 57*(3), 286–294. http://dx.doi.org/10.1021/acs.biochem.7b00874.

Sonnett, M., Yeung, E., & Wühr, M. (2018). Accurate, sensitive, and precise multiplexed proteomics using the complement reporter ion cluster. *Analytical Chemistry, 90*(8), 5032–5039. http://dx.doi.org/10.1021/acs.analchem.7b04713.

Sun, W., Dai, L., Yu, H., Puspita, B., Zhao, T., Li, F., et al. (2019). Monitoring structural modulation of redox-sensitive proteins in cells with MS-CETSA. *Redox Biology, 24*, 101168. http://dx.doi.org/10.1016/j.redox.2019.101168.

Tang, X., Munske, G. R., Siems, W. F., & Bruce, J. E. (2005). Mass spectrometry identifiable cross-linking strategy for studying Protein–Protein interactions. *Analytical Chemistry,* 77(1), 311–318. http://dx.doi.org/10.1021/ac0488762.

Tao, Z., Gao, P., & Liu, H. (2009). Studies of the expression of human poly(ADPribose) polymerase-1 in *Saccharomyces cerevisiae* and identification of PARP- 1 substrates by yeast proteome microarray screening. *Biochemistry, 48*(49), 11745–11754. http://dx.doi.org/10.1021/bi901387k.

Thompson, J. W., Nagel, J., Hoving, S., Gerrits, B., Bauer, A., Thomas, J. R., et al. (2014). Quantitative lys-ε-gly-gly (diGly) proteomics coupled with inducible RNAi reveals ubiquitin-mediated proteolysis of DNA damage-inducible transcript 4 (DDIT4) by the E3 ligase HUWE1. *Journal of Biological Chemistry, 289*(42), 28942–28955. http://dx.doi.org/10.1074/jbc.M114.573352.

Thompson, A., Schafer, J., Kuhn, K., Kienle, S., Schwarz, J., Schmidt, G., et al. (2003). Tandem mass tags: A novel quantification strategy for comparative analysis of complex protein mixtures by MS/MS. *Analytical Chemistry, 75*(8), 1895–1904. http://dx.doi.org/10.1021/ac0262560.

Ting, L., Rad, R., Gygi, S. P., & Haas, W. (2011). MS3 eliminates ratio distortion in isobaric multiplexed quantitative proteomics. *Nature Methods, 8*(11), 937–940.

Toennies, G. (1938). A sulfoxide OF methionine. *Science, 88*(2293), 545–546. http://dx.doi.org/10.1126/science.88.2293.545.

Tran, D. T., Adhikari, J., & Fitzgerald, M. C. (2014). Stable isotope labeling with amino acids in cell culture (SILAC)-based strategy for proteome-wide thermodynamic analysis of protein-ligand binding interactions. *Molecular & Cellular Proteomics: MCP, 13*(7), 1800–1813. http://dx.doi.org/10.1074/mcp.M113.034702.

Trinidad, A. G., Muller, P. A. J., Cuellar, J., Klejnot, M., Nobis, M., Valpuesta, J. M., et al. (2013). Interaction of p53 with the CCT complex promotes protein folding and wild-type p53 activity. *Molecular Cell, 50*(6), 805–817. http://dx.doi.org/10.1016/j.molcel.2013.05.002.

Tüchsen, E., & Woodward, C. (1985). Hydrogen kinetics of peptide amide protons at the bovine pancreatic trypsin inhibitor protein-solvent interface. *Journal of Molecular Biology, 185*(2), 405–419. http://dx.doi.org/10.1016/0022- 2836(85)90412-7.

Tyanova, S., Temu, T., & Cox, J. (2016). The MaxQuant computational platform for mass spectrometry-based shotgun proteomics. *Nature Protocols, 11*(12), 2301–2319. http://dx.doi.org/10.1038/nprot.2016.136.

Udeshi, N. D., Svinkina, T., Mertins, P., Kuhn, E., Mani, D. R., Qiao, J. W., et al. (2013). Refined preparation and use of anti-diglycine remnant (K-ε-GG) antibody enables routine quantification of 10,000s of ubiquitination sites in single proteomics experiments. *Molecular & Cellular Proteomics, 12*(3), 825–831. http://dx.doi.org/10.1074/mcp.O112.027094.

Unwin, R. D. (2010). Quantification of proteins by iTRAQ. In P. R. Cutillas, & J. F. Timms (Eds.), *LC-MS/MS in proteomics* (Vol. 658, pp. 205–215). http://dx.doi.org/10.1007/978-1-60761-780-8_12.

Vahidi, S., Stocks, B. B., Liaghati-Mobarhan, Y., & Konermann, L. (2012). Mapping pHIn-duced protein structural changes under equilibrium conditions by pulsed oxidative labeling and mass spectrometry. *Analytical Chemistry, 84*(21), 9124–9130. http://dx.doi.org/10.1021/ac302393g.

Varela, A. E., England, K. A., & Cavagnero, S. (2019). Kinetic trapping in protein folding. *Protein Engineering Design and Selection.* http://dx.doi.org/10.1093/protein/gzz018.

Vendruscolo, M., Knowles, T. P. J., & Dobson, C. M. (2011). Protein solubility and protein homeostasis: A generic view of protein misfolding disorders. *Cold Spring Harbor Perspectives in Biology, 3*(12), a010454. http://dx.doi.org/10.1101/cshperspect.a010454.

Walker, E. J., Bettinger, J. Q., Welle, K. A., Hryhorenko, J. R., & Ghaemmaghami, S. (2019). Global analysis of methionine oxidation provides a census of folding stabilities for the human proteome. *Proceedings of the National Academy of Sciences, 116*(13), 6081–6090. http://dx.doi.org/10.1073/pnas.1819851116.

Walters, B. T., Mayne, L., Hinshaw, J. R., Sosnick, T. R., & Englander, S. W. (2013). Folding of a large protein at high structural resolution. *Proceedings of the National Academy of Sciences, 110*(47), 18898–18903. http://dx.doi.org/10.1073/pnas.1319482110.

Walther, D. M., Kasturi, P., Zheng, M., Pinkert, S., Vecchi, G., Ciryam, P., et al. (2015). Widespread proteome remodeling and aggregation in aging *C. elegans*. *Cell, 161*(4), 919–932.

Wang, Y., Meriin, A. B., Zaarur, N., Romanova, N. V., Chernoff, Y. O., Costello, C. E., et al. (2009). Abnormal proteins can form aggresome in yeast: Aggresome-targeting signals and components of the machinery. *The FASEB Journal, 23*(2), 451–463. http://dx.doi.org/10.1096/fj.08-117614.

Ward, C. C., Kleinman, J. I., & Nomura, D. K. (2017). NHS-esters as versatile reactivity-based probes for mapping proteome-wide ligandable hotspots. *ACS Chemical Biology, 12*(6), 1478–1483. http://dx.doi.org/10.1021/acschembio.7b00125.

West, G. M., Tang, L., & Fitzgerald, M. C. (2008). Thermodynamic analysis of protein stability and ligand binding using a chemical modification- and mass spectrometry-based strategy. *Analytical Chemistry, 80*(11), 4175–4185. http://dx.doi.org/10.1021/ac702610a.

West, G. M., Tucker, C. L., Xu, T., Park, S. K., Han, X., Yates, J. R., et al. (2010). Quantitative proteomics approach for identifying protein-drug interactions in complex mixtures using protein stability measurements. *Proceedings of the National Academy of Sciences, 107*(20), 9078–9082. http://dx.doi.org/10.1073/pnas.1000148107.

White, K. M., De Jesus, P., Chen, Z., Abreu, P., Barile, E., Mak, P. A., et al. (2015). A potent anti-influenza compound blocks fusion through stabilization of the prefusion conformation of the hemagglutinin protein. *ACS Infectious Diseases, 1*(2), 98–109. http://dx.doi.org/10.1021/id500022h.

Wühr, M., Haas, W., McAlister, G. C., Peshkin, L., Rad, R., Kirschner, M. W., et al. (2012). Accurate multiplexed proteomics at the MS2 level using the complement reporter ion cluster. *Analytical Chemistry, 84*(21), 9214–9221. http://dx.doi.org/10.1021/ac301962s.

Xie, B., & Sharp, J. S. (2015). Hydroxyl radical dosimetry for high flux hydroxyl radical protein footprinting applications using a simple optical detection method. *Analytical Chemistry, 87*(21), 10719–10723. http://dx.doi.org/10.1021/acs.analchem.5b02865.

Xu, G., & Chance, M. R. (2007). Hydroxyl radical-mediated modification of proteins as probes for structural proteomics. *Chemical Reviews, 107*(8), 3514–3543. http://dx.doi.org/10.1021/cr0682047.

Xu, Y., Falk, I. N., Hallen, M. A., & Fitzgerald, M. C. (2011). Mass spectrometry- and lysine amidination-based protocol for thermodynamic analysis of protein folding and ligand binding interactions. *Analytical Chemistry, 83*(9), 3555–3562. http://dx.doi.org/10.1021/ac200211t.

Xu, Y., Wallace, M. A. G., & Fitzgerald, M. C. (2016). Thermodynamic analysis of the geldanamycin-Hsp90 interaction in a whole cell lysate using a mass spectrometry-based proteomics approach. *Journal of the American Society for Mass Spectrometry, 27*(10), 1670–1676. http://dx.doi.org/10.1007/s13361- 016-1457-2.

Yang, Z., Kuboyama, T., & Tohda, C. (2017). A systematic strategy for discovering a therapeutic drug for Alzheimer's disease and its target molecule. *Frontiers in Pharmacology, 8*, 340. http://dx.doi.org/10.3389/fphar.2017.00340.

Yoshiike, Y., Minai, R., Matsuo, Y., Chen, Y.-R., Kimura, T., & Takashima, A. (2008). Amyloid oligomer conformation in a group of natively folded proteins. *PLoS One, 3*(9), e3235. http://dx.doi.org/10.1371/journal.pone.0003235.

Yu, C., Huszagh, A., Viner, R., Novitsky, E. J., Rychnovsky, S. D., & Huang, L. (2016). Developing a multiplexed quantitative cross-linking mass spectrometry platform for comparative structural analysis of protein complexes. *Analytical Chemistry, 88*(20), 10301–10308. http://dx.doi.org/10.1021/acs.analchem.6b03148.

Zaarur, N., Xu, X., Lestienne, P., Meriin, A. B., McComb, M., Costello, C. E., et al. (2015). RuvbL1 and RuvbL2 enhance aggresome formation and disaggregate amyloid fibrils. *The EMBO Journal, 34*(18), 2363–2382. http://dx.doi.org/10.15252/embj.201591245.

Zauber, H., Kirchner, M., & Selbach, M. (2018). Picky: A simple online PRM and SRM method designer for targeted proteomics. *Nature Methods, 15*(3), 156–157. http://dx.doi.org/10.1038/nmeth.4607.

Zhang, Y., Fonslow, B. R., Shan, B., Baek, M.-C., & Yates, J. R. (2013). Protein analysis by shotgun/bottom-up proteomics. *Chemical Reviews, 113*(4), 2343–2394. http://dx.doi.org/10.1021/cr3003533.

Zhang, H., Gau, B. C., Jones, L. M., Vidavsky, I., & Gross, M. L. (2011). Fast photochemical oxidation of proteins for comparing structures of Protein–Ligand complexes: The Calmodulin–Peptide model system. *Analytical Chemistry, 83*(1), 311–318. http://dx.doi.org/10.1021/ac102426d.

Zhou, Y., & Vachet, R. W. (2013). Covalent labeling with isotopically encoded reagents for faster structural analysis of proteins by mass spectrometry. *Analytical Chemistry, 85*(20), 9664–9670. http://dx.doi.org/10.1021/ac401978w.

Zybailov, B., Coleman, M. K., Florens, L., & Washburn, M. P. (2005). Correlation of relative abundance ratios derived from peptide ion chromatograms and spectrum counting for quantitative proteomic analysis using stable isotope labeling. *Analytical Chemistry,* 77(19), 6218–6224. http://dx.doi.org/10.1021/ac050846.

CHAPTER FOUR

The intrinsic and extrinsic factors that contribute to proteostasis decline and pathological protein misfolding

Elise A. Kikis*
Biology Department, The University of the South, Sewanee, TN, United States
*Corresponding author: E-mail: eakikis@sewanee.edu

Contents

Abstract

Proteostasis refers to the ability of cells to maintain the health of the proteome. Highly conserved quality control mechanisms exist to maintain proteostasis. These include the heat shock response, the unfolded protein response, and protein clearance/degradation pathways. Together, these mechanisms and others comprise the proteostasis network. This network is under constant assault and is strikingly sensitive to changes in the protein folding environment, resulting in proteostasis collapse under certain conditions. Here, the intrinsic and extrinsic stresses experienced by the proteostasis network are explored. The intrinsic stresses include genetic background as well as transcriptional and translational fidelity. These cause changes in the abundance or amino acid sequence of cellular proteins. Extrinsic stresses refer to environmental perturbation of the proteome, such as those caused by temperature stress, oxidative stress, air pollution and cigarette smoke. As the stress to the proteome exceeds the capacity of the proteostasis network, progressive neurodegenerative diseases of aging, such as Alzheimer's disease and Huntington's disease are more likely to ensue.

Advances in Protein Chemistry and Structural Biology, Volume 118
ISSN 1876-1623
https://doi.org/10.1016/bs.apcsb.2019.07.001
© 2020 Elsevier Inc.
All rights reserved.

1. The proteostasis network

The term "proteostasis" was formally coined by William Balch, Richard Morimoto, Andrew Dillin, and Jeffery Kelly in a review article published in 2008 (Balch, Morimoto, Dillin, & Kelly, 2008). There, the authors defined proteostasis as "controlling the concentration, conformation, binding interactions (quaternary structure), and location of individual proteins making up the proteome by readapting the innate biology of the cell, often through transcriptional and translational changes." This definition is necessarily broad as it encompasses the entirety of protein metabolism—synthesis, maturation, localization, and turnover.

A healthy proteome is one in which proteins fold and function normally, without adverse consequences to the health of the cell. Maintaining the proteome in this healthy state is the purview of the large and diverse set of cellular factors that together comprise what has become known as the proteostasis network. These include proteins required for gene expression/protein biogenesis as well as molecular chaperones and other quality control factors that are required to maintain cellular proteins in their natively folded state, and, finally, the protein degradation machinery.

This chapter considers the various conditions that together stress the proteostasis network. Ultimately, the capacity of the proteostasis network to maintain a healthy proteome is surprisingly limited, meaning that the proteostasis network is easily overwhelmed by the accumulation of damaged and misfolded proteins (Gidalevitz, Kikis, & Morimoto, 2010). The less able a cell is to fold or clear damaged proteins, the more damaged proteins accumulate. This creates a vicious cycle, leading to catastrophic protein misfolding and disease. The remainder of this chapter will consider the cellular and environmental events that either alone or in combination have been shown to negatively affect the delicate proteostasis balance.

2. Genetic variation underlies differences in cellular proteostasis

The basal level of stress experienced by the proteostasis network is determined by the complement of proteins expressed in a specific cell type. As allele signature and genetic background varies between individuals, two unrelated but healthy individuals would be expected to express proteins that differ slightly in their amino acid sequences or to accumulate different steady-state protein levels. Depending on the thermodynamics of folding of the expressed proteins, the molecular chaperone components of the proteostasis machinery would be expected to be engaged to a greater or lesser extent. To test the hypothesis that genetic variation underlies differences in proteostasis, recent studies have utilized classical genetic model systems.

One of the many advantages to studying cellular processes in classical model systems, such as *Drosophila* or *Caenorhabditis elegans*, is that years of inbreeding in the laboratory has led to exquisite genetic uniformity. This has made it easier to study the effects of specific gene disruptions as a means to determine gene function. However, in uniform genetic strains, differences between individuals in a population — a fundamental feature of the natural world — is often overlooked. In recognition of this apparent limitation, recent studies have aimed to elucidate the extent to which genetic variation contributes in statistically significant ways to individual differences in gene expression and to cellular responses to stress. Many such studies still take advantage of the power of the classic genetic models, but instead of relying on single inbred isolates, different wild isolates and recombinant inbred lines have been compared to each other, revealing striking differences in proteostasis between genetically distinct individuals.

In one such study, the expression of 1353 genes was knocked down by RNAi in Bristol (N2) and Hawaiian (CB4856) isolates of *C. elegans* (Vu et al., 2015). While many gene inactivations caused stronger phenotypes in N2 animals, this was often attributed to stochastic differences in RNAi efficiency. However, ~250 gene inactivations showed stronger or weaker phenotypes in a manner dependent on the genetic background. To validate these findings and confirm that they were not simply an artifact of RNAi, they were recapitulated with bona fide genetic lesions. Ultimately, the observed phenotypic differences were shown to map closely to gene expression (Vu et al., 2015). Thus, two individuals whose genetic makeup differs in such a way as to affect gene expression levels will differ in their manifestation of mutant phenotypes. Extrapolating these data to the human population suggests that natural genetic variation might cause two carriers of the same disease-associated gene to have very different outcomes with respect to disease onset and severity of disease symptoms.

Human maladies that are associated with disruptions in proteostasis include neurodegenerative diseases, in which certain proteins misfold and overwhelm the ability of the proteostasis network to fold or clear damaged proteins. As damaged and misfolded proteins accumulate to toxic levels in the brain, neuronal cell death occurs, contributing to the onset of dementia. Overall changes in gene expression that translate into differences in cellular protein levels have been hypothesized to profoundly affect the quality control programs meant to monitor and manage protein (mis)folding. To determine whether genetic variation contributes to different mRNA transcription levels and different protein levels, microarrays and SILAC were employed to measure mRNA and protein, respectively, in Bristol and Hawaiian isolates of *C. elegans*. Although mRNA levels varied considerably between the two different wild type laboratory strains, only 5% of proteins accumulated to different levels (Kamkina et al., 2016). This illustrates the fact that mechanisms are in place to control protein levels even as mRNA levels fluctuate, suggesting that maintaining protein levels at precise steady-state amounts is of utmost importance. Of note, however, are the few proteins whose levels did differ between the two strains examined. Those proteins are enriched for an involvement in aging and stress responses (Kamkina et al., 2016). This is especially interesting because proteostasis declines precipitously in aging animals (Ben-Zvi, Miller, & Morimoto, 2009) and because stress responses represent the quality controls meant to maintain the delicate proteostasis balance.

While the aging brain is known to be under proteotoxic stress, often culminating in neurodegenerative disease, cancer cells are also under extreme proteotoxic stress. This is partly because of genomic instability affecting the proteome and also because of changes in gene expression and protein levels in cells that have undergone malignant transformation. As a result, the induction of cell stress responses, such as the heat shock response, is characteristic of many cancers (Dai & Sampson, 2016). To determine whether genetic background affects the steady-state levels of cancer-associated proteins, the abundance of proteins involved cell signaling pathways and apoptosis was compared between the Bristol (N2) and Hawaiian (CB4856) wild type strains of *C. elegans* (Singh et al., 2016). Overall, genetic variation had a negligible effect on the abundance of the aforementioned proteins, especially when comparing just the two parental strains. However, considerably more variation in protein levels was observed between recombinant inbred lines (RILs) generated by first crossing the N2 and CB4856 strains. These protein level differences were attributed to genetic variation between RILs (Singh et al., 2016).

Taken together, these data indicate that genetic variation has measurable effects on protein expression. Because studies done with *C. elegans* have revealed that genetic variation may have a particularly large effect on the cellular stress response pathways, this may be expected to contribute to survival differences for animals experiences proteotoxic stress. Such an outsized effect of genetic variation on stress resistance has also been demonstrated in *drosophila.* 114 lines from the *drosophila* Genetic Reference Panel were examined for genetic variation in resistance to ER stress (Chow, Wolfner, & Clark, 2013). The authors found a considerable amount of variation in the ability of each of these inbred wild isolates to survive tunicamycin-induced ER stress. They also found ER stress-responsive gene expression to be highly variable (Chow et al., 2013). This is consistent with findings with the Bristol (N2) and Hawaiian (CB4856) wild isolates of *C. elegans* that were shown to differ in their ability to survive heat stress (Rodriguez, Snoek, Riksen, Bevers, & Kammenga, 2012). Using recombinant inbred lines (RILs) created by crossing N2 and CB4856 followed by rounds of inbreeding, a QTL was identified on *C. elegans* chromosome 2 and shown to contribute to survival after heat shock (Rodriguez et al., 2012). This suggests that the differences in survival after heat shock, observed between the two wild isolates, are genetically determined (Rodriguez et al., 2012).

3. Errors in transcription and translation are a source of proteotoxic stress

The findings described above indicate that encoded differences between individuals in a population can have measurable effects on the ability of cells and organisms to maintain proteostasis. This raises the possibility that stochastic errors in normal cellular processes could give rise to similar variation in the proteome. This is predicated on the hypothesis that functional, biologically relevant, differences in the proteome need not be encoded within the genome itself. Instead, errors in transcription and translation, resulting in amino acid substitutions in proteins, may likewise impact protein structure and function.

To test the hypothesis that the transcriptional error rate is high enough to yield functional consequences, the first step was to measure that error rate. Doing so was complicated by the fact that the traditional RT-PCR-based methods of determining mRNA sequence are themselves inherently error-prone due to the low fidelity of reverse transcriptase. Therefore, a method called "circle sequencing" was utilized. This method involves circularizing RNA before synthesizing and sequencing any cDNA. This method therefore generates large cDNAs, such that each cDNA molecule contains many, presumably identical, copies of the same mRNA. Only point mutations that appear in every cDNA copy of the same mRNA can be considered legitimate transcriptional errors. In contrast, errors produced *ex post facto* by reverse transcriptase, would appear in only one of the many copies of the same mRNA. Utilizing this circle sequencing method, the error rate for transcription was determined to be 4 errors/1, 000, 000 bp, which is 100 times greater than the intrinsic error rate for DNA replication (Gout et al., 2017). This error rate is exacerbated under conditions of genotoxic stress during which RNA polymerase may transcribe over non-bulky lesions induced by DNA damaging agents, increasing the transcriptional error rate (Bregeon, Doddridge, You, Weiss, & Doetsch, 2003).

Such a high transcriptional error rate suggests the distinct possibility that some of the translated proteins would have incorrect amino acids and thereby be impaired in their folding pathway. We would expect that the magnitude of the load of misfolded proteins would be directly proportional to the stress exerted on the proteostasis balance. To determine whether transcriptional errors contribute to the misfolded protein load, the protein ADH1 was used as a model for proof of principle. It was determined that the vast majority of transcriptional errors in ADH1 mRNA led to deficiencies in protein folding (Gout et al., 2017). At face value, it may be surprising that the transcriptional errors led to folding defects rather than to more direct effects on catalytic activity or binding affinity, for example. However, as the authors of the study pointed out, the vast majority of amino acids within a protein are there to stabilize the native structure (Gout et al., 2017).

Ultimately, the findings with ADH1 demonstrate just how extraordinarily sensitive protein folding is to even slight changes in amino acid sequence. They are also consistent with previous thermodynamic models of protein folding that predicted that increasing numbers of amino acid substitutions should exponentially decrease the likelihood that a given protein will successfully fold (Bloom et al., 2005). This thermodynamic model does not make any presumptions regarding the origin of such substitutions. Theoretically, they could be genetically encoded, result as a consequence of transcriptional errors, or even result as a consequence of translational infidelity at the ribosome.

In summary, the combination of encoded genetic variation and transcriptional errors has marked effects on protein folding, induces cell stress responses, and alters the proteostasis balance. To test the hypothesis that translational infidelity likewise contributes to the misfolded protein load, Drummond and colleagues took both computational and experimental approaches. They demonstrated computationally that there is a strong selective pressure for translational fidelity, especially for highly expressed genes

(Drummond & Wilke, 2008). Ribosome-induced errors in protein sequence that result in protein misfolding are toxic, especially to neurons, which are hypersensitive to proteostasis imbalances. As such, natural selection has resulted in codon biases that seem to be optimized to prevent these ribosome-induced errors. Nonetheless, despite natural selection over an evolutionary timescale as well as cellular quality control mechanisms, 18% of translated proteins were shown to carry at least one missense substitution that was caused by ribosome infidelity (Drummond & Wilke, 2008). Such errors induced at the level of translation, even in just a single protein, were shown to carry a fitness cost and have molecular consequences (Geiler-Samerotte et al., 2011). This was determined experimentally with temperature sensitive, metastable, variants of the YFP protein expressed in yeast cells. Strikingly, misfolded YPF conferred a growth-rate defect and induced the expression of molecular chaperones, indicative of proteotoxic stress being sensed within the cellular milieu (Geiler-Samerotte et al., 2011).

4. Proteostasis and disease

This chapter has mostly considered the ways in which natural genetic variation and normal cellular mechanisms, such as transcription and translation, can alter the protein folding environment, putting more or less stress on the proteostasis network. While these sorts of low-level differences in misfolded protein load may be recognized by cell stress responses, leading to an upregulation of molecular chaperones (Geiler-Samerotte et al., 2011), cells can typically manage them successfully and thereby still maintain the overall integrity of the proteome. However, under extreme circumstances, proteostasis collapses and diseases of protein folding ensue. Such diseases include adult onset neurodegenerative diseases such as Alzheimer's disease (AD), Huntington's disease (HD), and amyotrophic lateral sclerosis (ALS). The fact that disease symptoms are associated with aging is consistent with a declining ability of cells to buffer against protein misfolding.

HD is an adult-onset, fatal, autosomal dominant disorder that affects the region of the brain known as the basal ganglia, leading to cognitive and motor impairment. The genetic lesion underlying HD was identified in 1993 as an expansion of a polyglutamine-encoding CAG repeat region of a gene on chromosome 4 (MacDonald et al., 1993). Eight additional neurodegenerative disorders have been shown to be caused by polyglutamine expansions in different proteins. In common with all of these polyglutamine expansions is the fact that they destabilize the folded state of the affected proteins. Additionally, the misfolded proteins negatively impact the entirety of the protein folding environment, leading to proteostasis collapse (Gidalevitz, Ben-Zvi, Ho, Brignull, & Morimoto, 2006). That a single mutant protein has the power to overwhelm the proteostasis network so completely as to cause the misfolding of other otherwise normal proteins is striking and evidence of the fragility of the proteostasis balance. However, even as the proteostasis balance is disrupted, some proteins seem to be mostly spared from misfolding. In fact, it is the metastable proteins, those whose amino acid sequence is such that the thermodynamics of folding are less than optimal, that are most likely to misfold. Because missense mutations that shift the thermodynamics of folding are common-place and caused by genetic variation as well as transcriptional and translational errors, an individual carrying more metastable proteins may be expected to have earlier or stronger symptoms of protein folding diseases.

This hypothesis was tested in a *C. elegans* model of Amyotrophic lateral sclerosis (ALS). Like HD, ALS is a neurodegenerative disease of protein misfolding. However, unlike HD, which is strictly a genetic disorder, ALS has both sporadic and familial counterparts. Familial ALS has been shown to be caused by the misfolding of a number of different proteins. 20% of ALS patients carry mutations in SOD1, destabilizing the folded state of the protein. Gidalevitz, Krupinski, Garcia, and Morimoto (2009) developed a *C. elegans* model of human SOD1 aggregation and toxicity, using SOD1 variants known to be determinants of familial forms of ALS. To determine whether genetic background affects mutant SOD1 aggregation and toxicity, they expressed different SOD1 mutant variants in the background of various metastable temperature sensitive alleles. The study revealed that the presence of folding-compromized temperature sensitive proteins exacerbated mutant SOD1 toxicity, and, further, that the phenotype of SOD1 expression was influenced by the identity of the specific temperature sensitive alleles (Gidalevitz et al., 2009).

These data indicate that slight changes not only in the overall load of misfolded protein, but also in the identities of those proteins themselves, contribute to the likelihood that mutant SOD1 will aggregate, in turn leading to the manifestation of disease symptoms. To test whether genetic backgrounds differ sufficiently with respect to the amount and identity of misfolded or metastable proteins as to trigger the aggregation of a disease-associated protein, the aggregation-prone and disease-associated polyglutamine (polyQ)-containing protein was introgressed into four different genetic backgrounds (N2, DR1350, JU258, and CB4856). The DR1350 genetic background significantly increased early onset polyQ aggregation and toxicity. In contrast, the CB4856 genetic background conferred a marked protection against polyQ aggregation and toxicity in *C. elegans*. Upon generating RILs with N2 and DR1350, the authors reported that aggregation and toxicity can be uncoupled through genetic variation (Gidalevitz, Wang, Deravaj, Alexander-Floyd, & Morimoto, 2013).

Like ALS, Alzheimer's disease (AD) is a neurodegenerative disease associated with protein misfolding, and has both sporadic and familial forms, with familial AD having earlier onset and being more aggressive than sporadic Late Onset Alzheimer's Disease (LOAD). One thing in common between early onset AD (EOAD) and LOAD is the presence of misfolded amyloid beta (Aβ) peptide accumulating as extracellular plaques in the brain, and the presence of tau neurofibrillary tangles. While LOAD has no single genetic cause, several risk factors have been identified. In one study aimed at identifying such risk factors, SNPs in 93 imprinted genes were examined and tested for whether any of them correlated with an increased risk for LOAD. The study cohort consisted of 1291 individuals with LOAD and 958 healthy control individuals. The study revealed that SNPs in four maternally imprinted genes affected, positively or negatively, LOAD risk (Chaudhry et al., 2015). Furthermore, cerebrospinal fluid of 690 individual people was harvested and tested for the presence of amyloid protein and/or tau. Individuals were grouped as being either amyloid positive, tau positive, or positive for both amyloid and tau. Ultimately, several single nucleotide polymorphisms (SNPs) were identified that influenced the likelihood that amyloid biomarkers would actually correlate with symptoms of neurodegeneration (Hohman, Koran, Thornton-Wells, & Alzheimer's Neuroimaging, 2014). These SNPs represent genetic variation between individuals. As such, these data can be interpreted to mean that genetic variation in the human population is an important driving force contributing to the likelihood that an individual will be diagnosed with LOAD.

Because protein misfolding is a hallmark of so many neurodegenerative diseases, genetic risk factors can be point mutations in proteins that play a functional role in the disease mechanism, or can be mutations that destabilize otherwise unrelated proteins. Such destabilizing mutations act by disrupting the proteostasis balance, triggering a cascade of events culminating in catastrophic and wide-spread misfolding of cellular proteins. Such catastrophic misfolding may be a normal part of aging that is exacerbated or accelerated in patients suffering from diseases of protein misfolding (Ben-Zvi et al., 2009). As such, the genetic background as well as errors in transcription and translation all work together to create a unique and individualized protein folding environment that may be more or less able to sustain proper protein folding.

5. The inflammatory response and proteostasis

The mechanisms by which aggregated and misfolded proteins exert their toxic effects on the brain are manifold, and an area of active investigation. One aspect of the mechanism of toxicity that we are starting to understand is the activation of the neuroinflammatory response. The neuroinflammatory response is initiated by ligand binding to Toll-like receptors (TLRs) on the cell surface of microglial cells within the brain. This leads to microglial activation and the resultant secretion of pro-inflammatory cytokines.

TLRs are type I transmembrane proteins consisting of an extracellular domain with leucine-rich repeats and a highly conserved cytoplasmic domain known as the TIR domain (Kawai & Akira, 2007). Such receptors were identified in *Drosophila* where they were shown to confer protection against fungal infection (Lemaitre, Nicolas, Michaut, Reichhart, & Hoffmann, 1996) and to have significant homology with human IL-1R (Gay & Keith, 1991). TLRs bind to pathogen-specific structures, such as those associated with the bacterial cell wall (lipopolysaccharide and peptidoglycan) as well as those associated with bacterial or viral genomes. Ligands expressed by foreign invaders are referred to as pathogen-associated microbial patterns (PAMPs) (Kawai & Akira, 2007). TLRs can also be activated by binding ligands that are derived from necrotic/dying cells. Such ligands are referred to as damage-associated molecular patterns (DAMPs) (Schaefer, 2014).

The initial steps of TLR activation include the recruitment of adapter proteins to their cytoplasmic domains. Upon adaptor protein binding, a series of kinases serve to activate the NF-κB family of transcription factors, by releasing them from the inhibitory IκB protein family (Sabroe, Parker, Dower, & Whyte, 2008). The many NF-κB family members are capable of homo- or hetero-dimerization, resulting in a wide range of transcriptional targets, such as cytokines that are secreted and induce a regional inflammatory response. Consistent with its role as a central regulator of the inflammatory response, NF-κB dysregulation is implicated in many autoimmune disease (Liu, Zhang, Joo, & Sun, 2017).

An important and timely question in the field of protein folding diseases relates to how changes in the proteostasis balance interface with the neuroinflammatory pathway to promote its activation. The last decade has seen a great deal of progress in elucidating the connection between proteostasis and inflammation. For example, it has been shown that when infection, insult, or injury disrupts the delicate balance of ER functions that allows for the maintenance of a healthy proteome, the resultant proteostasis imbalance is itself a trigger of the inflammatory response. The Unfolded Protein Response (UPR) is a well-characterized transcriptional response that allows the ER to enhance its ability to buffer against proteotoxic stress by increasing its protein folding capacity (Ron & Walter, 2007). One of the three established branches of the UPR involves the activation of the ER membrane-associated IRE1α by misfolded protein, resulting the non-canonical splicing of XBP1 mRNA. XBP1 is itself a transcription factor that controls the expression of ER molecular chaperones (Ron & Walter, 2007). XBP1 splicing can also be activated in macrophages downstream of TLR signaling in response to microbial stress (Martinon, Chen, Lee, & Glimcher, 2010). Importantly, activation of the UPR by ER stress leads to the degradation of IκB proteins, the resultant activation of NF-κB, and the secretion of proinflammatory cytokines. Likewise, the expression of proinflammatory cytokines can be induced in a JNK-dependent manner via the activation of the transcription factor AP1. This too is triggered by IRE1α signalizing in response to ER stress (Kaser et al., 2008).

While an acute traumatic event can trigger a sudden imbalance in ER proteostasis, neurodegenerative diseases are chronic, progressive, and associated with the misfolding and aggregation of specific disease-associated proteins. Several recent studies have implicated neurodegenerative disease-associated proteins in the activation of the neuroinflammatory response. Namely, aggregated protein has been shown to activate microglia via binding to TLRs on the cell surface. The resultant cytokine secretion leads to the permeabilization of the blood brain barrier, allowing for infiltration of the brain by T-cells that also have TLRs and are thereby also stimulated by aggregated protein. Furthermore, microglial activation also leads to the production of damaging reactive oxygen species (Chen, Zhang, & Huang, 2016). This leads to a downward spiral of protein damage and therefore increased stress on the proteostasis machinery. Importantly, Aβ is one such aggregated protein that has been shown to bind to TLR2, causing microglial activation and the secretion of pro-inflammatory cytokines (Liu et al., 2012). Additionally, the neuro-protective clearance of damaged/aggregated proteins by phagocytosis is suppressed by TLR2 (Liu et al., 2012), thereby further disrupting proteostasis and exacerbating disease progression.

6. Environmental effects on proteostasis and disease

With ER stress having been shown to trigger the inflammatory response, it is important to consider what triggers proteotoxic stress in the ER in the first place. Interestingly, several recent studies have pointed to cigarette smoke as a potent inducer of the UPR both *in vivo* and *in vitro*. In one study, lung tissue was obtained from smokers, ex-smokers, and non-smokers and was subjected to a large-scale proteomics analysis. Only twenty-six differentially expressed proteins were identified. Five of these are involved in the UPR and several others are known to affect protein synthesis or degradation (Kelsen et al., 2008). As such, ER proteostasis is significantly affected by cigarette smoke. Consistent with these findings, Jorgensen et al. (2008) performed a gene expression study to identify all the mRNA transcripts that are differentially expressed in response to cigarette smoke (Jorgensen et al., 2008). The most statistically significant differences were for genes involved in the UPR, although XBP1 splicing was not observed. It was ultimately concluded that vapors, not particulate matter, from cigarette smoke inhibited XBP1 through an unknown mechanism. However, it seems likely that the lack of particulate matter contribution was due to insufficient treatment time, and may therefore not be entirely representative of the *in vivo* situation (Jorgensen et al., 2008).

To study the effects of second hand smoke *in vivo*, Yuen-Shan Ho and colleagues exposed rats to cigarette smoke and measured makers of neuronal health (Ho et al., 2012). They found increased oxidative stress, reduced presynaptic proteins, impaired cellular transport, increased tau phosphorylation, and aberrant amyloid precursor protein (APP) processing leading to increased levels of the Aβ peptide in certain brain regions (Ho et al., 2012). Taken together with the data indicating that cigarette smoke induces the ER stress response, these findings on second hand smoke exposure suggest an important link between impaired proteostasis, neuroinflammation, and onset of neurodegenerative disease.

Consistent with cigarette smoke acting as a disrupter of the proteostasis balance and thus being a contributing factor in the onset of neurodegenerative disease, particulate air pollution has likewise been shown to exacerbate and accelerate the onset of neurodegenerative diseases including AD. To a large extent, this effect of air pollution is attributed to an induction of the inflammatory response. Samples of particulate matter harvested from the air of Riverside, CA, was subdivided based on size (coarse, fine, and ultrafine) and administered to rats in an exposure study. The authors found that course particulate matter was the most potent inducer of an inflammatory response, which was attributed to the resultant accumulation of Ni in the brain (Ljubimova et al., 2018). Nonetheless, nanoparticulate matter, the smallest particles found in the air, have been shown in other studies to be highly potent inducers of the inflammatory response (Woodward et al., 2017) and the oxidative stress response (Pomatto et al., 2018).

To identify genes and pathways that are responsive to nanoparticulate matter, mixed glial cell cultures were exposed to nanoparticulate matter and subjected to a microarray analysis. 2000 differentially expressed genes were identified. Interestingly, 530 of those genes are also sensitive to LPS exposure, meaning that they likely respond to air pollution via the induction of the inflammatory response. For example, a twofold upregulation of TLR4 mRNA was observed in response to nanoparticulate matter (Woodward et al., 2017). This gene induction was not further exacerbated by additional treatment with LPS, indicating that the particulate matter was apparently sufficient to induce a canonical inflammatory response (Woodward et al., 2017).

Evidence for particulate air pollution having a direct impact on the proteostasis balance came just last year with the finding that the oxidative stress responsive transcription factor, Nrf2, was upregulated in the heart and lung of young mice upon exposure to nanoparticulate matter. It was further shown that lon protease and proteasome activity was upregulated in these same tissues after 6 months of exposure to nanoparticulate matter (Pomatto et al., 2018). Ultimately, the authors concluded that a sublethal exposure to nanoparticulate matter likely triggers a stress response, which is attenuated in older, middle-aged, individuals (Pomatto et al., 2018). This is consistent with an earlier study that demonstrated that the olfactory epithelium is the first site of neuronal contact for inhaled air pollution, and that this tissue responds rapidly, with oxidative stress and inflammatory makers, well before the brain responds to exposure to air pollution (Cheng et al., 2016).

Suggesting yet a further connection between exposure to air pollution and neurodegenerative disease-associated impairment in protein processing and folding, a recent study demonstrated that particulate air pollution exacerbates AD in a manner dependent on the presence of the APOE ε4 allele (Cacciottolo et al., 2017). The authors first described an epidemiological study, revealing that women with the APOE ε4 allele were especially susceptible to the negative cognitive effects triggered by exposure to particulate air pollution. This also held true in a mouse model of AD. Namely, mice harboring the human APOE ε4 allele were exposed to particulate air pollution and displayed increased amyloid deposits as compared to non-exposed control animals (Cacciottolo et al., 2017).

7. Discussion and outlook

Significant progress has been made toward understanding the intrinsic and extrinsic factors that affect cellular proteostasis. First and foremost, the ability of cells and organisms to maintain proteostasis is limited. As such, slight perturbations to the protein folding environment can trigger proteostasis collapse, which has been shown to be a major contributor to diseases of protein misfolding. Intrinsic factors, such as genetic background differences between individuals, transcriptional errors, and errors in protein synthesis can together modulate proteostasis, with an increase in missense mutations typically stressing the proteostasis machinery.

The best understood environmental contributor to proteostasis decline is heat shock. However, this is generally shown in isolated cells in tissue culture rather than in animals that are able to regulate body temperature. Nonetheless, cells experiencing an acute heat shock event respond by upregulating molecular chaperone genes and thereby transiently bolstering the buffering capacity of the proteostasis network. More recently, studies have begun to look at other environmental conditions such as particulate matter in polluted air as possible chronic factors that contribute to a slow decline in proteostasis during aging. Exposure to air pollution has been shown to exacerbate molecular markers and symptoms of Alzheimer's disease, especially in individuals who are genetically predisposed. Whether this is due to direct effects on proteostasis is an open question. Nonetheless, findings that proteostasis network components are upregulated in response to exposure to nanoparticulate air pollution are tantalizing and open the door to further studies aimed at addressing the effect of air pollution on the protein folding environment and the proteostasis balance.

References

Balch, W. E., Morimoto, R. I., Dillin, A., & Kelly, J. W. (2008). Adapting proteostasis for disease intervention. *Science, 319*(5865), 916–919.

Ben-Zvi, A., Miller, E. A., & Morimoto, R. I. (2009). Collapse of proteostasis represents an early molecular event in *Caenorhabditis elegans* aging. *Proceedings of the National Academy of Sciences of the U S A, 106*(35), 14914–14919. http://dx.doi.org/10.1073/pnas.0902882106.

Bloom, J. D., Silberg, J. J., Wilke, C. O., Drummond, D. A., Adami, C., & Arnold, F. H. (2005). Thermodynamic prediction of protein neutrality. *Proceedings of the National Academy of Sciences of the U S A, 102*(3), 606–611. http://dx.doi.org/10.1073/pnas.0406744102.

Bregeon, D., Doddridge, Z. A., You, H. J., Weiss, B., & Doetsch, P. W. (2003). Transcriptional mutagenesis induced by uracil and 8-oxoguanine in *Escherichia coli. Molecular Cell, 12*(4), 959–970.

Cacciottolo, M., Wang, X., Driscoll, I., Woodward, N., Saffari, A., Reyes, J., et al. (2017). Particulate air pollutants, APOE alleles and their contributions to cognitive impairment in older women and to amyloidogenesis in experimental models. *Translational Psychiatry,* 7(1), e1022. http://dx.doi.org/10.1038/tp.2016.280.

Chaudhry, M., Wang, X., Bamne, M. N., Hasnain, S., Demirci, F. Y., Lopez, O. L., et al. (2015). Genetic variation in imprinted genes is associated with risk of late-onset Alzheimer's disease. *Journal of Alzheimer's Disease, 44*(3), 989–994. http://dx.doi.org/10.3233/JAD-142106.

Cheng, H., Saffari, A., Sioutas, C., Forman, H. J., Morgan, T. E., & Finch, C. E. (2016). Nanoscale particulate matter from urban traffic rapidly induces oxidative stress and inflammation in olfactory epithelium with concomitant effects on brain. *Environmental Health Perspectives, 124*(10), 1537–1546. http://dx.doi.org/10.1289/EHP134.

Chen, W. W., Zhang, X., & Huang, W. J. (2016). Role of neuroinflammation in neurodegenerative diseases (review). *Molecular Medicine Reports, 13*(4), 3391–3396. http://dx.doi.org/10.3892/mmr.2016.4948.

Chow, C. Y., Wolfner, M. F., & Clark, A. G. (2013). Using natural variation in Drosophila to discover previously unknown endoplasmic reticulum stress genes. *Proceedings of the National Academy of Sciences of the U S A, 110*(22), 9013–9018. http://dx.doi.org/10.1073/pnas.1307125110.

Dai, C., & Sampson, S. B. (2016). HSF1: Guardian of proteostasis in cancer. *Trends in Cell Biology, 26*(1), 17–28. http://dx.doi.org/10.1016/j.tcb.2015.10.011.

Drummond, D. A., & Wilke, C. O. (2008). Mistranslation-induced protein misfolding as a dominant constraint on coding-sequence evolution. *Cell, 134*(2), 341–352. http://dx.doi.org/10.1016/j.cell.2008.05.042.

Gay, N. J., & Keith, F. J. (1991). Drosophila Toll and IL-1 receptor. *Nature, 351*(6325), 355–356. http://dx.doi.org/10.1038/351355b0.

Geiler-Samerotte, K. A., Dion, M. F., Budnik, B. A., Wang, S. M., Hartl, D. L., & Drummond, D. A. (2011). Misfolded proteins impose a dosage-dependent fitness cost and trigger a cytosolic unfolded protein response in yeast. *Proceedings of the National Academy of Sciences of the U S A, 108*(2), 680–685. http://dx.doi.org/10.1073/pnas.1017570108.

Gidalevitz, T., Ben-Zvi, A., Ho, K. H., Brignull, H. R., & Morimoto, R. I. (2006). Progressive disruption of cellular protein folding in models of polyglutamine diseases. *Science, 311*(5766), 1471–1474.

Gidalevitz, T., Kikis, E. A., & Morimoto, R. I. (2010). A cellular perspective on conformational disease: The role of genetic background and proteostasis networks. *Current Opinion in Structural Biology, 20*(1), 23–32. http://dx.doi.org/10.1016/j.sbi.2009.11.001.

Gidalevitz, T., Krupinski, T., Garcia, S., & Morimoto, R. I. (2009). Destabilizing protein polymorphisms in the genetic background direct phenotypic expression of mutant SOD1 toxicity. *PLoS Genetics, 5*(3), e1000399. http://dx.doi.org/10.1371/journal.pgen.1000399.

Gidalevitz, T., Wang, N., Deravaj, T., Alexander-Floyd, J., & Morimoto, R. I. (2013). Natural genetic variation determines susceptibility to aggregation or toxicity in a *C. elegans* model for polyglutamine disease. *BMC Biology, 11*, 100. http://dx.doi.org/10.1186/1741-7007-11-100.

Gout, J. F., Li, W., Fritsch, C., Li, A., Haroon, S., Singh, L., et al. (2017). The landscape of transcription errors in eukaryotic cells. *Science Advances, 3*(10), e1701484. http://dx.doi.org/10.1126/sciadv.1701484.

Hohman, T. J., Koran, M. E., Thornton-Wells, T. A., & Alzheimer's Neuroimaging, I. (2014). Genetic variation modifies risk for neurodegeneration based on biomarker status. *Frontiers in Aging Neuroscience, 6*, 183. http://dx.doi.org/10.3389/fnagi.2014.00183.

Ho, Y. S., Yang, X., Yeung, S. C., Chiu, K., Lau, C. F., Tsang, A. W., et al. (2012). Cigarette smoking accelerated brain aging and induced pre-Alzheimer-like neuropathology in rats. *PLoS One,* 7(5), e36752. http://dx.doi.org/10.1371/journal.pone.0036752.

Jorgensen, E., Stinson, A., Shan, L., Yang, J., Gietl, D., & Albino, A. P. (2008). Cigarette smoke induces endoplasmic reticulum stress and the unfolded protein response in normal and malignant human lung cells. *BMC Cancer, 8*, 229. http://dx.doi.org/10.1186/1471-2407-8-229.

Kamkina, P., Snoek, L. B., Grossmann, J., Volkers, R. J., Sterken, M. G., Daube, M., et al. (2016). Natural genetic variation differentially affects the proteome and transcriptome in *Caenorhabditis elegans. Molecular & Cellular Proteomics, 15*(5), 1670–1680. http://dx.doi.org/10.1074/mcp.M115.052548.

Kaser, A., Lee, A. H., Franke, A., Glickman, J. N., Zeissig, S., Tilg, H., et al. (2008). XBP1 links ER stress to intestinal inflammation and confers genetic risk for human inflammatory bowel disease. *Cell, 134*(5), 743–756. http://dx.doi.org/10.1016/j.cell.2008.07.021.

Kawai, T., & Akira, S. (2007). TLR signaling. *Seminars in Immunology, 19*(1), 24–32. http://dx.doi.org/10.1016/j.smim.2006.12.004.

Kelsen, S. G., Duan, X., Ji, R., Perez, O., Liu, C., & Merali, S. (2008). Cigarette smoke induces an unfolded protein response in the human lung: A proteomic approach. *American Journal of Respiratory Cell and Molecular Biology, 38*(5), 541–550. http://dx.doi.org/10.1165/rcmb.2007-0221OC.

Lemaitre, B., Nicolas, E., Michaut, L., Reichhart, J. M., & Hoffmann, J. A. (1996). The dorsoventral regulatory gene cassette spatzle/Toll/cactus controls the potent antifungal response in Drosophila adults. *Cell, 86*(6), 973–983.

Liu, S., Liu, Y., Hao, W., Wolf, L., Kiliaan, A. J., Penke, B., et al. (2012). TLR2 is a primary receptor for Alzheimer's amyloid beta peptide to trigger neuroinflammatory activation. *The Journal of Immunology, 188*(3), 1098–1107. http://dx.doi.org/10.4049/jimmunol.1101121.

Liu, T., Zhang, L., Joo, D., & Sun, S. C. (2017). NF-kappaB signaling in inflammation. *Signal Transduction and Targeted Therapy, 2*. http://dx.doi.org/10.1038/sigtrans.2017.23.

Ljubimova, J. Y., Braubach, O., Patil, R., Chiechi, A., Tang, J., Galstyan, A., et al. (2018). Coarse particulate matter (PM2.5-10) in Los Angeles Basin air induces expression of inflammation and cancer biomarkers in rat brains. *Scientific Reports, 8*(1), 5708. http://dx.doi.org/10.1038/s41598-018-23885-3.

MacDonald, M. E., Ambrose, C. M., Duyao, M. P., Myers, R. H., Lin, C., Lakshmi, S., et al. (1993). A novel gene containing a trinucleotide repeat that is expanded and unstable on

Huntington's disease chromosomes. The Huntington's Disease Collaborative Research Group. *Cell, 72*(6), 971–983.

Martinon, F., Chen, X., Lee, A. H., & Glimcher, L. H. (2010). TLR activation of the transcription factor XBP1 regulates innate immune responses in macrophages. *Nature Immunology, 11*(5), 411–418. http://dx.doi.org/10.1038/ni.1857.

Pomatto, L. C. D., Cline, M., Woodward, N., Pakbin, P., Sioutas, C., Morgan, T. E., et al. (2018). Aging attenuates redox adaptive homeostasis and proteostasis in female mice exposed to traffic-derived nanoparticles ('vehicular smog'). *Free Radical Biology and Medicine, 121*, 86–97. http://dx.doi.org/10.1016/j.freeradbiomed.2018.04.574.

Rodriguez, M., Snoek, L. B., Riksen, J. A., Bevers, R. P., & Kammenga, J. E. (2012). Genetic variation for stress-response hormesis in *C. elegans* lifespan. *Experimental Gerontology, 47*(8), 581–587. http://dx.doi.org/10.1016/j.exger.2012.05.005.

Ron, D., & Walter, P. (2007). Signal integration in the endoplasmic reticulum unfolded protein response. *Nature Reviews Molecular Cell Biology, 8*(7), 519–529. http://dx.doi.org/10.1038/nrm2199.

Sabroe, I., Parker, L. C., Dower, S. K., & Whyte, M. K. (2008). The role of TLR activation in inflammation. *The Journal of Pathology, 214*(2), 126–135. http://dx.doi.org/10.1002/path.2264.

Schaefer, L. (2014). Complexity of danger: The diverse nature of damage-associated molecular patterns. *Journal of Biological Chemistry, 289*(51), 35237–35245. http://dx.doi.org/10.1074/jbc.R114.619304.

Singh, K. D., Roschitzki, B., Snoek, L. B., Grossmann, J., Zheng, X., Elvin, M., et al. (2016). Natural genetic variation influences protein abundances in *C. elegans* developmental signalling pathways. *PLoS One, 11*(3), e0149418. http://dx.doi.org/10.1371/journal.pone.0149418.

Vu, V., Verster, A. J., Schertzberg, M., Chuluunbaatar, T., Spensley, M., Pajkic, D., et al. (2015). Natural variation in gene expression modulates the severity of mutant phenotypes. *Cell, 162*(2), 391–402. http://dx.doi.org/10.1016/j.cell.2015.06.037.

Woodward, N. C., Levine, M. C., Haghani, A., Shirmohammadi, F., Saffari, A., Sioutas, C., et al. (2017). Toll-like receptor 4 in glial inflammatory responses to air pollution in vitro and in vivo. *Journal of Neuroinflammation, 14*(1), 84. http://dx.doi.org/10.1186/s12974-017-0858-x.

CHAPTER FIVE

Functional and dysfunctional folding, association and aggregation of caseins

John A. Carver[a,*] and Carl Holt[b]
[a]Research School of Chemistry, The Australian National University, Acton, ACT, Australia
[b]Institute of Molecular, Cell and Systems Biology, University of Glasgow, Glasgow, United Kingdom
*Corresponding author: E-mail: john.carver@anu.edu.au

Contents

Advances in Protein Chemistry and Structural Biology, Volume 118
ISSN 1876-1623
https://doi.org/10.1016/bs.apcsb.2019.09.002
© 2020 Elsevier Inc.
All rights reserved.

Abstract

Caseins are a group of closely related intrinsically disordered proteins (IDPs), best known for their occurrence in milk as stable, polydisperse, roughly spherical, amorphous particles, typically containing thousands of protein chains and hundreds of nanoclusters of calcium phosphate. The particles are called casein micelles though their structure bears no resemblance to detergent micelles. Caseins have an open and flexible conformation with a preponderance of poly-L-proline II secondary structure and hence cannot be described as hydrophobic proteins. Individually, and in combination, they associate through polar and non-polar interactions to form polydisperse fuzzy complexes (including the native casein micelle) while retaining their hydrated and flexible conformation to a large degree. Like many other IDPs, caseins are prone to form cytotoxic amyloid fibrils. However, they are also highly effective molecular chaperones so that a mixture of different caseins can form fuzzy complexes that are often self-limiting in size and, within which, amyloid fibril formation is suppressed. The remarkable ability of caseins to sequester nanoclusters of calcium phosphate in stable complexes is due to their flexible conformation and multiply-phosphorylated short sequences. These features combine to form a dense peptide shell around the calcium phosphate making the core-shell complex thermodynamically stable, even at high calcium and phosphate concentrations. Thus, the casein micelle provides a readily digested, high calcium food for the neonate. It also preserves the functional properties of caseins as IDPs and protects the mammary gland against amyloid formation and pathological calcification, dysfunctional processes that would reduce the future reproductive success of the mother.

Abbreviations

ACD α-Crystallin domain
ADH Alcohol dehydrogenase
CA *Corpora amylacea*

CaP	Calcium phosphate
CaP-SLiM	Short linear sequence motif that binds to calcium phosphate
CD	Circular dichroism
CMC	Critical micelle concentration
CTR	C-terminal region
HMW	High molecular weight
HO-SLiM	Short linear sequence motif of ≥3 hydrophobic and order-promoting residues
IDP	Intrinsically disordered protein
IDR	Intrinsically disordered region
K&D	Kyte and Doolittle
LFM	Linear functional motif
NTR	N-terminal region
PDS	Protein diagram of states
P,Q-rich	Pro and Gln-rich
PP-II	Poly-L-proline type II
RCM	Reduced and carboxymethylated
ROA	Raman optical activity
sHsp	Small heat-shock protein
SLiM	Short linear sequence motif
UHT	Ultra-heat treated

1. Introduction and overview

Casein proteins are found in all mammals and have a wide distribution among tissues, but they are chiefly known for their occurrence at high concentration ($\leq$150 g L^{-1}) in milk where they are mostly in the form of colloidal particles called casein micelles. Casein micelles are polydisperse and roughly spherical with an average diameter of about 150 nm and are mainly responsible, through their concentration and light scattering properties, for the white appearance of milk. Although called micelles they have no structural resemblance to soap micelles but typically comprise a water-rich matrix of several thousand disordered casein molecules within which several hundred nanoclusters of an amorphous form of calcium phosphate (CaP) are sequestered.

Caseins were among the first proteins to be recognized as having an unfolded "random coil" conformation (Halwer, 1954; Herskovits, 1966; Kresheck, 1965) which, it was thought, would facilitate proteolytic digestion in the stomach of the neonate. Notwithstanding this, in the last century caseins were widely regarded as hydrophobic proteins and likened to detergent molecules with polar "heads" and long hydrophobic "tails" that associated through the hydrophobic effect.

With recent advances in understanding intrinsically disordered proteins (IDPs) or intrinsically disordered regions (IDRs) in proteins, it became clear to us and others (Holt & Carver, 2012; Holt, Carver, Ecroyd, & Thorn, 2013; Thorn, Ecroyd, Carver, & Holt, 2015) that IDPs provide better models for understanding caseins than either detergent molecules or globular proteins. In promoting this change, we recognized that the ability of caseins to form amyloid fibrils, their action as molecular chaperones, the absence of a sharp denaturation transition, small NMR chemical shift dispersion and spectroscopic evidence of a high proportion of the poly-L-proline type-II (PP-II) secondary structure were characteristic properties of caseins and other IDPs. No unique hypotheses of protein behavior are required to explain either the conformation or association behavior of caseins. Accordingly, we proposed that caseins are best described by a dynamic ensemble of unfolded conformations and argued that in casein-casein interactions, the hydrophobic effect is not as important as formerly assumed. So, our analysis does not support the proposal that caseins have an open but rigid structure maintained by tensegrity (Farrell, Brown; Hoagland, & Malin, 2013) nor the proposal that the hydrophobic effect is the only (Horne, 1998) or main (Horne, 2017) driving force for casein-casein interactions (Carver, Thorn, Ecroyd, & Holt, 2017; Holt, Raynes, & Carver, 2019).

Herein, we review the recent work on CaP sequestration, amyloid structure formation and molecular chaperone action by caseins and the evidence accumulated over many years that caseins retain much of their disordered conformation and hydration in their associated states, including the native casein micelle. We present a coherent account of casein structures and interactions based upon modern concepts of protein science. Casein-casein and casein-CaP interactions are not just of biological importance as they determine the formation and properties of economically important food hydrogels including cheese and yoghurt.

2. Linear functional motifs in caseins

There are six casein genes known to science and of these four are expressed in the bovine mammary gland giving rise to forms of α_{S1}-, α_{S2}-, β- and κ-casein proteins. The subscripted "S" denotes a casein that is sensitive to precipitation by a low mM calcium ion concentration at neutral pH and temperature of 4 °C or higher. β-Casein is also sensitive to precipitation by calcium ions but only at slightly higher calcium ion concentrations and temperature, whereas κ-casein is relatively insensitive to precipitation by calcium ions. As a result, it is common in the literature to find bovine caseins and caseins from cognate species divided into two groups, namely the calcium-sensitive (α_{S1}, α_{S2}, β) and calcium-insensitive (κ) caseins.

Because of their mobile conformation, the functional properties of IDPs are often related to their sequence through linear functional motifs (LFMs) (van der Lee et al., 2014). The casein LFMs largely reflect the intron/exon structure of casein genes (Kawasaki, Lafont, & Sire, 2011) (Fig. 1). One LFM is the signal sequence at the N-terminus comprising 15 (calcium-sensitive caseins) or 21 (κ-casein) residues which is cleaved during secretion to give the mature caseins. The remaining LFMs in the mature caseins are (i) one or more CaP-binding motifs comprising a cluster of phosphorylated residues together with flanking sequences, all encoded by short exons, (ii) usually one, but in some α_{S2}-type caseins, two, Pro and Gln-rich (P,Q-rich) sequences encoded by long exons and (iii) in the C-terminal region (CTR) of κ-casein, a flexible mucin-like sequence, rich in Pro, Thr and Ser, known as the macropeptide. The CaP-binding LFMs facilitate the interaction of caseins with the CaP nanoclusters. The P,Q-rich regions are responsible for different types of polar and non-polar casein-casein interactions including disulphide bridged dimers of α_{S2}-casein and oligomers of κ-casein in bovine caseins. The macropeptide sequence, although it has a similar average hydropathy to the P,Q-rich sequences, is highly dynamic and forms few or weak protein-protein and protein-CaP interactions.

Short linear motifs (SLiMs) mediate the earliest stage of interaction between disordered regions of IDPs. These comprise 3–10 residues acting together to form transient, weak and reversible initial interactions with their target (Ren, Uversky, Chen, Dunker, & Obradovic, 2008). The clusters of phosphorylated residues in caseins have previously been called phosphate centers but to conform with IDP usage are here called CaP-SLiMs so that the CaP-binding LFM comprises a single CaP-SLiM and one or two mobile, hydrophilic flanking sequences (Holt, Sorensen, & Clegg, 2009). In addition to CaP-SLiMs, caseins contain SLiMs composed of three or more hydrophobic and order-promoting residues (HO-SLiMs) (Holt et al., 2019) and steric zipper SLiMs (Goldschmidt, Teng, Riek, & Eisenberg, 2010; Holt & Carver, 2012), which initially form reversible oligomers that may go on to form the refractory β-sheet core structure of amyloid fibrils (Wiltzius et al., 2008). Both steric zipper and HO-SLiMs are mainly, but not exclusively, found within the P,Q-rich LFMs (Holt & Carver, 2012; Holt et al., 2019). A cartoon representation of LFMs and

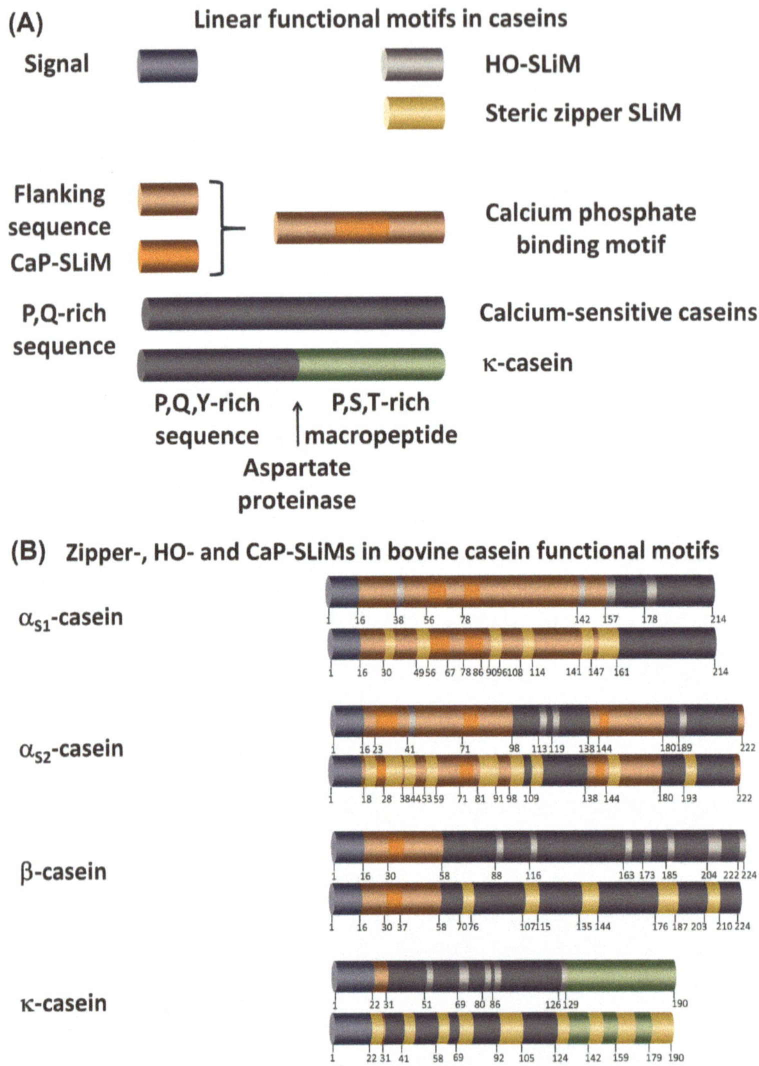

Fig. 1 Cartoon representation of linear functional motifs (LFMs) and SLiMs in bovine caseins. (A) Key explaining the color coding of LFMs and CaP-, HO- and steric zipper SLiMs in bovine caseins. (B) Approximate size and position of LFMs and SLiMs in the four bovine caseins. For each casein, a pair of diagrams is provided. Both contain the LFMs and CaP-SLiMs but for clarity, the HO-SLiMs (upper) and steric zipper SLiMs (lower) are given separately. Virtually all residues in the signal sequences are predicted to be HO-SLiMs.

SLiMs in bovine caseins is shown in Fig. 1. The identified CaP-SLiMs in each casein (Fig. 1) contain at least two phosphorylated residues and are relatively well-conserved but other CaP binding sites of lower affinity are present. Steric zipper SLiMs are common in caseins (Holt & Carver, 2012) but only a fraction of these are known to be involved in forming amyloid fibrils under physiological conditions (see Section 7).

3. Caseins are not hydrophobic proteins

A charge-hydropathy plot such as the protein diagram of states (PDS) (Fig. 2A–C) is very effective at discriminating between globular proteins and IDPs (Huang et al., 2014; Mao, Crick, Vitalis, Chicoine, & Pappu, 2010; Uversky, Gillespie, & Fink, 2000). Fig. 2A shows a plot of 50 IDPs or IDRs and 50 globular proteins or domains (Dosztányi, Csizmók, Tompa, & Simon, 2005) in a PDS using the Kyte and Doolittle (K&D) hydropathy scale (Kyte & Doolittle, 1982). In the PDS, the boundary planes efficiently separating the two groups intercept the hydropathy axis at its base value h_b

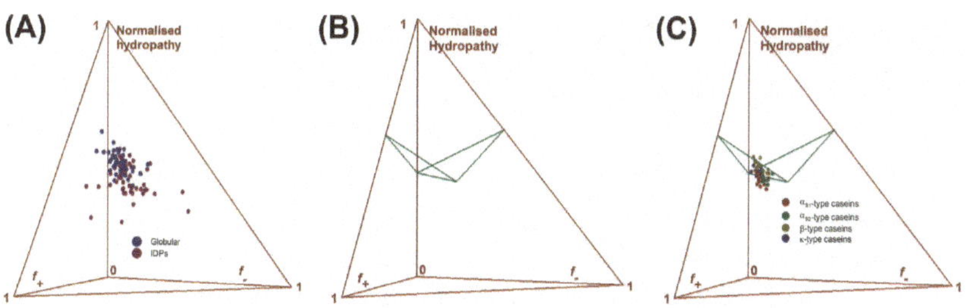

Fig. 2 The PDS is a charge-hydropathy diagram in three dimensions where the axes are the average normalized hydropathy, the fraction of positively charged residues (Arg and Lys, f_+) and the fraction of negatively charged residues (Asp and Glu, f_-). (A) Position of 50 globular proteins and domains compared to 50 IDPs. Data were selected from the tabulations of Dosztányi et al. (2005) and show that IDPs are more hydrophilic, in general, than globular proteins. (B) The Uversky-Pappu planes separating most folded from most unfolded proteins intercept the hydropathy axis at the base value h_b (Mao et al., 2010; Uversky et al., 2000). In this example, the normalized hydropathy scale of Guy has been used (Guy, 1985). (C) Position of 67 unphosphorylated mature caseins from 20 species in the PDS (Holt et al., 2019). Caseins lie in or close to the boundary between disordered and ordered proteins (below and above, respectively, the Uversky-Pappu plane). A very similar result is obtained with the Kyte and Doolittle scale (Carver, Ecroyd, Truscott, Thorn, & Holt, 2018; Kyte & Doolittle, 1982) and the disorder propensity scale (Holt et al., 2019; Huang et al., 2014). *Redrawn from Holt et al. (2019) and Carver et al. (2018).*

(Fig. 2B) At a given linear charge density, IDPs are less hydrophobic than globular proteins.

When 67 casein sequences from 20 species are plotted in the PDS (Fig. 2C) they cluster together close to the boundary between globular proteins and IDPs (Carver et al., 2018; Holt et al., 2019). As Fig. 2C conclusively shows, caseins cannot be described as hydrophobic proteins; for a given linear charge density, they are less hydrophobic than most globular proteins but more hydrophobic than most IDPs (Holt et al., 2019).

3.1 The average hydropathy of uncharged residues in the linear functional motifs of mature casein sequences is invariant

When the average hydropathy values of casein LFMs, i.e. CaP-binding motifs, macropeptides and P,Q-rich sequences, are plotted in the PDS (Fig. 3), they lie in or close to an inclined plane, consistent with an invariant average hydropathy of the uncharged residues (Holt et al., 2019). Fig. 3C shows an edge-on view of the fitted plane revealing the average hydropathy of the macropeptides and P,Q-rich regions to be very similar. The macropeptide CTR, however, contains few or no HO-SLiMs (Fig. 1B) and residues of higher average disorder propensity compared to the N-terminal region (NTR) of κ-casein and other P,Q-rich motifs.

The invariant average hydropathy of uncharged residues ($\overline{h}_0$) among the functional motifs of mature caseins is a property shared with the paralogous phosphoprotein amelogenin which, like β-casein, is a relatively hydrophobic, self-associating IDP (Holt et al., 2019). A possible explanation is that the need for intermolecular interactions sets a lower bound and the avoidance of conformational collapse sets an upper bound to the uncharged residue hydropathy and that these two bounds are fairly close together (Holt et al., 2019).

4. Caseins are intrinsically disordered proteins

Following the work of Tiffany and Krimm on the circular dichroism (CD) spectra of model peptides (Tiffany & Krimm, 1968a; 1968b), it was determined that the "random coil" spectrum was actually the PP-II helical secondary structure. Poly–proline can adopt an extended left-handed helical structure of three *trans* imide groups per turn. Any amino acid residue, not just proline, can adopt this conformation (Cubellis, Caillez, Blundell, & Lovell, 2005). The helix is stabilized by solvation (hydrogen bonds with

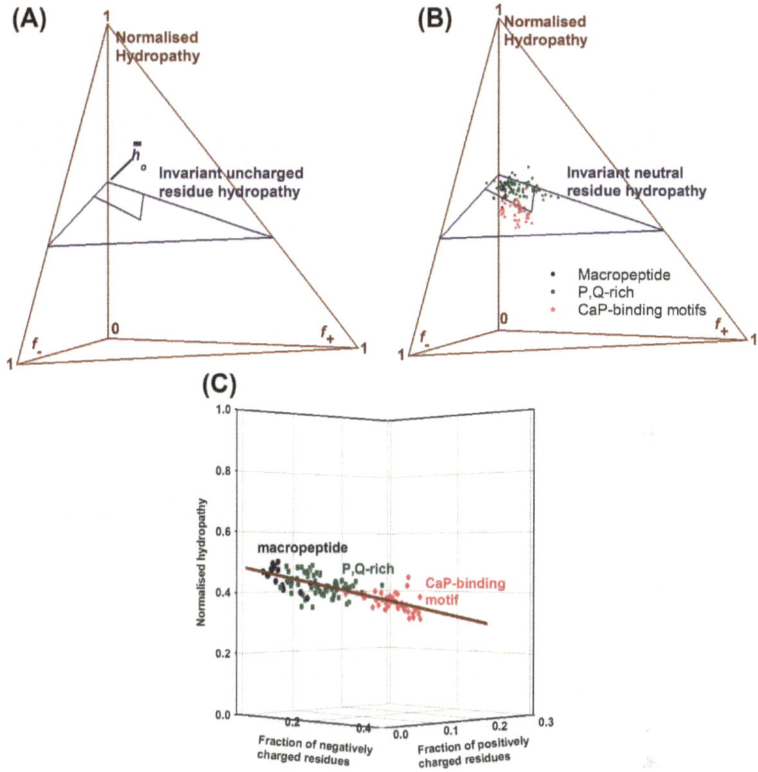

Fig. 3 Invariant average hydropathy of uncharged residues in the LFMs of 67 mature caseins from 20 species (Guy, 1985; Holt, 2016; Holt et al., 2019). (A) PDS showing an inclined plane intercepting the hydropathy axis at the invariant average hydropathy $\overline{h}_o$. A quadrilateral is formed at the apex of the plane by intersection with the Uversky-Pappu boundary planes provided $\overline{h}_o > h_b$. (B) Calculated average hydropathy of LMFs of 67 mature caseins from 20 species. Points fall in or near to an inclined regression plane with an invariant uncharged residue hydropathy ($\overline{h}_o$) of 0.498 ± 0.005. (C) Edge-on view of the invariant plane showing the similar average hydropathy of macropeptide and P,Q-rich sequences. *Redrawn from Holt et al. (2019).*

water) and weak dipolar inter-residue interactions (Maccallum, Poet, & Milner-White, 1995) but not by hydrogen bonding between residues so it is easily perturbed and gives rise to worm-like, flexible, polypeptide chains in many IDPs, including caseins. (Adzhubei, Sternberg, & Makarov, 2013; Baldwin, 2002; Barron, Blanch, & Hecht, 2002; Shi, Woody, & Kallenbach, 2002). The PP-II secondary structure is rarely present in the well-ordered domains of globular proteins (Shi et al., 2002).

Algorithms for secondary structure predictions based on the heuristic rules derived from globular proteins and the software used for the calculation of secondary structure from circular dichroism (CD) and infrared spectra with a basis set of globular protein reference spectra are likely to give erroneous results when applied to IDPs (Barron et al., 2002; Shi et al., 2002; van der Lee et al., 2014). This problem was not widely recognized, however, until the beginning of this century and many of the predictions and secondary structure calculations and conclusions made in the older work on caseins need to be interpreted with caution (Byler, Farrell, & Susi, 1988; Creamer, Richardson, & Parry, 1981; Graham, Malcolm, & McKenzie, 1984; Herskovits, 1966; Holt & Sawyer, 1988, 1993; Kumosinski, Brown, & Farrell, 1991a, 1991b, 1993; Qi et al., 2004; Raap, Kerling, Vreeman, & Visser, 1983; Tomatoda Ono, Yada, Yutani, & Nakai, 1987; Urbanová, Keiderling, & Pančoška, 1996).

It has been argued that Raman optical activity (ROA) provides the method of choice for identifying secondary structure in IDPs because the chiral effect is restricted more closely to individual peptide bonds than in other conformation-sensitive spectroscopies and therefore is less sensitive to the length or stability of ordered segments, temperature and other environmental effects (Barron et al., 2002). Application of ROA to caseins has shown (Fig. 4A) a predominance of PP-II in these proteins although some β-strands (not sheet) and turns are also present but little, if any, α-helix (Syme et al., 2002). Unfortunately, interpretation of ROA spectra to yield quantitative estimates of secondary structure is in the earliest stages of development (Mensch, Barron, & Johannessen, 2016; Mensch & Johannessen, 2019).

The CD spectra of bovine caseins (Creamer et al., 1981; Graham et al., 1984; Raap et al., 1983; Tomatoda Ono et al., 1987) show a prominent minimum at 195—205 nm which is likely to be due to a large contribution from PP-II structure. However, quantitative interpretation of these spectra relied on globular proteins without PP-II. As shown in Fig. 4B for β-casein and for β-casein truncated at its C-terminus by treatment with chymosin (Qi et al., 2005), the prominent ellipticity minimum for the latter is at a shorter wavelength than in the native protein. Qi et al. (2005) interpreted the shift to be due to the loss of some regular secondary structure in the truncated protein. Because PP-II is sensitive to temperature it transforms reversibly to a more disordered conformation with reduced ellipticity amplitudes at elevated temperatures (Drake et al., 1988). As shown in Fig. 4B, the ellipticity minimum of the β-casein spectra becomes less negative as temperature

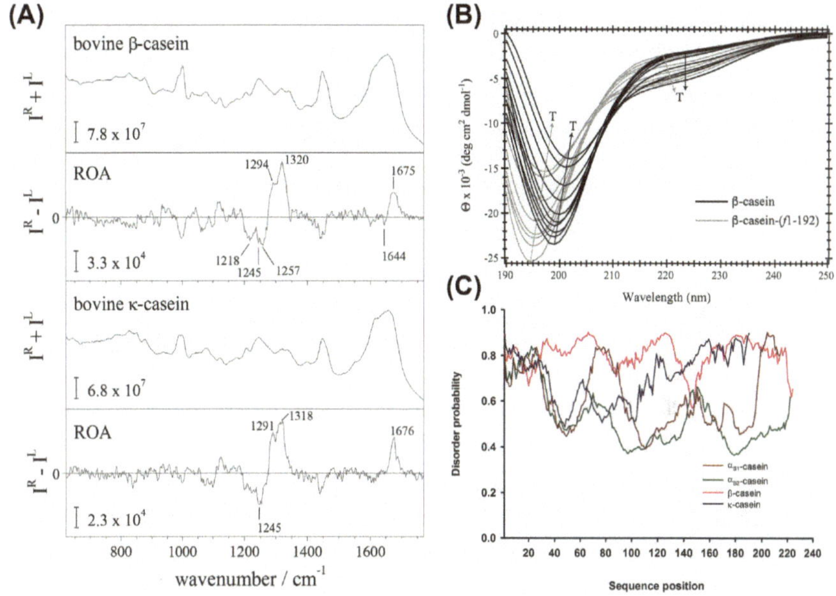

Fig. 4 Evidence that caseins have a preponderantly disordered conformation with PP-II as the dominant secondary structure. (A) Raman (top) and ROA (below) spectra of β- and κ-caseins (Syme et al., 2002) showing features characteristic of mainly PP-II secondary structure with some evidence of β-strands but little evidence of α-helix. Reproduced by permission of the FEBS Journal. (B) The CD spectra of whole β-casein (1-209) and β-casein 1-192, redrawn by Dr P. X. Qi from Qi, Wickham, and Farrell (2004), Qi, Wickham, Piotrowski, Fagerquist, and Farrell (2005). The strong temperature dependence of the CD spectrum of β-casein 1-192 is consistent with the spectral changes due to the interchange of PP-II with a more disordered conformation (Drake, Siligardi, & Gibbons, 1988). Reproduced by permission of Dr P. X. Qi and Springer. (C) Disorder predictions for bovine caseins from the intFOLD5 suite of predictors (McGuffin, Atkins, Salehe, Shuid, & Roche, 2015) (http://www.reading.ac.uk/bioinf/IntFOLD/IntFOLD5_form.html). Caseins are disordered throughout their length with no evidence of a stable conformation. *Redrawn from Supplementary Material A of Holt et al. (2019).*

increases. Although the spectrum was interpreted using globular protein reference spectra, without a PP-II component, Qi et al. (2005) also gave an analysis of the mid-infrared spectrum of β-casein that included a PP-II assignment and found 20% of this structure but also 25% α-helix, results that are very different from the qualitative conclusions from the ROA spectrum.

In summary, the CD and other conformation-sensitive spectra of IDPs in general, and caseins in particular, are dominated by PP-II but the use of globular protein models to interpret the spectra gives erroneous results.

Likewise, the application of heuristic rules of secondary structure prediction, derived from globular protein structures, fails to recognize the dominant importance of PP-II and tends to overpredict other regular secondary structures.

4.1 Other evidence of conformational disorder

Protein dynamics are being studied by a range of emerging inelastic and quasi-elastic neutron scattering techniques (Loupiac, 2016). Low frequency collective vibrational dynamics in amorphous materials give rise to the Boson peak, the position of which scales with local rigidity in structural motifs so that IDPs, for example, are softer than β-sheet proteins which are softer than α-helical proteins (Perticaroli, Nickels, Ehlers, & Sokolov, 2014). Caseins have been the subject of a number of studies to show that rigidity is increased by calcium binding and local diffusive motions can be larger in amplitude than in globular proteins (Gaspar, Appavou, Busch, Unruh, & Doster, 2008; Perticaroli, Nickels, Ehlers, Mamontov, & Sokoov, 2014). A similar increase in rigidity of the phosphoprotein and IDP, osteopontin, was noted when it was complexed to CaP nanoclusters (Lenton, Seydel, et al., 2015).

The specific heat of most caseins increases smoothly with temperature consistent with a gradual increase in mobility. Thus, no co-operative conformational transitions in caseins can be recognized by differential scanning calorimetry (de Kruif, Tuinier, Holt, Timmins, & Rollema, 2002; Mikheeva, Grinberg, Grinberg, Khokhlov, & de Kruif, 2003; Paulsson & Dejmek, 1990; Syme et al., 2002) except for a transition in β-casein at about 15 °C which coincides with endothermic oligomerization above this temperature (de Kruif & Grinberg, 2002; Qi et al., 2004, 2005).

If caseins are like other IDPs of ~200 residues in length, without disulphide bridges, they will have a hydrodynamic radius $R_h \sim$ 4.6 nm and a specific hydrodynamic volume of ~9.8 mLg^{-1}, but if they are typically globular, the R_h will be ~2.2 nm (Wilkins et al., 1999). At 4 °C, β-casein has $R_h \sim$ 4 nm (Evans, Irons, & Jones, 1971). Consistent with this, the ^{1}H NMR spectrum of the monomer has little chemical shift dispersion consistent with a "random coil" conformation. Scattering measurements on β-casein in 4 M guanidinium hydrochloride show it is a slightly stiff, worm-like, chain in a good solvent (Aschi, Gharbi, Daoud, Douillard, & Calmettes, 2007).

The initial conformation of reduced and carboxymethylated (RCM) κ-casein after dilution from concentrated urea was fully expanded with

$R_h = 4.6$ nm, but at a concentration too low to form micelles and in the absence of intermolecular disulphide bridges, it collapsed to a disordered more compact conformation with $R_h \sim 1.9$ nm (Jain, Bhattacharya, & Mukhopadhyay, 2011). The collapsed state was proposed to be a molten globule and an amyloid fibril precursor.

The α_{S1}-casein monomer at pH 12 or in 6-M guanidinium hydrochloride is fully expanded with $R_h = 4.2$ nm but at neutral pH $R_h = 3.3$ nm (Ho & Chen, 1967; Schmidt, Payens, van Markwijk, & Brinkhuis, 1967), indicating some more condensed conformers are present in its ensemble of conformations.

4.2 Caseins are predicted to have a dynamic ensemble of conformations

Predictor and metapredictor methods consistently show that caseins are disordered over the great majority of their mature sequence, if not all of it (Holt et al., 2009, 2019; Redwan, Xue, Almehdar, & Uversky, 2015), as illustrated by the disorder predictions for bovine caseins in Fig. 4C. Homology modeling has, however, been used to propose best-fit models for α_{S2}- and β-caseins with a significant amount of stable, regular secondary structure (Farrell, Malin, Brown, & Mora-Gutierrez, 2009; Zhou et al., 2019). In both studies, the authors recognized the limitations of using globular protein structures to model the structure of caseins. We reached a similar conclusion after applying the disorder and homology modeling algorithms in IntFOLD5 (http://www.reading.ac.uk/bioinf/IntFOLD/) (Atkins, Boateng, Sorensen, & McGuffin, 2015; McGuffin et al., 2015, 2018) to the four bovine casein sequences (Holt et al., 2019). Confidence and significance levels for the 50 best models, ranked in order of significance, indicated all were likely to be poor with little or no similarity to the native structure. In other words, no single conformation adequately describes casein structures.

5. Caseins bind to and sequester calcium phosphate

Arguably, the most remarkable property of caseins is their ability to bind CaP to form a stable complex. It is remarkable because mineral substances generally have a high lattice energy which means that their nucleation and growth from saturated solution to form a macroscopic crystalline phase should not be affected by weak interactions with proteins. When the physical chemistry of the micellar CaP was first examined, it became clear from its amorphous nature that there was no lattice energy contribution so it was at least feasible that complexing proteins could stabilize the CaP in a highly dispersed state to prevent it from converting into a more stable crystalline phase. Such non-stoichiometric complexes are described as kinetically stable but thermodynamically unstable. Further physico-chemical studies revealed, however, that certain peptides containing clusters of phosphorylated residues could not only bind to CaP nuclei as they formed from solution to slow the rate of precipitation, which was completely expected from prior work, but could also form an equilibrium complex (Holt, Timmins, Errington, & Leaver, 1998; Holt, Wahlgren, & Drakenberg, 1996). Moreover, in a type of reverse reaction, the peptides could disperse a pre-formed precipitate of amorphous CaP to form the same type of nanocluster complex (Little & Holt, 2004). In other words, the CaP nanocluster complexes are not just kinetically stable, they are reversible, of defined stoichiometry and thermodynamically stable. This finding has allowed the partition of calcium, inorganic phosphate and other ions between the micelle and milk serum to be calculated from thermodynamic principles (Bijl, Huppertz, van Valenberg, & Holt, 2019; Holt, 2004), a result which has important implications for understanding the limits to the stability of milk during secretion, storage and processing. Micellar CaP is a minor component, about 7% of the dry weight of micelles, but when the composition of the micelle was expressed in molar units, it became clear that the great majority of the calcium-sensitive caseins in the micelle are bound to the CaP nanoclusters (Bijl et al., 2019). The most recent review of the physical chemistry of CaP sequestration is that of Lenton, Nylander, Teixeira, and Holt (2015).

5.1 CaP-SLiMs in caseins

Sequences similar to the most highly phosphorylated CaP-SLiMs in caseins have been found in some other phosphoproteins involved in the control of biomineralization (Holt et al., 2009). In addition to the well-recognized CaP-SLiMs, other single sites of phosphorylation in the same protein may also have a supporting role in binding to CaP, particularly if they have flanking acidic residues. The most highly phosphorylated CaP-SLiMs, for example the sequence 14-EpSLpSpSpSEE-21 in mature β-casein, where pS is phosphoserine (Fig. 1), contain a high affinity binding site for calcium ions (Bijl et al., 2019) which probably means that they also have a high affinity for amorphous CaP as is also indicated by comparative studies of the strength of binding of phosphopeptides to hydroxyapatite (Clegg & Holt, 2009; Sleigh, Sculley, & Mackinlay, 1979). Similar conclusions have been reached from studies of the effect of the degree of phosphorylation of caseins on the rates of CaP precipitation and maturation from solution (van Kemenade & de Bruyn, 1989). Some evidence also exists of the binding of cationic residues and sequences in caseins to CaP (Peixoto et al., 2017).

In an improved thermodynamic description of the partition of milk salts (Bijl et al., 2019), all CaP-SLiMs on a protein molecule react with CaP once the initial binding event occurs, even if they have a lower affinity. One advantage of this assumption is that it predicts higher proportions of free (i.e. not bound to CaP) calcium-sensitive caseins, in better agreement with experimental findings from urea solubilization of caseins from casein micelles (Takayoshi Aoki, Sakamoto, & Kako, 1991; T. Aoki, Umeda, & Kako, 1992; T. Aoki, Yamada, & Kako, 1990; T. Aoki, Yamada, Tomita, Kako, & Imamura, 1987).

5.2 Physico-chemical aspects of CaP sequestration

Theoretical and experimental work on the CaP nanocluster complexes began with caseins from milk but has been generalized and further developed including work with other phosphoproteins and biofluids (Clegg & Holt, 2009; Cross, Huq, Palamara, Perich, & Reynolds, 2005; de Kruif, 2014; de Kruif, Huppertz, Urban, & Petukhov, 2012; Holt, 2013; Holt, Lenton, Nylander, Sorensen, & Teixeira, 2014; Holt et al., 2009; Ingham et al., 2015; Ingham et al., 2016; Lenton et al., 2017; Lenton et al., 2016; Lenton, Seydel, et al., 2015; Little & Holt, 2004; Rodriguez et al., 2014).

The use of simplified model CaP nanoclusters formed from a single pure casein phosphopeptide has been instrumental in clarifying the nature of the chemically more complex but physically similar micellar CaP nanoclusters. The most thoroughly studied CaP nanocluster complex is that formed with the N-terminal tryptic phosphopeptide from bovine β-casein. This has a core of amorphous CaP and a sequestering shell of phosphopeptides anchored to the core by the CaP-SLiM. The dimensions of the core and shell were determined by contrast-matched neutron scattering in the Guinier limit together with sedimentation equilibrium to determine the mass (Holt et al., 1998). In these calculations, the core was assumed to be an amorphous di-calcium phosphate and from this assumption, the number of peptides in the sequestering shell was calculated to be 48. Subsequently, the empirical chemical formula of the core calcium phosphate, determined from the invariant ion activity product, was shown to be $Ca(HPO_4)_{0.4}(PO_4)_{0.4}$. The anhydrous empirical chemical formula of a notional monomer containing a single peptide and associated CaP, consistent with the experimental molar ratios of Ca/phosphopeptide = 13.2 and P_i/phosphopeptide = 6.5 and electroneutral core and interface can be written

$$\underset{\text{Core}}{[Ca(HPO_4)_{0.4}(PO_4)_{0.4}]_{8.125}} - \underset{\text{Interface}}{[Ca_{5.075}(P_oO_4^{2-})_4(COO^-)_{1.825}]} - \underset{\text{Shell}}{[R]} \tag{1}$$

Where $P_oO_4^{2-}$ is the organic phosphate from the four phosphorylated residues and R is the peptide less the groups from sidechains forming the interface. These are assumed to be the phosphate moieties of the four phosphorylated residues and the side chain carboxyl groups needed to give overall charge neutrality. The average mass of the CaP nanocluster was 1.97×10^5 Da. Using the improved understanding of the chemical composition of the core reduces slightly the calculated number of monomers from 48 to 44.7. The anhydrous core molecular volume in nm is 101.3 nm^3. A sphere of the same volume has a radius of 2.43 nm, close to the experimental radius of 2.4 nm. The shell volume is 378.7 nm^3 which gives an outer radius of 4.71 nm compared to the experimental result of 4.0 nm. The core area per peptide is 1.62 nm^2. The area per peptide is comparable to that of the paralogous salivary IDP statherin in a monolayer on hydroxyapatite (Goobes et al., 2007).

The formation of a stable solution containing CaP nanocluster complexes can be understood qualitatively in the following terms. If a solution is supersaturated with respect to a certain phase of CaP then precipitation involves a negative change in free energy until, at equilibrium, the solution becomes exactly saturated and the change in free energy becomes zero. If instead of forming a precipitate, the CaP reacts with a competent phosphopeptide then there is an additional negative free energy contribution. Nanocluster formation can continue below the point of saturation, so CaP no longer precipitates from solution. A slight excess of the competent phosphopeptide is all that is needed to ensure the stability of a biofluid against physiological fluctuations in composition. In the milk of the cow, for example, the slight excess is in the μM-mM range but in other biofluids such as blood the total concentration of complexed and free peptide can be much lower and is determined by the affinity of the peptide for the CaP. From a balance of the free energy terms, expressions can be derived for the average core radius and the number distribution of core radii (Holt, 2013; Holt et al., 2009).

5.3 CaP nanoclusters in casein micelles

The evidence that CaP nanocluster complexes prepared with pure single phosphopeptides are comparable in composition, mass and structure to the complexes in casein micelles has been comprehensively reviewed (Holt, 1997; Lenton, Nylander, et al., 2015). The apparent mass of micellar CaP complexes isolated by tryptic digestion was estimated by gel filtration chromatography to be 1.7×10^5 Da, comparable to 1.97×10^5 Da for the artificial complexes prepared in the laboratory (Holt et al., 1998; T. Ono, Ohotawa, & Takagi, 1994). A much smaller mass of 1.745×10^4 Da was reported by Choi et al. (Choi, Horne, & Lucey, 2011) for CaP particles isolated from milk by a similar digestion procedure. No experimental data were given of the light scattering measurements, a particle refractive index for protein rather than CaP was assumed and the particles could not be shown to contain any peptides. Thus, these particles do not appear to be comparable to other CaP nanocluster complexes.

The core mass of 6.0×10^4 Da along with the mass fraction of CaP in casein micelles allows the number of CaP nanoclusters to be calculated. For a number average micelle radius of 75 nm (Le, Saveyn, Hoa, & Van der Meeren, 2008), this is about 300 and the average spacing is 18 nm (Holt, de Kruif, Tuinier, & Timmins, 2003). A study of casein micelles by resonant small-angle X-ray scattering near the L absorption edge of calcium conclusively identified key scattering features with the size of and the distance between CaP nanoclusters. (Ingham et al., 2015). However, the CaP nanoclusters by themselves do not account for all the scattering features arising from the substructure of bovine casein micelles (de Kruif, 2014; Ingham et al., 2016).

6. Caseins can act as molecular chaperones

6.1 Intra and extracellular molecular chaperones

Molecular chaperones encompass a large and diverse group of proteins that interact with partially folded or unfolded proteins to prevent their misfolding, incorrect association and aggregation (Ecroyd & Carver, 2008b). *In vivo*, chaperones are involved in a wide range of activities including facilitating protein folding, degradation, trafficking and stabilization (Cox, Carver, & Ecroyd, 2014). Because of these diverse capabilities, chaperones have a major role in proteostasis, i.e. ensuring that protein levels are contained within a narrow range to maintain cellular well-being (Carver et al., 2018). Herein, we discuss the chaperone action of caseins in preventing proteins, particularly caseins, from dysfunctional aggregation.

Small heat-shock proteins (sHsps) are the predominant intracellular chaperones that act in an ATP-independent manner to prevent proteins from unfolding. Their expression is elevated significantly under stress conditions. Accordingly, sHsps are the primary cellular defense mechanism against the adverse effects of stress. sHsps have a relatively small monomer mass ($\sim$20 kDa for mammalian sHsps) but mainly exist as heterogeneous oligomers that are dynamic and undergo rapid monomer exchange. Their NTR and CTR are IDRs. They flank the conserved, well-structured α-crystallin domain (ACD) (Buchner, 2019; Carver et al., 2018; Carver, Grosas, Ecroyd, & Quinlan, 2017; Treweek, Meehan, Ecroyd, & Carver, 2015). The dynamism, flexibility, heterogeneity and hence structural malleability of sHsps are of fundamental importance to their chaperone action.

Molecular chaperones are also found extracellularly where they stabilize proteins, particularly under stress conditions. They play an important role in transporting and degrading aged proteins and are therefore involved in the maintenance of extracellular proteostasis. The first extracellular chaperone to be discovered was clusterin (Humphreys, Carver, Easterbrook-Smith, & Wilson, 1999). Subsequently, other extracellular chaperones have been described (Wyatt, Yerbury, Ecroyd, & Wilson, 2013). As extracellular ATP levels are low, they operate ATP-independently to interact with, stabilize and often complex to, partially folded states of destabilized proteins. The mechanism of clusterin chaperone action bears many similarities to that of sHsps, even though there is no sequence similarity between them (Carver, Rekas, Thorn, & Wilson, 2003). Like sHsps, clusterin exists as a large, heterogeneous species, is very dynamic and IDRs are present, particularly in its NTR and CTR (Carver et al., 2003).

6.2 The molecular chaperone action of caseins against amorphous and amyloid fibrillar aggregation

There are two principal types of misfolded protein aggregates: (i) disordered or amorphous and (ii) well-structured amyloid fibrils (Hall et al., 2016). They each form via an alternative, off-folding aggregation pathway (more below).

Caseins are extracellular proteins. Caseins and their associated forms, including the casein micelle, share much in common with sHsps and clusterin; they are IDPs, they associate to form large, heterogeneous species, particularly when incorporated into the porous casein micelle, and individual casein subunits undergo exchange with the serum and other micelles. On the basis of these common features, it was postulated that caseins could exhibit ATP-independent chaperone activity to prevent unfolding and aggregation of a range of stressed target proteins. Indeed, Bhattacharyya & Das (1999) were the first to show this for bovine α_S-casein (composed of α_{S1}- and α_{S2}-caseins) whereby it inhibited the *in vitro* amorphous aggregation of a diverse group of unrelated proteins under different stress conditions (elevated temperature, reduction and ultraviolet light). The efficiency of the prevention of target protein aggregation, on a molar subunit basis, was significantly less than that of clusterin, and to a lesser degree, sHsps.

Subsequently, Morgan, Treweek, Lindner, Price, and Carver (2005) and Treweek, Thorn, Price, and Carver (2011) showed that each of the four bovine caseins exhibited chaperone action against a range of target proteins, including the whey proteins α-lactalbumin and β-lactoglobulin. As determined by real-time ^{1}H NMR spectroscopy, α_S-casein stabilized reduced apo α-lactalbumin when the latter was in a molten globule state in a comparable manner to that of the sHsp, α-crystallin (Carver et al., 2002), and clusterin (Poon, Treweek, Wilson, Easterbrook-Smith, & Carver, 2002). When elevated temperature and reduction stress were imposed upon β-lactoglobulin and apo α-lacatalbumin respectively in the presence of α_S-casein, a high-molecular weight (HMW) complex of very large mass formed by non-covalent interaction between the two proteins, as occurs with mixtures of amorphously aggregating target proteins and sHsps or clusterin (Carver et al., 2003). Furthermore in the presence of an inert crowding agent (dextran), reduced insulin and α-lactalbumin aggregated amorphously at a faster rate than in its absence, due to excluded volume effects of the dextran causing intermediately folded α-lacatalbumin molecules to associate more readily (Morgan et al., 2005). Under these crowded conditions, α_S-casein was a poorer chaperone compared to the non-crowded environment, as were α-crystallin and αA-crystallin (Ghahghaei, Rekas, Price, & Carver, 2007). Furthermore, conditions that cause rapid aggregation of target proteins lead to ineffective clusterin chaperone action compared to aggregation of the same target proteins under conditions of slow aggregation (Poon et al., 2002). Thus, α_S-casein (like sHsps and clusterin) is a more efficient chaperone when interacting with slowly aggregating target proteins to prevent their aggregation. Fig. 5 provides a schematic of the mechanism of casein chaperone interaction with amorphous and amyloid fibrillar aggregating proteins (Holt et al., 2013). The mechanism is analogous to that

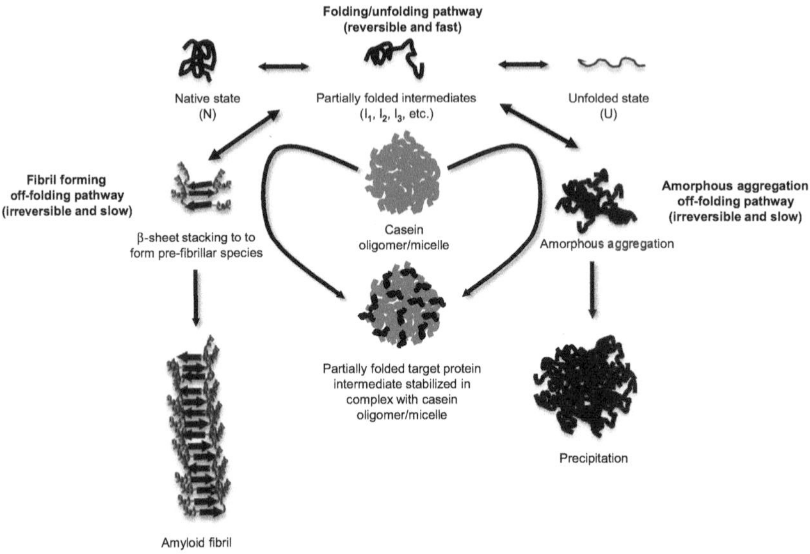

Fig. 5 The protein folding/unfolding and off-folding pathways and the chaperone interaction of caseins to prevent protein aggregation (Holt et al., 2013; Treweek et al., 2011). *Reproduced from Holt et al. (2013).*

for sHsp and clusterin chaperone action (Carver et al., 2003, 2018; Poon et al., 2002). Caseins specifically interact with proteins that are in a destabilized, intermediately folded state that have entered the irreversible, off-folding aggregation pathways. The interaction occurs early along these pathways. For amorphously aggregating target proteins, a HMW complex forms. The casein micelle and individual casein oligomers are dynamic entities. Their monomer exchange may contribute to chaperone interaction with target proteins, as occurs in sHsp chaperone action (Carver et al., 2003).

There have been many other investigations of the chaperone ability of caseins against both amorphous and amyloid fibrillar aggregation. Much of the literature up to 2009 relating to casein chaperone ability against amorphous aggregation, including the potential use of this property in stabilizing food proteins, was reviewed by Yong & Foegeding (2010).

Zhang et al. (2005) showed that α_S- and β-casein were both efficient chaperones at inhibiting the stress-induced, amorphous aggregation of a variety of target proteins. Hassanisadi et al. (2008) confirmed that β-casein interacted with a partially folded intermediate of heated alcohol dehydrogenase (ADH) and formed a HMW complex whilst not maintaining ADH enzyme activity. The inability to retain enzyme activity is also a characteristic of α_S-casein when stabilizing stressed ADH, catalase and glutathione-S-transferase (Treweek et al., 2011). Clusterin and sHsps exhibit the same properties during their stabilization of stressed enzymes (Carver et al., 2003). Hassanisadi et al. (2008) also showed that bovine β-casein is a significantly more efficient chaperone than camel β-casein in preventing the heat-induced aggregation of ADH.

The ability of β-casein to inhibit the temperature-induced, amorphous aggregation of α-lactalbumin and β-lactoglobulin was investigated by Foegeding and co-workers (Kehoe & Foegeding, 2011; Yong & Foegeding, 2010), i.e. conditions that are relevant to pasteurization and ultra-heat treated (UHT) milk. They observed that the chaperone ability of β-casein was pH dependent and decreased in the presence of cations, implying that electrostatic factors are involved in β-casein chaperone action. O'Kennedy and Mounsey (2006) observed similar behavior for an α_S/β-casein mixture and whole casein in preventing the heat-induced precipitation of whey proteins. β-Casein and whole casein are similarly effective at minimizing the precipitation of carbonic anhydrase after rapid dilution from denaturant, and at inhibiting amyloid fibril formation of ovalbumin at 80 °C (Khodarahmi, Beyrami, & Soori, 2008).

Matsudomi, Kanda, Yoshika, and Moriwaki (2004) found that α_S-casein inhibited the aggregation of ovotransferrin at 80 °C. Furthermore, the presence of salt reduced the chaperone ability of α_S-casein as did its dephosphorylation. The latter observation was confirmed for α_S- and β-casein by Koudelka, Hoffmann, and Carver (2009) for a variety of amorphously aggregating target proteins, and for a recombinant but unphosphorylated form of β-casein (Yousefi et al., 2009). On the other hand, engineered dimeric β-casein (linked via its C-terminal region) was more chaperone active implying that the greater overall hydrophilicity of the dimer enhanced β-casein chaperone ability (Yousefi et al., 2009). For α_S and β-casein, removal of the negatively charged phosphate groups via phosphatase action led to slightly more structured caseins, implying that the IDP nature of caseins, and hence their malleable structure, are integral to their chaperone ability (Koudelka, Hoffmann, & Carver, 2009). The role of electrostatic interactions in casein chaperone action may also be factor, for example involving the post-translationally added phosphate groups. Thus, the chaperone action of caseins is most likely a combination of hydrophobic and electrostatic interactions, in contrast to that of sHsps and clusterin in which hydrophobic interactions are considered to be dominant (Carver et al., 2003, 2018). It is a conclusion that may explain why Hsp70, coupled with ATP hydrolysis, is unable to refold and reactivate destabilized enzymes (ADH and catalase) bound to α_S-casein, in contrast to the ability of sHsps and clusterin to do so (Ehrnsperger, Graber, Gaestel, & Buchner, 1997; Poon et al., 2002). Thus, the electrostatic component of binding between the target protein and α_S-casein in the HMW complex, perhaps also with the target protein buried therein, does not enable ready access by Hsp70 and subsequent release upon refolding. For sHsps under reduction stress, this is not the case as the target protein is loosely held in the outer, accessible region of the HMW complex (Regini et al., 2010).

β-Casein inhibits the conversion of amyloid β (the principal component of the extracellular deposits (plaques) in Alzheimer's disease) into amyloid fibrils, possibly via interaction with the hydrophobic, aggregation-prone region of amyloid β (Hojati, Ghahghaei, & Lagzian, 2018). κ-Casein forms amyloid fibrils readily at neutral pH (Thorn et al., 2005) (Section 7). α_S-Casein and its individual components, α_{S1} and α_{S2}-casein, exhibited variable effectiveness at inhibiting κ-casein fibril formation with α_S and α_{S2}-casein being the most effective (Treweek et al., 2011), most likely because α_{S2}-casein itself is amyloidogenic (Thorn, Ecroyd, Sunde, Poon, & Carver, 2008). Dephosphorylation of α_s-casein and β-casein had no effect on their chaperone ability to prevent κ-casein fibril formation, possibly because of the non-standard mechanism of fibril formation of κ-casein (Koudelka et al., 2009). From an investigation of the interaction of α_S-casein with insulin at pH 2, i.e. very non-physiological conditions but those in which insulin forms amyloid fibrils readily, Librizzi, Carrotta, Spigolon, Bulone, and San Biagio (2014) concluded that α_S-casein prevents insulin secondary nucleation and thereby delays its fibril formation. No well-defined chaperone interaction or binding sites have been identified in sHsps. Most likely, there are many such sites, reflecting the mainly unstructured nature of sHsps, which provides a range of chaperone interacting sites. In general, the unstructured NTR interacts with amorphously aggregating target proteins whereas the β-sheet structured, central ACD interacts with fibril-forming target proteins (Treweek et al., 2015). The plurality of chaperone interacting sites in sHsps is consistent with their promiscuous chaperone ability, i.e. to stabilize a range of amorphously and fibrillar aggregating target proteins. Little has been published of similar studies to determine the region(s) involved in casein chaperone action. Koudelka (2010) undertook limited proteolysis studies of the complex formed between reduced α-lactalbumin and β-casein and identified by mass spectrometry that the central region of β-casein (A91 to W143) was protected from cleavage by proteases, implying its involvement in chaperone interaction. As caseins are IDPs, they are most likely to have many sites that facilitate their chaperone interaction with a range of destabilized target proteins.

6.3 Calcium phosphate sequestration competes with molecular chaperone action in milk

Understanding CaP nanocluster sequestration is of fundamental importance in describing the biological and food functional properties of milk. Fig. 6A shows that increasing sequestration of CaP nanoclusters reduces the concentration of α_{S1}-, α_{S2}- and β-caseins that are free to exchange with the milk serum. It is reasonable to suppose that the surface composition of casein micelles and the net charge reflect the composition of the free caseins and determine aspects of casein micelle stability. The surface potential of the micelles also depends on pH and free divalent cation concentration. Fig. 6B shows how the average net charge on bound and free caseins varies with the free concentration of divalent cations from individual milks at their natural pH. The individual cow milk samples on which these results are based show a very wide range of variation in the total concentrations of casein and milk salts but because of the complex equilibria between the two, the properties of casein micelles can only be appreciated by detailed modeling of the partition of salts and caseins as has recently been demonstrated (Bijl et al., 2019).

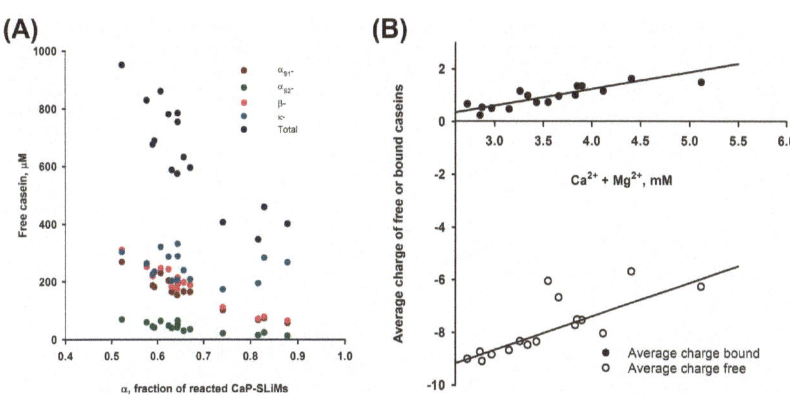

Fig. 6 Effect of CaP nanocluster complex formation on the properties of casein micelles in the milk from 18 individual cows. (A) Calculated concentrations of individual and total caseins plotted as a function of the fraction of CaP-SLiMs that are bound to the CaP nanoclusters. (B) Effect of divalent metal ion binding on the average charge on caseins that either are or are not complexed to CaP nanoclusters. *Reproduced from Bijl et al. (2019).*

In addition to the direct effect of CaP sequestration on the free concentrations of calcium-sensitive caseins and surface properties of casein micelles, there are further consequences for the ability of caseins to act as molecular chaperones in stabilizing whey proteins against denaturation and aggregation.

The CaP sequestration and molecular chaperone roles of caseins are schematically summarized in Fig. 7 (Carver et al., 2018; Holt et al., 2013). Bovine α_{S1}-, α_{S2}- and β-caseins contain CaP-SLiMs that can sequester CaP, whereas κ-casein does not (Fig. 1B). A large proportion of κ-casein and smaller proportions of the other caseins lie on the surface of the casein micelle and regulate micelle size to prevent large-scale amorphous casein aggregation and precipitation (Fig. 7). The chaperone action of α_{S1} and β-caseins prevents amyloid fibril formation of κ- and α_{S2}-caseins (Fig. 7), and the inappropriate aggregation of other milk proteins, for example β-lactoglobulin and α-lactalbumin which are less stress (e.g. heat) stable

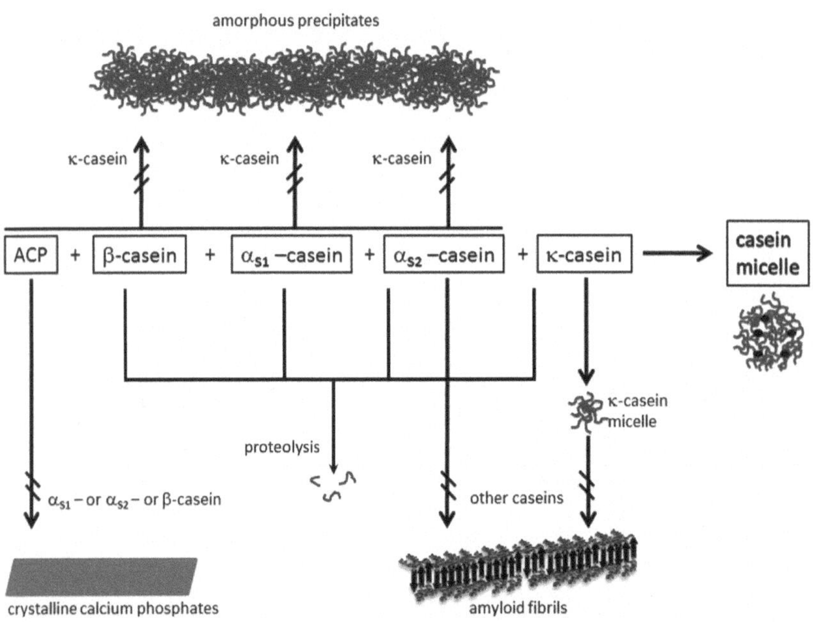

Fig. 7 The central reaction is the functional formation of casein micelles whereas the off-axis reactions lead to dysfunctional outcomes such as pathological calcification, premature proteolysis, uncontrolled amorphous aggregation or amyloid fibril formation (Carver et al., 2018; Holt et al., 2013). Side reactions are inhibited by interaction with the relevant casein as indicated by a double strike-through arrow. ACP is amorphous calcium phosphate. *Reproduced from Carver et al. (2018).*

than the caseins. Within milk, casein chaperone action is probably restricted to the proportion of caseins not involved in CaP sequestration. Recent work with individual cow milk samples (Bijl et al., 2019) has highlighted the profound differences between individual milks in the concentrations of caseins that are free to act as chaperones which may be an important consideration in determining the physico-chemical and processing properties of the milk.

In traditional cheese-making, cleavage of one peptide bond (F105-M106) of bovine κ-casein by the protease chymosin produces the C-terminal macropeptide and the N-terminal para-κ-casein peptide. Deletion of the macropeptide from the casein micelle removes a source of steric stabilization provided by the highly disordered macropeptide, which contains no HO-SLiMs (Fig. 1B) and interacts weakly, if at all, with other caseins. Similar sequences in other IDPs have been called entropic chains because of the contribution of conformational entropy to repulsive interactions with other proteins (van der Lee et al., 2014). Highly flexible IDRs at the N- and/or C-termini of mammalian sHsps and eye lens crystallins perform similar roles, for example in contributing to the solubility of crystallins and thereby maintenance of lens transparency (Carver et al., 2017; 2018).

After removing the entropic chains, the casein micelles can begin to aggregate at a rate determined by the pH and free divalent cation concentration in the milk. Again, these are properties that can be calculated from the model of salt and casein partition in the milk (Bijl et al., 2019).

7. Caseins can form amyloid fibrils

Individual κ- and α_{S2}-caseins form amyloid fibrils under physiological conditions (Farrell, Cooke, Wickham, Piotrowski, & Hoagland, 2003; Léonil et al., 2008; Thorn et al., 2005, 2008). Each is about 10% of total bovine casein and the remaining bovine caseins, β- and α_{S1}- are approximately 40% of the total although considerable variation is found in individuals (Bijl, van Valenberg, Huppertz, & van Hooijdonk, 2013). The formation of amyloid fibrils by κ-casein is readily suppressed by an excess of α_S- or β-casein whereas α_{S2}-casein fibril formation is potently inhibited by α_{S1}-casein but not by β-casein (Thorn et al., 2008; Treweek et al., 2011). The potent *in vitro* inhibition by α_{S1}-casein of amyloid formation by α_{S2}-casein is evidence that these two proteins are preferentially associated in the native casein micelle. In total, therefore, the chaperone action and molar proportions of the individual caseins are sufficient to inhibit fibril formation in the native bovine casein micelle (Thorn et al., 2005). It was concluded, therefore, that an important biological function of α_{S1}- and β-casein within the native bovine casein micelle is to suppress undesirable amyloid fibril formation by κ- and α_{S2}-caseins (Holt & Carver, 2012; Holt et al., 2013; Treweek et al., 2011).

7.1 Amorphous aggregates and amyloid fibrils

Fig. 5 represents schematically the reversible folding/unfolding pathway of a protein between its unfolded (U) and folded, native (N) states via intermediates (I) (Ecroyd & Carver, 2008b). If the intermediates linger for too long during protein folding or are generated due to unfolding from the native state as a result of elevated stress, the protein can associate via their exposed hydrophobic regions and enter the off-folding pathways to form either amorphous (disordered) or amyloid fibrillar (ordered) aggregates. Molecular chaperones (such as caseins and sHsps) interact with these intermediates as they enter the off-folding pathways to stabilize them and suppress their aggregation (Fig. 5). The driving force for amorphous and amyloid fibrillar aggregation is that both are lower in free energy than their natively folded counterparts. Hence, it is generally believed that all proteins can form amyloid fibrils, under the right conditions (Knowles, Vendruscolo, & Dobson, 2014).

Amyloid fibrils, in contrast to the disordered arrangement of amorphous aggregates, are comprised of well-ordered fibrils of nano to micrometer in length of 4–15 nm in diameter. Their polypeptide chains are arranged in a cross β-sheet conformation with the β-stands oriented perpendicular to the fibril axis. The distance between the β-stands is around 4 to 5 Å and that between the β-sheets is 10–11 Å (Fig. 8). The amyloid fibril is a generic structure as it is mainly stabilized by hydrogen bonding involving backbone amide and carbonyl groups, with inter-digitation of sidechains contributing. Of late, atomic-level structures have been determined by cryoelectron microscopy and solid-state NMR spectroscopy for various fibril-forming peptides and proteins, for example α-synuclein associated with the Lewy body deposits in Parkinson's disease (B. Li et al., 2018; Y. Li et al., 2018). The structures are polymorphs with variations in the arrangements of the β-strands. The intermediates from which amyloid fibrils form arise from either the protein partially unfolding from its globular state or, if an IDP, from the protein adopting a partially folded conformation that is aggregation prone. In relation to disease, the intermediately-folded, prefibrillar oligomers are the putative cytotoxic species (Fig. 5).

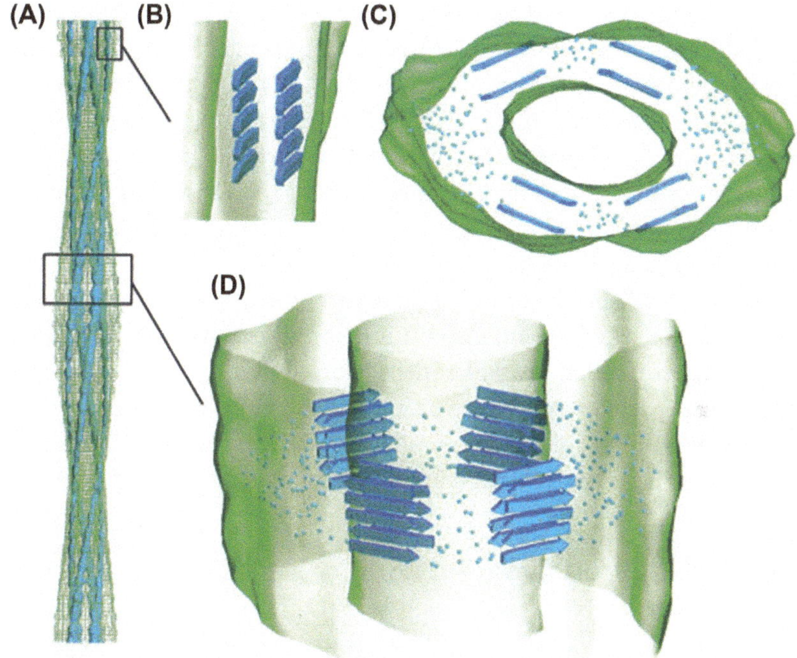

Fig. 8 Model of an amyloid fibril of a SH3 domain, as determined by cryoelectron microscopy and molecular modeling (Jimenez et al., 1999). (A) Four protofibrils wrap around each other to form the fibril. (B) The arrangement of the β-strands and β-sheets within one of the protofibrils. (C) Cross-section showing the central cavity. (D) Overall arrangement of the protofibrils. The disconnected dots indicate unresolved polypeptide regions. *Reproduced from Jimenez et al. (1999).*

7.2 Amyloid fibril formation of κ-casein

IDPs sample a wide range of conformational space which predisposes them to adopt a β-strand conformation, at least momentarily. Potentially, this entity can associate with a similarly structured species to form a relatively stable dimer. Further association of this dimeric template structure with other monomers leads to oligomer formation and eventually an amyloid fibril. Often, IDPs contain amyloidogenic steric zipper sequences (Goldschmidt et al., 2010), which cannot be buried within a protein core. Furthermore, they can adopt the PP-II helix, a conformation whose backbone torsion angles are not far removed from those of the β-strand such that it is the potential intermediate conformation on the pathway to fibril formation (Blanch et al., 2000). Hence, many IDPs form amyloid fibrils under physiological conditions, including those associated with the deposits in diseases such as Alzheimer's (amyloid β) and Parkinson's (α-synuclein).

Since caseins are IDPs and contain significant PP-II helix and steric zipper SLiMs primarily in their P,Q-rich regions (Fig. 1), they are primed to form amyloid fibrils. Indeed, Farrell et al. (2003) were the first to observe that RCM bovine κ-casein readily formed amyloid fibrils under physiological conditions, a property that was also confirmed for the native and reduced forms of κ-casein by Thorn et al. (2005). Both studies highlighted that reduction of the disulphide bonds in κ-casein promoted its fibril formation.

Most amyloid fibril-forming peptides and proteins associate via a nucleation-dependent mechanism. Association of intermediates during the lag phase leads to formation of an oligomeric nucleus which, when of sufficient size, sequesters destabilized monomers to form prefibrillar and then fibrillar species during an exponential elongation phase. Ultimately, this leads to the plateau phase characteristic of the fully formed amyloid fibril (Fig. 9) (Ecroyd

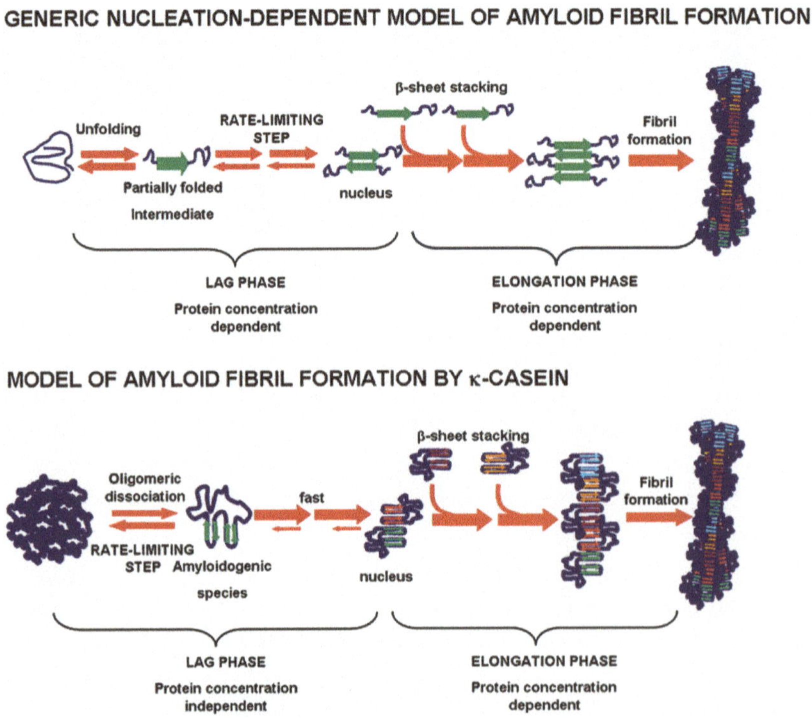

Fig. 9 A comparison of the mechanism of amyloid fibril formation by proteins undergoing standard nucleation-dependent amyloid fibril formation and that for κ-casein (Ecroyd et al., 2008). The rate-limiting step in κ-casein fibril formation is dissociation from the oligomer, which is followed by rapid aggregation of the amyloidogenic monomer. *Reproduced from Ecroyd et al. (2008).*

& Carver, 2008b; Ecroyd et al., 2008). By contrast, RCM κ-casein has a different and unusual mechanism of amyloid fibril formation. It, and reduced κ-casein, exhibit a very short lag phase before aggregating, independent of protein concentration. Furthermore, κ-casein aggregation cannot be seeded by the addition of preformed fibrils (Ecroyd et al., 2008). In standard nucleation-dependent aggregation, the opposite of both of these properties occurs.

On its own, κ-casein (and its RCM form) exist as a dynamic, oligomeric ensemble. Under physiological conditions, RCM κ-casein exhibits a critical micelle concentration (CMC) of ~29 μM (Ecroyd, Thorn, Liu, & Carver, 2010; Vreeman, 1979). Ecroyd et al. (2008, 2010) determined that monomeric κ-casein is the amyloidogenic species and that dissociation from the κ-casein oligomer is the rate-determining step in fibril formation. Thus, κ-casein aggregation is dependent on the presence of the monomer and hence addition of seeds and increased κ-casein concentration do not affect the lag phase. Fig. 9 summarizes schematically the mechanism of κ-casein fibril formation and compares it with the standard nucleation-dependent mechanism. By contrast, Léonil et al. (2008) proposed that the oligomeric (micelle-like) state of κ-casein is the amyloidogenic precursor. Consistent with the monomer being an amyloidogenic-prone species, Jain et al. (2011) determined that the RCM κ-casein monomer (i.e. at a concentration below its CMC) adopted a collapsed, molten globule-like conformation, one that would predispose it to aggregation and amyloid fibril formation.

Ecroyd et al. (2008) determined by limited proteolysis of RCM κ-casein fibrils and subsequent sequencing of the peptide fragments by mass spectrometry that Y25-K86 encompassed the κ-casein amyloid fibril core. The fibril core corresponds well with that predicted by a variety of amyloid fibril-predicting algorithms (Ecroyd et al., 2008) and with the presence of potential fibril-forming, steric zipper SLiMs in κ-casein (Fig. 1B) (Holt & Carver, 2012). The κ-casein fibril core also contains eight proline and seven glutamine residues (i.e. 24% of the core). It is part of the P,Q-rich sequence within para-κ-casein that has a propensity to adopt PP-II secondary structure and is predisposed to fibril formation (Holt et al., 2013, 2019). The κ-casein fibril core is in the middle of its primary sequence. The fibril cores of many other fibril-forming peptides and proteins are similarly located; it is hypothesized that the unstructured and flexible flanking regions inhibit the sequence-specific interactions of the core region and thereby minimize the possibility of association and amyloid fibril formation (Carver, Grosas, et al., 2017).

7.3 Amyloid fibril formation by α_{S2}-casein

α_{S2}-Casein does not form amyloid fibrils as rapidly as κ-casein, although like κ-casein, it has a very short lag phase in its aggregation profile. Unlike κ-casein, α_{S2}-casein aggregation is inhibited to a significant extent upon reduction of monomeric intra- or dimeric inter-molecular disulphide bonds involving C30 and C34 (Rasmussen et al., 1999). The morphology of α_{S2}-casein amyloid fibrils is very different from that of κ-casein fibrils; α_{S2}-casein fibrils are twisted and ribbon-like with occasional formation of rings, whereas κ-casein forms straight fibrils (Thorn et al., 2008; 2014). Increased temperature promotes α_{S2}- and κ-casein fibril formation and does not affect α_{S2}-casein fibril morphology significantly (Thorn, Ecroyd, & Carver, 2014).

7.4 Amyloid fibril formation by α_{S1}- and β-caseins under non-physiological conditions

Whereas the two cysteine-containing bovine caseins (κ and α_{S2}) form amyloid fibrils under physiological conditions, the α_{S1}- and β-caseins do not. The presence in the latter two caseins of steric zipper SLiMs (Fig. 1B) (Holt & Carver, 2012; Holt et al., 2019), their IDP nature and PP-II helical formation implies that they may form amyloid fibrils under other conditions, for example low pH, shear and/or elevated temperature. Indeed, Pan & Zhong (2015) made amyloid fibrils from β-casein, but not α_S-casein, solutions by incubation at pH 2.0 and 90 °C for over 30 h.

7.5 Cytotoxicity of casein amyloid

Inhibition of κ- and α_{S2}-casein amyloid fibril formation by α_{S1}- and β-casein also prevents the cellular toxicity of the former two proteins. Prefibrillar and fibrillar RCM κ-casein are more cytotoxic to model rat neuronal pheochromocytoma (PC12) cells than native κ-casein (Dehle, Ecroyd, Musgrave, & Carver, 2010; Hudson, Ecroyd, Dehle, Musgrave, & Carver, 2009; Koudelka, Dehle, Musgrave, Hoffmann, & Carver, 2012). Co-incubation with α_S- or β-casein prevented this toxicity (Dehle et al., 2010). Similarly, amyloid fibrils of α_{S2}-casein are toxic to PC12 cells and to epithelial colon carcinoma cells (Elias, Scanlon, Musgrave, & Carver, 2014). The cytotoxicity of casein fibrils has disease relevance. The mammary glands of lactating cows can develop proteinaceous, calcified deposits known as *Corpora amylacea* (CA) that contain amyloid fibrils and CaP. CA formation can block the mammary gland and affect milk production. Peptide fragments of α_{S2}-casein have been detected in the amyloid deposits of bovine CA (Niewold, Murphy, Hulskamp-Koch, Tooten, & Gruys, 1999). They arise from 32, 33 and 45 amino acid-long α_{S2}-casein fragments commencing at A81. The longest fragment (A81-R125) corresponds almost exactly to one of the P,Q-rich regions of α_{S2}-casein (Holt et al., 2019) and contains the most fibril-prone, predicted steric zipper SLiM in all the bovine caseins, E99-Y104 (Holt & Carver, 2012).

7.6 Effects of oxidation, milk constituents and macromolecular crowding on κ-casein fibril formation

As caseins are extracellular proteins they exist in an oxidative environment. In milk, they interact with a range of other molecules, large and small, for example salts, lactose, lipids and other proteins. The casein micelle is a crowded environment. All these factors may affect casein structure and function. Various experimental studies have explored these aspects.

Oxidation has been linked to diseases associated with protein misfolding and aggregation. Methionine oxidation alters amyloid fibril formation for a variety of peptides and proteins. Accordingly, the effect on κ-casein amyloid fibril formation of selective oxidation (to form the sulfoxide) of its two methionine residues at positions 95 and 106, i.e. outside its fibril core region, was investigated (Koudelka et al., 2012). The structure of oxidized κ-casein was very similar to that of the native protein, but it had an enhanced rate of fibril formation and exhibited greater toxicity toward PC12 cells. Also, it was a slightly better chaperone to prevent the amorphous aggregation of target proteins but the chaperone ability of β-casein to prevent oxidized κ-casein fibril formation was reduced relative to unmodified β-casein (Koudelka et al., 2012).

Phosphatidylcholine lipids (as found in milk fat globules), in contrast to shorter phospholipids and triglycerides, promoted RCM κ-casein amyloid fibril formation (J. Liu et al., 2016). Heparan sulfate, which is a component of the milk fat globule membrane and is associated with the *in vivo* fibril formation of peptides and proteins, had little effect on RCM κ-casein fibril formation. Likewise, the presence of calcium ions and lactose, the sugar in milk, had minimal influence on RCM κ-casein aggregation. Inert crowding agents did influence RCM κ-casein fibril formation but in an opposite manner to other fibril-forming peptides and proteins that aggregate via the 'standard' nucleation-dependent mechanism (Fig. 9). The presence of a crowding agent (polyethylene glycol) reduced the rate and extent of RCM κ-casein aggregation, consistent with stabilization of the κ-casein oligomer, i.e. promotion of interactions between the RCM κ-casein monomers. Since the rate-limiting step in κ-casein fibril formation is dissociation from the oligomer (Fig. 9), stabilization of the oligomer reduces monomer concentration and hence fewer fibrils are formed, an observation that is consistent with stabilization being inherent in co-association of caseins within the micelle to prevent amyloidosis.

7.7 Applications and consequences of casein amyloid fibril formation and chaperone action

The chaperone ability and the highly reproducible amyloid fibril-forming propensity of κ- and α_{S2}-caseins provide opportunities for their application in a diversity of ways.

1. Thorn et al. (2014) discussed the potential use of κ-casein fibrils as bionanomaterials, for example via functionalization to incorporate small and large (e.g. enzyme) molecules, nanowires and as composite materials. The attraction of κ-casein (and the other casein) amyloid fibrils as bionanomaterials arises from their ready availability, biocompatibility, well-ordered self-assembly at the nanometer level, rigidity, strength, stability and resistance to proteolysis. Indeed, fibrils generated from β- and κ-casein under acid conditions and high temperature have relatively high stiffness with Young's moduli of ~2 and 4 GPa respectively (Pan & Zhong, 2015).
2. Ultra-heat treatment enables long-term storage of milk without refrigeration. However, gel formation of UHT milk also often occurs which has been proposed to arise from association of κ-casein and the major whey protein β-lactoglobulin, for example via inter-molecular disulphide bond formation (Holt et al., 2013). *In vitro* incubation of both proteins at elevated temperature gave rise to amyloid fibrils containing both proteins co-associated (Raynes, Day, Crepin, Horrocks, & Carver, 2017), a phenomenon that has been observed rarely between two structurally unrelated proteins. Gel formation in UHT milk may therefore arise from co-amyloid fibril formation between κ-casein and β-lactoglobulin.
3. The high reproducibility of RCM κ-casein amyloid fibril formation (Ecroyd et al., 2008, 2010), and its ready availability, mean that RCM κ-casein has potential use for high-throughput, large-scale screening of small molecule fibril inhibitors such as polyphenols (Carver et al., 2010). In a similar vein, RCM κ-casein has been used as a model fibril-forming protein to assess the amyloid fibril-inhibitory capabilities of a variety of small molecule natural products, such as

(-)-epigallocatechin-3-gallate (EGCG) (Hudson, Ecroyd, Dehle, et al., 2009), gallic acid (Y. Liu et al., 2013) and hemin (Y. Liu et al., 2014), along with monitoring the effect of differently charged amino acids on fibril formation (Ecroyd & Carver, 2008a). RCM κ-casein has also been used successfully to assess the efficacy of the novel fluorescent dye TPE-TPP to monitor the formation of prefibrillar oligomers during amyloid fibril formation (Kumar, Hong, Thorn, Ecroyd, & Carver, 2017). In doing so, TPE-TPP exhibits superior characteristics compared to the routinely and widely used dye, thioflavin T. The reproducible nature of RCM κ-casein aggregation was also exploited to determine that the absorptive and fluorescent properties of fibril-inhibiting polyphenols such as curcumin can interfere with the thioflavin T fluorescence and therefore bias the assay (Hudson, Ecroyd, Kee, & Carver, 2009).

4. Each molecule of α_{S1}-casein binds two molecules of curcumin (Sneharani, Singh, & Rao, 2009), a natural product from the plant, *Curcuma longa*, and the principal component of turmeric. Curcumin has antioxidant and anti-inflammatory properties and hence has potential use in the treatment of diseases such as cancer, arthritis and Alzheimer's. Curcumin is relatively unstable and insoluble in water. Its ready binding to α_{S1}-casein provides a means for enhancing curcumin's solubility, availability and therefore utilization, for example as a food additive or therapeutic.
5. In an intriguing approach, Chun et al. (2012) proposed the use of κ-casein amyloid fibrils as a drug-delivery agent via formation of a hydrogel to incorporate small molecules, in this case retinoic acid which was released in the presence of salt to promote neuronal cell growth.

8. Caseins can form dynamic, fuzzy complexes

Formation of disordered (fuzzy) complexes through multivalent and promiscuous interactions is characteristic of many IDPs (Fuxreiter & Tompa, 2012). Extreme fuzzy complexes are ones where all the interacting IDPs retain the high dynamic disorder of the monomeric proteins to a large degree. At the other extreme are polymorphic complexes where the disorder is more static. Homotypic oligomers, called micelles, are formed by β- and κ-casein under physiological conditions above their CMC. Casein micelles contain additional conformational constraints arising from the sequestration of CaP nanoclusters, disulphide bridges and the binding of divalent cations. Comprehensive reviews are available (Dalgleish & Corredig, 2012;

Huppertz, Fox, & Kelly, 2018; Rollema, 1992). Only selected evidence is reviewed here to show that caseins form fuzzy complexes in which the dynamic disorder and high specific volume of the monomers are retained.

8.1 Endothermic conformational transition and self-association of β-casein

Above its CMC, an increasing proportion of β-casein forms polydisperse oligomers (Cragnell et al., 2017) as the temperature is raised, and the mass of the oligomers also increases. At 20 °C the micelle has a radius of gyration (R_g) of 7.3 nm (compared to a R_g value of 9.4 nm assuming a homogeneous sphere) and contains 14 monomers, corresponding to a specific volume of about 8.8 mLg^{-1}, only slightly smaller than that of the monomeric protein at 4 °C. However, small-angle neutron scattering data show that the particle is not a homogeneous sphere but can be modeled by a denser core and very diffuse shell of worm-like chains in a good solvent (Leclerc & Calmettes, 1997). The core size and density increased with temperature but remained more highly solvated than the core of a globular protein. ^{1}H NMR experiments showed that about half of residues H50 to V209 were mobile at 20 °C but the mobile fraction increased with temperature (de Kruif et al., 2002) even though the oligomer size increased. ROA measurements on the oligomer revealed spectra dominated by the PP-II regular secondary structure (Fig. 4A) as was also apparent in the CD spectra (Fig. 4B). As expected, PP-II structure decreased with increasing temperature. Differential scanning calorimetry shows a broad endothermic transition with a maximum at about 15 °C, only slightly dependent on concentration (de Kruif & Grinberg, 2002). Isothermal calorimetry of the enthalpy of dilution of oligomers at pH 7.0 has been reported over the temperature range 16–45 °C. The enthalpy passed through a minimum at about 28 °C. At lower temperatures, association was by a stepwise mechanism but became more co-operative at the higher temperatures.

The removal of as few as two residues from the C-terminus of β-casein reduces the small proportion of regular secondary structure and postpones micelle formation to higher temperatures (Qi et al., 2005) (Fig. 4B). In the CD spectrum of β-casein, a sharp but minor transition was observed at 10 °C, in which the position of the ellipticity minimum increased by 1 nm (Fig. 10) (Qi et al., 2004). It has been proposed (Qi et al., 2004, 2005) that this transition is to a more ordered (local) conformation that favors the formation of oligomers at about 15 °C (folding prior to binding rather than folding on binding). This proposal has merit in identifying a

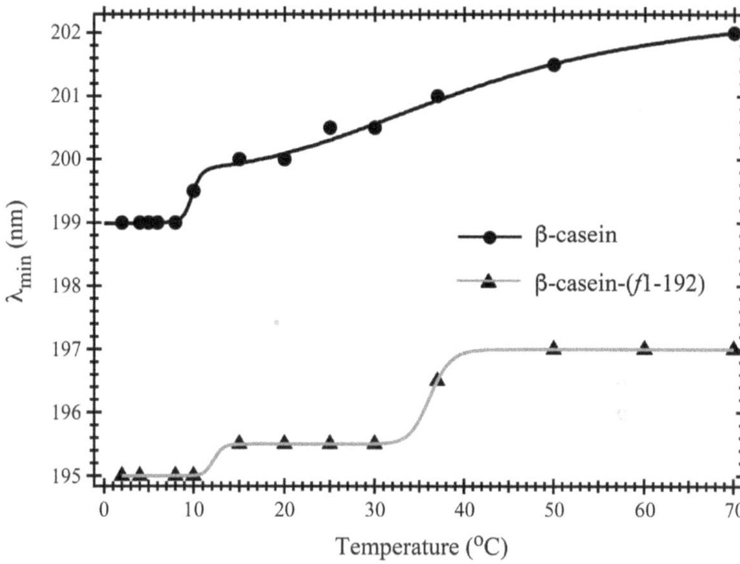

Fig. 10 Effect of temperature (2–70 °C) on the wavelength of the ellipticity minimum in the CD spectrum of whole β-casein (1-209) and β-casein (1-192). *Redrawn by Dr P.X. Qi from the original published versions in Qi et al. (2004) and Qi et al. (2005).*

unifying factor in a range of temperature-dependent phenomena in which β-casein dissociates on cooling from homo and heterotypic casein particles.

8.2 Athermal self-association of reduced and carboxymethylated κ-casein

Disulfide-linked oligomers of κ-casein are found in milk (Groves, Wickham, & Farrell, 1998; Rasmussen et al., 1999) but most studies have been made on RCM κ-casein which associates to form a spherical 30-mer. The diameter of 23 nm (Vreeman, 1979; Vreeman, Brinkhuis, & Vanderspek, 1981) gives an average specific volume of 6.4 mLg^{-1}. The oligomer has been modeled as a core-shell structure with a denser and less mobile core and a mobile diffuse shell (de Kruif et al., 2002; Rollema, Brinkhuis, & Vreeman, 1988; Vreeman et al., 1981) of thickness 4.5 nm and core specific volume of 2.3 mLg^{-1}. Although the core is denser than the β-casein micelle, differential scanning calorimetry demonstrated only a smooth increase of specific heat with temperature, consistent with an amorphous structure (de Kruif et al., 2002). Simulation of the ^{1}H NMR spectrum showed that it can be fitted by a broader component and a sharper component corresponding in composition to the mobile CTR (macropeptide) of κ-casein.

ROA and CD measurements on the oligomer generated spectra dominated by PP-II secondary structure (Fig. 4A) (Raap et al., 1983; Tomatoda Ono et al., 1987).

8.3 Limited self-association of α_{S1}-casein

Self-association of α_{S1}-casein is stepwise producing a series of oligomers (dimers, tetramers, hexamers etc.) depending strongly on ionic strength and pH but hardly at all on temperature (reviewed by Huppertz (2013); Swaisgood (2003)). A CD and ^{1}H NMR study of the peptide I136 to M196 (Alaimo, Farrell, & Germann, 1999), which self-associates like the whole protein, provided evidence of the involvement of a Tyr-rich region near P162 in the formation of dimers and of W164 in the formation of higher oligomers. W164 occurs in the well-conserved HO-SLIM 163-AWYYV-167 and another well-conserved HO-SLiM 142-LAYFY-146 (Holt et al., 2019) is tyrosine rich (Fig. 1B). The 1D NMR spectra were broadened by association and the 2D spectra contained few cross peaks but spectral features sharpened at elevated temperatures consistent with an increase in residue mobility.

8.4 Limited self-association of α_{S2}-casein

The concentration-dependent self-association of reduced α_{S2}-casein at pH 6.7 was studied as a function of ionic strength (Snoeren, van Markwijk, & van Montfort, 1980). The specific volume of the oligomers, assuming a spherical shape, was almost constant at 4.5−5.5 mL g^{-1}, comparable to other casein oligomers and the casein micelle. Oligomer size did not level-off with increasing protein concentration up to the limit studied of 1 g L^{-1} and the free energy of successive association steps was roughly constant at -38 kJ mol^{-1} and 20 °C. The dependence of oligomer aggregation number on ionic strength was unusual in showing a maximum at about 0.2. Thus, they calculated an aggregation number of about 5 at a protein concentration of 1 g L^{-1} in both 0.02 and 0.6 M NaCl and in 0.2 M NaCl it was about twice as large.

9. Casein micelles are non-stoichiometric, fuzzy complexes

Although casein micelles are found in all milks, their composition, size and structure are not constant. The composition of casein micelles varies greatly among species and to a lesser extent among individuals of the same species (Bijl et al., 2013; Martin, Cebo, & Miranda, 2013;

Oftedal, 2012) and in a given milk with the size of the micelle (Dalgleish, Horne, & Law, 1989; Donnelly, McNeill, Buchheim, & McGann, 1984). In human casein micelles there are three casein gene products, and in the African elephant micelles only two (M. Madende, Kemp, Stoychev, & Osthoff, 2018; Madende & Osthoff, 2019; M. Madende et al., 2015); in both species a β-casein ortholog predominates. In horse casein micelles, the predominant casein is also a β-casein ortholog but the casein fraction is remarkable for the very low or undetectable proportion of κ-casein (Ochirkhuyag, Chobert, Dalgalarrondo, & Haertle, 2000; Potocnik, Gantner, Kuterovac, & Cividini, 2011; Uniacke-Lowe, Huppertz, & Fox, 2010). It had been thought that κ-casein was the one essential protein needed for forming stable micelles, but this may not be so. The function of κ-casein in the horse casein micelle may be supplemented or even completely replaced by β-caseins with a low level of phosphorylation (Ochirkhuyag et al., 2000). In rabbit casein micelles, there are five casein gene products, the main one being an ortholog of bovine α_{S1}-casein (Martin et al., 2013). In bovine casein micelles, the four caseins are present in the approximate average molar proportions of 15.6% κ-, 37.0% α_{S1}-, 37.2% β- and 10.3% α_{S2}-casein, after allowing for partial plasmin digestion of β-casein (Bijl et al., 2019; Davies & Law, 1980).

The specific volume of bovine casein micelles, calculated from the intrinsic viscosity (de Kruif & Holt, 2003; Jeurnink & De kruif, 1993), is about 4.2 mLg^{-1} which lies between values for the β-casein micelle and the core of the κ-casein micelle. Three ^{1}H NMR studies (Griffin & Roberts, 1985; Rollema, 1992; Rollema & Brinkhuis, 1989; Rollema et al., 1988) show a broad spectrum with sharper resonances possibly arising from the mobile CTR macropeptide of κ-casein. The peak areas of the sharper resonances could be explained if half the κ-casein macropeptide content of the micelles is mobile at room temperature. The surface composition of casein micelles, however, is more diverse (Dalgleish et al., 1989; Donnelly et al., 1984) suggesting that there are non-κ-casein peptides contributing to the mobile NMR component. An overall rise in the intensity of the sharper resonances, including aromatic ring protons, is consistent with a general and reversible increase in mobility with temperature, particularly above 60 °C. The native casein micelle is therefore like the κ- and β-casein micelles in being a mixture of static and dynamically disordered sequences in a highly hydrated state. In none is there evidence of protein condensation and desolvation to form structures like the core of globular proteins.

10. Functional and dysfunctional interactions of caseins in the casein micelle

The elevated calcium, phosphate and casein concentrations in milk and the evident risk of cytotoxic amyloidosis and pathological calcification endanger lactation, neonatal nutrition and, crucially, the future reproductive success of the mother (Holt & Carver, 2012). The avoidance of dysfunctional processes, we suggest, imposes stricter constraints on the composition and structure of caseins than the provision of adequate amino acid nutrition for the neonate, important as this factor is. The constraints are (i) a flexible casein structure that can sequester amorphous calcium phosphate to form thermodynamically stable core-shell structures or substructures, (ii) a casein composition that allows amorphous aggregation to occur in preference to toxic amyloid fibrils and (iii) a fuzzy casein micelle complex that can accommodate compositional variation to give stable colloidal particles and a free-flowing fluid rather than a gel or protein precipitate. These ideas are summarized in Figs. 7 and 11.

10.1 Structural requirements for calcium phosphate sequestration

The peptides forming the CaP binding motif are tightly packed in the sequestering shell around CaP nanoclusters to maximize the free energy

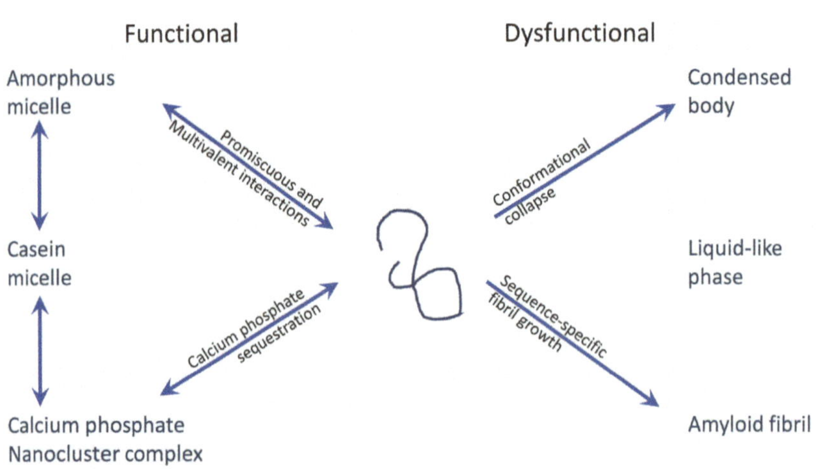

Fig. 11 Functional folding and aggregation of caseins leading to stable and reversible structures or dysfunctional processes generating unstable structures, condensed particles, toxic amyloid fibrils or phase separation.

of sequestration per unit of core surface area. An unfolded conformation is therefore an advantage in forming a thermodynamically stable complex. The affinity with which the phosphorylated sequence binds to the surface of the CaP depends on the sequence and structure of the CaP-SLiM. A sequence containing three consecutive phosphorylated seryl residues generates a single high affinity binding site and two sites of normal affinity for calcium ions (Bijl et al., 2019). The association constant of the high affinity site was an order of magnitude greater than that of an isolated phosphorylated residue and this suggests that the clustered arrangement of phosphorylated residues in the CaP-SLiMs of caseins and other phosphoproteins involved in the control of biocalcification is an important requirement for CaP sequestration.

10.2 Amorphous aggregation versus amyloid fibril formation

A recognized mechanism for slowing or eliminating the specific interactions leading to amyloid structures is through competing promiscuous interactions leading to amorphous structures (Hall, Kardos, Edskes, Carver, & Goto, 2015; Lee, 2009), as occurs for α-synuclein interacting with the sHsp, αB-crystallin (Rekas et al., 2004) and for the prevention of RCM κ-casein fibril formation by the polyphenol EGCG (Hudson, Ecroyd, Dehle, et al., 2009). We previously discussed this idea as a way of explaining why casein micelles are formed from a mixture of caseins such that their action as molecular chaperones counters their ability to form amyloid fibrils (Holt & Carver, 2012). In general, IDPs interact with other proteins mainly through charged and hydrophilic residues with a supporting contribution from the hydrophobic effect (Fitzpatrick, Knowles, Waudby, Vendruscolo, & Dobson, 2011; van der Lee et al., 2014; Wong, Na, & Gsponer, 2013). The largest contribution to binding free energy, according to the calculations of Wong et al. (2013), comes from the electrostatic attraction of complementary charges at the binding interface. Accordingly, we suggest that similar principles are involved in the formation of fuzzy complexes by caseins. For example, if caseins can form a range of different and transient hydrophobic clusters involving the HO-SLiMs, there will be complementary charged residue interactions from flanking sequences (Holt et al., 2019).

Charge segregation is important in determining preferred conformations in IDPs (Beveridge et al., 2019; Holehouse & Pappu, 2018) and in the caseins there is, according to Qi (2007), at least one prominent cluster of basic residues in each bovine casein. We have identified the sequences: 1–RPKHPIKHQ-10 and 100-RLKKYK-105 in α_{S1}-; 165-KKISQR-170 and 181-KTVYQHQKAMK-191 in α_{S2}-; 97–KVK-99 and 105-KHKEMPFPK-113 in β- and 97-RHPHPH-102 and 111-KKNQDK-116 in κ-casein. They are candidates for attractive electrostatic interactions among caseins and fuzzy casein complexes with the highly negatively charged CaP-SLiMs that are not involved in CaP sequestration. Moreover, between the two basic clusters in α_{S2}-casein is the well-conserved HO-SLiM 174-FAL-176, between the two basic clusters in β-casein is the HO-SLiM 101-AMA-103 and between the two basic clusters in κ-casein is the well-conserved chymosin cleavage site and HO-SLiM 105-FMAI-109 (Fig. 1B).

Suppression of amyloid structures may not be total. Evidence has been reported of proto-amyloid structures in casein micelles (Lencki, 2007). They involve interactions via hydrogen bonding of β-strands to form β-sheets and the inter-digitation of side chains and desolvation to stack the β-sheets into the cross-β-sheet structure at the core of amyloid fibrils.

Beside the electron dense shell around the CaP nanocluster and the amyloid protofibrils of the native casein micelle, there may be other, less dense and possibly transient structures such as hydrophobic clusters and elements of regular secondary structure such as turns or β-sheet that play some part in determining the nature of casein intra- or inter-molecular interactions. One example is the proposed folding at the C-terminus of β-casein which predisposes the protein to form oligomers (Qi et al., 2005).

10.3 Control of micelle size and milk fluidity

Traditionally, κ-casein has been assigned the role of acting as a schutzcolloid or steric stabilizer of the hydrophobic core of casein micelles (Holt & Horne, 1996; Walstra, Bloomfield, Wei, & Jenness, 1981; Waugh & Talbot, 1971). A protein science description of the way casein micelle size is determined has not yet been reported but, in principle, all caseins that are not bound to the CaP nanoclusters (Bijl et al., 2019) can act as molecular chaperones to limit casein aggregation by binding to the micelle surface (Morgan et al., 2005). Thus, the surface composition should reflect the composition of free caseins rather than κ-casein alone which would be in better agreement with the measured composition of the micelle surface (Dalgleish et al., 1989; Donnelly et al., 1984).

11. Concluding remarks

We have summarized in this chapter our views on the relationship between the structure of caseins and their biological functions; the latter include the prevention of dysfunctional processes of amyloidosis and calcification that potentially affect the mammary gland, and the promotion of functional processes that generate a stable, high calcium and high protein food for the neonate. In doing so, we have made use of the extensive new knowledge and understanding of the structures and biological functions of IDPs that supplement and, to a degree, supplant the colloid science views of caseins and the casein micelle that predominated in the last century. Caseins are undoubtedly specific examples of the broad class of IDPs, exhibiting typical properties such as a flexible and expanded range of conformations with a predominance of PP-II secondary structure. They can form fuzzy complexes in which the hydration and dynamic properties of the monomer are largely retained, and like many other IDPs, they can form amyloid fibrils and act as molecular chaperones. Less typical of IDPs is their ability to sequester and stabilize nanoclusters of CaP though this may be a property shared with a number of other IDPs that function to control aspects of biocalcification. One particular insight that has come from the comparison with other IDPs is the first rationale for the composition and structure of the native casein micelle as a means of limiting the tendency of caseins to form amyloid fibrils.

All of the implications and ramifications of comparing caseins to IDPs are not fully explored and will require the continued application of the concepts and methods developed for characterizing other IDPs. The enhanced understanding of caseins and the casein micelle that results will inevitably translate into improvements in dairy science and technology.

Acknowledgments

We thank Dr Phoebe Qi for redrawing Figs. 4B and 10 and thank her and Dr Thom Huppertz for reviewing and commenting on the whole chapter. JC's research was supported by a grant (#1068087) from the National Health and Medical Research Council of Australia.

References

Adzhubei, A. A., Sternberg, M. J. E., & Makarov, A. A. (2013). Polyproline-ii helix in proteins: Structure and function. *Journal of Molecular Biology, 425*(12), 2100–2132. http://dx.doi.org/10.1016/j.jmb.2013.03.018.

Alaimo, M. H., Farrell, H. M., & Germann, M. W. (1999). Conformational analysis of the hydrophobic peptide alpha(s1)-casein(136-196). *Biochimica et Biophysica Acta (BBA) -*

Protein Structure and Molecular Enzymology, 1431(2), 410–420. http://dx.doi.org/10.1016/s0167-4838(99)00061-8.

Aoki, T., Sakamoto, H., & Kako, Y. (1991). Cross-linking of caseins by colloidal calcium phosphate in the presence of urea. *International Dairy Journal, 1*(1), 67–75. http://dx.doi.org/10.1016/0958-6946(91)90028-7.

Aoki, T., Umeda, T., & Kako, Y. (1992). The least number of phosphate groups for cross-linking of casein by colloidal calcium phosphate. *Journal of Dairy Science, 75*(4), 971–975.

Aoki, T., Yamada, N., & Kako, Y. (1990). Relation between colloidal calcium phosphate cross-linkage and release of beta-casein from bovine casein micelles on cooling. *Agricultural & Biological Chemistry, 54*(9), 2287–2292. http://dx.doi.org/10.1080/00021369.1990.10870320.

Aoki, T., Yamada, N., Tomita, I., Kako, Y., & Imamura, T. (1987). Caseins are cross-linked through their ester phosphate groups by colloidal calcium-phosphate. *Biochimica et Biophysica Acta, 911*(2), 238–243. http://dx.doi.org/10.1016/0167-4838(87)90013-6.

Aschi, A., Gharbi, A., Daoud, M., Douillard, R., & Calmettes, P. (2007). Study of chemically unfolded beta-casein by means of small-angle neutron scattering. *Physica B: Condensed Matter, 387*(1–2), 179–183. http://dx.doi.org/10.1016/j.physb.2006.04.010.

Atkins, J. D., Boateng, S. Y., Sorensen, T., & McGuffin, L. J. (2015). Disorder prediction methods, their applicability to different protein targets and their usefulness for guiding experimental studies. *International Journal of Molecular Sciences, 16*(8), 19040–19054. http://dx.doi.org/10.3390/ijms160819040.

Baldwin, R. L. (2002). A new perspective on unfolded proteins. *Unfolded Proteins, 62*, 361–367.

Barron, L. D., Blanch, E. W., & Hecht, L. (2002). Unfolded proteins studied by Raman optical activity. *Unfolded Proteins, 62*, 51–90.

Beveridge, R., Migas, L. G., Das, R. K., Pappu, R. V., Kriwacki, R. W., & Barran, P. E. (2019). Ion mobility mass spectrometry uncovers the impact of the patterning of oppositely charged residues on the conformational distributions of intrinsically disordered proteins. *Journal of the American Chemical Society, 141*(12), 4908–4918. http://dx.doi.org/10.1021/jacs.8b13483.

Bhattacharyya, J., & Das, K. P. (1999). Molecular chaperone-like properties of an unfolded protein, alpha(s)-casein. *Journal of Biological Chemistry, 274*(22), 15505–15509.

Bijl, E., Huppertz, T., van Valenberg, H., & Holt, C. (2019). A quantitative model of the bovine casein micelle: Ion equilibria and calcium phosphate sequestration by individual caseins in bovine milk. *European Biophysics Journal, 48*(1), 45–59. http://dx.doi.org/10.1007/s00249-018-1330-2.

Bijl, E., van Valenberg, H. J. F., Huppertz, T., & van Hooijdonk, A. C. M. (2013). Protein, casein, and micellar salts in milk: Current content and historical perspectives. *Journal of Dairy Science, 96*(9), 5455–5464. http://dx.doi.org/10.3168/jds.2012-6497.

Blanch, E. W., Morozova-Roche, L. A., Cochran, D. A. E., Doig, A. J., Hecht, L., & Barron, L. D. (2000). Is polyproline II helix the killer conformation? A Raman optical activity study of the amyloidogenic prefibrillar intermediate of human lysozyme. *Journal of Molecular Biology, 301*(2), 553–563.

Buchner, J. (2019). Molecular chaperones and protein quality control: An introduction to the JBC reviews thematic series. *Journal of Biological Chemistry, 294*(6), 2074–2075. http://dx.doi.org/10.1074/jbc.REV118.006739.

Byler, D. M., Farrell, H. M., & Susi, H. (1988). Raman-spectroscopic study of casein structure. *Journal of Dairy Science, 71*(10), 2622–2629.

Carver, J. A., Duggan, P. J., Ecroyd, H., Liu, Y., Meyer, A. G., & Tranberg, C. E. (2010). Carboxymethylated-kappa-casein: A convenient tool for the identification of polyphenolic inhibitors of amyloid fibril formation. *Bioorganic & Medicinal Chemistry, 18*(1), 222–228. http://dx.doi.org/10.1016/j.bmc.2009.10.063.

Carver, J. A., Ecroyd, H., Truscott, R. J. W., Thorn, D. C., & Holt, C. (2018). Proteostasis and the regulation of intra- and extracellular protein aggregation by ATP-independent molecular chaperones: Lens α-crystallins and milk caseins. *Accounts of Chemical Research, 51*(3), 745–752. http://dx.doi.org/10.1021/acs.accounts.7b00250.

Carver, J. A., Grosas, A. B., Ecroyd, H., & Quinlan, R. A. (2017). The functional roles of the unstructured N- and C-terminal regions in alpha B-crystallin and other mammalian small heat-shock proteins. *Cell Stress & Chaperones, 22*(4), 627–638. http://dx.doi.org/10.1007/s12192-017-0789-6.

Carver, J. A., Lindner, R. A., Lyon, C., Canet, D., Hernandez, H., Dobson, C. M., et al. (2002). The interaction of the molecular chaperone alpha-crystallin with unfolding alpha-lactalbumin: A structural and kinetic spectroscopic study. *Journal of Molecular Biology, 318*(3), 815–827. http://dx.doi.org/10.1016/s0022-2836(02)00144-4.

Carver, J. A., Rekas, A., Thorn, D. C., & Wilson, M. R. (2003). Small heat-shock proteins and clusterin: Intra- and extracellular molecular chaperones with a common mechanism of action and function? *IUBMB Life, 55*(12), 661–668. http://dx.doi.org/10.1080/15216540310001640498.

Carver, J. A., Thorn, D. C., Ecroyd, H., & Holt, C. (2017). Letter to the editor: A response to Horne and Lucey (2017). *Journal of Dairy Science, 100*(7), 5121–5124. http://dx.doi.org/10.3168/jds.2017-13053.

Choi, J., Horne, D. S., & Lucey, J. A. (2011). Determination of molecular weight of a purified fraction of colloidal calcium phosphate derived from the casein micelles of bovine milk. *Journal of Dairy Science, 94*(7), 3250–3261. http://dx.doi.org/10.3168/jds.2010-3762.

Chun, J., Bhak, G., Lee, S.-G., Lee, J.-H., Lee, D., Char, K., et al. (2012). Kappa-casein-based hierarchical suprastructures and their use for selective temporal and spatial control over neuronal differentiation. *Biomacromolecules, 13*(9), 2731–2738. http://dx.doi.org/10.1021/bm300692k.

Clegg, R. A., & Holt, C. (2009). An *E. coli* over-expression system for multiply-phosphorylated proteins and its use in a study of calcium phosphate sequestration by novel recombinant phosphopeptides. *Protein Expression and Purification, 67*(1), 23–34. http://dx.doi.org/10.1016/j.pep.2009.04.004.

Cox, D., Carver, J. A., & Ecroyd, H. (2014). Preventing alpha-synuclein aggregation: The role of the small heat-shock molecular chaperone proteins. *Biochimica et Biophysica Acta - Molecular Basis of Disease, 1842*(9), 1830–1843. http://dx.doi.org/10.1016/j.bbadis.2014.06.024.

Cragnell, C., Choi, J., Segad, M., Lee, S., Nilsson, L., & Skepö, M. (2017). Bovine β-casein has a polydisperse distribution of equilibrium micelles. *Food Hydrocolloids, 70*, 65–68. http://dx.doi.org/10.1016/j.foodhyd.2017.03.021.

Creamer, L. K., Richardson, T., & Parry, D. A. D. (1981). Secondary structure of bovine αS1- and β-casein in solution. *Archives of Biochemistry and Biophysics, 211*(2), 689–696. http://dx.doi.org/10.1016/0003-9861(81)90505-1.

Cross, K. J., Huq, N. L., Palamara, J. E., Perich, J. W., & Reynolds, E. C. (2005). Physicochemical characterization of casein phosphopeptide-amorphous calcium phosphate nanocomplexes. *Journal of Biological Chemistry, 280*(15), 15362–15369. http://dx.doi.org/10.1074/jbc.M413504200.

Cubellis, M. V., Caillez, F., Blundell, T. L., & Lovell, S. C. (2005). Properties of polyproline II, a secondary structure element implicated in protein-protein interactions. *Proteins-Structure Function and Bioinformatics, 58*(4), 880–892. http://dx.doi.org/10.1002/prot.20327.

Dalgleish, D. G., & Corredig, M. (2012). The structure of the casein micelle of milk and its changes during processing. In M. P. Doyle, & K.T.R. (Eds.), *Annual review of food science and technology* (Vol. 3, pp. 449–467).

Dalgleish, D. G., Horne, D. S., & Law, A. J. R. (1989). Size-related differences in bovine casein micelles. *Biochimica et Biophysica Acta, 991*(3), 383–387.
Davies, D. T., & Law, A. J. R. (1980). Content and composition of protein in creamery milks in southwest scotland. *Journal of Dairy Research, 47*(1), 83–90.
de Kruif, C. G. (2014). The structure of casein micelles: A review of small-angle scattering data. *Journal of Applied Crystallography, 47*, 1479–1489.
de Kruif, C. G., & Grinberg, V. Y. (2002). Micellisation of beta-casein. *Colloids and Surfaces A-Physicochemical and Engineering Aspects, 210*(2–3), 183–190.
de Kruif, C. G., & Holt, C. (2003). Casein micelle structure, functions and interactions. In P. F. Fox, & P. L. H. McSweeney (Eds.), *Advanced dairy chemistry: Vol. 1A. Proteins* (pp. 675–698). New York: Kluwer Academic/Plenum.
de Kruif, C. G., Huppertz, T., Urban, V. S., & Petukhov, A. V. (2012). Casein micelles and their internal structure. *Advances in Colloid and Interface Science, 171*, 36–52. http://dx.doi.org/10.1016/j.cis.2012.01.002.
de Kruif, C. G., Tuinier, R., Holt, C., Timmins, P. A., & Rollema, H. S. (2002). Physicochemical study of kappa- and beta-casein dispersions and the effect of cross-linking by transglutaminase. *Langmuir, 18*(12), 4885–4891. http://dx.doi.org/10.1021/la025543w.
Dehle, F. C., Ecroyd, H., Musgrave, I. F., & Carver, J. A. (2010). Alpha B-Crystallin inhibits the cell toxicity associated with amyloid fibril formation by kappa-casein and the amyloid-beta peptide. *Cell Stress & Chaperones, 15*(6), 1013–1026. http://dx.doi.org/10.1007/s12192-010-0212-z.
Donnelly, W. J., McNeill, G. P., Buchheim, W., & McGann, T. C. A. (1984). A comprehensive study of the relationship between size and protein composition in natural bovine casein micelles. *Biochimica et Biophysica Acta, 789*(2), 136–143. http://dx.doi.org/10.1016/0167-4838(84)90197-3.
Dosztányi, Z., Csizmók, V., Tompa, P., & Simon, I. (2005). The pairwise energy content estimated from amino acid composition discriminates between folded and intrinsically unstructured proteins. *Journal of Molecular Biology, 347*(4), 827–839. http://dx.doi.org/10.1016/j.jmb.2005.01.071.
Drake, A. F., Siligardi, G., & Gibbons, W. A. (1988). Reassessment of the electronic circular dichroism criteria for random coil conformations of poly(l-lysine) and the implications for protein folding and denaturation studies. *Biophysical Chemistry, 31*(1), 143–146. http://dx.doi.org/10.1016/0301-4622(88)80019-X.
Ecroyd, H., & Carver, J. A. (2008a). The effect of small molecules in modulating the chaperone activity of alpha B-crystallin against ordered and disordered protein aggregation. *FEBS Journal, 275*(5), 935–947. http://dx.doi.org/10.1111/j.1742-4658.2008.06257.x.
Ecroyd, H., & Carver, J. A. (2008b). Unraveling the mysteries of protein folding and misfolding. *IUBMB Life, 60*(12), 769–774. http://dx.doi.org/10.1002/iub.117.
Ecroyd, H., Koudelka, T., Thorn, D. C., Williams, D. M., Devlin, G., Hoffmann, P., et al. (2008). Dissociation from the oligomeric state is the rate-limiting step in fibril formation by kappa-casein. *Journal of Biological Chemistry, 283*(14), 9012–9022. http://dx.doi.org/10.1074/jbc.M709928200.
Ecroyd, H., Thorn, D. C., Liu, Y., & Carver, J. A. (2010). The dissociated form of kappa-casein is the precursor to its amyloid fibril formation. *Biochemical Journal, 429*, 251–260. http://dx.doi.org/10.1042/bj20091949.
Ehrnsperger, M., Graber, S., Gaestel, M., & Buchner, J. (1997). Binding of non-native protein to Hsp25 during heat shock creates a reservoir of folding intermediates for reactivation. *The EMBO Journal, 16*(2), 221–229. http://dx.doi.org/10.1093/emboj/16.2.221.
Elias, A. K., Scanlon, D., Musgrave, I. F., & Carver, J. A. (2014). SEVI, the semen enhancer of HIV infection along with fragments from its central region, form amyloid fibrils that

are toxic to neuronal cells. *Biochimica et Biophysica Acta (BBA) - Proteins & Proteomics, 1844*(9), 1591–1598. http://dx.doi.org/10.1016/j.bbapap.2014.06.006.

Evans, M. T. A., Irons, L., & Jones, M. (1971). Physicochemical properties of beta-casein and some carboxyacyl derivatives. *Biochimica et Biophysica Acta, 229*(2), 411–&.

Farrell, H. M., Brown, E. M., Hoagland, P. D., & Malin, E. L. (2013). Higher order structures of the caseins: A paradox. In P. F. Fox, & P. L. H. McSweeney (Eds.) (4 ed,*Advanced dairy chemistry: Vol. 1A. Proteins: Basic aspects* (pp. 161–184). New York, NY, USA: Springer.

Farrell, H. M., Cooke, P. H., Wickham, E. D., Piotrowski, E. G., & Hoagland, P. D. (2003). Environmental influences on bovine κ-casein: Reduction and conversion to fibrillar (amyloid) structures. *Journal of Protein Chemistry, 22*(3), 259–273. http://dx.doi.org/10.1023/a:1025020503769.

Farrell, H. M., Malin, E. L., Brown, E. M., & Mora-Gutierrez, A. (2009). Review of the chemistry of alpha(S2)-casein and the generation of a homologous molecular model to explain its properties. *Journal of Dairy Science, 92*(4), 1338–1353. http://dx.doi.org/10.3168/jds.2008-1711.

Fitzpatrick, A. W., Knowles, T. P. J., Waudby, C. A., Vendruscolo, M., & Dobson, C. M. (2011). Inversion of the balance between hydrophobic and hydrogen bonding interactions in protein folding and aggregation. *PLoS Computational Biology,* 7(10), e1002169. http://dx.doi.org/10.1371/journal.pcbi.1002169.

Fuxreiter, M., & Tompa, P. (2012). Fuzzy complexes: A more stochastic view of protein function. In M. Fuxreiter, & P. Tompa (Eds.), *Fuzziness: Structural disorder in protein complexes* (Vol. 725, pp. 1–14). New York: Springer.

Gaspar, A. M., Appavou, M. S., Busch, S., Unruh, T., & Doster, W. (2008). Dynamics of well-folded and natively disordered proteins in solution: A time-of-flight neutron scattering study. *European Biophysics Journal with Biophysics Letters, 37*(5), 573–582. http://dx.doi.org/10.1007/s00249-008-0266-3.

Ghahghaei, A., Rekas, A., Price, W. E., & Carver, J. A. (2007). The effect of dextran on subunit exchange of the molecular chaperone alpha A-crystallin. *Biochimica et Biophysica Acta (BBA) - Proteins & Proteomics, 1774*(1), 102–111. http://dx.doi.org/10.1016/j.bbapap.2006.10.002.

Goldschmidt, L., Teng, P. K., Riek, R., & Eisenberg, D. (2010). Identifying the amylome, proteins capable of forming amyloid-like fibrils. *Proceedings of the National Academy of Sciences, 107*(8), 3487–3492. http://dx.doi.org/10.1073/pnas.0915166107.

Goobes, R., Goobes, G., Shaw, W. J., Drobny, G. P., Campbell, C. T., & Stayton, P. S. (2007). Thermodynamic roles of basic amino acids in statherin recognition of hydroxyapatite. *Biochemistry, 46*(16), 4725–4733. http://dx.doi.org/10.1021/bi602345a.

Graham, E. R. B., Malcolm, G. N., & McKenzie, H. A. (1984). On the isolation and conformation of bovine beta-casein A^1. *International Journal of Biological Macromolecules, 6*(3), 155–161. http://dx.doi.org/10.1016/0141-8130(84)90058-8.

Griffin, M. C. A., & Roberts, G. C. K. (1985). A H^{-1}-NMR study of casein micelles. *Biochemical Journal, 228*(1), 273–276.

Groves, M. L., Wickham, E. D., & Farrell, H. M. (1998). Environmental effects on disulfide bonding patterns of bovine kappa-casein. *Journal of Protein Chemistry, 17*(2), 73–84. http://dx.doi.org/10.1023/a:1022518613574.

Guy, H. R. (1985). Amino-acid side-chain partition energies and distribution of residues in soluble proteins. *Biophysical Journal, 47*(1), 61–70.

Hall, D., Kardos, J., Edskes, H., Carver, J. A., & Goto, Y. (2015). A multi-pathway perspective on protein aggregation: Implications for control of the rate and extent of amyloid formation. *FEBS Letters, 589*(6), 672–679. http://dx.doi.org/10.1016/j.febslet.2015.01.032.

Hall, D., Zhao, R., Dehlsen, I., Bloomfield, N., Williams, S. R., Arisaka, F., et al. (2016). Protein aggregate turbidity: Simulation of turbidity profiles for mixed aggregation reactions. *Analytical Biochemistry, 498*, 78–94. http://dx.doi.org/10.1016/j.ab.2015.11.021.

Halwer, M. (1954). Light-scattering study of effect of electrolytes on alpha-casein and beta-casein solutions. *Archives of Biochemistry and Biophysics, 51*(1), 79–87. http://dx.doi.org/10.1016/0003-9861(54)90455-5.

Hassanisadi, M., Barzear, A., Yousefi, R., Dalgalarrondo, M., Chobert, J.-M., Haertle, T., et al. (2008). Chemometric study of the aggregation of alcohol dehydrogenase and its suppression by beta-caseins: A mechanistic perspective. *Analytica Chimica Acta, 613*(1), 40–47. http://dx.doi.org/10.1016/j.aca.2008.02.036.

Herskovits, T. T. (1966). On the conformation of caseins. Optical rotatory properties. *Biochemistry, 5*(3), 1018–1026. http://dx.doi.org/10.1021/bi00867a030.

Ho, C., & Chen, A. H. (1967). Polymerisation of bovine alpha-S1-casein B. *Journal of Biological Chemistry, 242*(3), 551.

Hojati, S., Ghahghaei, A., & Lagzian, M. (2018). The potential inhibitory effect of beta-casein on the aggregation and deposition of A(1-42) fibrils in alzheimer's disease: Insight from in-vitro and in-silico studies. *Journal of Biomolecular Structure & Dynamics, 36*(8), 2118–2130. http://dx.doi.org/10.1080/07391102.2017.1345326.

Holehouse, A. S., & Pappu, R. V. (2018). Collapse transitions of proteins and the interplay among backbone, sidechain, and solvent interactions. In K. A. Dill (Ed.), *Annual review of biophysics* (Vol. 47, pp. 19–39). Palo Alto: Annual Reviews.

Holt, C. (1997). The milk salts and their interaction with casein. In P. F. Fox (Ed.), *Advanced dairy chemistry: Vol. 3. Lactose, Water, Salts and Vitamins* (pp. 233–254). London: Chapman and Hall.

Holt, C. (2004). An equilibrium thermodynamic model of the sequestration of calcium phosphate by casein micelles and its application to the calculation of the partition of salts in milk. *European Biophysics Journal with Biophysics Letters, 33*(5), 421–434. http://dx.doi.org/10.1007/s00249-003-0377-9.

Holt, C. (2013). Unfolded phosphopolypeptides enable soft and hard tissues to coexist in the same organism with relative ease. *Current Opinion in Structural Biology, 23*(3), 420–425. http://dx.doi.org/10.1016/j.sbi.2013.02.010.

Holt, C. (2016). Casein and casein micelle structures, functions and diversity in 20 species. *International Dairy Journal, 60*, 2–13. http://dx.doi.org/10.1016/j.idairyj.2016.01.004.

Holt, C., & Carver, J. A. (2012). Darwinian transformation of a 'scarcely nutritious fluid' into milk. *Journal of Evolutionary Biology, 25*(7), 1253–1263. http://dx.doi.org/10.1111/j.1420-9101.2012.02509.x.

Holt, C., Carver, J. A., Ecroyd, H., & Thorn, D. C. (2013). Invited review: Caseins and the casein micelle: Their biological functions, structures, and behavior in foods. *Journal of Dairy Science, 96*(10), 6127–6146. http://dx.doi.org/10.3168/jds.2013-6831.

Holt, C., de Kruif, C. G., Tuinier, R., & Timmins, P. A. (2003). Substructure of bovine casein micelles by small-angle X-ray and neutron scattering. *Colloids and Surfaces A-Physicochemical and Engineering Aspects, 213*(2–3), 275–284.

Holt, C., & Horne, D. S. (1996). The hairy casein micelle: Evolution of the concept and its implications for dairy technology. *Netherlands Milk and Dairy Journal, 50*(2), 85–111.

Holt, C., Lenton, S., Nylander, T., Sorensen, E. S., & Teixeira, S. C. M. (2014). Mineralisation of soft and hard tissues and the stability of biofluids. *Journal of Structural Biology, 185*(3), 383–396. http://dx.doi.org/10.1016/j.jsb.2013.11.009.

Holt, C., Raynes, J. K., & Carver, J. A. (2019). Sequence characteristics responsible for protein-protein interactions in the intrinsically disordered regions of caseins, amelogenins, and small heat-shock proteins. *Biopolymers, 0*(0), e23319. http://dx.doi.org/10.1002/bip.23319 [Epub ahead of print].

Holt, C., & Sawyer, L. (1988). Primary and predicted secondary structures of the caseins in relation to their biological functions. *Protein Engineering, 2*(4), 251–259. http://dx.doi.org/10.1093/protein/2.4.251.

Holt, C., & Sawyer, L. (1993). Caseins as rheomorphic proteins - interpretation of primary and secondary structures of the alpha-S1-caseins, beta-caseins and kappa-caseins. *Journal of the Chemical Society Faraday Transactions, 89*(15), 2683–2692.

Holt, C., Sorensen, E. S., & Clegg, R. A. (2009). Role of calcium phosphate nanoclusters in the control of calcification. *FEBS Journal, 276*(8), 2308–2323. http://dx.doi.org/10.1111/j.1742-4658.2009.06958.x.

Holt, C., Timmins, P. A., Errington, N., & Leaver, J. (1998). A core-shell model of calcium phosphate nanoclusters stabilized by beta-casein phosphopeptides, derived from sedimentation equilibrium and small-angle X-ray and neutron-scattering measurements. *European Journal of Biochemistry, 252*(1), 73–78.

Holt, C., Wahlgren, N. M., & Drakenberg, T. (1996). Ability of a beta-casein phosphopeptide to modulate the precipitation of calcium phosphate by forming amorphous dicalcium phosphate nanoclusters. *Biochemical Journal, 314*, 1035–1039.

Horne, D. S. (1998). Casein interactions: Casting light on the black boxes, the structure in dairy products. *International Dairy Journal, 8*(3), 171–177. http://dx.doi.org/10.1016/s0958-6946(98)00040-5.

Horne, D. S. (2017). A balanced view of casein interactions. *Current Opinion in Colloid & Interface Science, 28*, 74–86. http://dx.doi.org/10.1016/j.cocis.2017.03.009.

Huang, F., Oldfield, C. J., Xue, B., Hsu, W. L., Meng, J. W., Liu, X. W., et al. (2014). Improving protein order-disorder classification using charge-hydropathy plots. *BMC Bioinformatics, 15*. http://dx.doi.org/10.1186/1471-2105-15-s17-s4.

Hudson, S. A., Ecroyd, H., Dehle, F. C., Musgrave, I. F., & Carver, J. A. (2009). (-)-Epigallocatechin-3-Gallate (EGCG) maintains kappa-casein in its pre-fibrillar state without redirecting its aggregation pathway. *Journal of Molecular Biology, 392*(3), 689–700. http://dx.doi.org/10.1016/j.jmb.2009.07.031.

Hudson, S. A., Ecroyd, H., Kee, T. W., & Carver, J. A. (2009). The thioflavin T fluorescence assay for amyloid fibril detection can be biased by the presence of exogenous compounds. *FEBS Journal, 276*(20), 5960–5972. http://dx.doi.org/10.1111/j.1742-4658.2009.07307.x.

Humphreys, D. T., Carver, J. A., Easterbrook-Smith, S. B., & Wilson, M. R. (1999). Clusterin has chaperone-like activity similar to that of small heat shock proteins. *Journal of Biological Chemistry, 274*(11), 6875–6881. http://dx.doi.org/10.1074/jbc.274.11.6875.

Huppertz, T. (2013). Chemistry of the caseins. In P. F. Fox, & P. L. H. McSweeney (Eds.), *Advanced dairy chemistry: Vol. 1A. Proteins: Basic aspects* (pp. 135–160). New York, NY, USA: Springer.

Huppertz, T., Fox, P. F., & Kelly, A. L. (2018). 3 - the caseins: Structure, stability, and functionality. In R. Y. Yada (Ed.), *Proteins in food processing (Second edition)* (2 ed., pp. 49–92). Cambridge, UK: Woodhead Publishing.

Ingham, B., Erlangga, G. D., Smialowska, A., Kirby, N. M., Wang, C., Matia-Merino, L., et al. (2015). Solving the mystery of the internal structure of casein micelles. *Soft Matter, 11*(14), 2723–2725. http://dx.doi.org/10.1039/c5sm00153f.

Ingham, B., Smialowska, A., Erlangga, G. D., Matia-Merino, L., Kirby, N. M., Wang, C., et al. (2016). Revisiting the interpretation of casein micelle SAXS data. *Soft Matter, 12*(33), 6937–6953. http://dx.doi.org/10.1039/C6SM01091A.

Jain, N., Bhattacharya, M., & Mukhopadhyay, S. (2011). Chain collapse of an amyloidogenic intrinsically disordered protein. *Biophysical Journal, 101*(7), 1720–1729. http://dx.doi.org/10.1016/j.bpj.2011.08.024.

Jeurnink, T. J. M., & De kruif, C. G. (1993). Changes in milk on heating- viscosity measurements. *Journal of Dairy Research, 60*(2), 139–150.

Jimenez, J. L., Guijarro, J. L., Orlova, E., Zurdo, J., Dobson, C. M., Sunde, M., et al. (1999). Cryo-electron microscopy structure of an SH3 amyloid fibril and model of the molecular packing. *The EMBO Journal, 18*(4), 815–821. http://dx.doi.org/10.1093/emboj/18.4.815.

Kawasaki, K., Lafont, A.-G., & Sire, J.-Y. (2011). The evolution of casein genes from tooth genes before the origin of mammals. *Molecular Biology and Evolution, 28*(4), 2053–2061. http://dx.doi.org/10.1093/molbev/msr020.

Kehoe, J. J., & Foegeding, E. A. (2011). Interaction between beta-casein and whey proteins as a function of pH and salt concentration. *Journal of Agricultural and Food Chemistry, 59*(1), 349–355. http://dx.doi.org/10.1021/jf103371g.

van Kemenade, M. J. J. M., & de Bruyn, P. L. (1989). The influence of casein on the kinetics of hydroxyapatite precipitation. *Journal of Colloid and Interface Science, 129*(1), 1–14.

Khodarahmi, R., Beyrami, M., & Soori, H. (2008). Appraisal of casein's inhibitory effects on aggregation accompanying carbonic anhydrase refolding and heat-induced ovalbumin fibrillogenesis. *Archives of Biochemistry and Biophysics, 477*(1), 67–76. http://dx.doi.org/10.1016/j.abb.2008.04.028.

Knowles, T. P. J., Vendruscolo, M., & Dobson, C. M. (2014). The amyloid state and its association with protein misfolding diseases. *Nature Reviews Molecular Cell Biology, 15*(6), 384–396. http://dx.doi.org/10.1038/nrm3810.

Koudelka, T. (2010). *Casein proteins: Investigating their chaperone activity and amyloid fibril formation* (PhD). Adelaide, Australia: The University of Adelaide.

Koudelka, T., Dehle, F. C., Musgrave, I. F., Hoffmann, P., & Carver, J. A. (2012). Methionine oxidation enhances kappa-casein amyloid fibril formation. *Journal of Agricultural and Food Chemistry, 60*(16), 4144–4155. http://dx.doi.org/10.1021/jf205168t.

Koudelka, T., Hoffmann, P., & Carver, J. A. (2009). Dephosphorylation of alpha(s)- and beta-caseins and its effect on chaperone activity: A structural and functional investigation. *Journal of Agricultural and Food Chemistry, 57*(13), 5956–5964. http://dx.doi.org/10.1021/jf9008372.

Kresheck, G. C. (1965). Conformation of casein in aqueous solution. *Acta Chemica Scandinavica, 19*(2), 375–382. http://dx.doi.org/10.3891/acta.chem.scand.19-0375.

Kumar, M., Hong, Y. N., Thorn, D. C., Ecroyd, H., & Carver, J. A. (2017). Monitoring early-stage protein aggregation by an aggregation induced emission fluorogen. *Analytical Chemistry, 89*(17), 9322–9329. http://dx.doi.org/10.1021/acs.analchem.7b02090.

Kumosinski, T. F., Brown, E. M., & Farrell, H. M. (1991a). 3-Dimensional molecular modeling of bovine caseins - alpha-S1-casein. *Journal of Dairy Science, 74*(9), 2889–2895. http://dx.doi.org/10.3168/jds.S0022-0302(91)78470-1.

Kumosinski, T. F., Brown, E. M., & Farrell, H. M. (1991b). 3-Dimensional molecular modeling of bovine caseins - kappa casein. *Journal of Dairy Science, 74*(9), 2879–2889. http://dx.doi.org/10.3168/jds.S0022-0302(91)78469-5.

Kumosinski, T. F., Brown, E. M., & Farrell, H. M. (1993). Three-dimensional molecular modeling of bovine caseins: An energy-minimized beta-casein structure. *Journal of Dairy Science, 76*(4), 931–945. http://dx.doi.org/10.3168/jds.S0022-0302(93)77420-2.

Kyte, J., & Doolittle, R. F. (1982). A simple method for displaying the hydropathic character of a protein. *Journal of Molecular Biology, 157*(1), 105–132. http://dx.doi.org/10.1016/0022-2836(82)90515-0.

Le, T. T., Saveyn, P., Hoa, H. D., & Van der Meeren, P. (2008). Determination of heat-induced effects on the particle size distribution of casein micelles by dynamic light scattering and nanoparticle tracking analysis. *International Dairy Journal, 18*(12), 1090–1096. http://dx.doi.org/10.1016/j.idairyj.2008.06.006.

Leclerc, E., & Calmettes, P. (1997). Structure of beta-casein micelles. *Physica B: Condensed Matter, 241*, 1141–1143. http://dx.doi.org/10.1016/s0921-4526(97)00850-8.

Lee, C. F. (2009). Self-assembly of protein amyloids: A competition between amorphous and ordered aggregation. *Physical Review E, 80*(3). http://dx.doi.org/10.1103/PhysRevE.80.031922.

van der Lee, R., Buljan, M., Lang, B., Weatheritt, R. J., Daughdrill, G. W., Dunker, A. K., et al. (2014). Classification of intrinsically disordered regions and proteins. *Chemical Reviews, 114*(13), 6589–6631. http://dx.doi.org/10.1021/cr400525m.

Lencki, R. W. (2007). Evidence for fibril-like structure in bovine casein micelles. *Journal of Dairy Science, 90*(1), 75–89.

Lenton, S., Grimaldo, M., Roosen-Runge, F., Schreiber, F., Nylander, T., Clegg, R., et al. (2017). Effect of phosphorylation on a human-like osteopontin peptide. *Biophysical Journal, 112*(8), 1586–1596. http://dx.doi.org/10.1016/j.bpj.2017.03.005.

Lenton, S., Nylander, T., Holt, C., Sawyer, L., Härtlein, M., Müller, H., et al. (2016). Structural studies of hydrated samples of amorphous calcium phosphate and phosphoprotein nanoclusters. *European Biophysics Journal with Biophysics Letters, 45*(5), 405–412. http://dx.doi.org/10.1007/s00249-015-1109-7.

Lenton, S., Nylander, T., Teixeira, S. C. M., & Holt, C. (2015). A review of the biology of calcium phosphate sequestration with special reference to milk. *Dairy Science & Technology, 95*(1), 3–14. http://dx.doi.org/10.1007/s13594-014-0177-2.

Lenton, S., Seydel, T., Nylander, T., Holt, C., Hartlein, M., Teixeira, S., et al. (2015). Dynamic footprint of sequestration in the molecular fluctuations of osteopontin. *Journal of the Royal Society Interface, 12*(110), 12. http://dx.doi.org/10.1098/rsif.2015.0506.

Léonil, J., Henry, G., Jouanneau, D., Delage, M.-M., Forge, V., & Putaux, J.-L. (2008). Kinetics of fibril formation of bovine κ-casein indicate a conformational rearrangement as a critical step in the process. *Journal of Molecular Biology, 381*(5), 1267–1280. http://dx.doi.org/10.1016/j.jmb.2008.06.064.

Li, B., Ge, P., Murray, K. A., Sheth, P., Zhang, M., Nair, G., et al. (2018). Cryo-EM of full-length alpha-synuclein reveals fibril polymorphs with a common structural kernel. *Nature Communications, 9*. http://dx.doi.org/10.1038/s41467-018-05971-2.

Librizzi, F., Carrotta, R., Spigolon, D., Bulone, D., & San Biagio, P. L. (2014). Alpha-casein inhibits insulin amyloid formation by preventing the onset of secondary nucleation processes. *Journal of Physical Chemistry Letters, 5*(17), 3043–3048. http://dx.doi.org/10.1021/jz501570m.

Little, E. M., & Holt, C. (2004). An equilibrium thermodynamic model of the sequestration of calcium phosphate by casein phosphopeptides. *European Biophysics Journal with Biophysics Letters, 33*(5), 435–447. http://dx.doi.org/10.1007/s00249-003-0376-x.

Liu, Y., Carver, J. A., Ho, L. H., Elias, A. K., Musgrave, I. F., & Pukala, T. L. (2014). Hemin as a generic and potent protein misfolding inhibitor. *Biochemical and Biophysical Research Communications, 454*(2), 295–300. http://dx.doi.org/10.1016/j.bbrc.2014.10.062.

Liu, J., Dehle, F. C., Liu, Y., Bahraminejad, E., Ecroyd, H., Thorn, D. C., et al. (2016). The effect of milk constituents and crowding agents on amyloid fibril formation by kappa-casein. *Journal of Agricultural and Food Chemistry, 64*(6), 1335–1343. http://dx.doi.org/10.1021/acs.jafc.5b04977.

Liu, Y., Pukala, T. L., Musgrave, I. F., Williams, D. M., Dehle, F. C., & Carver, J. A. (2013). Gallic acid is the major component of grape seed extract that inhibits amyloid fibril formation. *Bioorganic & Medicinal Chemistry Letters, 23*(23), 6336–6340. http://dx.doi.org/10.1016/j.bmcl.2013.09.071.

Li, Y., Zhao, C., Luo, F., Liu, Z., Gui, X., Luo, Z., et al. (2018). Amyloid fibril structure of alpha-synuclein determined by cryoelectron microscopy. *Cell Research, 28*(9), 897–903. http://dx.doi.org/10.1038/s41422-018-0075-x.

Loupiac, C. (2016). How neutron scattering experiments can target the structure and dynamics of milk proteins? *Current Opinion in Food Science, 9*, 93–97. http://dx.doi.org/10.1016/j.cofs.2016.10.005.

Maccallum, P. H., Poet, R., & Milner-White, E. J. (1995). Coulombic attractions between partially charged main-chain atoms stabilize the right-handed twist found in most beta-strands. *Journal of Molecular Biology, 248*(2), 374–384. http://dx.doi.org/10.1006/jmbi.1995.0228.

Madende, M., Kemp, G., Stoychev, S., & Osthoff, G. (2018). Characterisation of african elephant beta casein and its relevance to the chemistry of caseins and casein micelles. *International Dairy Journal, 85*, 112–120. http://dx.doi.org/10.1016/j.idairyj.2018.05.013.

Madende, M., & Osthoff, G. (2019). Comparative genomics of casein genes. *Journal of Dairy Research*, 1–8. http://dx.doi.org/10.1017/S0022029919000414.

Madende, M., Osthoff, G., Patterton, H. G., Patterton, H. E., Martin, P., & Opperman, D. J. (2015). Characterization of casein and alpha lactalbumin of African elephant (*Loxodonta africana*) milk. *Journal of Dairy Science, 98*(12), 8308–8318. http://dx.doi.org/10.3168/jds.2014-9195.

Mao, A. H., Crick, S. L., Vitalis, A., Chicoine, C. L., & Pappu, R. V. (2010). Net charge per residue modulates conformational ensembles of intrinsically disordered proteins. *Proceedings of the National Academy of Sciences of the United States of America, 107*(18), 8183–8188. http://dx.doi.org/10.1073/pnas.0911107107.

Martin, P., Cebo, C., & Miranda, G. (2013). Interspecies comparison of milk proteins: Quantitative variability and molecular diversity. In P. L. H. McSweeney, & P. F. Fox (Eds.) (4th ed.,*Advanced dairy chemistry: Vol. 1A. Proteins: Basic aspects* (pp. 386–429). New York, NY, USA: Springer.

Matsudomi, N., Kanda, Y., Yoshika, Y., & Moriwaki, H. (2004). Ability of alpha s-casein to suppress the heat aggregation of ovotransferrin. *Journal of Agricultural and Food Chemistry, 52*(15), 4882–4886. http://dx.doi.org/10.1021/jf030802o.

McGuffin, L. J., Atkins, J. D., Salehe, B. R., Shuid, A. N., & Roche, D. B. (2015). IntFOLD: An integrated server for modelling protein structures and functions from amino acid sequences. *Nucleic Acids Research, 43*(W1), W169–W173. http://dx.doi.org/10.1093/nar/gkv236.

McGuffin, L. J., Shuid, A. N., Kempster, R., Maghrabi, A. H. A., Nealon, J. O., Salehe, B. R., et al. (2018). Accurate template-based modeling in CASP12 using the IntFOLD4-TS, ModFOLD6, and ReFOLD methods. *Proteins: Structure, Function, and Bioinformatics, 86*(S1), 335–344. http://dx.doi.org/10.1002/prot.25360.

Mensch, C., Barron, L. D., & Johannessen, C. (2016). Ramachandran mapping of peptide conformation using a large database of computed Raman and Raman optical activity spectra. *Physical Chemistry Chemical Physics, 18*(46), 31757–31768. http://dx.doi.org/10.1039/c6cp05862k.

Mensch, C., & Johannessen, C. (2019). The influence of the amino acid side chains on the Raman optical activity spectra of proteins. *ChemPhysChem, 20*(1), 42–54. http://dx.doi.org/10.1002/cphc.201800924.

Mikheeva, L. M., Grinberg, N. V., Grinberg, V. Y., Khokhlov, A. R., & de Kruif, C. G. (2003). Thermodynamics of micellization of bovine beta-casein studied by high-sensitivity differential scanning calorimetry. *Langmuir, 19*(7), 2913–2921. http://dx.doi.org/10.1021/la026702e.

Morgan, P. E., Treweek, T. M., Lindner, R. A., Price, W. E., & Carver, J. A. (2005). Casein proteins as molecular chaperones. *Journal of Agricultural and Food Chemistry, 53*(7), 2670–2683. http://dx.doi.org/10.1021/jf048329h.

Niewold, T. A., Murphy, C. L., Hulskamp-Koch, C. A. M., Tooten, P. C. J., & Gruys, E. (1999). Casein related amyloid, characterization of a new and unique amyloid protein isolated from bovine corpora amylacea. *Amyloid: International Journal of Experimental & Clinical Investigation, 6*(4), 244–249.

O'Kennedy, B. T., & Mounsey, J. S. (2006). Control of heat-induced aggregation of whey proteins using casein. *Journal of Agricultural and Food Chemistry, 54*(15), 5637–5642. http://dx.doi.org/10.1021/jf0607866.

Ochirkhuyag, B., Chobert, J. M., Dalgalarrondo, M., & Haertle, T. (2000). Characterization of mare caseins. Identification of alpha(S1)- and alpha(S2)-caseins. *Le Lait, 80*(2), 223–235.

Oftedal, O. T. (2012). The evolution of milk secretion and its ancient origins. *Animal, 6*(3), 355–368. http://dx.doi.org/10.1017/s1751731111001935.

Ono, T., Ohotawa, T., & Takagi, Y. (1994). Complexes of casein phosphopetides and calcium phosphate prepared from casein micelles by tryptic digestion. *Bioscience Biotechnology and Biochemistry, 58*(8), 1376–1380.

Ono, T., Yada, R., Yutani, K., & Nakai, S. (1987). Comparison of conformations of κ-casein, para-κ-casein and glycomacropeptide. *Biochimica et Biophysica Acta (BBA) - Protein Structure and Molecular Enzymology, 911*(3), 318–325. http://dx.doi.org/10.1016/0167-4838(87)90072-0.

Pan, K., & Zhong, Q. (2015). Amyloid-like fibrils formed from intrinsically disordered caseins: Physicochemical and nanomechanical properties. *Soft Matter, 11*(29), 5898–5904. http://dx.doi.org/10.1039/c5sm01037c.

Paulsson, M., & Dejmek, P. (1990). Thermal-denaturation of whey proteins in mixtures with caseins studied by differential scanning calorimetry. *Journal of Dairy Science, 73*(3), 590–600. http://dx.doi.org/10.3168/jds.S0022-0302(90)78707-3.

Peixoto, P. D. S., Silva, J. V. C., Laurent, G., Schmutz, M., Thomas, D., Bouchoux, A., et al. (2017). How high concentrations of proteins stabilize the amorphous state of calcium orthophosphate: A solid-state nuclear magnetic resonance (NMR) study of the casein case. *Langmuir, 33*(5), 1256–1264. http://dx.doi.org/10.1021/acs.langmuir.6b04235.

Perticaroli, S., Nickels, J. D., Ehlers, G., Mamontov, E., & Sokoov, A. P. (2014). Dynamics and rigidity in an intrinsically disordered protein, beta-casein. *Journal of Physical Chemistry B, 118*(26), 7317–7326. http://dx.doi.org/10.1021/jp503788r.

Perticaroli, S., Nickels, J. D., Ehlers, G., & Sokolov, A. P. (2014). Rigidity, secondary structure, and the universality of the Boson peak in proteins. *Biophysical Journal, 106*(12), 2667–2674. http://dx.doi.org/10.1016/j.bpj.2014.05.009.

Poon, S., Treweek, T. M., Wilson, M. R., Easterbrook-Smith, S. B., & Carver, J. A. (2002). Clusterin is an extracellular chaperone that specifically interacts with slowly aggregating proteins on their off-folding pathway. *FEBS Letters, 513*(2–3), 259–266. http://dx.doi.org/10.1016/s0014-5793(02)02326-8.

Potocnik, K., Gantner, V., Kuterovac, K., & Cividini, A. (2011). Mare's milk: Composition and protein fraction in comparison with different milk species. *Mljekarstvo, 61*(2), 107–113.

Qi, P. X. (2007). Studies of casein micelle structure: The past and the present. *Le Lait, 87*(4–5), 363–383. http://dx.doi.org/10.1051/lait:2007026.

Qi, P. X., Wickham, E. D., & Farrell, H. M. (2004). Thermal and alkaline denaturation of bovine beta-casein. *The Protein Journal, 23*(6), 389–402. http://dx.doi.org/10.1023/B:JOPC.0000039553.66233.3f.

Qi, P. X., Wickham, E. D., Piotrowski, E. G., Fagerquist, C. K., & Farrell, H. M. (2005). Implication of C-terminal deletion on the structure and stability of bovine beta-casein. *The Protein Journal, 24*(7–8), 431–444. http://dx.doi.org/10.1007/s10930-005-7639-6.

Raap, J., Kerling, K. E. T., Vreeman, H. J., & Visser, S. (1983). Peptide substrates for chymosin (rennin)- conformational studies of kappa-casein and some kappa-casein-related oligopeptides by circular dichroism and secondary structure prediction. *Archives of Biochemistry and Biophysics, 221*(1), 117–124. http://dx.doi.org/10.1016/0003-9861(83)90127-3.

Rasmussen, L. K., Johnsen, L. B., Tsiora, A., Sorensen, E. S., Thomsen, J. K., Nielsen, N. C., et al. (1999). Disulphide-linked caseins and casein micelles. *International Dairy Journal, 9*(3–6), 215–218.

Raynes, J. K., Day, L., Crepin, P., Horrocks, M. H., & Carver, J. A. (2017). Coaggregation of kappa-casein and beta-lactoglobulin produces morphologically distinct amyloid fibrils. *Small, 13*(14). http://dx.doi.org/10.1002/smll.201603591.

Redwan, E. M., Xue, B., Almehdar, H. A., & Uversky, V. N. (2015). Disorder in milk proteins: Caseins, intrinsically disordered colloids. *Current Protein & Peptide Science, 16*(3), 228–242.

Regini, J. W., Ecroyd, H., Meehan, S., Bremmell, K., Clarke, M. J., Lammie, D., et al. (2010). The interaction of unfolding alpha-lactalbumin and malate dehydrogenase with the molecular chaperone alpha B-crystallin: A light and X-ray scattering investigation. *Molecular Vision, 16*(260–62), 2446–2456.

Rekas, A., Adda, C. G., Aquilina, J. A., Barnham, K. J., Sunde, M., Galatis, D., et al. (2004). Interaction of the molecular chaperone alpha B-crystallin with alpha-synuclein: Effects on amyloid fibril formation and chaperone activity. *Journal of Molecular Biology, 340*(5), 1167–1183. http://dx.doi.org/10.1016/jmb.2004.05.054.

Ren, S., Uversky, V. N., Chen, Z., Dunker, A. K., & Obradovic, Z. (2008). Short Linear Motifs recognized by SH2, SH3 and Ser/Thr Kinase domains are conserved in disordered protein regions. *BMC Genomics, 9*. http://dx.doi.org/10.1186/1471-2164-9-s2-s26.

Rodriguez, D. E., Thula-Mata, T., Toro, E. J., Yeh, Y.-W., Holt, C., Holliday, L. S., et al. (2014). Multifunctional role of osteopontin in directing intrafibrillar mineralization of collagen and activation of osteoclasts. *Acta Biomaterialia, 10*(1), 494–507. http://dx.doi.org/10.1016/j.actbio.2013.10.010.

Rollema, H. S. (1992). Casein association and micelle formation. In P. F. Fox (Ed.), *Advanced dairy chemistry* (Vol. 1, pp. 111–140). Barking, Essex, UK: Elsevier Science Publishers.

Rollema, H. S., & Brinkhuis, J. A. (1989). A H^1-NMR study of bovine casein micelles: influence of pH, temperature and calcium ions on micellar structure. *Journal of Dairy Research, 56*(3), 417–425.

Rollema, H. S., Brinkhuis, J. A., & Vreeman, H. J. (1988). H^1-NMR studies of bovine kappa casein and casein micelles. *Netherlands Milk and Dairy Journal, 42*(2), 233–248.

Schmidt, D. G., Payens, T. A. J., van Markwijk, B. W., & Brinkhuis, J. A. (1967). On the subunit of alpha-S1-casein. *Biochemical and Biophysical Research Communications, 27*(4), 448. http://dx.doi.org/10.1016/s0006-291x(67)80005-6.

Shi, Z. S., Woody, R. W., & Kallenbach, N. R. (2002). Is polyproline II a major backbone conformation in unfolded proteins? *Unfolded Proteins, 62*, 163–240.

Sleigh, R. W., Sculley, T. B., & Mackinlay, A. G. (1979). Binding of beta casein to hydroxyapatite - effect of phosphate content and location. *Journal of Dairy Research, 46*(2), 337–342. http://dx.doi.org/10.1017/s0022029900017271.

Sneharani, A. H., Singh, S. A., & Rao, A. G. A. (2009). Interaction of alpha(S1)-casein with curcumin and its biological implications. *Journal of Agricultural and Food Chemistry, 57*(21), 10386–10391. http://dx.doi.org/10.1021/jf902464p.

Snoeren, T. H. M., van Markwijk, B., & van Montfort, R. (1980). Some physicochemical properties of bovine αs2-casein. *Biochimica et Biophysica Acta (BBA) - Protein Structure, 622*(2), 268–276. http://dx.doi.org/10.1016/0005-2795(80)90037-9.

Swaisgood, H. E. (2003). *Chemistry of the caseins* (Vol. 1A). New York, NY, USA: Kluwer/Plenum.

Syme, C. D., Blanch, E. W., Holt, C., Jakes, R., Goedert, M., Hecht, L., et al. (2002). A Raman optical activity study of rheomorphism in caseins, synucleins and tau - new insight into the structure and behaviour of natively unfolded proteins. *European Journal of Biochemistry, 269*(1), 148–156. http://dx.doi.org/10.1046/j.0014-2956.2001.02633.x.

Thorn, D. C., Ecroyd, H., & Carver, J. A. (2014). Chapter 30 - polymorphism in casein protein aggregation and amyloid fibril formation. In V. N. Uversky, & Y. L. Lyubchenko (Eds.), *Bio-nanoimaging* (pp. 323–331). Boston: Academic Press.

Thorn, D. C., Ecroyd, H., Carver, J. A., & Holt, C. (2015). Casein structures in the context of unfolded proteins. *International Dairy Journal, 46*, 2–11. http://dx.doi.org/10.1016/j.idairyj.2014.07.008.

Thorn, D. C., Ecroyd, H., Sunde, M., Poon, S., & Carver, J. A. (2008). Amyloid fibril formation by bovine milk alpha(s2)-casein occurs under physiological conditions yet is prevented by its natural counterpart, alpha(s1)-casein. *Biochemistry, 47*(12), 3926–3936. http://dx.doi.org/10.1021/bi701278c.

Thorn, D. C., Meehan, S., Sunde, M., Rekas, A., Gras, S. L., MacPhee, C. E., et al. (2005). Amyloid fibril formation by bovine milk kappa-casein and its inhibition by the molecular chaperones alpha(s-) and beta-casein. *Biochemistry, 44*(51), 17027–17036. http://dx.doi.org/10.1021/bi051352r.

Tiffany, M. L., & Krimm, S. (1968a). Circular dichroism of poly-L-proline in an unordered conformation. *Biopolymers, 6*(12), 1767–&. http://dx.doi.org/10.1002/bip.1968.360061212.

Tiffany, M. L., & Krimm, S. (1968b). New chain conformations of poly(glutamic acid) and polylysine. *Biopolymers, 6*(9), 1379–&. http://dx.doi.org/10.1002/bip.1968.360060911.

Treweek, T. M., Meehan, S., Ecroyd, H., & Carver, J. A. (2015). Small heat-shock proteins: Important players in regulating cellular proteostasis. *Cellular and Molecular Life Sciences, 72*(3), 429–451. http://dx.doi.org/10.1007/s00018-014-1754-5.

Treweek, T. M., Thorn, D. C., Price, W. E., & Carver, J. A. (2011). The chaperone action of bovine milk alpha(S1)- and alpha(S2)-caseins and their associated form alpha(S)-casein. *Archives of Biochemistry and Biophysics, 510*(1), 42–52. http://dx.doi.org/10.1016/j.abb.2011.03.012.

Uniacke-Lowe, T., Huppertz, T., & Fox, P. F. (2010). Equine milk proteins: Chemistry, structure and nutritional significance. *International Dairy Journal, 20*(9), 609–629. http://dx.doi.org/10.1016/j.idairyj.2010.02.007.

Urbanová, M., Keiderling, T. A., & Pančoška, P. (1996). Conformational study of some milk proteins. Comparison of the results of electronic circular dichroism and vibrational circular dichroism. *Bioelectrochemistry and Bioenergetics, 41*(1), 77–80. http://dx.doi.org/10.1016/0302-4598(96)01932-0.

Uversky, V. N., Gillespie, J. R., & Fink, A. L. (2000). Why are "natively unfolded" proteins unstructured under physiologic conditions? *Proteins Structure Function and Genetics, 41*(3), 415–427. http://dx.doi.org/10.1002/1097-0134(20001115)41:3<415::aid-prot130>3.0.co;2-7.

Vreeman, H. J. (1979). Association of bovine SH-kappa casein at pH 7.0. *Journal of Dairy Research, 46*(2), 271–276.

Vreeman, H. J., Brinkhuis, J. A., & Vanderspek, C. A. (1981). Some association properties of bovine SH-kappa-casein. *Biophysical Chemistry, 14*(2), 185–193. http://dx.doi.org/10.1016/0301-4622(81)85018-1.

Walstra, P., Bloomfield, V. A., Wei, G. J., & Jenness, R. (1981). Effect of chymosin action on the hydrodynamic diameter of casein micelles. *Biochimica et Biophysica Acta, 669*(2), 258–259.

Waugh, D. F., & Talbot, B. (1971). Equilibrium casein micelle systems. *Biochemistry, 10*(23), 4153–4162.

Wilkins, D. K., Grimshaw, S. B., Receveur, V., Dobson, C. M., Jones, J. A., & Smith, L. J. (1999). Hydrodynamic radii of native and denatured proteins measured by pulse field gradient NMR techniques. *Biochemistry, 38*(50), 16424–16431. http://dx.doi.org/10.1021/bi991765q.

Wiltzius, J. J. W., Sievers, S. A., Sawaya, M. R., Cascio, D., Popov, D., Riekel, C., et al. (2008). Atomic structure of the cross-β spine of islet amyloid polypeptide (amylin). *Protein Science, 17*(9), 1467–1474. http://dx.doi.org/10.1110/ps.036509.108.

Wong, E. T. C., Na, D., & Gsponer, J. (2013). On the importance of polar interactions for complexes containing intrinsically disordered proteins. *PLoS Computational Biology, 9*(8), e1003192. http://dx.doi.org/10.1371/journal.pcbi.1003192.

Wyatt, A. R., Yerbury, J. J., Ecroyd, H., & Wilson, M. R. (2013). Extracellular chaperones and proteostasis. In R. D. Kornberg (Ed.), *Annual review of biochemistry* (Vol. 82, pp. 295–322).

Yong, Y. H., & Foegeding, E. A. (2010). Caseins: Utilizing molecular chaperone properties to control protein aggregation in foods. *Journal of Agricultural and Food Chemistry, 58*(2), 685–693. http://dx.doi.org/10.1021/jf903072g.

Yousefi, R., Shchutskaya, Y. Y., Zimny, J., Gaudin, J.-C., Moosavi-Movahedi, A. A., Muronetz, V. I., et al. (2009). Chaperone-like activities of different molecular forms of beta-casein. Importance of polarity of N-terminal hydrophilic domain. *Biopolymers, 91*(8), 623–632. http://dx.doi.org/10.1002/bip.21190.

Zhang, X. F., Fu, X. M., Zhang, H., Liu, C., Jiao, W. W., & Chang, Z. Y. (2005). Chaperone-like activity of beta-casein. *The International Journal of Biochemistry & Cell Biology, 37*(6), 1232–1240. http://dx.doi.org/10.1016/j.biocel.2004.12.004.

Zhou, M., Xia, Y., Cao, F., Li, N., Hemar, Y., Tang, S., et al. (2019). A theoretical and experimental investigation of the effect of sodium dodecyl sulfate on the structural and conformational properties of bovine β-casein. *Soft Matter, 15*(7), 1551–1561. http://dx.doi.org/10.1039/C8SM01967C.

CHAPTER SIX

Protein misfolding in endoplasmic reticulum stress with applications to renal diseases

Samera Nademi and Jeffrey G. Dickhout*
Department of Medicine, Hamilton Centre for Kidney Research, McMaster University and St. Joseph's Healthcare Hamilton, Hamilton, ON, Canada
*Corresponding author: Email: jdickhou@stjosham.on.ca

Contents

Abstract

Protein misfolding may be the result of a variety of different processes that disrupt the ability of a protein to form a thermodynamically stable tertiary structure that allows it to perform its proper function. In this chapter, we explore the nature of a protein's form that allows it to have a stable tertiary structure, and examine specific mutation that are known to occur in the coding regions of DNA that disrupt a protein's ability to be folded into a thermodynamically stable tertiary structure. We examine the

Advances in Protein Chemistry and Structural Biology, Volume 118
ISSN 1876-1623
https://doi.org/10.1016/bs.apcsb.2019.08.001
© 2020 Elsevier Inc.
All rights reserved.

consequences of these protein misfoldings in terms of the endoplasmic reticulum stress response and resulting unfolded protein response. These conditions are specifically related to renal diseases. Further, we explore novel therapeutics, pharmacological chaperones, that are being developed to alleviate the disease burden associated with protein misfolding caused by mutations. These interventions aim to stabilize protein folding intermediates and allow proper folding to occur as well as prevent protein aggregation and the resulting pathophysiological consequences.

Abbreviations

ACE Angiotensin-converting enzyme
ADTKD-UMOD Autosomal Dominant Tubulointerstitial Kidney Disease Uromodulin-Related
AQP2 Aquaporin 2
ARB Angiotensin receptor blockers
AS Alport Syndrome
ASK1 Apoptosis Signal Regulating Kinase-1
ATF6 Activating transcription factor 6
AVP Arginine vasopressin
AVPR2 Arginine vasopressin receptor type 2
BiP Binding immunoglobulin protein
cAMP cyclic adenosine monophosphate
CHOP C/EBP homologous protein
CKD chronic kidney disease
CLDN14 claudin-14
CLIC2 chloride intracellular channel 2
CNDI congenital nephrogenic Diabetes Insipidus
DI Diabetes Insipidus
eIF2α eukaryotic translation initiation factor 2 alpha
ER Endoplasmic reticulum
ERAD ER-associated degradation
ESRD end-stage renal disease
FD Fabry disease
GADD34 growth arrest and DNA damage-inducible protein 34
GBM glomerular basement membrane
GFB glomerular filtration barrier
GRP78 78-kDa glucose-regulated protein
HIF-α hypoxia-inducible factor 1 alpha
IRE1 inositol requiring enzyme 1
JNK c- Jun N-terminal Kinase NDI
NS Nephrotic Syndrome
PERK Protein kinase R(PKR)-like endoplasmic reticulum kinase
PRKAG2 Protein Kinase AMP-Activated Non-Catalytic Subunit Gamma 2
ROS reactive oxygen species
S1P Site-1 protease
S2P Site-2 protease
TAL thick ascending limb
TRAF2 Tumor Necrosis Factor Receptor-associated Factor 2

UGGT	UDP-glucose/glycoprotein glucosyltransferase
UMOD	Uromodulin
UPR	Unfolded Protein Response
VEGF	vascular endothelial growth factor
VHL	Von Hippel-Lindau disease
XBP1	X-box-binding protein-1
XLID	X-linked Intellectual Disability

1. Introduction

Protein misfolding is becoming recognized as a key mediator of pathology (Hartl, 2017). Proteins are macromolecules that comprise our bodies and play a critical role in cell function, since they are the components of the nanomachines that allow all living processes to function. Proteins act to acquire energy for living organisms either through harvesting it from light or through the breakdown of chemical bonds in metabolism. This input of energy is required to maintain all complex life against the entropic forces that affect all complex systems and drive them to decay. To provide these critical functions, however, proteins themselves are required to maintain their own structure, and this structure is also subject to entropic decay (Hartl, 2017). The information that has been stored in DNA, arrived at through countless millennia of evolution, has developed a program for the generation of proteins, which allows for their critical function to maintain life, but also provides this function in a state that is thermodynamically stable. This stability is achieved by allowing the proper conformation of a protein to metaphorically sit in a low energy trough between higher energy transitions to unfolded and misfolded states. Such an elegant program allows some less complex proteins to spontaneously fold whereas others require additional molecular machines, they themselves proteins that are referred to as chaperones (Hartl, 2017), to reach this trough state (Fig. 1). Protein misfolding is a low energy conformation for the polypeptide chain that interferes with function. This misfolded state is thermodynamically preferred but gated off by high activation energy of misfolding. Thus, the program found in the DNA for protein structure incorporates information regarding the preferred folded state to allow function as well as a protective barrier to prevent degradation during function. To understand these concepts in detail, we must first delve into protein structure.

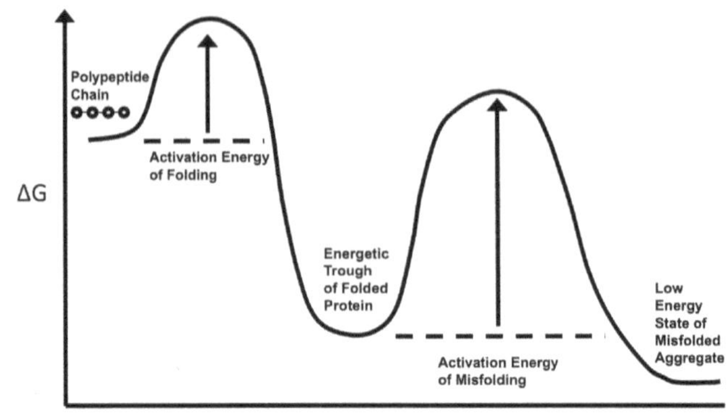

Fig. 1 ***Change in Free Energy (ΔG) associated with Protein Folding.*** Description of the change in free energy (ΔG) associated with protein folding in a complex protein that does not spontaneously fold. The nascent polypeptide chain is aided in acquiring the necessary energy by protein folding chaperones to reach the activation energy of folding. Folding can then proceed by release of free energy to the surroundings. This then places the properly folded protein in an energetic trough giving the protein thermodynamic stability. The protein will only reach a misfolded state if additional free energy is added to the system to raise the protein above the activation energy of misfolding, ie. heat induced denaturation of albumin, that will cause the protein to reach a new lower free energy state as a misfolded aggregate.

1.1 Protein structure

The three-dimensional conformation of a protein is critical for its proper functioning and is formed by four levels of protein structures: primary, secondary, tertiary, and quaternary. Quaternary structure represents an assemblage of properly folded individual proteins to allow function (Yu, Wang, & Li, 2006). Primary structure is the linear sequence of amino-acids held together by peptide bonds to form a polypeptide chain. The peptide bond is formed between the carboxyl group of one amino acid and the amino group of the next amino acid in the sequence. The rest of the amino acid molecule, not included in the peptide bond, can be refereed to as the rest or R-group of the amino acid. Secondary structures are formed within a polypeptide chain due to formation of hydrogen bonds between the amide and carbonyl groups of amino acids (Dill & MacCallum, 2012). Two types of secondary structures are the α-helix and β-pleated sheets. Alpha-helices are formed when one amino acid binds to another that is four down the chain recurrently; pulling the polypeptide into a helical structure that was modeled to contains 3.7 amino acids per turn of the helix (Pauling, Corey, & Branson, 1951). In α-helices, the amino acids' R-groups are directed outwards to interact with R-groups of other amino acids. Beta-pleated sheets are shaped as two or more segments of a polypeptide chain hydrogen-bond to form either parallel (both strands end in C- or N-terminus) or anti-parallel (N-terminus of one strand matches C-terminus of another) sheets alongside each other. In β-pleated sheets, the amino acids' R-groups extend above or below the plane of the sheets and interact with R groups of other amino acids (Eisenberg, 2003). Tertiary structure is the three-dimensional conformation of a polypeptide that is shaped as a result of hydrophobic, non-covalent, and covalent interactions. The primary driving forces for folding a polypeptide are hydrophobic and hydrophilic interactions between amino acids' R-groups and the environment (Dubey, Kini, Balaji, & Kumar, 2018). For instance, in an aqueous environment such as the cytosol of the cell, non-polar R-groups cluster together in the core of the polypeptide leaving the hydrophilic R-groups outside of the polypeptide to interact with the surrounding water molecules. Non-covalent (ie. hydrogen bonding, ionic bonding, dipole-dipole interactions, and London dispersion forces) and covalent disulfide bonds also determine the ultimate tertiary structure of a polypeptide. Some proteins also have a quaternary structure, which is formed by assembly of two or more polypeptides (also called subunits). Protein structures can become disrupted due to environmental factors (ie. salt concentration, solvents, pH, temperature) or genetic mutations that disrupt the genetic plan to create a thermodynamically stable properly folded protein, leading to dysfunctional proteins.

1.2 Protein production

The endoplasmic reticulum (ER) is the site for the production and folding of secretory, transmembrane, and ER luminal resident proteins. This folding mechanism involves a series of steps to exert protein folding quality control. Glycosylated proteins undergo the covalent addition of an N-linked glycan to the protein. In this process, the two outermost glucose residues are enzymatically removed by glucosidase I and II sequentially (Araki & Nagata, 2011). N-glycosylated proteins are recognized by the ER resident chaperones calnexin and calreticulin. These chaperones aid in the folding of the glycoprotein and prevent its aggregation. When the innermost glucose residual is removed by glucosidase II the protein is released from the calnexin and calreticulin chaperones. The protein is then recognized by UDP-glucose/glycoprotein glucosyltransferase (UGGT), which senses the folding state of the protein. If the protein is misfolded, then UGGT reattaches a glucose allowing the protein to be bond again by calnexin and calreticulin and retained in the ER. This cycle continues until the protein is sensed to be in the correctly folded state by UGGT, thus acting as a quality control mechanism to prevent the release of misfolded proteins (Araki & Nagata, 2011). Other chaperones in the ER that aid in protein folding include 78-kDa glucose-regulated protein (GRP78). GRP78 acts as a chaperone like calnexin and calreticulin and contains a hydrophobic region that allows it to bind to unfolded proteins. GRP78 activity to fold proteins is aided by a number of co-chaperones that contain a J-domain with a histidine-proline-aspartic acid region to allow GRP78 binding (Araki & Nagata, 2011). One of these co-chaperones, ERdj4, has been shown to associate with the ER-associated degradation (ERAD) machinery and may lead to misfolded protein degradation (Lai, Otero, Hendershot, & Snapp, 2012). Misfolded protein degradation occurs in the ERAD and consists of transport of the misfolded protein out of the ER, and finally ubiquitin-proteasome mediated degradation (Araki & Nagata, 2011).

1.3 Mutations that result in protein misfolding

Coding regions (also called protein-coding regions) are segments of the DNA that are translated into proteins. About 2% of human DNA are coding regions, while the rest does not translate into amino acids that provide the protein with its structure. Even though noncoding DNA does not code for proteins, it has important regulatory functions including determining which genes are "turned-on" to code proteins, these regions form the centromeres and telomeres needed for DNA replication, and are involved in the epigenetic control of gene expression (Kellis et al., 2014; Patrushev & Kovalenko, 2014). This chapter focuses on disorders that result from mutations that lead to protein misfolding resulting from changes in the DNA sequence of the coding regions. Mutations are predominantly due to insertion, deletion, or substitution of a DNA segment, leading to beneficial, harmful, or neutral (neither beneficial nor detrimental to the organism) consequences depending on its effect on protein functioning (Stefl, Nishi, Petukh, Panchenko, & Alexov, 2013). Even changing a single amino acid in a protein sequence (called missense mutation) can be detrimental to the function of the protein and survival of the organism by affecting a protein's function through modification of its structure (Stefl et al., 2013). Coding mutations affect a protein's function by changing the protein's localization, catalytic activity, hydrogen bonding network, stability, and/or folding; each of which is described below. Miss-localization of proteins may occur due to mutations in their nuclear localization sequences preventing the protein from entering the cell nucleus for instance. As a result, a protein may be located in an incorrect environment or unfold due to the chemical changes in their cellular environment such as differences in pH, or salt concentrations, in different cellular compartments (Stefl et al., 2013). Additionally, mutations may affect the catalytic activities of a protein including changes to the active site residues, protein-protein interactions sites, or allosteric regulation sites (Stefl et al., 2013). These mutations can dramatically affect catalytic reactions of proteins (Stefl et al., 2013). Moreover, changing amino acid residues can also affect the hydrogen-bonding network of a protein and expose their hydrophobic core to the aqueous environment, which frequently causes aggregation of the proteins through a process of misfolding (Fig. 2). Mutations in the proteins' hydrogen bonding network can also affect their catalytic activity, pH-dependent characteristics, stability, and folding. Furthermore, protein stability can be determined by the difference in the free energy (ΔG) of the unfolded to the folded state and into the misfolded state of a protein, as illustrated in Fig. 1. The activation energy of misfolding provides a thermodynamic barrier to maintain the protein in its folded state. Coding mutations can either destabilize or stabilize the protein by changing the activation energy of misfolding, both of which may compromise the function of a protein. For instance, in X-linked Intellectual Disability (XLID), one causative mutation occurs in the protein that codes for chloride intracellular channel 2 (CLIC2). This mutation increases its stability, but inhibits its dissociation from the ryanodine receptor (Witham, Takano, Schwartz, & Alexov, 2011). This prevents the channel from being transported to the neural membranes, leading to developmental delays, seizures, and cardiac abnormalities (Takano et al., 2012). Another significant consequence of coding mutation is protein misfolding, which leads to various pathologies through several different mechanisms some of which are described below (Stefl et al., 2013; Valastyan & Lindquist, 2014). Firstly, mutations can stabilize an intermediate during the folding process, favoring this misfolded-intermediate over the fully-folded functional protein. An example of this is a familiar form of amyotrophic lateral sclerosis caused by a mutation in the profilin-1 gene that stabilizes this protein in an intermediate folding state (Del Poggetto et al., 2018). Secondly, in spite of having partial functionality, misfolded proteins are usually destroyed through ER-mediated degradation processes. As an example, mutant ß-glucosidase (a lysosomal enzyme) in Gaucher disease retains partial functionality but is degraded prematurely by ER-mediated degradation factors, contributing to the severity of the disease (Ron & Horowitz, 2005). Thirdly, misfolded

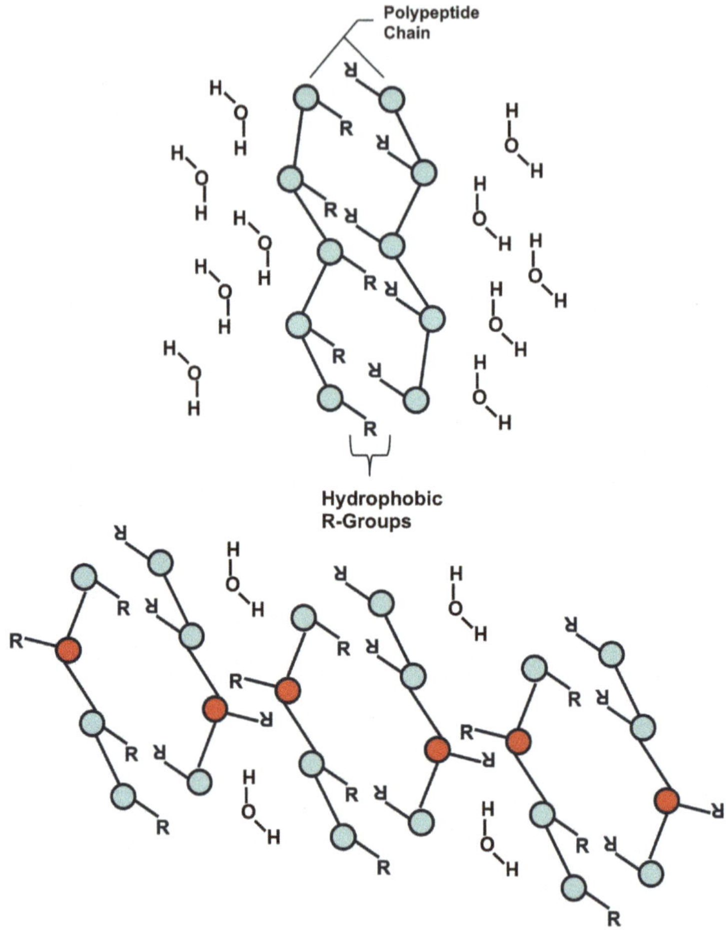

Fig. 2 ***Mechanism of Mutation induced Misfolding.*** The properly folded protein is stabilized in its native conformation via hydrogen bonding allowing hydrophobic residues to be stabilized inside the three-dimensional structure of the protein away from the aqueous environment. Mutations that change amino acids within the protein (new amino acid residue caused by coding mutation, red circle) disrupt the hydrogen bonding network causing expose of hydrophobic amino acid residues to the aqueous environment during protein folding. This results in bonding of hydrophobic R-groups of partially folded intermediates through hydrophobic interactions (minimum free energy state) causing misfolded protein aggregates.

proteins frequently fail to be transported to their final destination and accumulate in the ER (the site of synthesis for secretory and transmembrane proteins). This mislocalization and accumulation in the ER can have detrimental consequences for the cells involved in secretion. For instance in α-1 antitrypsin deficiency, mutant α-1 antitrypsin is trapped inside the ER of liver cells and fails to be transported to the lungs. The accumulation of this misfolded protein in the liver impairs liver function, while the deficiency of it in the lungs causes pulmonary emphysema and chronic obstructive pulmonary disease (Ghouse, Chu, Wang, & Perlmutter, 2014). Fourthly, misfolded proteins may become constitutively active as in the case of a secondary form of hypertension that is inherited with a mendelian character (Geller et al., 2000). This mutation results in a mineralocorticoid receptor that is constitutively active producing a form of hypertension that resembles aldosterone excess. Fifthly, misfolded mutant proteins may aggregate with the wild type protein leading to a loss of protein function even in a heterozygous individual. An example of this is Epidermolysis Bullosa Simplex (EBS), which occurs due to a mutation in keratin, a component of cytoskeleton intermediate filaments. The mutant keratin joins wild type keratin to make defective intermediate filaments that fragment upon mechanical stress, causing the skin to form painful wounds and blisters upon small frictions (Coulombe, Kerns, & Fuchs, 2009; Valastyan & Lindquist, 2014). Lastly, misfolded proteins may aggregate with each other to form toxic oligomers (a polymer with repeating units (Crews, Tsigelny, Hashimoto, & Masliah, 2009)), amyloid pores (pore-like assemblies that behave like ion channels (Di Scala et al., 2016)), and amyloid fibrils (insoluble fibrils resistant to degradation (Rambaran & Serpell, 2008)); disrupting cellular functions. To illustrate, Alzheimer's and Parkinson's diseases are caused by mutations in amyloid-β and α-synuclein respectively. These mutant proteins aggregate to form amyloid oligomers that later construct calcium-permeable amyloid pores which are dysregulated, contributing to the main culprit of these diseases (Di Scala et al., 2016). Another example is amyloid fibrils in cataracts, which are formed due to aggregation of mutant crystallin, a structural protein in the lens and cornea of the eye (Surguchev & Surguchov, 2010; Valastyan & Lindquist, 2014). Amyloid fibrils, oligomers, and amyloid pores are examples of misfolded and aggregated proteins that accumulate in the lumen of the ER (and other cellular organelles) and activate a cellular stress response called Unfolded Protein Response (UPR) (LaFerla, Green, & Oddo, 2007; Zheng et al., 2012).

1.4 Unfolded protein response

The ER is a membranous network that is continuous with the nuclear envelope and is a site of calcium storage, lipid synthesis, and protein homeostasis (synthesis, folding, modifications, transport, and degradation)

(Cybulsky, 2010; Xu et al., 2016). Secretory and membrane resident proteins are synthesized in the ER through translation by ribosomes to produce a nascent polypeptide chain that is then folded by ER resident chaperones, modified post-translationally (ie. glycosylation, disulfide bond formation, lipidation) by ER enzymes, and transported to other destinations (Cybulsky, 2010; Xu et al., 2016). Before proteins are secreted, the ER ensures that the proteins are correctly folded by ER chaperones (Xu et al., 2016). Glucose Regulated Protein 78 (GRP78), also called BiP, is an ER-resident chaperone that is normally attached to three transmembrane proteins: Protein kinase R(PKR)-like endoplasmic reticulum kinase (PERK), inositol requiring enzyme 1 (IRE1), and activating transcription factor 6 (ATF6), keeping them inactivated (Dickhout, Carlisle, & Austin, 2011a; Dickhout & Krepinsky, 2009). When there are misfolded proteins in the ER, GRP78 detaches from these transmembrane proteins and binds to the misfolded proteins in order to fold them. This dissociation mediates the activity of three downstream signaling pathways: PERK signaling pathway, IRE1 signaling pathway, and ATF6 signaling pathway, leading to a cellular response called Unfolded Protein Response (UPR) (Dickhout & Krepinsky, 2009). It has also been demonstrated that unfolded proteins can directly bond with IRE1 causing activation (Gardner & Walter, 2011). The PERK signaling pathway is initiated when PERK dimerizes and autophosphorylates, transforming to a kinase that phosphorylates eukaryotic translation initiation factor 2 alpha (eIF2a), a subunit of ribosomes. Phosphorylated-eIF2a (p-eIF2a) reduces protein translation and thus the ER's burden to fold proteins. PERK activation also leads to the preferential translation of activating transcription factor 4 (ATF4) and CCAAT-enhancer-binding protein homologous protein (CHOP), both of which are transcription factors that translocate to the nucleus and increase expression of chaperones (ie. GRP78), Growth Arrest and DNA Damage Protein 34 (GADD34) (a phosphatase that removes eIF2a phosphate to restore protein translation), or apoptotic genes (occurs in case of prolonged PERK signaling activation) (Amin-Wetzel et al., 2017; Dickhout et al., 2011a). The IRE1 signaling pathway is instigated when GRP78 dissociates from IRE1 or unfolded proteins bind directly to IRE1 (Gardner & Walter, 2011), an endoribonuclease, causing it to dimerize and become activated. IRE1 activation leads to three main events. Firstly, the dimerized IRE1 splices the X-box-binding protein-1 (sXBP-1) mRNA leading to its translation to XBP1 protein, a potent transcription factor (Xu et al., 2016). XBP1 protein increases transcription of chaperones (ie. GRP78) and ERAD factors in an attempt to refold or degrade misfolded proteins (Amin-Wetzel et al., 2017; Xu et al., 2016). Secondly, the dimerized IRE1 prevents translation of new proteins by cleaving mRNAs that are to be translated into the ER (Amin-Wetzel et al., 2017). Thirdly, the IRE1 dimer reveals a binding site to Tumor Necrosis Factor Receptor-associated Factor 2 (TRAF2), which phosphorylates Apoptosis Signal Regulating Kinase-1 (ASK1) and activates c- Jun N-terminal Kinase (JNK) signaling pathway, leading to apoptosis (Xu et al., 2016). The third wing of UPR is the ATF6 signaling pathway. The ATF6 protein has a Golgi localization signal, which is normally covered by GRP78. Dissociation of GRP78 uncovers this signal and translocates ATF6 to the Golgi where ATF6 is cleaved by Site-1 (S1P) and Site-2 (S2P) protease (Dickhout, Carlisle, & Austin, 2011b). Upon this cleavage, ATF6 is released from the Golgi and moves to the nucleus where it acts as a transcription factor and upregulates transcription of ER chaperones and enzymes that promote protein folding or degradation, XBP1, CHOP (Cybulsky, 2010; Xu et al., 2016). Initially, UPR activation aims at folding or degrading unfolded proteins in order to prevent their problematic accumulation and aggregation. However, prolonged UPR activation unleashes apoptotic pathways through all three wings of the UPR and destroys the cell. For this reason, UPR activation has been described as a double-edged sword that can restore proteostasis or demolish the cell leading to a diverse range of pathologies (Kitamura, 2008).

2. ER stress in renal diseases

Endoplasmic reticulum (ER) stress is the state of misfolded proteins being accumulated in the ER lumen, activating the UPR (Cybulsky, 2010). ER stress has been associated with proteopathies such as Alzheimer's, Parkinson's, and Huntington's diseases, as well as other pathologies such as ischemia/reperfusion injury, diabetes, and atherosclerosis (Inagi, 2009). In the kidneys, ER stress has been observed in conditions such as acute and chronic kidney disease (CKD) (Inagi, 2009), glomerulosclerosis, Heymann nephritis (an experimental rat model of membranous nephropathy), mesangioproliferative glomerulonephritis, diabetic nephropathy, and minimal change disease (Inagi, 2009). ER stress and UPR pathways in the kidney can be induced in podocytes, mesangial cells, or tubular epithelium due to environmental insults such as glucose deprivation, hypoxia, oxidative stress, cellular lipid concentrations, exposure to toxins, or genetic mutations

(Inagi, 2009; Pineau et al., 2009). For instance in podocytes, accumulation of misfolded proteins in the ER is associated with ER stress and seen in several pathologies such as Passive Heymann nephritis, Puromycin nephrosis, and megsin overexpression (Inagi et al., 2005). Likewise, retention of misfolded proteins in the ER of mesengium and the tubular epithelium has been observed in mesangioproliferative glomerulonephritis and diabetic nephropathy respectively (Inagi et al., 2005).

2.1 Heritability of renal diseases

Many of these diseases including diabetic nephropathy have an inherited component (Brennan, McEvoy, Sadlier, Godson, & Martin, 2013). Heritability of a disease is the ratio of genetic variation over phenotypic variation, and is an indicator of how well genetic differences in a population explain phenotypic variation in that population (Tenesa & Haley, 2013). Heritability score ranges from zero to one. A score close to one means that almost all the variation in a phonotypic trait is due to genetic differences, suggesting the presence of a Mendelian disorder such as autosomal dominant polycystic kidney disease. On the contrary, a heritability score close to zero shows that all the variations in a trait are due to environmental factors. Heritability of several kidney diseases has been demonstrated through genome-wide association studies (Freedman, Bostrom, Daeihagh, & Bowden, 2007). These studies have identified genomic loci that are associated with kidney diseases and measures of kidney function (Freedman et al., 2007). Namely, common variants in the uromodulin (UMOD) and Protein Kinase AMP-Activated Non-Catalytic Subunit Gamma 2 (PRKAG2) genes were shown to be associated with chronic kidney disease (CKD) (Kottgen, 2010) while variants in Claudin 14 (CLDN14) was linked to kidney stone disease (Thorleifsson et al., 2009). Additionally, heritability of the primary outcome measurement of kidney function; Glomerular Filtration Rate (GFR), was estimated to be 0.33 to 0.82 which means that between 33% and 82% of the phenotypic variations in GFR values can be attributed to genetic factors (Kottgen, 2010). Moreover, heritability of end-stage renal disease (ESRD) was investigated by a group of Swedish scientists who studied ESRD in Swedish-born adopted children as well as their biological and adoptive parents. They reported the heritability of ESRD to be as high as 59.5% ± 18.2% (Akrawi et al., 2018). Nevertheless, complex diseases such as CKD are thought to be due to environmental insults affecting genetically susceptible individuals, making both genetic and environmental factors critical in the occurrence of these diseases (Freedman et al., 2007). Mendelian disorders (also called monogenic diseases) are genetically determinant since they are a result of mutation in a single gene inherited according to Mendel's Laws, or occur through spontaneous mutations (with no family history). Many of these Mendelian disorders involve generation of dysfunctional proteins that may misfold, leading to ER stress and activation of UPR (Dombroski et al., 2010). This chapter focuses on coding region mutations that lead to renal diseases through ER stress.

2.2 Renal diseases that involve coding region mutations and ER stress

2.2.1 *Fabry disease*

Fabry disease (FD) is an X-linked lysosomal storage disease involving multiple organ-systems (Di Toro, Favalli, & Arbustini, 2018). FD is caused by mutations in α-galactosidase (α-GAL) which normally catalyzes removal of a terminal α-galactose from glycolipids, glycopeptides, and polysaccharides, leading to accumulation of these macromolecules inside lysosomes (Guce et al., 2010). It has been suggested that point mutations in α-GAL can be classified to active-site or folding mutations. Active site mutations reduce the catalytic activity but may not affect the folding of the enzyme. Folding mutations, on the other hand, disrupt hydrophobic interactions of the constituent amino acids, compromising the stability of the enzyme (Garman, 2007). The resulting misfolded proteins may be retained and degraded prematurely in the ER, contributing to pathogenesis of the disease (Cigna et al., 2013; Fan & Ishii, 2003). The severity of FD symptoms is directly related to the level of residual enzyme activity with early-onset forms having almost no remaining enzyme activity and multi-system symptoms including chronic pain, circulatory problems, liver and kidney defects, ocular deterioration, and/or cardiac abnormalities (Cigna et al., 2013; Garman, 2007). Milder late-onset forms typically show some residual enzyme activities and single-organ (ie. kidney or heart) involvement (Cigna et al., 2013; Ishii et al., 2009). Accumulation of the enzyme's substrates, particularly a glycosphingolipids globotriaosylceramide (Gb3), in body organs such as the kidneys produces reactive oxygen species (ROS) and inflammatory molecules, contributing to the symptoms of the disease (Cigna et al., 2013). Renal-specific symptoms of FD experienced by some patients include presence of protein and blood in the urine, and progressive reduction in kidney function that eventually leads to kidney failure. Enzyme replacement therapy has been approved for FD treatment for about a decade. However, other therapies, such as chaperone therapy, substrate reduction therapy, and gene therapy, are rapidly being developed and tested in clinical trials. Nevertheless, FD symptoms often manifest differently in different patients, requiring the therapies to be individualized. For example, FD patients commonly receive multiple treatments for their organ-specific symptoms such as CKD, cardiac, or neurological complications (Wanner et al., 2018).

2.2.2 Von Hippel-Lindau disease

Von Hippel-Lindau disease (VHL) is an autosomal-dominant inherited disease that is as a result of a germline mutation in VHL gene on chromosome 3 (Ganeshan et al., 2018). Around 80% of VHL patients inherit one allele from an affected parent, and about 20% acquire it through random germline mutations (Ganeshan et al., 2018). All patients with hereditary VHL inherit one affected allele and over time a second mutation occur in their wild-type allele, leading to multi-organ benign or malignant tumors (Ganeshan et al., 2018). The most common germline mutations in the VHL gene are missense mutations (27%–52%). But less commonly, other types of mutations including frame-shift mutations, nonsense mutations, large or microdeletions, gene rearrangements, in-frame deletions or insertions, and splice site mutations also occur (Ganeshan et al., 2018). The second random somatic inactivation in the wild-type allele occur typically due to allelic loss, hypermethylation, or point mutations (Ganeshan et al., 2018). More than 90% of patients develop the disease by the age of 65, with the mean age of diagnosis being 26 years old (Ganeshan et al., 2018). VHL patients develop tumors in their eyes, brain, spinal cord, pancreas, adrenal glands, and kidneys. Renal manifestations of VHL mutation are benign cysts and/or renal clear cell carcinoma, the most common type of kidney cancer (Feldman, Thulasiraman, Ferreyra, & Frydman, 1999; Stebbins, Kaelin, & Pavletich, 1999). VHL gene codes for a tumor suppressor protein, VHL, that normally hydroxylases hypoxia-inducible factor 1 alpha (HIF-α) under normoxic conditions. Mutations in VHL gene often translate into a dysfunctional protein, leading to elevation of HIF-α levels and activation of several hypoxia-inducible genes such as vascular endothelial growth factor (VEGF), platelet-derived growth factor, transforming growth factor, glucose transporter 1, and phosphofructokinase-1 (Ganeshan et al., 2018).

2.2.3 Autosomal dominant tubulointerstitial kidney disease uromodulin-related (ADTKD-UMOD)

UMOD, also called Tamm-Horsfall protein, is the most abundant glycoprotein in normal urine, being excreted into the urine at a rate of 50–150 mg per day in humans (Reindl, Gröne, Wolf, & Busch, 2018). UMOD is synthesized exclusively in the kidney by epithelial cells of the thick ascending limb (TAL) of the Loop of Henle and distal convoluted tubules (DCT, 10% of TAL expression) (Tokonami et al., 2018). UMOD's tertiary structure contains 48 cysteine residues (form disulfide bonds), three epidermal growth factor domains, one zona pellucida-like domain, one D8C domain with an unknown function, and a chain of hydrophobic amino acids at the C-terminus which anchors UMOD to the plasma membrane (Rampoldi, Scolari, Amoroso, Ghiggeri, & Devuyst, 2011; Serafini-Cessi, Malagolini, & Cavallone, 2003; Van Rooijen, Voskamp, Kamerling, & Vliegenthart, 1999). At the plasma membrane, Hepsin, a member of type II transmembrane serine proteases, cleaves UMOD releasing its external hydrophobic patch and allows the internal hydrophobic patch to homopolymerize into filaments that are excreted in the urine where they play important roles (Jovine, Qi, Williams, Litscher, & Wassarman, 2004; Schaeffer, Santambrogio, Perucca, Casari, & Rampoldi, 2009). UMOD's critical functions were revealed through knockout mice and patient studies and include protection against urinary tract infections, innate immunity, preventing kidney stone formation, and salt-electrolyte balance (Devuyst, Olinger, & Rampoldi, 2017; Gudbjartsson et al., 2010; Tokonami et al., 2018). UMOD non-coding and promoter variants are associated with CKD (Olden et al., 2014), while UMOD coding mutations cause autosomal dominant tubulointerstitial kidney disease uromodulin-related (ADTKD-UMOD) (Bleyer, Hart, & Kmoch, 2016); previously also called familial juvenile hyperuricemic nephropathy type 1, medullary cystic kidney disease type 2, or uromodulin-associated kidney disease (Bleyer et al., 2016). Symptoms of ADTKD-UMOD include hyperuricemia, gout, and progressive loss of renal function (C. Schaeffer, Merella, Pasqualetto, Lazarevic, & Rampoldi, 2017). Over 100 mutations in the UMOD gene have been identified and the majority of them are missense mutations in N-terminal end of the UMOD protein (Devuyst et al., 2017). UMOD mutation has been shown to cause ER stress and activate the UPR. Mutant UMOD has increased interaction with ER chaperones such as GRP78, calnexin, and PDI; shows enhanced degradation by ER proteases through ERAD; and increases XBP1 splicing by the IRE1 enzyme (Schaeffer et al., 2017). These findings suggest that ER stress plays a critical role in the pathogenesis of ADTKD-UMOD and this pathology could be targeted for developing therapeutic strategies (Schaeffer et al., 2017). Unfortunately, currently there is no direct treatment for ADTKD-UMOD (Reindl et al., 2018). Treatments for ADTKD-UMOD are limited to treating symptoms of the disease such as gout, hypertension, CKD progression, and ESRD with renal replacement therapy (Hamada et al., 2008; Reindl et al., 2018; Sampson, Singer, & Walters, 2017).

2.2.4 Alport Syndrome

Alport Syndrome (AS) is a rare genetic disorder that affects about 1 in 50,000 births in the United States (Hicks, Mierau, Wartchow, & Eldin, 2012; Watson & Bush, 2018). Around 80% of AS cases are due to X-linked mutations in Collagen Type IV Alpha (COL4A) 5 gene, whereas 20% are due to autosomal recessive or dominant mutations in COL4A3 or COL4A4 genes (Kashtan, 2018; Savige, 2018; Vos, Zietse, van Geel, Brooks, & Cransberg, 2018; Watson & Bush, 2018). COL4A5, COL4A3, and COL4A4 genes encode for three out of the six subunits that construct type-IV collagen (Boutaud et al., 2000; Watson & Bush, 2018). Normally, the six subunits combine with each other in the ER to make three heterotrimers that self-assemble to construct type-IV collagen (Brown, Cummings, Vanacore, & Hudson, 2017). Mutations in any of these subunits can induce ER stress and activate the UPR (Pieri et al., 2014; Wang et al., 2017). To illustrate, mutant COL4A5 and COL4A3 have been shown to upregulate ER stress markers (ie. GRP78, calnexin, calreticulin, GADD34, spliced XBP1, and CHOP) and ERAD markers (ie. ER degradation-enhancing α-mannosidase-like proteins, EDEM proteins), suggesting that mutated type-IV collagen is misfolded and targeted for degradation (Pieri et al., 2014; Wang et al., 2017). Mutations in type-IV collagen cause hearing loss, eye abnormalities, and renal disease; as type-IV collagen is a component of the organ of Corti (in the inner ear), eyes (ie. lens, retinal pigment epithelium), and the glomerular basement membrane (GBM) in the kidneys (Bailey, 1987; Funk, Lin, & Miner, 2018; Watson & Bush, 2018). GBM contains five out of six subunits of type-IV collagen (COL4A1-A5 (Gunwar et al., 1998). COL4A1 and COL4A2 subunits assemble during fetal development, but COL4A3, COL4A4, and COL4A5 are formed later in life as the glomeruli mature (Kalluri, Shield, Todd, Hudson, & Neilson, 1997). In AS patients, mutations in any of the latter three subunits prevent maturation of type-IV collagen in the glomeruli. The resulting defective GBM abnormally passes proteins and blood into the urine (termed proteinuria and hematuria respectively), leading to progressive renal disease over a period of 10–20 years (Boutaud et al., 2000; Kalluri et al., 1997). At initial stages of AS, GBM may be abnormally thinned but as the disease progresses, GBM becomes focally and segmentally thickened (a condition known as focal segmental glomerulosclerosis) until a complete sclero-hyalinosis of the glomeruli is formed, leading to a total loss the glomeruli function and ESRD (Funk et al., 2018; Heidet & Gubler, 2009). About 0.2% of adults with ESRD and 3% of children with CKD are diagnosed with AS (Hicks et al., 2012). Unfortunately currently there is no cure for AS. Treatments are limited to improving the symptoms of the disease and slowing its progression to ESRD. Angiotensin-converting enzyme (ACE) inhibitors or Angiotensin receptor blockers (ARBs) are used to control the blood pressure, diuretics, and dietary salt restriction have been shown to successfully reduce blood pressure and proteinuria (Gross et al., 2003; Watson & Bush, 2018). Nevertheless, AS in most patients eventually progresses to ESRD, requiring dialysis or renal transplantation as renal replacement therapies (Funk et al., 2018). New *in-vitro* and *in-vivo* human and mice studies respectively have also shown potential therapeutic avenues for using chemical chaperones (ie. Sodium 4-phenylbutyrate, 4-PBA) and intracardiac injection of amniotic fluid stem cells in the treatment of AS (Sedrakyan et al., 2012; Wang et al., 2017).

2.2.5 Nephrogenic Diabetes Insipidus

Arginine vasopressin (AVP) (also called vasopressin or antidiuretic hormone (ADH)) is a hormone that is synthesized by the hypothalamus and released by posterior pituitary gland in response to hyperosmolality of the plasma and low arterial blood volume (Park & Yoo, 2017). AVP then travels through the circulation and binds to type 2 vasopressin receptor (AVPR2) on the basolateral membrane of principal cells in distal convoluted tubules and collecting ducts of the kidneys (Moeller, Rittig, & Fenton, 2013; Robben, Knoers, & Deen, 2004). This binding activates a downstream signaling pathway that results in phosphorylation of aquaporin 2 (AQP2) water channels in intracellular vesicles. This leads to translocation of AQP2 channels to the apical plasma membrane, where they facilitate osmotic pressure-driven water flow from the lumen to the bloodstream, concentrating the urine (Katsura, Gustafson, Ausiello, & Brown, 1997; Kwon, Frøkiær, & Nielsen, 2013; Moeller, MacAulay, Knepper, & Fenton, 2009). Diabetes Insipidus (DI) is an inherited or acquired condition that results in extreme urine excretion (called polyuria) and thirst (called polydipsia) through impairing the AVP system (Kalra et al., 2016). DI pathophysiology is very different from diabetes mellitus even though both involve ample urine production and thirst (Klibanski, Schlechte, & Tritos, 2013). There are 4 types of DI: (a) gestational DI occurs during pregnancy due to excessive vasopressinase (a cysteine aminopeptidase) production by placental trophoblasts which cleaves AVP (Marques, Gunawardana, & Grossman, 2015); (b) dipsogenic DI involves undue thirst that results in excessive fluid intake (Sarma, 2013); (c) central or neurogenic DI (CDI) is the most common type of DI and is characterized by inadequate synthesis or release of AVP (Crawford & Harris, 2011; Makaryus & McFarlane, 2006); and (d) nephrogenic DI (NDI) is an acquired or congenital condition in which the kidneys are unable to respond to AVP due to mutations predominantly in AVPR2 (X-linked recessive mutations) or AQP2 (autosomal recessive or dominant mutations) (Castaño et al., 2015; Kalra et al., 2016; Loonen, Knoers, van Os, & Deen, 2008). According to a Japanese nationwide study, congenital NDI (CNDI) occurs more commonly (about 80%) than acquired NDI (about 20%), and most CNDI cases are due to mutations in AVPR2 (about 75%), with AQP2 mutations accounting for only 10% of the cases (Fujimoto et al., 2014). AVPR2 mutations can be grouped into "loss of function" or "gain of function" mutations. "Loss of function" mutations can also be classified into 5 categories based on mutation types and subcellular localization (Moeller et al., 2013). Class 1 mutations may occur due to promoter alterations, exon skipping, or aberrant splicing; leading to an unstable mRNA that is rapidly degraded and not translated into a protein (Robben, Knoers, & Deen, 2006). Class 2 mutations, the most common type, are due to missense, insertion, or deletion of triplet nucleotides that result in a fully expressed but misfolded proteins that are retained inside the ER lumen and not transported to the plasma membrane (Bichet & Bockenhauer, 2016; Robben et al., 2006). Class 3 mutated AVPR2 proteins are folded properly, translocated to the membrane, bound to AVP, but have reduced ability to activate adenylate cyclase and start the cyclic adenosine monophosphate (cAMP) singling response (Moeller et al., 2013). Class 4 mutated AVPR2 proteins reach the membrane but have reduced AVP binding due to mutations close to AVP binding pocket such as V2R-ΔR202 (Ala et al., 1998). Class 5 mutated AVPR2 may reach the plasma membrane but they have an incorrect localization signal which misroutes them to the wrong organelle. An example is NDI R137H mutation which results in constitutive internalization of AVPR2 from the membrane, leading to only a brief presence of AVPR2 on the membrane (Andersen-Beckh et al., 1999; Barak, Oakley, Laporte, & Caron, 2001). AVPR2 "gain of function" is classically due to substitution of arginine with cysteine/leucine at number 137 amino acid, leading to a constitutively active receptor and inappropriate antidiuretic production (Hayat, Florkowski, Kolla, & Potter,

2019; Morin, Tenenbaum, Ranchin, & Durroux, 2012). Unlike AVPR2 mutations, almost all AQP2 mutations lead to misfolded proteins. AQP2 mutations that result in autosomal recessive NDI cause the misfolded AQP2 proteins to be trapped inside the ER lumen and degraded by proteasomes (Kortenoeven & Fenton, 2014; Miranda, Lee, Chou, & Knepper, 2013). In contrast, AQP2 mutations that result in autosomal dominant NDI are typically due to alterations in carboxy terminus of the AQP2 which causes the mutated protein to aggregate with the correctly-folded wild-type AQP2, obliterating the protein function (de Mattia et al., 2004; Kamsteeg et al., 2003; Loonen et al., 2008; Marr et al., 2002; Mulders et al., 1998). Chemical chaperones that aid in protein folding (ie. 4-PBA, Tolvaptan, Mozavaptan) have been shown to increase expression and correct localization of AQP2 and AVPR2 proteins, and reduce ER stress and UPR activation (Robben, Sze, Knoers, & Deen, 2007; Zheng et al., 2016). These molecular and pharmacological chaperones can be potentially used as new therapeutic strategies to treat congenital NDI (Mouillac & Mendre, 2018; Zheng et al., 2016). Other treatments currently in-use are thiazide to inhibit co-transport of sodium and chloride at distal convoluted tubules (Abraham, Rao, Price, & Choong, 2014; Makaryus & McFarlane, 2006); amiloride and hydrochlorothiazide to reduce water intake, urinary potassium loss, hypokalemia, and alkalosis (Alon & Chan, 1985); and indomethacin along with desmopressin to reduce polyuria (Kalra et al., 2016; Weinstock & Moses, 1990).

2.2.6 Congenital nephrotic syndrome

Nephrotic Syndrome (NS) is when the GFB of the kidneys abnormally pass proteins from plasma into the urine due to being damaged (Cattran et al., 2012). NS has four major characteristics through which it is diagnosed: proteinuria (protein in urine, $\geq$40 mg/m^2/hour), hypoalbuminemia (low plasma albumin, <25 g/L), edema, and hyperlipidemia (excessive lipids in blood) (Andolino & Reid-Adam, 2015; Cattran et al., 2012; Ranganathan, 2016). NS can be classified to congenital (occurring in the first 3 months of life), infantile (between 3 months and 1 year), or childhood (after 1 year of age) NS depending on the age of symptom presentation (Andolino & Reid-Adam, 2015; Downie, Gallibois, Parekh, & Noone, 2017). Congenital nephrotic syndrome (CNS) predominantly occurs due to mutations in proteins constituting the GFB (autosomal dominant and recessive mutations), and less commonly due to congenital infections (ie. syphilis, toxoplasmosis, and cytomegalovirus) (Andolino & Reid-Adam, 2015; Holmberg & Jalanko, 2014). GFB consists of a fenestrated endothelium layer that is covered by a dense structure of extracellular matrix components called the GBM. The endothelium and GBM layers are surrounded by podocytes that have long cytoplasmic extensions called foot processes, which are interconnected. The interconnected foot processes leave slits in between them called filtration slits or the slit diaphragm through which blood is filtered (Andolino & Reid-Adam, 2015; Löwik, Groenen, Levtchenko, Monnens, & Van Den Heuvel, 2009). These slit diaphragms are covered by nephrin proteins interacting with each other and forming zip-like structures. The intracellular domain of nephrin connects the slit diaphragm to the podocyte cytoskeleton through interactions with many molecules including podocin (Patrakka & Tryggvason, 2007; Yu, Nissaisorakarn, Husain, & Jim, 2018). Although since 1998, a growing number of mutations have been linked to CNS, mutations in the genes encoding for nephrin and podocin (NPHS1 and NPHS2 genes respectively) account for about 75% of the CNS cases (Hinkes et al., 2007; Kestilä et al., 1998; Wang & Mao, 2016). Other proteins critical for maintaining integrity of the GFB can also be mutated including Wilms tumor 1 (WT1, a transcription factor that binds to nephrin promoter), Laminin β2 chain (LAMB2, a component of GBM), and Phospholipse C epsilon 1 (PLCE1, an enzyme involved in podocyte growth and differentiation) (Cil et al., 2015; Domogatskaya, Rodin, & Tryggvason, 2012; Guo, Morrison, Licht, & Quaggin, 2004; Wang & Mao, 2016; Wing, Bourdon, & Harden, 2003). Depending on the mutated protein involved, CNS can be inherited as an autosomal recessive or dominant mutation (Wang & Mao, 2016). More than 200 mutations in nephrin have been reported including deletions, insertions, nonsense, missense, splice site, and promoter mutations; and most of them lead to severe CNS (Beltcheva, Martin, Lenkkeri, & Tryggvason, 2001; Gigante et al., 2002; Holmberg & Jalanko, 2014; Koziell et al., 2002; Liu et al., 2004). Missense mutations account for the majority of nephrin alterations (Liu et al., 2004). These mutations lead to protein misfolding depicted by increased interaction of the mutant proteins with ER chaperones (ie. calnexin), retention in the ER, initiation of ER-associated degradation, decreased translocation to the membrane, and activation of ER stress markers (ie. CHOP, GRP94) (Drozdova, Papillon, & Cybulsky, 2013; Liu et al., 2004). Missense mutations in LAMB2 also have been reported to induce ER stress and pro-apoptotic markers leading to podocyte injury and proteinuria (Chen et al., 2013). Misfolding in LAMB2 and nephrin have been shown to improve upon treatments with chemical chaperones taurodeoxycholic acid and 4-PBA respectively (Chen et al., 2013; Liu et al., 2004). CNS that are due to congenital infections can be cured by antibiotics depending on the type of the infection (Downie et al., 2017; Wang & Mao, 2016). Nonetheless, CNS that are caused by genetic defects can only be cured by renal transplantation with usually excellent prognosis (Holmberg & Jalanko, 2014). Medications such as angiotensin-converting enzyme (ACE) inhibitors and indomethacin may also be useful in the management of proteinuria (Jalanko & Holmberg, 2009, pp. 601–619).

3. Therapeutic options for the treatment of kidney diseases involving protein misfolding

A new emerging therapeutic avenue for the treatment of protein misfolding diseases is the use of molecular chaperones and proteasome regulators (Gámez et al., 2018; Mu, Ong, et al., 2008). Molecular chaperones are proteins that bind transitionally to unfolded polypeptides in order to fold them (Gámez et al., 2018; Hartl, 1996), whereas proteasome regulators are molecules that enhance proteostasis by inducing the UPR response, increasing transcription and translation of UPR components (ie. resident chaperones), and altering calcium signaling proteostasis pathways (Gámez et al., 2018; Mu, Fowler, & Kelly, 2008). Therapies that use chaperones and proteasome regulators synergistically may prove more beneficial than using either one alone. This is because proteasome regulators have the ability to further increase expression of resident chaperones and properly fold the mutated proteins (Lindquist & Kelly, 2011; Mu, Ong, et al., 2008). Drugs that restore protein folding can be classified into 3 different categories based on their effects on the mutated proteins: protein stabilizers, aggregation inhibitors, and proteostasis enhancers (Denny, Gavrin, & Saiah, 2013) (Table 1). First, protein stabilizers are typically small pharmacological chaperones that directly bind to unfolded proteins and direct them toward a lower free energy minimum, folding and stabilizing the unfolded proteins. For example, misfolded glucocerebrosidase in Gaucher disease (predominantly due to missense mutations) can be improved by using a small pharmacological chaperone, isofagomine, that facilitates folding of the mutated enzymes and increases its availability and functionality in the lysosomes. This type of treatment has proved to be more effective than the currently available enzyme replacement and substrate reduction therapies as isofagomine crosses blood brain barrier and improves the neurological aspects of the disorder (Denny et al., 2013). Second, aggregation inhibitors prevent formation of oligomers and amyloid clumps by binding to the misfolded proteins, this binding is

Table 1 Categorizing drugs that restore protein folding in genetic kidney diseases.

Category	Kidney disease	Mutated protein	Potential therapies	Cite
protein stabilizer	Fabry Disease	α-galactosidase	Migalastat	Cigna et al., 2013 (Hughes et al., 2017; Zhou et al., 2018)
proteostasis enhancer	von Hippel-Lindau disease (kidney cancer)	Von Hippel-Lindau protein	IDF-1174 reduces HSP70 chaperone activity promoting HIF-1 misfolding and degradation.	(Ban et al., 2017; Ganeshan et al., 2018)
proteostasis enhancer and protein stabilizers	Alport Syndrome	Type-IV collagen	4-PBA	Watson & Bush, 2018(Sedrakyan et al., 2012; Wang et al., 2017).
protein stabilizers	Nephrogenic Diabetes Insipidus	AVPR2 or AQP2	4-PBA, Tolvaptan, Mozavaptan	Moeller et al., 2013(Robben et al., 2007; Zheng et al., 2016).
protein stabilizers and proteostasis enhancer	Congenital Nephrotic Syndrome	Glomerular Filtration Barrier proteins (ie. nephrin, podocin)	TUDCA and 4-PBA	Wang & Mao, 2016(Chen et al., 2013; Liu et al., 2004)

Protein stabilizers are molecules that prevent the irreversible aggregation of misfolded protein intermediates giving appropriate chaperones an opportunity to continue to fold the protein or the ERAD system to degrade it. Proteostasis enhances modify the inherent system that regulates protein folding or degradation to produce an outcome in terms of mutant protein production that limits the disease state. *4-PBA*, Sodium 4-phenylbutyrate; *AQP2*, aquaporin 2; *AVPR2*, type 2 vasopressin receptor; *TUDCA*, taurodeoxycholic acid.

reversible and can lead to completely folding the protein, and stabilizing this new completely or partially folded structure. For instance, aggregation of prion proteins and Superoxide dismutase 1 (SOD1) proteins can be prevented by partially folding these proteins. Third, proteostasis enhancers act on the UPR pathway to upregulate chaperones and other protein quality control components such as heat shock transcription factor 1 (HSF1) and heat shock proteins (HSPs) (Denny et al., 2013). Molecular chaperones may be mutation-specific or they may work for many types of mutations in the target enzyme (Suzuki, 2014). For instance, several types of mutant AVPR2 and rhodopsin can be successfully rescued in cell culture by a single pharmacochaperone, making these compounds more applicable to treating patients (Conn & Janovick, 2009). Moreover, proteinuria, albuminuria, and hypertension are frequent complications in kidney diseases involving GFB injuries (Doris, 2012; Freedman & Sedor, 2008; Yu et al., 2018). For this reason, mutation-induced kidney diseases are often managed based on the patient's individual symptoms. This management includes use of ACE inhibitors, ARBs, and diuretics such as in the cases of Alport syndrome and CNS as previously described.

References

Abraham, M. B., Rao, S., Price, G., & Choong, C. S. (2014). Efficacy of hydrochlorothiazide and low renal solute feed in neonatal central diabetes insipidus with transition to oral desmopressin in early infancy. *International Journal of Pediatric Endocrinology, 2014*(1), 11.

Akrawi, D. S., PirouziFard, M., Fjellstedt, E., Sundquist, J., Sundquist, K., & Zoller, B. (2018). Heritability of end-stage renal disease: A Swedish adoption study. *Nephron, 138*(2), 157–165.

Ala, Y., Morin, D., Mouillac, B., Sabatier, N., Vargas, R., Cotte, N., et al. (1998). Functional studies of twelve mutant V2 vasopressin receptors related to nephrogenic diabetes insipidus: Molecular basis of a mild clinical phenotype. *Journal of the American Society of Nephrology, 9*(10), 1861–1872.

Alon, U., & Chan, J. C. (1985). Hydrochlorothiazide-amiloride in the treatment of congenital nephrogenic diabetes insipidus. *American Journal of Nephrology, 5*(1), 9–13.

Amin-Wetzel, N., Saunders, R. A., Kamphuis, M. J., Rato, C., Preissler, S., Harding, H. P., et al. (2017). A J-protein Co-chaperone recruits BiP to monomerize IRE1 and repress the unfolded protein response. *Cell, 171*(7), 1625–1637. e1613.

Andersen-Beckh, B., Dehe, M., Schülein, R., Wiesner, B., Rutz, C., Liebenhoff, U., et al. (1999). Polarized expression of the vasopressin V2 receptor in Madin-Darby canine kidney cells. *Kidney International, 56*(2), 517–527.

Andolino, T. P., & Reid-Adam, J. (2015). Nephrotic syndrome. *Pediatrics in Review, 36*(3), 117–125. quiz 126, 129.

Araki, K., & Nagata, K. (2011). Protein folding and quality control in the ER. *Cold Spring Harbor Perspectives in Biology, 3*(11). a007526.

Bailey, A. J. (1987). Structure, function and ageing of the collagens of the eye. *Eye, 1*(2), 175.

Ban, H. S., Kim, B. K., Lee, H., Kim, H. M., Harmalkar, D., & Nam, M. (2017). The novel hypoxia-inducible factor-1alpha inhibitor IDF-11774 regulates cancer metabolism, thereby suppressing tumor growth. *Cell Death Disease, 8*(6), e2843.

Barak, L. S., Oakley, R. H., Laporte, S. A., & Caron, M. G. (2001). Constitutive arrestin-mediated desensitization of a human vasopressin receptor mutant associated with nephrogenic diabetes insipidus. *Proceedings of the National Academy of Sciences, 98*(1), 93–98.

Beltcheva, O., Martin, P., Lenkkeri, U., & Tryggvason, K. (2001). Mutation spectrum in the nephrin gene (NPHS1) in congenital nephrotic syndrome. *Human Mutation, 17*(5), 368–373.

Bichet, D. G., & Bockenhauer, D. (2016). Genetic forms of nephrogenic diabetes insipidus (NDI): Vasopressin receptor defect (X-linked) and aquaporin defect (autosomal recessive and dominant). *Best Practice & Research Clinical Endocrinology & Metabolism, 30*(2), 263–276.

Bleyer, A. J., Hart, P. S., & Kmoch, S. (2016). *Autosomal dominant tubulointerstitial kidney disease, UMOD-related GeneReviews® [Internet]*. Seattle: University of Washington.

Boutaud, A., Borza, D.-B., Bondar, O., Gunwar, S., Netzer, K.-O., Singh, N., et al. (2000). Type IV collagen of the glomerular basement membrane evidence that the chain specificity of network assembly is encoded by the noncollagenous nc1 domains. *Journal of Biological Chemistry, 275*(39), 30716–30724.

Brennan, E., McEvoy, C., Sadlier, D., Godson, C., & Martin, F. (2013). The genetics of diabetic nephropathy. *Genes (Basel), 4*(4), 596–619.

Brown, K. L., Cummings, C. F., Vanacore, R. M., & Hudson, B. G. (2017). Building collagen IV smart scaffolds on the outside of cells. *Protein Science, 26*(11), 2151–2161.

Castaño, A. G., De Nanclares, G. P., Madariaga, L., Aguirre, M., Chocron, S., Madrid, A., et al. (2015). Novel mutations associated with nephrogenic diabetes insipidus. A clinical-genetic study. *European Journal of Pediatrics, 174*(10), 1373–1385.

Cattran, D. C., Feehally, J., Cook, H. T., Liu, Z. H., Fervenza, F. C., Mezzano, S. A., et al. (2012). Kidney disease: Improving global outcomes (KDIGO) glomerulonephritis work group. KDIGO clinical practice guideline for glomerulonephritis. *Kidney International Supplements, 2*(2), 139–274.

Chen, Y. M., Zhou, Y., Go, G., Marmerstein, J. T., Kikkawa, Y., & Miner, J. H. (2013). Laminin β2 gene missense mutation produces endoplasmic reticulum stress in podocytes. *Journal of the American Society of Nephrology, 24*(8), 1223–1233.

Cigna, D., D'Anna, C., Zizzo, C., Francofonte, D., Sorrentino, I., Colomba, P., et al. (2013). Alteration of proteomic profiles in PBMC isolated from patients with fabry disease: Preliminary findings. *Molecular BioSystems, 9*(6), 1162–1168.

Cil, O., Besbas, N., Duzova, A., Topaloglu, R., Peco-Antić, A., Korkmaz, E., et al. (2015). Genetic abnormalities and prognosis in patients with congenital and infantile nephrotic syndrome. *Pediatric Nephrology, 30*(8), 1279–1287.

Conn, P. M., & Janovick, J. A. (2009). Drug development and the cellular quality control system. *Trends in Pharmacological Sciences, 30*(5), 228–233.

Coulombe, P. A., Kerns, M. L., & Fuchs, E. (2009). Epidermolysis bullosa simplex: A paradigm for disorders of tissue fragility. *Journal of Clinical Investigation, 119*(7), 1784–1793.

Crawford, A., & Harris, H. (2011). Water world PART 1: Understanding the syndrome of inappropriate antidiuretic hormone secretion. *Nursing2018 Critical Care, 6*(6), 24–30.

Crews, L., Tsigelny, I., Hashimoto, M., & Masliah, E. (2009). Role of synucleins in Alzheimer's disease. *Neurotoxicity Research, 16*(3), 306–317.

Cybulsky, A. V. (2010). Endoplasmic reticulum stress in proteinuric kidney disease. *Kidney International, 77*(3), 187–193.

Del Poggetto, E., Toto, A., Aloise, C., Di Piro, F., Gori, L., Malatesta, F., et al. (2018). Stability of an aggregation-prone partially folded state of human profilin-1 correlates with aggregation propensity. *Journal of Biological Chemistry, 293*(26), 10303–10313.

Denny, R. A., Gavrin, L. K., & Saiah, E. (2013). Recent developments in targeting protein misfolding diseases. *Bioorganic & Medicinal Chemistry Letters, 23*(7), 1935–1944.
Devuyst, O., Olinger, E., & Rampoldi, L. (2017). Uromodulin: From physiology to rare and complex kidney disorders. *Nature Reviews Nephrology, 13*(9), 525–544.
Di Scala, C., Yahi, N., Boutemeur, S., Flores, A., Rodriguez, L., Chahinian, H., et al. (2016). Common molecular mechanism of amyloid pore formation by Alzheimer's beta-amyloid peptide and alpha-synuclein. *Scientific Reports, 6*, 28781.
Di Toro, A., Favalli, V., & Arbustini, E. (2018). Anderson-Fabry disease. *Journal of Cardiovascular Medicine, 19*(Suppl. 1), e1–e5.
Dickhout, J. G., Carlisle, R. E., & Austin, R. C. (2011a). Inter-relationship between cardiac hypertrophy, heart failure and chronic kidney disease – endoplasmic reticulum stress as a mediator of pathogenesis. *Circulation Research, 108*(5), 629–642.
Dickhout, J. G., Carlisle, R. E., & Austin, R. C. (2011b). Interrelationship between cardiac hypertrophy, heart failure, and chronic kidney disease: Endoplasmic reticulum stress as a mediator of pathogenesis [Research Support, Non-U.S. Gov't Review] *Circulation Research, 108*(5), 629–642.
Dickhout, J. G., & Krepinsky, J. C. (2009). Endoplasmic reticulum stress and renal disease [Research Support, Non-U.S. Gov't Review] *Antioxidants and Redox Signaling, 11*(9), 2341–2352.
Dill, K. A., & MacCallum, J. L. (2012). The protein-folding problem, 50 years on. *Science, 338*(6110), 1042–1046.
Dombroski, B. A., Nayak, R. R., Ewens, K. G., Ankener, W., Cheung, V. G., & Spielman, R. S. (2010). Gene expression and genetic variation in response to endoplasmic reticulum stress in human cells. *The American Journal of Human Genetics, 86*(5), 719–729.
Domogatskaya, A., Rodin, S., & Tryggvason, K. (2012). Functional diversity of laminins. *Annual Review of Cell and Developmental Biology, 28*, 523–553.
Doris, P. A. (2012). Genetic susceptibility to hypertensive renal disease. *Cellular and Molecular Life Sciences, 69*(22), 3751–3763.
Downie, M. L., Gallibois, C., Parekh, R. S., & Noone, D. G. (2017). Nephrotic syndrome in infants and children: Pathophysiology and management. *Paediatrics and International Child Health, 37*(4), 248–258.
Drozdova, T., Papillon, J., & Cybulsky, A. V. (2013). Nephrin missense mutations: Induction of endoplasmic reticulum stress and cell surface rescue by reduction in chaperone interactions. *Physiological Reports, 1*(4), e00086.
Dubey, S. P. N., Kini, N. G., Balaji, S., & Kumar, M. S. (2018). A review of protein structure prediction using lattice model. *Critical Reviews in Biomedical Engineering, 46*(2), 147–162.
Eisenberg, D. (2003). The discovery of the alpha-helix and beta-sheet, the principal structural features of proteins. *Proceedings of the National Academy of Sciences of the U S A, 100*(20), 11207–11210.
Fan, J. Q., & Ishii, S. (2003). Cell-based screening of active-site specific chaperone for the treatment of Fabry disease. *Methods in Enzymology, 363*, 412–420.
Feldman, D. E., Thulasiraman, V., Ferreyra, R. G., & Frydman, J. (1999). Formation of the VHL-elongin BC tumor suppressor complex is mediated by the chaperonin TRiC. *Molecular Cell, 4*(6), 1051–1061.
Freedman, B. I., Bostrom, M., Daeihagh, P., & Bowden, D. W. (2007). Genetic factors in diabetic nephropathy. *Clinical Journal of the American Society of Nephrology, 2*(6), 1306–1316.
Freedman, B. I., & Sedor, J. R. (2008). Hypertension-associated kidney disease: Perhaps no more. *Journal of the American Society of Nephrology, 19*(11), 2047–2051.
Fujimoto, M., Okada, S.-i., Kawashima, Y., Nishimura, R., Miyahara, N., Kawaba, Y., et al. (2014). Clinical overview of nephrogenic diabetes insipidus based on a nationwide survey in Japan. *Yonago Acta Medica, 57*(2), 85.

Funk, S. D., Lin, M.-H., & Miner, J. H. (2018). Alport syndrome and pierson syndrome: Diseases of the glomerular basement membrane. *Matrix Biology, 71*, 250–261.

Gámez, A., Yuste-Checa, P., Brasil, S., Briso-Montiano, Á., Desviat, L. R., Ugarte, M., et al. (2018). Protein misfolding diseases: Prospects of pharmacological treatment. *Clinical Genetics, 93*(3), 450–458.

Ganeshan, D., Menias, C. O., Pickhardt, P. J., Sandrasegaran, K., Lubner, M. G., Ramalingam, P., et al. (2018). Tumors in von Hippel-lindau syndrome: From head to toe-comprehensive state-of-the-art review. *RadioGraphics, 38*(3), 982.

Gardner, B. M., & Walter, P. (2011). Unfolded proteins are Ire1-activating ligands that directly induce the unfolded protein response. *Science, 333*(6051), 1891–1894.

Garman, S. C. (2007). Structure-function relationships in alpha-galactosidase A. *Acta Paediatrica, 96*(455), 6–16.

Geller, D. S., Farhi, A., Pinkerton, N., Fradley, M., Moritz, M., Spitzer, A., et al. (2000). Activating mineralocorticoid receptor mutation in hypertension exacerbated by pregnancy. *Science, 289*(5476), 119–123.

Ghouse, R., Chu, A., Wang, Y., & Perlmutter, D. H. (2014). Mysteries of alpha1-antitrypsin deficiency: Emerging therapeutic strategies for a challenging disease. *Dis Model Mech*, 7(4), 411–419.

Gigante, M., Monno, F., Roberto, R., Laforgia, N., Assael, M. B., Livolti, S., et al. (2002). Congenital nephrotic syndrome of the Finnish type in Italy: A molecular approach. *Journal of Nephrology, 15*(6), 696–702.

Gross, O., Beirowski, B., Koepke, M.-L., Kuck, J., Reiner, M., Addicks, K., et al. (2003). Preemptive ramipril therapy delays renal failure and reduces renal fibrosis in COL4A3-knockout mice with Alport syndrome. *Kidney International, 63*(2), 438–446.

Guce, A. I., Clark, N. E., Salgado, E. N., Ivanen, D. R., Kulminskaya, A. A., Brumer, H., 3rd, et al. (2010). Catalytic mechanism of human alpha-galactosidase. *Journal of Biological Chemistry, 285*(6), 3625–3632.

Gudbjartsson, D. F., Holm, H., Indridason, O. S., Thorleifsson, G., Edvardsson, V., Sulem, P., et al. (2010). Association of variants at UMOD with chronic kidney disease and kidney stones—role of age and comorbid diseases. *PLoS Genetics, 6*(7), e1001039.

Gunwar, S., Ballester, F., Noelken, M. E., Sado, Y., Ninomiya, Y., & Hudson, B. G. (1998). Glomerular basement membrane identification of a novel disulfide-cross-linked network of A3, A4, and A5 chains of type IV collagen and its implications for the pathogenesis of alport syndrome. *Journal of Biological Chemistry, 273*(15), 8767–8775.

Guo, G., Morrison, D. J., Licht, J. D., & Quaggin, S. E. (2004). WT1 activates a glomerular-specific enhancer identified from the human nephrin gene. *Journal of the American Society of Nephrology, 15*(11), 2851–2856.

Hamada, T., Ichida, K., Hosoyamada, M., Mizuta, E., Yanagihara, K., Sonoyama, K., et al. (2008). Uricosuric action of losartan via the inhibition of urate transporter 1 (URAT 1) in hypertensive patients. *American Journal of Hypertension, 21*(10), 1157–1162.

Hartl, F. U. (1996). Molecular chaperones in cellular protein folding. *Nature, 381*(6583), 571.

Hartl, F. U. (2017). Protein misfolding diseases. *Annual Review of Biochemistry, 86*, 21–26.

Heidet, L., & Gubler, M.-C. (2009). The renal lesions of Alport syndrome. *Journal of the American Society of Nephrology, 20*(6), 1210–1215.

Hicks, J., Mierau, G., Wartchow, E., & Eldin, K. (2012). Renal diseases associated with hematuria in children and adolescents: A brief tutorial. *Ultrastructural Pathology, 36*(1), 1–18.

Hinkes, B. G., Mucha, B., Vlangos, C. N., Gbadegesin, R., Liu, J., Hasselbacher, K., et al. (2007). Nephrotic syndrome in the first year of life: Two thirds of cases are caused by mutations in 4 genes (NPHS1, NPHS2, WT1, and LAMB2). *Pediatrics, 119*(4), e907–e919.

Holmberg, C., & Jalanko, H. (2014). Congenital nephrotic syndrome and recurrence of proteinuria after renal transplantation. *Pediatric Nephrology, 29*(12), 2309–2317.

Hayat, A., Florkowski, C., Kolla, A., & Potter, H. (2019). Nephrogenic syndrome of inappropriate antidiuresis. *Journal of Internal Medicine, 49*(5), 680–681.
Hughes, D. A., Nicholls, K., Shankar, S. P., Sunder-Plassmann, G., Koeller, D., & Nedd, K. (2017). Oral pharmacological chaperone migalastat compared with enzyme replacement therapy in Fabry disease: 18-month results from the randomised phase III ATTRACT study. *Journal of Medical Genetics, 54*(4), 288–296.
Inagi, R. (2009). Endoplasmic reticulum stress in the kidney as a novel mediator of kidney injury. *Nephron Experimental Nephrology, 112*(1), e1–9.
Inagi, R., Nangaku, M., Onogi, H., Ueyama, H., Kitao, Y., Nakazato, K., et al. (2005). Involvement of endoplasmic reticulum (ER) stress in podocyte injury induced by excessive protein accumulation. *Kidney International, 68*(6), 2639–2650.
Ishii, S., Chang, H. H., Yoshioka, H., Shimada, T., Mannen, K., Higuchi, Y., et al. (2009). Preclinical efficacy and safety of 1-deoxygalactonojirimycin in mice for Fabry disease. *Journal of Pharmacology and Experimental Therapeutics, 328*(3), 723–731.
Jalanko, H., & Holmberg, C. (2009). Congenital nephrotic syndrome. In *Pediatric nephrology: Sixth completely revised*. Updated and Enlarged Edition.
Jovine, L., Qi, H., Williams, Z., Litscher, E. S., & Wassarman, P. M. (2004). A duplicated motif controls assembly of zona pellucida domain proteins. *Proceedings of the National Academy of Sciences, 101*(16), 5922–5927.
Kalluri, R., Shield, C. F., Todd, P., Hudson, B. G., & Neilson, E. G. (1997). Isoform switching of type IV collagen is developmentally arrested in X-linked Alport syndrome leading to increased susceptibility of renal basement membranes to endoproteolysis. *The Journal of Clinical Investigation, 99*(10), 2470–2478.
Kalra, S., Zargar, A. H., Jain, S. M., Sethi, B., Chowdhury, S., Singh, A. K., et al. (2016). Diabetes insipidus: The other diabetes. *Indian Journal of Endocrinology and Metabolism, 20*(1), 9.
Kamsteeg, E.-J., Bichet, D. G., Konings, I. B., Nivet, H., Lonergan, M., Arthus, M.-F., et al. (2003). Reversed polarized delivery of an aquaporin-2 mutant causes dominant nephrogenic diabetes insipidus. *The Journal of Cell Biology, 163*(5), 1099–1109.
Kashtan, C. E. (2018). Renal transplantation in patients with Alport syndrome: Patient selection, outcomes, and donor evaluation. *International Journal of Nephrology and Renovascular Disease, 11*, 267.
Katsura, T., Gustafson, C. E., Ausiello, D. A., & Brown, D. (1997). Protein kinase A phosphorylation is involved in regulated exocytosis of aquaporin-2 in transfected LLC-PK1 cells. *American Journal of Physiology - Renal Physiology, 272*(6), F816–F822.
Kellis, M., Wold, B., Snyder, M. P., Bernstein, B. E., Kundaje, A., Marinov, G. K., et al. (2014). Defining functional DNA elements in the human genome. *Proceedings of the National Academy of Sciences of the U S A, 111*(17), 6131–6138.
Kestilä, M., Lenkkeri, U., Männikkö, M., Lamerdin, J., McCready, P., Putaala, H., et al. (1998). Positionally cloned gene for a novel glomerular protein—nephrin—is mutated in congenital nephrotic syndrome. *Molecular Cell, 1*(4), 575–582.
Kitamura, M. (2008). Endoplasmic reticulum stress and unfolded protein response in renal pathophysiology: Janus faces. *American Journal of Physiology - Renal Physiology, 295*(2), F323–F334.
Klibanski, A., Schlechte, J., & Tritos, N. (2013). Diabetes insipidus. *The Journal of Clinical Endocrinology & Metabolism, 98*(7), 35A–36A.
Kortenoeven, M. L., & Fenton, R. A. (2014). Renal aquaporins and water balance disorders. *Biochimica et Biophysica Acta (BBA) - General Subjects, 1840*(5), 1533–1549.
Kottgen, A. (2010). Genome-wide association studies in nephrology research. *American Journal of Kidney Diseases, 56*(4), 743–758.
Koziell, A., Grech, V., Hussain, S., Lee, G., Lenkkeri, U., Tryggvason, K., et al. (2002). Genotype/phenotype correlations of NPHS1 and NPHS2 mutations in nephrotic

syndrome advocate a functional inter-relationship in glomerular filtration. *Human Molecular Genetics, 11*(4), 379–388.

Kwon, T.-H., Frøkiær, J., & Nielsen, S. (2013). Regulation of aquaporin-2 in the kidney: A molecular mechanism of body-water homeostasis. *Kidney Research and Clinical Practice, 32*(3), 96–102.

LaFerla, F. M., Green, K. N., & Oddo, S. (2007). Intracellular amyloid-beta in Alzheimer's disease. *Nature Reviews Neuroscience, 8*(7), 499–509.

Lai, C. W., Otero, J. H., Hendershot, L. M., & Snapp, E. (2012). ERdj4 protein is a soluble endoplasmic reticulum (ER) DnaJ family protein that interacts with ER-associated degradation machinery. *Journal of Biological Chemistry, 287*(11), 7969–7978.

Lindquist, S. L., & Kelly, J. W. (2011). Chemical and biological approaches for adapting proteostasis to ameliorate protein misfolding and aggregation diseases–progress and prognosis. *Cold Spring Harbor Perspectives in Biology, 3*(12). a004507.

Liu, X. L., Doné, S. C., Yan, K., Kilpeläinen, P., Pikkarainen, T., & Tryggvason, K. (2004). Defective trafficking of nephrin missense mutants rescued by a chemical chaperone. *Journal of the American Society of Nephrology, 15*(7), 1731–1738.

Loonen, A. J., Knoers, N. V., van Os, C. H., & Deen, P. M. (2008). *Aquaporin 2 mutations in nephrogenic diabetes insipidus*. Paper presented at the Seminars in nephrology.

Löwik, M., Groenen, P., Levtchenko, E., Monnens, L., & Van Den Heuvel, L. (2009). Molecular genetic analysis of podocyte genes in focal segmental glomerulosclerosis—a review. *European Journal of Pediatrics, 168*(11), 1291.

Makaryus, A. N., & McFarlane, S. I. (2006). Diabetes insipidus: Diagnosis and treatment of a complex disease. *Cleveland Clinic Journal of Medicine, 73*(1), 65.

Marques, P., Gunawardana, K., & Grossman, A. (2015). Transient diabetes insipidus in pregnancy. *Endocrinology, Diabetes & Metabolism Case Reports, 2015*(1).

Marr, N., Bichet, D. G., Hoefs, S., Savelkoul, P. J., Konings, I. B., De Mattia, F., et al. (2002). Cell-biologic and functional analyses of five new Aquaporin-2 missense mutations that cause recessive nephrogenic diabetes insipidus. *Journal of the American Society of Nephrology, 13*(9), 2267–2277.

de Mattia, F., Savelkoul, P. J., Bichet, D. G., Kamsteeg, E.-J., Konings, I. B., Marr, N., et al. (2004). A novel mechanism in recessive nephrogenic diabetes insipidus: Wild-type aquaporin-2 rescues the apical membrane expression of intracellularly retained AQP2-P262L. *Human Molecular Genetics, 13*(24), 3045–3056.

Miranda, C. A., Lee, J. W., Chou, C.-L., & Knepper, M. A. (2013). Tolvaptan as a tool in renal physiology. *American Journal of Physiology - Renal Physiology, 306*(3), F359–F366.

Moeller, H. B., MacAulay, N., Knepper, M. A., & Fenton, R. A. (2009). Role of multiple phosphorylation sites in the COOH-terminal tail of aquaporin-2 for water transport: Evidence against channel gating. *American Journal of Physiology - Renal Physiology, 296*(3), F649–F657.

Moeller, H. B., Rittig, S., & Fenton, R. A. (2013). Nephrogenic diabetes insipidus: Essential insights into the molecular background and potential therapies for treatment. *Endocrine Reviews, 34*(2), 278–301.

Morin, D., Tenenbaum, J., Ranchin, B., & Durroux, T. (2012). Nephrogenic syndrome of inappropriate antidiuresis. *International Journal of Pediatric Otorhinolaryngology, 2012*, 937175.

Mouillac, B., & Mendre, C. (2018). Pharmacological chaperones as potential therapeutic strategies for misfolded mutant vasopressin receptors. In A. Ulloa-Aguirre, & Y.-X. Tao (Eds.), *Targeting trafficking in drug development* (pp. 63–83). Cham: Springer International Publishing.

Mu, T.-W., Fowler, D. M., & Kelly, J. W. (2008). Partial restoration of mutant enzyme homeostasis in three distinct lysosomal storage disease cell lines by altering calcium homeostasis. *PLoS Biology, 6*(2), e26.

Mulders, S. M., Bichet, D. G., Rijss, J. P., Kamsteeg, E.-J., Arthus, M.-F., Lonergan, M., et al. (1998). An aquaporin-2 water channel mutant which causes autosomal dominant nephrogenic diabetes insipidus is retained in the Golgi complex. *The Journal of Clinical Investigation, 102*(1), 57–66.

Mu, T.-W., Ong, D. S. T., Wang, Y.-J., Balch, W. E., Yates, J. R., Segatori, L., et al. (2008). Proteostasis regulators and pharmacologic chaperones synergize to correct protein misfolding diseases. *Cell, 134*(5), 769.

Olden, M., Corre, T., Hayward, C., Toniolo, D., Ulivi, S., Gasparini, P., et al. (2014). Common variants in UMOD associate with urinary uromodulin levels: A meta-analysis. *Journal of the American Society of Nephrology, 25*(8), 1869–1882.

Park, K. S., & Yoo, K. Y. (2017). Role of vasopressin in current anesthetic practice. *Korean Journal of Anesthesiology, 70*(3), 245.

Patrakka, J., & Tryggvason, K. (2007). Nephrin—a unique structural and signaling protein of the kidney filter. *Trends in Molecular Medicine, 13*(9), 396–403.

Patrushev, L. I., & Kovalenko, T. F. (2014). Functions of noncoding sequences in mammalian genomes. *Biochemistry (Mosc), 79*(13), 1442–1469.

Pauling, L., Corey, R. B., & Branson, H. R. (1951). The structure of proteins; two hydrogen-bonded helical configurations of the polypeptide chain. *Proceedings of the National Academy of Sciences of the U S A, 37*(4), 205–211.

Pieri, M., Stefanou, C., Zaravinos, A., Erguler, K., Stylianou, K., Lapathitis, G., et al. (2014). Evidence for activation of the unfolded protein response in collagen IV nephropathies. *Journal of the American Society of Nephrology, 25*(2), 260–275.

Pineau, L., Colas, J., Dupont, S., Beney, L., Fleurat-Lessard, P., Berjeaud, J. M., et al. (2009). Lipid-induced ER stress: Synergistic effects of sterols and saturated fatty acids. *Traffic, 10*(6), 673–690.

Rambaran, R. N., & Serpell, L. C. (2008). Amyloid fibrils: Abnormal protein assembly. *Prion, 2*(3), 112–117.

Rampoldi, L., Scolari, F., Amoroso, A., Ghiggeri, G., & Devuyst, O. (2011). The rediscovery of uromodulin (Tamm-Horsfall protein): From tubulointerstitial nephropathy to chronic kidney disease. *Kidney International, 80*(4), 338–347.

Ranganathan, S. (2016). Pathology of podocytopathies causing nephrotic syndrome in children. *Frontiers in Pediatrics, 4*, 32.

Reindl, J., Gröne, H.-J., Wolf, G., & Busch, M. (2018). Uromodulin-related autosomal-dominant tubulointerstitial kidney disease—pathogenetic insights based on a case. *Clinical Kidney Journal, 12*(2), 172–179.

Robben, J., Knoers, N., & Deen, P. (2004). Regulation of the vasopressin V2 receptor by vasopressin in polarized renal collecting duct cells. *Molecular Biology of the Cell, 15*(12), 5693–5699.

Robben, J. H., Knoers, N. V., & Deen, P. M. (2006). Cell biological aspects of the vasopressin type-2 receptor and aquaporin 2 water channel in nephrogenic diabetes insipidus. *American Journal of Physiology - Renal Physiology, 291*(2), F257–F270.

Robben, J., Sze, M., Knoers, N., & Deen, P. M. (2007). Functional rescue of vasopressin V2 receptor mutants in MDCK cells by pharmacochaperones: Relevance to therapy of nephrogenic diabetes insipidus. *American Journal of Physiology - Renal Physiology, 292*(1), F253–F260.

Ron, I., & Horowitz, M. (2005). ER retention and degradation as the molecular basis underlying Gaucher disease heterogeneity. *Human Molecular Genetics, 14*(16), 2387–2398.

Sampson, A. L., Singer, R. F., & Walters, G. D. (2017). Uric acid lowering therapies for preventing or delaying the progression of chronic kidney disease. *Cochrane Database of Systematic Reviews, 10*.

Sarma, K. (2013). Algorithmic approach for the diagnosis of polyuria. *Medicine Update, 23*, 311–313.

Savige, J. (2018). Alport syndrome: Deducing the mode of inheritance from the presence of haematuria in family members. *Pediatric Nephrology*, 1–8.
Schaeffer, C., Merella, S., Pasqualetto, E., Lazarevic, D., & Rampoldi, L. (2017). Mutant uromodulin expression leads to altered homeostasis of the endoplasmic reticulum and activates the unfolded protein response. *PLoS One, 12*(4), e0175970.
Schaeffer, C., Santambrogio, S., Perucca, S., Casari, G., & Rampoldi, L. (2009). Analysis of uromodulin polymerization provides new insights into the mechanisms regulating ZP domain-mediated protein assembly. *Molecular Biology of the Cell, 20*(2), 589–599.
Sedrakyan, S., Da Sacco, S., Milanesi, A., Shiri, L., Petrosyan, A., Varimezova, R., et al. (2012). Injection of amniotic fluid stem cells delays progression of renal fibrosis. *Journal of the American Society of Nephrology, 23*(4), 661–673.
Serafini-Cessi, F., Malagolini, N., & Cavallone, D. (2003). Tamm-horsfall glycoprotein: Biology and clinical relevance. *American Journal of Kidney Diseases, 42*(4), 658–676.
Stebbins, C. E., Kaelin, W. G., Jr., & Pavletich, N. P. (1999). Structure of the VHL-ElonginC-ElonginB complex: Implications for VHL tumor suppressor function. *Science, 284*(5413), 455–461.
Stefl, S., Nishi, H., Petukh, M., Panchenko, A. R., & Alexov, E. (2013). Molecular mechanisms of disease-causing missense mutations. *Journal of Molecular Biology, 425*(21), 3919–3936.
Surguchev, A., & Surguchov, A. (2010). Conformational diseases: Looking into the eyes. *Brain Research Bulletin, 81*(1), 12–24.
Suzuki, Y. (2014). Emerging novel concept of chaperone therapies for protein misfolding diseases. *Proceedings of the Japan Academy, Series B, 90*(5), 145–162.
Takano, K., Liu, D., Tarpey, P., Gallant, E., Lam, A., Witham, S., et al. (2012). An X-linked channelopathy with cardiomegaly due to a CLIC2 mutation enhancing ryanodine receptor channel activity. *Human Molecular Genetics, 21*(20), 4497–4507.
Tenesa, A., & Haley, C. S. (2013). The heritability of human disease: Estimation, uses and abuses. *Nature Reviews Genetics, 14*(2), 139–149.
Thorleifsson, G., Holm, H., Edvardsson, V., Walters, G. B., Styrkarsdottir, U., Gudbjartsson, D. F., et al. (2009). Sequence variants in the CLDN14 gene associate with kidney stones and bone mineral density. *Nature Genetics, 41*(8), 926–930.
Tokonami, N., Takata, T., Beyeler, J., Ehrbar, I., Yoshifuji, A., Christensen, E. I., et al. (2018). Uromodulin is expressed in the distal convoluted tubule, where it is critical for regulation of the sodium chloride cotransporter NCC. *Kidney International, 94*(4), 701–715.
Valastyan, J. S., & Lindquist, S. (2014). Mechanisms of protein-folding diseases at a glance. *Disease Models & Mechanisms,* 7(1), 9–14.
Van Rooijen, J. J., Voskamp, A. F., Kamerling, J. P., & Vliegenthart, J. F. (1999). Glycosylation sites and site-specific glycosylation in human Tamm-Horsfall glycoprotein. *Glycobiology, 9*(1), 21–30.
Vos, P., Zietse, R., van Geel, M., Brooks, A. S., & Cransberg, K. (2018). Diagnosing Alport syndrome: Lessons from the pediatric ward. *Nephron, 140*(3), 203–210.
Wang, J.-J., & Mao, J.-H. (2016). The etiology of congenital nephrotic syndrome: Current status and challenges. *World Journal of Pediatrics, 12*(2), 149–158.
Wang, D., Mohammad, M., Wang, Y., Tan, R., Murray, L. S., Ricardo, S., et al. (2017). The chemical chaperone, PBA, reduces ER stress and autophagy and increases collagen IV α5 expression in cultured fibroblasts from men with X-linked Alport syndrome and missense mutations. *Kidney International Reports, 2*(4), 739–748.
Wanner, C., Arad, M., Baron, R., Burlina, A., Elliott, P. M., Feldt-Rasmussen, U., et al. (2018). European expert consensus statement on therapeutic goals in Fabry disease. *Molecular Genetics and Metabolism, 124*(3), 189–203.
Watson, S., & Bush, J. S. (2018). *Alport syndrome StatPearls [Internet].* StatPearls Publishing.

Weinstock, R., & Moses, A. (1990). Desmopressin and indomethacin therapy for nephrogenic diabetes insipidus in patients receiving lithium carbonate. *Southern Medical Journal, 83*(12), 1475–1477.

Wing, M. R., Bourdon, D. M., & Harden, T. K. (2003). PLC-ε: A shared effector protein in ras-, rho-, and Gαβγ-mediated signaling. *Molecular Interventions, 3*(5), 273.

Witham, S., Takano, K., Schwartz, C., & Alexov, E. (2011). A missense mutation in CLIC2 associated with intellectual disability is predicted by in silico modeling to affect protein stability and dynamics. *Proteins, 79*(8), 2444–2454.

Xu, Y., Guo, M., Jiang, W., Dong, H., Han, Y., An, X. F., et al. (2016). Endoplasmic reticulum stress and its effects on renal tubular cells apoptosis in ischemic acute kidney injury. *Renal Failure, 38*(5), 831–837.

Yu, S. M.-W., Nissaisorakarn, P., Husain, I., & Jim, B. (2018). Proteinuric kidney diseases: A podocyte's slit diaphragm and cytoskeleton approach. *Frontiers in Medicine, 5.*

Yu, X., Wang, C., & Li, Y. (2006). Classification of protein quaternary structure by functional domain composition. *BMC Bioinformatics, 7*, 187.

Zheng, L., Cedazo-Minguez, A., Hallbeck, M., Jerhammar, F., Marcusson, J., & Terman, A. (2012). Intracellular distribution of amyloid beta peptide and its relationship to the lysosomal system. *Translational Neurodegeneration, 1*(1), 19.

Zheng, P., Lin, Y., Wang, F., Luo, R., Zhang, T., Hu, S., et al. (2016). 4-PBA improves lithium-induced nephrogenic diabetes insipidus by attenuating ER stress. *American Journal of Physiology - Renal Physiology, 311*(4), F763–F776.

Zhou, W., Ni, Z., & Zhang, M. (Sep 2018). *Hemizygous Fabry disease associated with membranous nephropathy: A rare case report.*

CHAPTER SEVEN

Misfolding of vasopressin receptors: biased agonist pharmacochaperones as potential therapeutics

Christiane Mendre and Bernard Mouillac*
Institut de Génomique Fonctionnelle, CNRS, INSERM, Université de Montpellier, Montpellier, France
*Corresponding author: E-mail: bernard.mouillac@igf.cnrs.fr

Contents

Abstract

Biased agonists and pharmacological chaperones have demonstrated their potential to harness G protein-coupled receptor signaling and trafficking, and have collectively opened new possibilities in G protein-coupled receptor drug discovery. Combining pharmacological chaperoning and biased agonism properties into a unique given molecule would be of high therapeutic interest in many human diseases resulting from G protein-coupled receptor mutation and misfolding. This strategy perfectly fits to congenital Nephrogenic Diabetes Insipidus which is a typical conformational disease. In most of the cases, it is associated to inactivating mutations of the renal arginine-vasopressin V2 receptor leading to misfolding and intracellular retention of the receptor, causing the inability of patients to concentrate their urine in response to

Advances in Protein Chemistry and Structural Biology, Volume 118
ISSN 1876-1623
https://doi.org/10.1016/bs.apcsb.2019.07.002
© 2020 Elsevier Inc.
All rights reserved.

the antidiuretic hormone. Cell-permeable pharmacological chaperones have been successfully challenged to restore plasma membrane localization of the receptor mutants and to rescue their function. Interestingly, different classes of pharmacological chaperones of the V2 receptor have proven their usefulness and efficacy, such as antagonists, agonists as well as biased agonists. These compounds, particularly small-molecule biased agonists which elicit the V2-induced Gs protein-dependent signaling pathway, but not V2-related arrestin-dependent cell responses, represent a potential therapeutic treatment of this X-linked genetic pathology.

1. Biased ligands and pharmacological chaperones for G protein-coupled receptors

1.1 Biased agonists

The discovery and the characterization of ligand-biased signaling at G protein-coupled receptors (GPCRs) has completely changed the classical concepts of receptor pharmacology (Galandrin, Oligny-Longpré, & Bouvier, 2007; Lutrell, 2014; Luttrell, Maudsley, & Bohn, 2015; Reiter, Ahn, Shukla, & Lefkowitz, 2012). For decades, the activity of each GPCR was associated to only one major signaling pathway, in general a specific coupling to a unique G protein and generation of a second messenger (inositol phosphates, calcium, cAMP for instance). If considering this single downstream outcome, GPCR ligands were classified as agonists (able to generate a maximal response), partial agonists (able to induce a submaximal response at saturating concentration), antagonists (inhibitors of agonist response but without intrinsic activity), or inverse agonists (ligands that can decrease or suppress basal receptor activity). This view evolved rapidly, taking into account that GPCRs can couple to several different G proteins, allowing a single receptor to engage multiple signaling pathways simultaneously (Laugwitz et al., 1996; Offermanns et al., 1994), and that structurally distinct ligands can activate the same GPCR in different ways (Holloway et al., 2002; Sagan, Chassaing, Pradier, & Lavielle, 1996; Takasu, Gardella, Luck, Potts, & Bringhurst, 1999). In other words, a GPCR possesses different active states, and ligand structure can "bias" downstream signaling. Moreover, it also became evident that ligand-activated GPCRs can engage G protein-independent signaling pathways, for instance through activation of β-arrestins (Ferguson, 2001; Rajagopal, Rajagopal, & Lefkowitz, 2010; Shenoy & Lefkowitz, 2011) whose primarily role was clearly defined in desensitization of G protein-dependent signaling. Whereas second messengers generated via G protein-dependent activation of enzymatic effectors account for most of the classical short-term consequences of GPCR activation, arrestin-mediated signals appear to perform numerous functions on a much longer time-scale (up to several hours) among them protein synthesis, cell migration, cytoskeletal rearrangement, and cell proliferation to apoptosis (Luttrell & Gesty-Palmer, 2010). Thus, discovering that GPCRs can elicit separable G protein- and arrestin-dependent signaling pathways and that ligands can differentially activate or inhibit one or the other process, opened the way to a complete pharmacological reassessment of known compounds and to the development of novel ligands with unique properties, able to selectively modulate GPCR activity and associated downstream cellular events. These properties have been referred to as "biased agonism" or functional selectivity.

Either G protein-biased or β-arrestin-biased ligands have been identified. A full G protein-biased agonist leads to robust coupling and activation of G proteins but not recruitment of β-arrestins, whereas a complete β-arrestin-biased agonist does not promote G protein coupling but induces robust β-arrestin interaction. These notions uncovered a new paradigm in drug discovery that relies on the pluridimensionality of GPCR signaling, with the aim to develop potential therapeutics with better efficacy and fewer adverse effects. Proof-of-concept studies have demonstrated that both G protein and β-arrestin pathway-selective ligands can promote beneficial effects *in vivo* while simultaneously antagonizing deleterious ones. Some drugs with biased properties targeting β-adrenergic receptors (Carter & Hill, 2005; Wisler et al., 2007), dopamine receptors (Chen et al., 2012), histamine receptors (Thurmond et al., 2004), opioid receptors (Groer et al., 2007) or angiotensin II receptors (Violin et al., 2010) are listed in Table 1.

1.2 Pharmacological chaperones

Many genetic and neurodegenerative diseases in humans result from protein misfolding and/or aggregation. These pathologies are classified as conformational diseases (Chaudhuri & Paul, 2006; Cohen & Kelly, 2003) in which misfolded proteins are rejected and misrouted by the cellular quality control system, consequently being unable to play their physiological roles. Many of the misfolded protein mutants responsible for these pathologies are sequestered in an intracellular compartment (usually the endoplasmic reticulum, ER), but can be rescued by chemical chaperones (Sato, Ward, Krouse, Wine, & Kopito, 1996) or pharmacological chaperones (PC). Contrary to chemical chaperones which are nonspecific small organic compounds (glycerol, dimethyl sulfoxide, trimethyl-N-oxide), PC (also termed

Table 1 Several biased reference ligands for GPCRs: their receptor target, their *in vitro* activity and their *in vivo* beneficial physiological effects.

Receptor	Biased ligand	Target pathology	*in vitro* activity	Therapeutic advantages
β1-AR	carvedilol	cardiac arrhythmia, heart failure	β-arrestin biased agonist, no G protein signaling	better cardioprotection
β2-AR	salmeterol	respiratory diseases, asthma	Gs biased agonist, very low β-arrestin signaling	long-acting bronchodilatation
D2R	aripiprazole derivatives	psychiatric disorders, schizophrenia	β-arrestin biased agonist, no Gs protein signaling	better antipsychotic activity (mice)
H4R	JNJ7777120, VUF10214	allergies, asthma	G protein antagonist, non-selective β-arrestin biased agonist	suppression of cough through reduction airway inflammation (guinea pig)
μ-OR	herkinorin, TRV130	pain	Gi biased agonist, no β-arrestin signaling	analgesic with less adverse effects (constipation and respiratory suppression)
AT1AR	TRV027, SII	hypertension, acute heart failure	partial β-arrestin biased agonist, full G protein antagonist	decreased blood pressure, improved cardiomyocyte contraction

β1-AR, β1-adrenergic receptor; *β2-AR*, β2-adrenergic receptor; *AT1AR*, angiotensin II type 1A receptor; *D2R*, dopamine type 2 receptor; *FFA*, free fatty acid; *H4R*, histamine type 4 receptor; *μ-OR*, μ-opioid receptor.

pharmacochaperones or pharmacoperones) are specific small-molecules that bind to their protein targets to facilitate biogenesis and/or prevent/correct misfolding (Loo & Clarke, 1997; Morello et al., 2000a; Morello, Petäjä-Repo, Bichet, & Bouvier, 2000b). These are usually hydrophobic structures that enter cells by diffusion, and rescue protein localization and function. The discovery and activity of PC have also been extensively reviewed and discussed elsewhere (Bernier, Bichet, & Bouvier, 2004a; Bernier et al., 2006; Conn et al., 2014; Conn & Ulloa-Aguirre, 2010; Leidenheimer & Ryder, 2014). Most of the PC target secretory pathway proteins including enzymes, transporters, receptors (among them, many GPCRs) and ion channels. In addition, while most PC have been used *in vitro*, the demonstration of their efficacy in animal models and humans established that their use holds great promise as novel therapeutic strategy.

Lysosomal enzymes are best examples of protein targets that can be functionally rescued *in vivo* by PC. Lysosomal storage disorders (LSDs) are metabolic diseases caused by mutations in genes that encode proteins involved in different lysosomal functions, in most cases enzymes, including acid-β-glucosidase (Gaucher disease), α-galactosidase A (Fabry disease) and many other acidic hydrolases (Karageorgos et al., 1997; Parkinson-Lawrence et al., 2010). The biological and clinical interest of LSD is high and different therapeutic approaches have been developed to treat these disorders (Parenti, Andria, & Valenzano, 2015). The therapeutic approach that has been most successful is enzyme replacement therapy (Brady, 2006). This strategy is based on the periodic intravenous administration of a manufactured enzyme that is taken up into cells, delivered to lysosomes and can reduce substrate storage. Alternative strategies also exist and are directed toward reducing the synthesis of substrates by enhancing clearance of substrates from cells and tissues (Platt & Jeyakumar, 2008). Recently, PC therapy (PCT) for a number of LSDs has been evaluated in the clinic (Parenti et al., 2015). This is an emerging approach based on small-molecule ligands that selectively bind and stabilize mutant enzymes, increase their cellular levels and improve lysosomal trafficking and activity. The PC migalastat for treating Fabry disease (Germain et al., 2012) and afegostat or ambroxol for treating Gaucher disease (Zimran, Altarescu, & Elstein, 2013) are very promising therapeutic avenues. Indeed very recently, migalastat has been approved in Europe by the European Medicines Agency as the first PC therapeutic molecule (Germain et al., 2016).

Membrane proteins like receptors, transporters and ion channels for which three-dimensional (3D) folding is tightly controlled by the cellular quality control system and that are targeted to the plasma membrane through the secretory pathway are major targets for PCT. PCT is of particular interest for GPCRs, since mutations in GPCRs are responsible for many human pathologies and GPCRs constitute the largest class of membrane targets for a majority of currently marketed drugs (more that 30% of FDA (Food and Drug Administration)-approved drugs). For instance, the human gonadotropin-releasing hormone receptor (GnRHR) has been a central focus of drug development and many useful compounds (agonists and antagonists) have been characterized for the treatment of reproduction disorders (Conn, Ulloa-Aguire, Ito, & Janovick, 2007; Conn & Ulloa-Aguirre, 2011; Janovick, Maya-Nunez, & Conn, 2002). PC of the GnRHR have shown efficacy in cell culture systems but also in a small animal model, a knock-in mouse expressing the GnRHR E90K mutant which causes hypogonadotropic hypogonadism in humans (Janovick et al., 2013). Indeed, pulsatile PCT rescued the E90K receptor plasma membrane localization and responsiveness of the endogenous natural ligand gonadotropin releasing hormone. Spermatogenesis, proteins associated with steroid transport and steroidogenesis, and androgen levels were restored in mutant male mice following the PCT. A PC action can be generalized to many intracellular-retained misfolded mutant receptors from many GPCR families. A few examples of PC targeting GnRHR (Janovick et al. 2013), vasopressin receptors (Bernier et al., 2006; Jean-Alphonse et al., 2009; Morello et al., 2000a), calcium-sensing receptor (White, McKenna, Cavanaugh, & Breitwieser, 2009), luteinizing and follicle-stimulating hormone receptors (Janovick et al., 2009; Newton et al., 2011) are shown in Table 2. Most of these PC have been shown to be useful in cellular systems, and still have to confirm proof-of-concept for *in vivo* protein rescue.

2. Misfolded arginine-vasopressin V2 receptors and congenital nephrogenic diabetes insipidus (cNDI): an avenue for pharmacological chaperone therapy

Combining biased agonist properties with pharmacochaperone activity would be a fantastic strategy to develop small-molecule compounds for treating diseases related to protein misfolding and for which drug-induced beneficial versus deleterious effect ratio has to be improved. The kidney arginine-vasopressin (AVP) type 2 receptor (V2R) which regulates water homeostasis, constitutes a major target for voiding disorders and water imbalance. It also possesses a large variety of pharmacological ligands and

Table 2 Representative pharmacological chaperones for GPCRs: their specific receptor and target pathology, their *in vitro* efficacy and *in vivo* effects.

Receptor	PC	Target pathology	In vitro efficacy	In vivo effects
GnRHR	IN3 (antagonist)	hypogonadotropic hypogonadism	plasma membrane rescue, restoration of GnRH responsiveness	restoration of testis function (mouse model)
V2R	SR121463 (inverse agonist)	cNDI	plasma membrane rescue, restoration of AVP responsiveness	§
	SR49059 (antagonist)	cNDI	equivalent to SR121463	decrease in urine volume and water intake (humans)
	MCF compounds (biased agonists)	cDNI	plasma membrane rescue, direct activation of V2R	not tested
CaSR	NPS R-568 (allosteric agonist)	hypocalciuric hypercalcemia	plasma membrane and functional rescue	*
LHR	Org 42599 (allosteric agonist)	reproductive disorders, infertility due to Leydig cell hypoplasia	plasma membrane and functional rescue	not tested
FSHR	Org 41841 (allosteric agonist)	reproductive disorders, infertility	plasma membrane and functional rescue	not tested

CaSR, calcium-sensing receptor; *cNDI*, congenital nephrogenic diabetes insipidus; *FSHR*, follicle-stimulating hormone receptor; *LHR*, luteinizing hormone receptor.
§SR121463 is a compound of the vaptan family, an is named satavaptan. The ligand is efficient in patients with dilutional hyponatremia by increasing serum sodium concentrations (DILIPO study, Aronson et al., 2011). The vaptans may also have therapeutic potential for heart failure and polycystic kidney disease (Torres et al., 2012).
*The calcimimetic cinacalcet which has been developed through optimization of ligands such as NPS R-568 and NPS R-467 is widely used in clinic for treating hyperparathyroidism. Org 42599 and Org 41841 are thienopyrimidine gonadotropin analog compounds.

therapeutic compounds, and can be considered as a model GPCR system for challenging this strategy.

2.1 Pathology of cNDI

Regulated water excretion by the kidney is crucial to preserve water homeostasis of our body. The adjustment of water renal reabsorption, to respond to increase in blood osmolality (hypernatremia) or decrease in blood volume (hypovolemia), mainly depends on the release of the antidiuretic hormone AVP from the pituitary (Feinstein et al., 2013; Moeller, Rittig, & Fenton, 2013; Treschan & Peters, 2006). Binding of AVP to the V2R, a Gs protein-coupled receptor localized at the basolateral membrane of the principal cells of the kidney collecting duct, results in an intracellular cyclic adenosine monophosphate (cAMP)-dependent signaling cascade of events. Among them, phosphorylation of the water channel aquaporin-2 (AQP2) and its translocation from storage compartments to the apical surface of the cells are of primary importance. Water from pro-urine which enters the cells exits via aquaporin-3 and aquaporin-4 at the basolateral side, leading to water reabsorption and urine concentration. Upon restoration of water balance, the level of plasma AVP drops and AQP2 is internalized, leaving the apical membrane watertight again.

Disorders that interfere with proper urine concentration can be life-threatening, especially in children. One such disease is cNDI (Morello & Bichet, 2001). Indeed, cNDI is a rare inherited disease, characterized by insensivity of the kidney to AVP and absence of water reabsorption. It results in polyuria and compensatory polydipsia, and may lead to severe dehydration and electrolyte imbalance (hypernatremia) in the case of inadequate water supply.

The X-linked form of cDNI is caused by mutations in the V2R gene (Bichet et al., 1994). To date, over 200 different V2R mutations have been described (Spanakis, Milord, & Gragnoli, 2008). V2R mutations are divided into five different classes. Class II mutations (the most prevalent, more than 50%) are most frequently missense mutations (amino acid substitutions). These mutations result into misfolded, trafficking-deficient receptors that do not reach plasma membrane of the basolateral side of the principal cells of the kidney. Indeed, most of the mutants are retained in the endoplamic reticulum (ER) and in the ER-Golgi intermediate compartment. Consequently, V2R mutants are unable to interact with circulating AVP (Tsukagushi et al., 1995). These V2R mutants, rather than resulting in nonfunctional proteins (mutants from classes III and IV) are intrinsically functional, as demonstrated by overexpression in heterologous cell expression systems (Ala et al., 1998).

Untreated adult cNDI patients may have a daily output of 15–20 L of highly dilute urine. Newborn infants often suffer from hypernatremic dehydration with symptoms of irritability, poor feeding and weight gain. In addition, repeated periods of brain dehydration may result in permanent brain damage, mental retardation and seizures can occur. The main strategy for treating cNDI patients consists of a sufficient water supply to replace the urinary water loss, but this can seriously impact on the quality of life due to excessive drinking and urine voiding. Some diuretics, like hydrochlorothiazide, amiloride or the cyclooxygenase inhibitor indomethacin, have been proven effective to reduce urine output by up to 50% (Bockenhauer & Bichet, 2014). However, diuretics may affect the sodium and potassium balance in patients and therefore these treatments require tight monitoring of serum electrolytes and osmolality.

Although understanding of cNDI from molecular and cell biological point of view has largely increased since the cloning of the V2R gene (Birnbaumer et al., 1992; Lolait et al., 1992; Rosenthal et al., 1992), developing alternative strategies to manage water homeostasis and induce antidiuresis in cNDI patients is still obvious. The V2R is a "natural" target for establishing new forms of therapies, and PC rescue of its function is a very elegant and specific approach.

2.2 Treatment with antagonist pharmacological chaperones

Chemical chaperones, like glycerol and DMSO, were shown for instance to correct mutants of the AQP2 water channel, as assessed by protein maturation, cellular targeting and water permeability (Tamarappoo & Verkman, 1998). Taking the concept of chemical chaperones further, artificial mutants of the multidrug resistance P-glycoprotein-1, a cell surface transporter which interacts with a panel of cytotoxic agents, were functionally rescued. Indeed, ER-retained mutants were targeted to the plasma membrane and their functional rescue was demonstrated using specific substrates or inhibitors like vinblastin, cyclosporin or verapamil (Loo & Clarke, 1997). These compounds were proposed to stabilize a specific native-like conformation of the transporter, allowing its release from the ER quality control cell system.

The concept was applied to the mutants of V2R responsible for cNDI, based on the idea that pharmacological ligands act by binding to and stabilizing specific conformations of their receptors. Selective cell-permeant nonpeptidic V2R antagonists (which block the V2R in an inactive conformation) were assessed to check whether they could facilitate the folding of mutant receptors that are retained in ER and unable to interact with AVP (Morello et al., 2000a). Given that these antagonists are specific to the V2R and that they perform chaperone-like activity, these compounds were named PC for the first time (Morello et al., 2000b). The first antagonist (or inverse agonist) to be used was SR121463 (satavaptan), a selective high-affinity V2R ligand (Serradeil-Le Gal et al., 1996). An overnight treatment of the cells retaining different V2R mutants into an intracellular compartment converted precursor forms into fully glycosylated mature receptors that were targeted to the cell surface. Once correctly localized at the plasma membrane, these mutants were able to differentially bind AVP and produce a correlated cAMP intracellular signal (Morello et al., 2000a). Interestingly, V2R membrane-impermeable peptidic antagonists were unable to mimic the SR121463 effect, indicating a PC intracellular effect. The PC-driven V2R mutant rescue was not limited to SR121463, because another nonpeptidic antagonist, VPA985 (lixivaptan), reproduced equivalent results. Since the publication of this study, SR121463 was used to rescue a larger panel of V2R mutants (Bockenhauer et al., 2010; Janovick, Park, & Conn, 2011; Tan, Nickols, & Limbird, 2003; Wüller et al., 2004). Different other antagonists (or inverse agonists) were tested for their PC properties, such as the V1AR-selective SR49059 (relcovaptan) (Wüller et al., 2004; Bernier et al., 2004b), the mixed V1A/V2R YM087 (conivaptan) (Bernier et al., 2006) and the two V2R-selective OPC31260 (mozavaptan) and OPC41061 (tolvaptan) (Robben, Sze, Knoers, & Deen, 2007). Very importantly in some cases, the PC effect was also reproducible in polarized renal cells where V2R mutants were appropriately targeted to the basolateral membrane, the natural localization of the wild-type V2R (Robben, Sze, Knoers, & Deen, 2006). In addition, because of their target specificity, the PC compounds are active at nanomolar concentrations on cultured cells (Robben et al., 2007). This is a tremendous advantage, compared to chemical chaperones which are active at micromolar concentrations or even more. Some prototypic PC for the V2R are presented in Table 3, as well as the different cNDI mutants for which their beneficial effect was proven. Indeed, studies with the antagonists SR121463 (Bockenhauer et al., 2010; Janovick et al., 2011; Jean-Alphonse et al., 2009; Morello et al., 2000a; Robben et al., 2006; Tan et al., 2003; Wüller et al., 2004), SR49059 (Bernier et al., 2004b), OPC41061 and OPC31260 (Robben et al., 2007), the agonists OPC51803, VA999088 and VA999089 (Robben

Table 3 Different classes of V2R pharmacological chaperones: their intrinsic activity, specificity and efficiency in rescuing mutant receptor plasma membrane localization and function.

PC	Intrinsic activity	Specificity	V2R mutants with plasma membrane and functional rescue
SR121463 (Satavaptan, Aquilda®)	inverse agonist	V2R	L44P, L59P, Δ62–64, L83Q, V88M, Y128S, S167L, S167T, V206D, L292P, A294P, P322H, R337X
SR49059 (Relcovaptan)	antagonist	V1AR	L44P, Δ62-64, I130F, R137H, W164S, S167T, C319Y
OPC41061 (Tolvaptan, Samsca®, Jinarc®)	antagonist	V2R	L44P, I130F, S167T
OPC31260 (Mozavaptan)	inverse agonist	V2R	L44P, I130F, S167T
OPC51803*	agonist	V2R	L44P, Y128S, I130F, S167T, Y280C, R337X
VA999088* and VA999089*	agonists	V2R	L44P, Y128S, I130F, S167T, Y280C, R337X
MCF14, MCF18 and MCF57	biased agonists	V2R	L44P, A294P, R337X

Apart from SR49059 which is a selective antagonist for the AVP V1A receptor subtype, all other ligands have been classified regarding their activity toward the V2R-induced Gs protein-dependent signaling pathway (linked to the antidiuretic effect of AVP). Interestingly, MCF compounds are Gs-biased agonists with no β-arrestin signaling.

*In these studies, the different V2R mutants were not targeted to the plasma membrane but were functionally rescued.

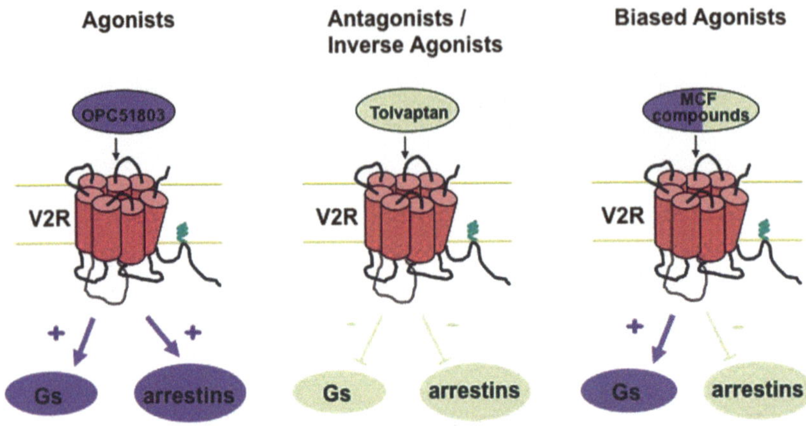

Fig. 1 ***The different classes of V2R pharmacological chaperones.*** Agonist, antagonist/inverse agonist and biased agonist pharmacochaperones of the V2R are compared. All pharmacological chaperones are small molecule compounds that can cross the cellular membranes, due to their hydrophobic nature. Agonists like OPC51803 directly stimulate the V2R and induce both Gs-dependent and arrestin-associated signaling pathways. Antagonists such as Tolvaptan bind to the V2R and inhibit all AVP-induced signaling pathways of the V2R whereas inverse agonists are even able to decrease basal activity of the different V2R signaling outcomes. Biased agonists like the MCF compounds are mixed ligands, being activators of the V2R-induced Gs-dependent cell responses and inhibitors of the arrestin-associated pathways.

et al., 2009), or the biased agonists MCF14, MCF18 and MCF57 (Jean-Alphonse et al., 2009) are listed. As shown in Fig. 1, these different classes of V2R pharmacological chaperones are compared to each others.

Interestingly, the concept of PC developed for V2R mutants responsible for cNDI was also studied with the V1A and V1B subtypes of the vasopressin receptor family. SSR149415, which is a specific nonpeptidic antagonist for V1B was demonstrated to rescue plasma membrane localization and function of the V1B mutant 341FN$(X)_2$LL$(X)_3$L350 (Auzan, Ventura, & Clauser, 2005). In addition, the selective nonpeptidic antagonist SR49059 was shown to rescue the functional properties of surface-impaired D148A/N/E mutants of the V1A receptor subtype (Hawtin, 2006). To date, no more study was investigated using PC to rescue other artificial mutants of V1AR, V1BR or of the vasopressin-related oxytocin (OT) receptor.

Using antagonists for rescuing function of the V2R and more generally GPCR mutants responsible for inherited conformational diseases is somehow paradoxical. First of all, because these antagonists specifically block (inhibit) their receptors (see Fig. 1), they can not directly stimulate receptor-associated signaling pathways. Regarding patients who suffer from cNDI, the therapeutic beneficial effect would be antidiuresis, through activation of a cAMP-dependent signaling cascade and particularly membrane translocation of AQP2. Indeed, using the PC antagonist strategy, functional rescue of mutants of the V2R is a subtle balance between the ability of the ligand to target cell-surface expression of the mutants and its possibility to be displaced by endogenous AVP for receptor activation (Mendre & Mouillac, 2010). In this regard, considering the antagonist affinity is an important feature for this challenge, and therefore low-affinity antagonists (those which are easily displaced by AVP) may possess a highest clinical value (Los, Deen, & Robben, 2010). However, the efficiency of such low-affinity antagonist ligands in rescuing receptor function is lower than that of high-affinity ligands (the higher the affinity, the better the rescue). Moreover, high concentrations of low-affinity antagonists to be administered for clinical efficiency might lead to unwanted side effects in patients (see paragraph below). In addition, compound-intrinsic factors other than affinity may influence their capacity to confer functional rescue and their extent to be displaced by AVP, like their localization in the binding pocket of the V2R, their intrinsic activity or their lipophilic value. Overall, it seems that concerning cNDI patients, the high-affinity OPC31260 (mozavaptan) and OPC41061 (tolvaptan) nonpeptidic antagonists would combine best clinical potential, in terms of cell surface rescue, low concentration to be used, and efficient displacement by AVP (Robben et al., 2007; Wesche, Deen, & Knoers, 2012).

Interestingly, a small-scale short-term clinical trial has been set up using SR49059, a low-affinity antagonist for V2R (it is specific for the V1A receptor subtype) (Bernier et al., 2006). Different patients with R137H, W164S or Δ62-64 mutations were treated with the nonpeptidic antagonist. Interestingly, SR49059 significantly decreased the 24-h urine volume and water intake, demonstrating a successful PC behavior *in vivo*. However, because of a potential interference with the cytochrome P450 metabolic pathway (hepatic toxicity), the development of this molecule was discontinued during clinical phase II. To date, no other clinical trial has been developed for cNDI patients. Recently, the cell-permeable nonpeptidic antagonist OPC41061 (tolvaptan, Samsca®, Jinarc®) has been approved in USA and Europe for the treatment of hyponatremia for instance in the syndrome of inappropriate antidiuretic hormone secretion and congestive heart failure (Schrier, Gross, & Gheorghiade, 2006) and more recently for the treatment of polycystic kidney disease (Torres et al., 2012). This ligand may be of high therapeutic value for novel clinical trials for treating cNDI in the future, provided that it can be displaced by AVP *in vivo*.

3. Biased agonist pharmacochaperones: promising therapeutics for treating cNDI

3.1 Agonists versus antagonists

Other pharmacological approaches to treat cNDI patients may have a higher potential that the antagonist PC strategy. Indeed, agonists and biased agonists of the V2R may prove to be of higher clinical value. Theoretically, PC agonists possess advantages over antagonists since they are able to directly stimulate the V2R and induce receptor-associated signaling pathways (see Fig. 1). In this case, V2R-selective high-affinity ligands are likely to be the most appropriate to efficiently rescue plasma membrane targeting of the mutants and their direct activation. However, agonists also promote V2R internalization, and consequently a decrease in the cAMP signal, a phenomenon that could reduce the beneficial effects of these compounds.

Recent investigations indicated that ER-retained but intrinsically functional V2R mutants can be activated intracellularly by different agonists (Robben et al., 2009). The activation surprisingly leads to sufficient cAMP increase to induce AQP2 to be translocated to the apical membrane of renal polarized cells. The recently developed nonpeptidic agonist OPC51803 and two novel agonists VA999088 and VA999089 were used in this study to rescue function of a panel of different cNDI V2R mutants (see Table 3). In contrast to classical PC-assisted receptor folding and rescue, the localization and maturation of the cNDI mutants did not change upon ligand binding and receptor activation. Due to their structural features (small lipophilic molecule compounds which have the ability to penetrate cell membranes and reach the ER), it would be conceivable to classify these three ligands as PC but they do not induce plasma membrane targeting of the different V2R mutants. Consequently, they are somehow different from the reference pharmacological chaperones but still constitute promising future therapeutic candidates for cNDI clinical studies (Moeller et al., 2013). Like for PC-based approaches, V2R rescue and intracellular receptor activation by OPC51803 and both VA compounds may only work for mutations that affect folding or proper intracellular transport, but neither for highly truncated receptors, nor those that lost their

Gs-dependent coupled signaling pathways. These ligands may also be tested on a larger panel of cNDI mutants, in order to check their PC potential properties.

3.2 Biased agonists versus agonists

The discovery of biased agonist PC of the V2R is a novel therapeutic opportunity for cNDI patients (Jean-Alphonse et al., 2009). The V2R nonpeptidic agonists MCF14 (previously described as OPC23h), MCF18 (previoulsy known as VNA932) and MCF57 possess only part of the AVP signaling properties. Indeed (see Fig. 1), these molecules are full agonists of the Gs-dependent cAMP signal (which is responsible for AQP2 translocation and water reabsorption) but do not induce receptor internalization and arrestin-related signaling pathways (antagonists for arrestin recruitment and associated events). On a therapeutic point of view, these particular properties may lead to additional beneficial effects for cNDI patients, as compared to common agonists. The ligands MCF14, MCF18 and MCF57 are high-affinity agonists for V2R and are capable of inducing cNDI mutant receptor maturation, translocation to the plasma membrane and can directly initiate a cAMP response (they may act in synergy with circulating AVP in the case of *in vivo* clinical trials). Functional rescue with MCF14, MCF18 and MCF57 was demonstrated for different mutants of the V2R (see Table 3). These ligands are however not totally selective for the V2R, possessing a significant affinity for the V1AR and the OT receptor. Their pharmacological profile has thus to be improved yet. They however constitute a novel class of PC. Indeed, the V2R biased agonist PC, able to generate a cAMP signal and acting as noninternalizing ligands, potentially providing a long-lasting cellular response during drug administration, may constitute ideal therapeutic compounds for treating cNDI (Mouillac & Mendre, 2014).

4. Insights from pharmacochaperone high-throughput screening studies and V2 structural studies

To develop new V2R biased ligands with unique beneficial therapeutic effects and no adverse effects (no activity on other V2R-induced signaling pathways but also no activation of other AVP/OT receptors), rational drug development may be based on biophysical and structural studies. In theory, to fully understand the structural basis of biased signaling would be to crystallize a given receptor not only in its inactive, active and biased active conformations but also in complex with both G protein and β-arrestin. In addition, cristallographic approaches have to be complemented with dynamic studies of ligand-receptor-G protein/arrestin complexes like NMR spectroscopy. Today, we are far from having such a complete set of data for any given receptor and no V2R 3D structure has been described to date. Therefore, there are different alternatives to develop original PC for treating cNDI.

The first one may be based on high-throughput screening (HTS) assays to analyze thousands of small molecules from various known chemical libraries. This is a strategy that has been developed by P. Michael Conn and his team during the last few years (Conn, Smith, Hodder, Janovick, & Smithson, 2013; Janovick et al., 2011). Most of the ligands that were used as PC for the V2R before this screening programme were only classified considering their activity toward the V2R-induced Gs protein-dependent signaling pathway (agonists or antagonists), but not on their capacity to increase receptor trafficking. Consequently, many valuable molecules may have been overlooked because of this limitation. The screen approach was based on targeting V2R mutants to the plasma membrane by incubating cultured cells with a potential pharmacochaperone and measuring the capacity of the rescued receptor to trigger an AVP-stimulated cAMP production using a luminescence-based system (cAMP-Glo Max luminescence detection system from Promega). The screening assay was miniaturized (up to 1536-well format) and automated using a robotic platform (Smithson, Janovick, & Conn, 2013). The procedure allowed to screen first the Library of Pharmacologically Active Compouds (LOPACs) from Sigma Aldrich (Conn et al., 2014) and then to screen the Scripps Drug Diversity Library (more than 650,000 compounds, Conn, Spicer, Scampavia, & Janovick, 2015). The reference PC compound, SR121463, was systematically included in the HTS assays as a positive control. The discovery campaigns were applied to a well-known V2R mutation responsible for cNDI, L83Q (Morello et al., 2000a; Bernier et al., 2006), which induces retention of the receptor within an intracellular compartment and loss-of-function. Once the mutant is rescued to the cell surface upon PC treatment, AVP is still able to bind to the mutant (albeit with a lower affinity) and to trigger a cAMP response. Following this large-scale PC screening programme for the L83Q V2R, hit compounds from different chemical families were identified (Smith et al., 2016). The campaign initially identified 3734 positive modulators of cAMP in the HTS screen. The development of counter-screens to eliminate false positive molecules confirmed 147 active ligands with an EC50 ≤ 5 μM. The most promising compounds (83 active small-molecules) were broken down into several main structural classes: amino-benzothiazole benzamides, aryl aminothiazole amides, heteroaryl amides (in which the core heterocycle was neither a thiazole nor a benzothiazole), amide sulfonamides and bis-sulfonamides, dihydropyridines, triazines, and some "singleton" structures, like an azabenzothiazole, an acyl urea, a cyanoacrylamide, or a pyrazole amide. In each class, some PC were also characterized as antagonists or inverse agonists for the V2R (Janovick et al., 2016; Janovick, Spicer, Bannister, Scampavia, & Conn, 2017). Unfortunately, none of the tested ligands were agonists. Affinity and V2R selectivity of the different leads have now to be defined, and their therapeutic potential tested *in vivo*. In addition, their pharmacochaperone properties will have to be evaluated onto many other representative cNDI V2R mutants.

The second alternative to get information on biased signaling molecular mechanisms and to develop new PC may be derived from biophysical and structural approaches complementary to X-ray crystallography. Indeed, we used tryptophane fluorescence and lanthanide resonance energy transfer fluorescence to study ligand-induced structural changes of the purified human V2R (Rahmeh et al., 2012). We compared the effects of the reference unbiased agonist AVP to those of SR121463, a PC with inverse agonist and partial agonist properties toward Gs and arrestin respectively, and of MC14, a PC with full Gs biased properties. Decrease and increase in the overall intrinsic tryptophan fluorescence were recorded using the SR121463 or AVP/MCF14 respectively, indicating the existence of at least two different receptor populations in response to inverse agonist or agonists of the V2R-dependent Gs protein signaling pathway. Moreover, introduction of lanthanide fluorescence resonance energy transfer-based intramolecular sensors at transmembrane (TM) 6 and 7 and at the C-terminus of the V2R clearly demonstrated that β-arrestin- and Gs-biased ligands differentially affected the average life-time constant of the major population of the receptor, indicating the existence of different conformational states. Conformational movements of functional domains of the V2R relative to each others are different depending on the biased ligands, and consequently may explain why different signaling pathways are activated or not. Indeed, movements of the V2R TM 6 are involved in Gs signaling whereas those of TM7 and helix 8 are involved in β-arrestin recruitment.

Some other methodologies can also be used to study the structural basis of biased signaling at GPCRs. For instance, bimane fluorescence in the ghrelin receptor GHS-R1a was used to study ligand-induced conformational changes (Mary et al., 2012). Indeed, sensors were introduced in the second or the third intracellular loop respectively to detect specific effects of biased and unbiased ligands on receptor conformations. Interestingly, the JMV 3002 and JMV 3018 biased ligands for the Gq protein induced smaller changes in bimane fluorescence compared to unbiased ligands whereas the SPA inverse agonist triggered a significant increase in fluorescence. These results are in agreement with the presence of distinct conformations stabilized by functionally different ligands. Chemical labeling coupled to ^{19}F-NMR spectroscopy is also a methodology that can help to decipher structural information in biased signaling. It has been shown that the β-arrestin-biased ligand carvedilol induces a unique conformational signature of the β2-AR (Liu, Horst, Katritch, Stevens, & Wüthrich, 2012). Indeed, introducing site-specific ^{19}F-NMR labels in the structure of the β2-AR at cytoplasmic ends of TM 6 and 7 allowed to demonstrate that these regions adopt two major conformational states and that carvedilol predominantly impact the conformation of TM 7.

5. Conclusions

Over the last two decades, V2R-selective biased ligands and PC compounds have been identified and pharmacologically characterized. Developing molecules which combine PC properties and bias for Gs signaling pathway is a promising strategy for treating cNDI and more generally of particular clinical interest in GPCR research. In principle, V2R-specific PC have more desirable properties as therapeutics than current nonspecific treatments like thiazides with indomethacine. Because the rescuing properties of these ligands have been analyzed to only a few misfolded receptors, it would be important to investigate their PC properties on a larger panel of V2R mutants (most of the mutants from class II are potential candidates to be treated). Additionally, their Gs protein bias is also a major criteria to select compounds that do not display β-arrestin recruitment, in order to favor beneficial effects and abolish adverse effects. We anticipate that biased agonist PC are novel small-molecules in the toolbox that will become promising therapeutics and pharmacological ligands useful for selectively modulating GPCR activity.

Acknowledgments

This work was supported, in part, by INSERM and CNRS.

References

Ala, Y., Morin, D., Mouillac, B., Sabatier, N., Vargas, R., Cotte, N., ... Jard, S. (1998). Functional studies of twelve mutant V2 vasopressin receptors related to nephrogenic diabetes insipidus: molecular basis of a mild clinical phenotype. *Journal of the American Society of Nephrology, 9*(10), 1861–1872.

Aronson, D., Verbalis, J. G., Mueller, M., Krum, H., ... DILIPO Investigators. (2011). Short- and long-term treatment of dilutional hyponatraemia with satavaptan, a selective arginine-vasopressin V2 receptor antagonist: The DILIPO study. *European Journal of Heart Failure, 13*(3), 327–336.

Auzan, R. J., Ventura, M. A., & Clauser, E. (2005). Mechanisms of cell-surface rerouting of an endoplasmic reticulum-retained mutant of the vasopressin V1b/V3 receptor by a pharmacological chaperone. *The Journal of Biological Chemistry, 280*(51), 42198–42206.

Bernier, V., Bichet, D. G., & Bouvier, M. (2004a). Pharmacological chaperone action on G protein-coupled receptors. *Current Opinion in Pharmacology, 4*(5), 528–533.

Bernier, V., Lagacé, M., Lonergan, M., Arthus, M. F., Bichet, D. G., & Bouvier, M. (2004b). Functional rescue of the constitutively internalized V2 vasopressin receptor mutant R137H by the pharmacological chaperone action of SR49059. *Molecular Endocrinology, 18*(8), 2074–2084.

Bernier, V., Morello, J. P., Zarruk, A., Debrand, N., Salahpour, A., Lonergan, M., ... Bichet, D. G. (2006). Pharmacologic chaperones as a potential treatment for X-linked nephrogenic diabetes insipidus. *Journal of the American Society of Nephrology, 17*(1), 233–243.

Bichet, D. G., Birnbaumer, M., Lonergan, M., Arthus, M. F., Rosenthal, W., Goodyer, P., ... Morgan, K. (1994). Nature and recurrence of AVPR2 mutations in X-linked nephrogenic diabetes insipidus. *American Journal of Human Genetics, 55*(2), 278–286.

Birnbaumer, M., Seibold, A., Gilbert, S., Ishido, M., Barberis, C., Antaramian, A., ... Rosenthal, W. (1992). Molecular cloning of the receptor for human antidiuretic hormone. *Nature, 357*(6376), 333–335.

Bockenhauer, D., & Bichet, D. G. (2014). Urinary concentration: Different ways to open and close the tap. *Pediatric Nephrology, 29*(8), 1297–1303.

Bockenhauer, D., Carpentier, E., Rochdi, D., Van't Hoff, W., Breton, B., Bernier, V., ... Bichet, D. G. (2010). Vasopressin type 2 receptor V88M mutation: Molecular basis of partial and complete nephrogenic diabetes insipidus. *Nephron Physiology, 114*(1), 1–10.

Brady, R. O. (2006). Enzyme replacement for lysosomal diseases. *Annual Review of Medicine, 57*, 283–296.

Carter, A. A., & Hill, S. J. (2005). Characterization of isoprenaline- and salmeterol-stimulated interactions between beta2-adrenoceptors and beta-arrestin 2 using beta-galactosidase complementation in C2C12 cells. *The Journal of Pharmacology and Experimental Therapeutics, 315*(2), 839–848.

Chaudhuri, T. K., & Paul, S. (2006). Protein-misfolding diseases and chaperone-based therapeutic approaches. *FEBS Journal, 273*(7), 1331–1349.

Chen, X., Sassano, M. F., Zheng, L., Setola, V., Chen, M., Bai, X., ... Jin, J. (2012). Structure-functional selectivity relationship studies of β-arrestin-biased dopamine D2 receptor agonists. *Journal of Medicinal Chemistry, 55*(16), 7141–7153.

Cohen, F. E., & Kelly, L. W. (2003). Therapeutic approaches to protein-misfolding diseases. *Nature, 426*(6968), 905–909.

Conn, P. M., Smith, E., Hodder, P., Janovick, J. A., & Smithson, D. C. (2013). High-throughput screen for pharmacoperones of the vasopressin type 2 receptor. *Journal of Biomolecular Screening, 18*(8), 930–937.

Conn, P. M., Smithson, D. C., Hodder, P. S., Stewart, M. D., Behringer, R. R., Smith, E., ... Janovick, J. A. (2014). Transitioning pharmacoperones to therapeutic use: In vivo proof-of-principle and design of high throughput screens. *Pharmacological Research, 83*, 38–51.

Conn, P. M., Spicer, T. P., Scampavia, L., & Janovick, J. A. (2015). Assay strategies for identification of therapeutic leads that target protein trafficking. *Trends in Pharmacological Sciences, 36*(8), 498–505.

Conn, P. M., Ulloa-Aguire, A., Ito, J., & Janovick, J. A. (2007). G protein-coupled receptor trafficking in health and disease: Lessons learned to prepare for therapeutic mutant rescue in vivo. *Pharmacological Reviews, 59*(3), 225–250.

Conn, P. M., & Ulloa-Aguirre, A. (2010). Trafficking of G protein-coupled receptors to the plasma membrane: Insights from pharmacoperone drugs. *Trends in Endocrinology and Metabolism, 21*(3), 190–197.

Conn, P. M., & Ulloa-Aguirre, A. (2011). Pharmacological chaperones for misfolded gonadotropin-releasing hormone receptors. *Advances in Pharmacology, 62*, 109–141.

Feinstein, T. N., Yui, N., Webber, M. J., Wehbi, V. L., Stevenson, H. P., King, J. D., Jr., ... Vilardaga, J. P. (2013). Noncanonical control of vasopressin receptor type 2 signaling by retromer and arrestin. *The Journal of Biological Chemistry, 288*(39), 27849–27860.

Ferguson, S. S. (2001). Evolving concepts in G protein-coupled receptor endocytosis: The role in receptor desensitization and signaling. *Pharmacological Reviews, 53*(1), 1–24.

Galandrin, S., Oligny-Longpré, G., & Bouvier, M. (2007). The evasine nature of drug efficacy: Implications for drug discovery. *Trends in Pharmacological Sciences, 28*(8), 423–430.

Germain, D. P., Giugliani, R., Hughes, D. A., Mehta, A., Nicholls, K., Barisoni, L., ... Boudes, P. F. (2012). Safety and pharmacodynamic effects of a pharmacological chaperone on α-galactosidase a activity and globotriaosylceramide clearance in Fabry disease: Report from two phase 2 clinical studies. *Orphanet Journal of Rare Diseases, 7*, 91.

Germain, D. P., Hughes, D. A., Nicholls, K., Bichet, D. G., Giugliani, R., Wilcox, W. R., ... Schiffmann, R. (2016). Treatment of Fabry's disease with the pharmacologic chaperone Migalastat. *New England Journal of Medicine, 375*(6), 545–555.

Groer, C. E., Tidgewell, K., Moyer, R. A., Harding, W. W., Rothman, R. B., Prisinzano, T. E., et al. (2007). An opioid agonist that does not induce mu-opioid receptor-arrestin interactions or receptor internalization. *Molecular Pharmacology, 71*(2), 549–557.

Hawtin, S. R. (2006). Pharmacological chaperone activity of SR49059 to functionally recover misfolded mutations of the vasopressin V1a receptor. *The Journal of Biological Chemistry, 281*(21), 14604–14614.

Holloway, A. C., Qian, H., Pipolo, L., Ziogas, J., Miura, S., Karnik, S., ... Thomas, W. G. (2002). Side-chain substitutions within angiotensin II reveal different requirements for signaling, internalization, and phosphorylation of type 1A angiotensin receptors. *Molecular Pharmacology, 61*(4), 768–777.

Janovick, J. A., Maya-Nunez, G., & Conn, P. M. (2002). Rescue of hypogonadotropic hypogonadism-causing and manufactured GnRH receptor mutants by a specific protein-folding template: Misrouted proteins as a novel disease etiology and therapeutic target. *The Journal of Clinical Endocrinology & Metabolism, 87*(7), 3255–3262.

Janovick, J. A., Maya-Nunez, G., Ulloa-Aguire, A., Huhtaniemi, I. T., Dias, J. A., Verbost, P., et al. (2009). Increased plasma membrane expression of human follicle-stimulating hormone receptor by a small molecule thienopyr(im)idine. *Molecular and Cellular Endocrinology, 298*(1–2), 84–88.

Janovick, J. A., Park, B. S., & Conn, P. M. (2011). Therapeutic rescue of misfolded mutants: Validation of primary high throughput screens for identification of pharmacoperone drugs. *PLoS One, 6*(7), e22784.

Janovick, J. A., Spicer, T. P., Bannister, T. D., Scampavia, L., & Conn, P. M. (2017). Pharmacoperone rescue of vasopressin 2 receptor mutants reveals unexpected constitutive activity and coupling bias. *PLoS One, 12*(8), e0181830.

Janovick, J. A., Spicer, T. P., Smith, E., Bannister, T. D., Kenakin, T., Scampavia, L., et al. (2016). Receptor antagonism/agonism can be can be uncoupled from pharmacoperone activity. *Molecular and Cellular Endocrinology, 434*, 176–185.

Janovick, J. A., Stewart, M. D., Jacob, D., Martin, L. D., Deng, J. M., Stewart, C. A., ... Conn, P. M. (2013). Restoration of testis function in hypogonadotropic hypogonadal mice harboring a misfolded GnRHR mutant by pharmacoperone drug therapy. *Proceedings of the National Academy of Sciences of the United States of America, 110*(52), 21030–21035.

Jean-Alphonse, F., Perkovska, S., Frantz, M. C., Durroux, T., Méjean, C., Morin, D., et al. (2009). Biased agonist pharmacochaperones of the AVP V2 receptor may treat congenital Nephrogenic Diabetes Insipidus. *Journal of the American Society of Nephrology, 20*(10), 2190–2203.

Karageorgos, L. E., Isaac, E. L., Brooks, D. A., Ravenscroft, E. M., Davey, R., Hopwood, J. J., et al. (1997). Lysosomal biogenesis in lysosomal storage disorders. *Experimental Cell Research, 234*(1), 85–97.

Laugwitz, K. L., Allgeier, A., Offermanns, S., Spicher, K., Van Snade, J., Dumont, J. E., et al. (1996). The human thyrotropin receptor: A heptahelical receptor capable of stimulating members of all four G protein families. *Proceedings of the National Academy of Sciences of the United States of America, 93*(1), 116–120.

Leidenheimer, N. J., & Ryder, K. G. (2014). Pharmacological chaperoning: A primer on mechanism and pharmacology. *Pharmacological Research, 83*, 10–19.

Liu, J. J., Horst, R., Katritch, V., Stevens, R. C., & Wüthrich, K. (2012). Biased signaling pathways in β2-adrenergic receptor characterized by 19F-NMR. *Science, 335*(6072), 1106–1110.

Lolait, S. J., O'Carroll, A. M., McBride, O. W., Konig, M., Morel, A., & Brownstein, M. J. (1992). Cloning and characterization of a vasopressin V2 receptor and possible link to nephrogenic diabetes insipidus. *Nature, 357*(6376), 526–529.

Loo, T. W., & Clarke, D. M. (1997). Correction of defective protein kinesis of human P-glycoprotein mutants by substrates and modulators. *The Journal of Biological Chemistry, 272*(2), 709–712.

Los, E. L., Deen, P. M. T., & Robben, J. H. (2010). Potential of nonpeptide (ant)agonists to rescue vasopressin V2 receptor mutants for the treatment of X-linked nephrogenic diabetes insipidus. *Journal of Neuroendocrinology, 22*(5), 393–399.

Lutrell, L. M. (2014). More than just a hammer : Ligand "bias" and pharmaceutical discovery. *Molecular Endocrinology, 28*(3), 281–294.

Luttrell, L. M., & Gesty-Palmer, D. (2010). Beyond desensitization: Physiological relevance of arrestin-dependent signaling. *Pharmacological Reviews, 62*(2), 305–330.

Luttrell, L. M., Maudsley, S., & Bohn, L. M. (2015). Fulfilling the promise of "biased" G protein-coupled receptor agonism. *Molecular Pharmacology, 88*(3), 579–588.

Mary, S., Damian, M., Louet, M., Floquet, N., Fehrentz, J. A., Marie, J., et al. (2012). Ligands and signaling proteins govern the conformational landscape explored by a G protein-coupled receptor. *Proceedings of the National Academy of Sciences of the United States of America, 109*(21), 8304–8309.

Mendre, C., & Mouillac, B. (2010). Pharmacological chaperones: A potential therapeutic treatment for conformational diseases. *Médecine/Sciences (Paris), 26*(6–7), 627–635.

Moeller, H. B., Rittig, S., & Fenton, R. A. (2013). Nephrogenic diabetes insipidus: Essential insights into the molecular background and potential therapies for treatment. *Endocrine Reviews, 34*(2), 278–301.

Morello, J. P., & Bichet, D. G. (2001). Nephrogenic diabetes insipidus. *Annual Review of Physiology, 63*, 607–630.

Morello, J. P., Petäjä-Repo, U. E., Bichet, D. G., & Bouvier, M. (2000b). Pharmacological chaperones: A new twist on receptor folding. *Trends in Pharmacological Sciences, 21*(12), 466–469.

Morello, J. P., Salahpour, A., Laperrière, A., Bernier, V., Arthus, M. F., Lonergan, M., et al. (2000a). Pharmacological chaperones rescue cell-surface expression and function of misfolded V2 vasopressin receptor mutants. *The Journal of Clinical Investigation, 105*(7), 887–895.

Mouillac, B., & Mendre, C. (2014). Vasopressin receptors and pharmacological chaperones: From functional rescue to promising therapeutic strategies. *Pharmacological Research, 83*, 74–78.

Newton, C. L., Whay, A. M., McArdle, C. A., Zhang, M., Van Koppen, C. J., Van de Lagemaat, R., et al. (2011). Rescue of expression and signaling of human luteinizing hormone G protein-coupled receptor mutants with an allosterically binding small-molecule agonist. *Proceedings of the National Academy of Sciences of the United States of America, 108*(17), 7172–7176.

Offermanns, S., Wieland, T., Homann, D., Sandmann, J., Bombien, E., Spicher, K., et al. (1994). Transfected muscarinic acetylcholine receptors selectively couple to Gi-type G proteins and Gq/11. *Molecular Pharmacology, 45*(5), 890–898.

Parenti, G., Andria, G., & Valenzano, K. J. (2015). Pharmacological chaperone therapy: Preclinical development, clinical translation, and prospects for the treatment of lysosomal storage disorders. *Molecular Therapy, 23*(7), 1138–1148.

Parkinson-Lawrence, E. J., Shandala, T., Prodoehl, M., Plew, R., Borlace, G. N., & Brooks, D. A. (2010). Lysosomal storage disease: Revealing lysosomal function and physiology. *Physiology (Bethesda), 25*(2), 102–115.

Platt, F. M., & Jeyakumar, M. (2008). Substrate reduction therapy. *Acta Paediatrica, 97*(457), 88–93.

Rahmeh, R., Damian, M., Cottet, M., Orcel, H., Mendre, C., Durroux, T., et al. (2012). Structural insights into biased G protein-coupled receptor signaling revealed by fluorescence spectroscopy. *Proceedings of the National Academy of Sciences of the United States of America, 109*(17), 6733–6738.

Rajagopal, S., Rajagopal, K., & Lefkowitz, R. J. (2010). Teaching old receptors new tricks: Biasing seven-transmembrane receptors. *Nature Reviews Drug Discovery, 9*(5), 373–386.

Reiter, E., Ahn, S., Shukla, A. K., & Lefkowitz, R. J. (2012). Molecular mechanism of β-arrestin-biased agonism at seven-transmembrane receptors. *Annual Review of Pharmacology and Toxicology, 52*, 179–197.

Robben, J. H., Kortenoeven, M. L. A., Sze, M., Yae, C., Milligan, G., Oorschot, V. M., et al. (2009). Intracellular activation of vasopressin V2 receptor mutants in nephrogenic diabetes insipidus by nonpeptide agonists. *Proceedings of the National Academy of Sciences of the United States of America, 106*(29), 12195–12200.

Robben, J. H., Sze, M., Knoers, N. V., & Deen, P. M. (2006). Rescue of vasopressin V2 receptor mutants by chemical chaperones: Specificity and mechanism. *Molecular Biology of the Cell, 17*(1), 379–386.

Robben, J. H., Sze, M., Knoers, N. V., & Deen, P. M. (2007). Functional rescue of vasopressin V2 receptor mutants in MDCK cells by pharmacochaperones: Relevance to therapy of nephrogenic diabetes insipidus. *American Journal of Physiology-Renal Physiology, 292*(1), F253–F260.

Rosenthal, W., Seibold, A., Antaramian, A., Lonergan, M., Arthus, M. F., Hendy, G. N., et al. (1992). Molecular identification of the gene responsible for congenital nephrogenic diabetes insipidus. *Nature, 359*(6392), 233–235.

Sagan, S., Chassaing, G., Pradier, L., & Lavielle, S. (1996). Tachykinin peptides affect differently the second messenger pathways after binding to CHO-expressed human NK-1 receptors. *The Journal of Pharmacology and Experimental Therapeutics, 276*(3), 1039–1048.

Sato, S., Ward, C. L., Krouse, M. E., Wine, J. J., & Kopito, R. R. (1996). Glycerol reverses the misfolding phenotype of the most common cystic fibrosis mutation. *The Journal of Biological Chemistry, 271*(2), 635–638.

Schrier, R. W., Gross, P., & Gheorghiade, M. (2006). Tolvaptan, a selective oral vasopressin V2-receptor antagonist, for hyponatremia. *New England Journal of Medicine, 355*(20), 2099–2112.

Serradeil-Le Gal, C., Lacour, C., Valette, G., Garcia, G., Foulon, L., Galindo, G., et al. (1996). Characterization of SR 121463A, a highly potent and selective, orally active vasopressin V2 receptor antagonist. *The Journal of Clinical Investigation, 98*(12), 2729–2738.

Shenoy, S., & Lefkowitz, R. J. (2011). β-arrestin-mediated receptor trafficking and signal transduction. *Trends in Pharmacological Sciences, 32*(9), 521–533.

Smith, E., Janovick, J. A., Bannister, T. D., Shumate, J., Scampavia, L., Conn, P. M., et al. (2016). Identification of potential pharmacoperones capable of rescuing the functionality of misfolded vasopressin 2 receptor involved in Nephrogenic Diabetes Insipidus. *Journal of Biomolecular Screening, 21*(8), 824–831.

Smithson, D. C., Janovick, J. A., & Conn, P. M. (2013). Therapeutic rescue of misfolded.-mistrafficked mutants: Automation-friendly high-throughput assays for identification of pharmacoperone drugs of GPCRs. *Methods in Enzymology, 521*, 3–16.

Spanakis, E., Milord, E., & Gragnoli, C. (2008). AVPR2 variants and mutations in Nephrogenic Diabetes Insipidus ; review and missense mutation significance. *Journal of Cellular Physiology, 217*(3), 605–617.

Takasu, H., Gardella, T. J., Luck, M. D., Potts, J. T., Jr., & Bringhurst, F. R. (1999). Amino-terminal modifications of human parathyroid hormone (PTH) selectively alter phospholipase C signaling via the type 1 PTH receptor: Implications for design of signal-specific PTH ligands. *Biochemistry, 38*(41), 13453–13460.

Tamarappoo, B. K., & Verkman, A. S. (1998). Defective aquaporin-2 trafficking in nephrogenic diabetes insipidus and correction by chemical chaperones. *The Journal of Clinical Investigation, 101*(10), 2257–2267.

Tan, C. M., Nickols, H. H., & Limbird, L. E. (2003). Appropriate polarization following pharmacological rescue of V2 vasopressin receptors encoded by X-linked nephrogenic diabetes insipidus alleles involves a conformation of the receptor that also attains mature glycosylation. *The Journal of Biological Chemistry, 278*(37), 35678–35686.

Thurmond, R. L., Desai, P. J., Dunford, P. J., Fung-Leung, W. P., Hofstra, C. L., Jiang, W., et al. (2004). A potent and selective histamine H4 receptor antagonist with anti-inflammatory properties. *The Journal of Pharmacology and Experimental Therapeutics, 309*(1), 404–413.

Torres, V. E., Chapman, A. B., Devuyst, O., Gansevoort, R. T., Grantham, J. J., Higashihara, E., et al., TEMPO 3:4 Trial investigators. (2012). Tolvaptan in patients with autosomal dominant polycystic kidney disease. *New England Journal of Medicine, 367*(25), 2407–2418.

Treschan, T. A., & Peters, J. (2006). The vasopressin system. *Anesthesiology, 105*(3), 599–612.

Tsukagushi, H., Matsubara, H., Taketani, S., Mori, Y., Seido, T., & Inada, M. (1995). Binding, intracellular transport and biosynthesis-defective mutants of vasopressin type 2 receptor in patients with X-linked nephrogenic diabetes insipidus. *The Journal of Clinical Investigation, 96*(4), 2043–2050.

Violin, J. D., DeWire, S. M., Yamashita, D., Rominger, D. H., Nguyen, L., Schiller, K., et al. (2010). Selectively engaging β-arrestins at the angiotensin II type 1 receptor reduces blood pressure and increases cardiac performance. *The Journal of Pharmacology and Experimental Therapeutics, 335*(3), 572–579.

Wesche, D., Deen, P. M. T., & Knoers, N. V. (2012). Congenital nephrogenic diabetes insipidus : The current state of affairs. *Pediatric Nephrology, 27*(12), 2183–2204.

White, E., McKenna, J., Cavanaugh, A., & Breitwieser, G. E. (2009). Pharmacochaperone-mediated rescue of calcium-sensing receptor loss-of-function mutants. *Molecular Endocrinology, 23*(7), 1115–1123.

Wisler, J. W., DeWire, S. M., Whalen, E. J., Violin, J. D., Drake, M. T., Ahn, S., et al. (2007). A unique mechanism of beta-blocker action : Carvedilol stimulates beta-arrestin signaling. *Proceedings of the National Academy of Sciences of the United States of America, 104*(42), 16657–16662.

Wüller, S., Wiesner, B., Loffler, A., Furkert, J., Krause, G., Hermosilla, R., et al. (2004). Pharmacochaperones post-translationally enhance cell surface expression by increasing conformational stability of wild-type and mutant vasopressin V2 receptors. *The Journal of Biological chemistry, 279*(45), 47254–47263.

Zimran, A., Altarescu, G., & Elstein, D. (2013). Pilot study using ambroxol as a pharmacological chaperone in type 1 Gaucher disease. *Blood Cells, Molecules, and Diseases, 50*(2), 134–137.

CHAPTER EIGHT

LOXL1 folding in exfoliation glaucoma

Audrey M. Bernstein[a], Robert Ritch[b] and J. Mario Wolosin[c,*]
[a]SUNY Upstate Medical University, Department of Ophthalmology and Visual Sciences, Syracuse, NY, United States
[b]Einhorn Clinical Research Center, New York Eye and Ear Infirmary of Mount Sinai, New York, NY, United States
[c]Eye and Vision Research Institute, Department of Ophthalmology, Icahn School of Medicine at Mount Sinai, New York, NY, United States
*Corresponding author: E-mail: jmario.wolosin@gmail.com

Contents

Abstract

Exfoliation syndrome (XFS) is an age-related disease defined by the deposition of aggregated fibrous material (XFM) in the peri-cellular space. Principal morbidity occurs in the eye, where XFM accumulates on the anterior ocular tissues. *GWAS* have found that certain genetic variants of lysyl oxidase-like 1 (*LOXL1*), a matrix cross-linking enzyme that is required for elastic fiber formation confer risk for the development of XFS, but are not a single causative factor as many genetically affected individuals do not develop XFS or subsequent glaucoma (XFG). We have found that XFG cells display defects in lysosomes, microtubules, autophagy, and mitochondria resembling defects found in cells from age-related syndromes, such as the main neurodegenerative diseases. In the majority of these diseases, the determining cellular factor is a protein containing intrinsically disordered regions (IDRs) and displaying a high propensity for aggregation. We have found that in XFG patient-derived cells, LOXL1 protein is actively subjected to autophagic clearance, suggesting that LOXL1 is undergoing aggregation. *In silico* analysis demonstrates that LOXL1's first 369 aa constitute an IDR with the highest disorder

Advances in Protein Chemistry and Structural Biology, Volume 118
ISSN 1876-1623
https://doi.org/10.1016/bs.apcsb.2019.09.005
© 2020 Elsevier Inc.
All rights reserved.

probability peak centering around the known risk positions. Experimentally, we have found over-expression of either unmodified LOXL1 or fluorescent chimeras preserving the well-structured N-terminus cause copious intracellular aggregation and that aggregation wanes when the high IDR peaks are deleted. Overall, our work suggests that XFS/G results from the aggregation of the LOXL1 protein coupled with a reduction of cellular proteostasis capabilities in aging, resulting in a chronic build-up of LOXL1-containing protein aggregates.

1. Introduction

The age-dependent Exfoliation Syndrome (XFS) is a systemic disease characterized by the formation of aggregated protein deposits called Exfoliation Material (XFM) in all tissues that can synthesize elastin fibers (Nazarali, Damji, & Damji, 2018). While there are reports showing XFS health impact in multiple organs, in particular in peri-cardiac vessels (Anastasopoulos, Founti, & Topouzis, 2015; Prince & Ritch, 1986), the most significant morbidity occurs in the eye. Flaky material accumulates on the ciliary body, iris, lens, and trabecular meshwork. When glaucoma develops in the presence of these aggregates, it is referred to as Exfoliation Glaucoma (XFG)(Aboobakar, Johnson, Stamer, Hauser, & Allingham, 2016; Henry et al., 1987; Ritch & Schlotzer-Schrehardt, 2001; Ritch, Schlotzer-Schrehardt, & Konstas, 2003). XFG displays a marked resistance to classical pharmacological treatments for primary open-angle glaucoma (POAG) and induces a much faster vision loss. The XFG contribution to the incidence of open-angle glaucoma shows a strong dependence on the ethnic background, geography, and environmental variables (Pasquale et al., 2014). It constitutes between 20 and 60% of glaucoma cases in many regions of the world, including several Scandinavian countries, Russia, Iran, Ethiopia, and the South African Bantu tribe (Bialasiewicz, Wali, Shenoy, & Al-Saeidi, 2005; Forsius, 1988; Konstas & Ringvold, 2018; Ringvold, 1996), and overall, it accounts for 20%–25% of the estimated 70 million worldwide glaucoma sufferers. Because the earliest onset of POAG is around age 40, but for XFG is around age 60, both XFG prevalence and its fractional contribution to total glaucoma can be projected to increase as worldwide lifespan extends. For instance, in rural Finland XFS prevalence in the 60–69, 70–79 and 80 + age cohorts are, 14.2%, 21.9% and 34.7%, respectively. Projecting these numbers forward, yields a >90% probability of XFS incidence for a presently hypothetical, 100–110 cohort (Krause et al., 1988).

2. Macroscopic, structural and biochemical features of XFG

Whereas elsewhere the XFS material will manifest as extracellular deposits within an extracellular matrix not accessible to the naked eye, the accumulations in the eye occur within the aqueous space of the eye, resulting in their easy recognition as white flaky deposits (Fig. 1-A and B). Ultrastructural studies have shown that the material originates from the surface of tissues facing the posterior aqueous chamber. In the ciliary body, putative aggregated material has been visualized at invaginations of the aqueous facing, basolateral membrane of the non-pigmented epithelial layer (Schlotzer-Schrehardt, 2009). The pigment-rich pseudostratified epithelium covering the inner surface of the iris may also secrete XFM. Whether XFM is made there or not, one of the more dramatic manifestations of XFG is the release of pigment granules from these cells, indicating that XFG causes cellular stress. Yet, the most visible accumulations occur at the equator of the crystalline lens. The cellular source of these deposits remains to be

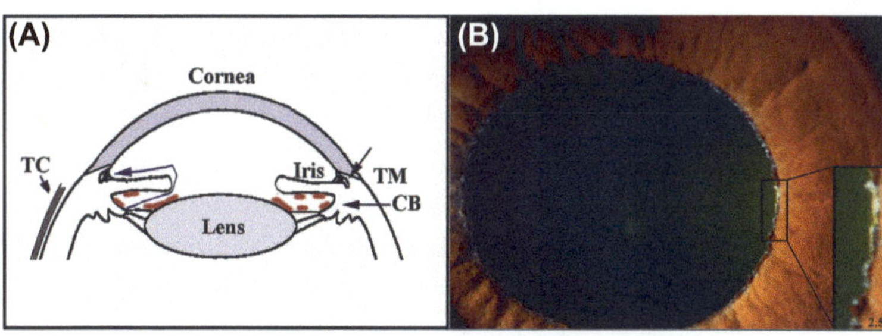

Fig. 1 (A) ***Schematic description of the cellular components of the anterior segments of the eye in relation to aqueous flow.*** Fluid (blue line) is generated at the inner surface of the epithelium of the ciliary body CB) in the posterior aqueous chamber. From there it flows through the space between the iris and lens into the anterior aqueous chamber to exit the eye through the trabecular meshwork (TM) also referred to as the outflow facility. Increased, abnormal flow resistance of the facility causes elevation of the intraocular pressure (IOP). The increased pressure lead to the development of glaucoma, loss of retinal visual function due to retinal neural death. In individuals affected by exfoliation syndrome (XSF), proteinaceous material (shown tan marks in sketch) forms or deposits on the surface of the epithelial facing the posterior aqueous chamber, including the equator of the lens. The deposition is frequently associated with a type of glaucoma, exfoliation glaucoma, characterized by drug-unresponsive IOP increases and rapid progression of glaucoma. The Tenon's capsule (TC) is a thin, elastic collagenous lining covering the sclera. (B) **Flaky white deposits in the equator of the lens.** The pupil has been dilated.

elucidated. Since the permeability of the human lens capsule has an MW cut off of < 170 Kd (Kastner et al., 2013; Lee, Vroom, Fishman, & Bent, 2006) this XFM may arrive at the lens from the CB-Iris carried up by the aqueous flow. Nevertheless, ultrastructural studies suggest that the deposit may be generated by the lens cells and transferred through the capsule by an impaling-type mechanical force. The glaucoma-promoting effect of XFS may be related to the iris epithelium release of pigment granules, which carried away by the fluid flow to the outflow facility, deposit there and thereby increase outflow resistance elevating the intraocular pressure. This secondary consequence of XFS may combine with a putative effect on the integrity of the lamina cribrosa, the protective cushion for the retinal neurons at the optic nerve head; XFM accumulation there may weaken the lamina resulting in damage to the optic nerve fibers at lower pressures than those needed to elicit the damage in POAG. Ultrastructural examination (atomic force microscopy) reveals features of organized structures within XFM. Short beaded stretches crosslink at variable angles (Fig. 2A). These structures do not resemble the structure of fibrillin fibrils (Fig. 2B).

At the biochemical level, mass spectra proteomic studies have shown that XFM incorporates an inordinate number of different proteins. The initial analysis identified, fibrillin-1, microfibril-associated glycoprotein, and clusterin/ApoJ and elastin, between other components (Ronci, Sharma, Martin, Craig, & Voelcker, 2013; Schlotzer-Schrehardt, 2009; Sharma et al., 2009, 2018). The first two proteins are logical candidates to contribute to a fibrous structure. The presence of Clusterin/ApoJ is unsurprising; this protein is secreted by cells when a proximal extracellular aggregate forms in an attempt

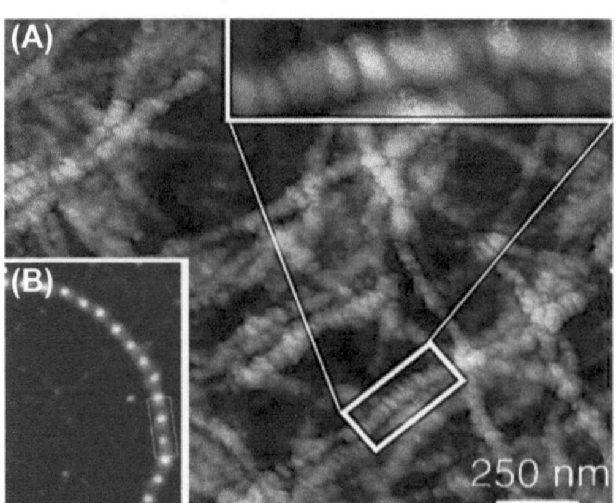

Fig. 2 (A) ***XFM structure.*** (B) ***Fibrillin microfibrils.***

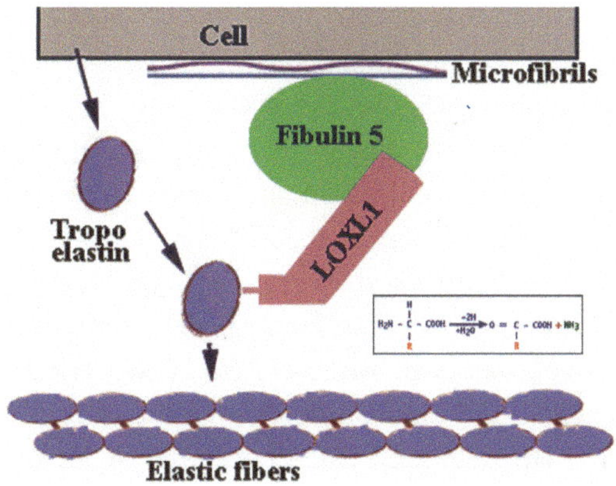

Fig. 3 ***Schematic description of elastic fiber development.*** At the peri-cellular space the adapter protein Fibulin 5 anchoring at Fibrillin 1, binds the deaminase LOXL1 setting it in the proper spatial configuration to cause deamination of secreted Tropoelastin monomers. The reactive aldehyde groups allow homologous polymerization of Tropoelastin to form elastic fibers.

to dissociate it. Other more unexpected components such as basement membrane and blood proteins may merely become sequestered in the forming meshwork. LOXL1, the main subject of this review was identified as an XFM component only after its relationship to XFG was indicated by the genomic wide association studies (*GWAS*; described below). Since LOXL1's primary function is to catalyze the weaving of elastic fibers from the tropoelastin monomers through its deaminase activity, the presence of both elastin and LOXL1 peptides in the aggregates suggests the involvement of the elaborate extracellular process for elastin weaving in aggregate formation (Fig. 3). Supporting this hypothesis, it has been recently shown that mutations in fibulin 5, the scaffold protein that spatially positions LOXL1 for tropoelastin deamination, fosters the formation of extracellular aggregates (Padhy, Kapuganti, Hayat, Mohanty, & Alone, 2019).

3. LOXL1 variants and XFG risk

A major advance toward the elucidation of the molecular drivers of XFG was provided by a *GWAS* that found that in the Swedish population the XFG risk was markedly affected by three sequence variants in LOXL1. One risk variant occurs in a long non-coding (lnc) region of the LOXL1 gene. The other two SNPs are CDS variants involving arginine (R) or leucine at the amino acid position (aa) 141 and glycine (G) or aspartic acid (D) at aa 153 (Hauser et al., 2015; Malukiewicz et al., 2011; Thorleifsson et al., 2007). The two risk-linked variants, R141 and G153 are not rare; in most world populations 153G incidence is about 70%. The strong statistical link results from the fact that in most studies including both regional LOXL1-focused genetic investigation and a very large worldwide *GWAS* (Abu-Amero et al., 2010; Chen et al., 2010; de Juan-Marcos et al., 2016; Hewitt et al., 2008; Thorleifsson et al., 2007) the incidence of G153 in XFS patients is 99% or 100 % (Fan et al., 2008). Similar numbers, though less accentuated, support the link of 141R to XFG. These links have been challenged by confounding results obtained with two high XFS prevalence populations. In the Japanese high risk is conferred by leucine occupation of aa 141, whereas in the Bantu, high XFS risk is associated with aspartic acid at aa 151 (Chen et al., 2010; Hayashi, Gotoh, Ueda, Nakanishi, & Yoshimura, 2008; Rautenbach, Bardien, Harvey, & Ziskind, 2011; Williams et al., 2010). Given that presently no high XFS prevalence population in which the variants at either of these two positions display risk indifference has been identified, seeing these divergent observations as voiders of the presumptive LOXL1 variant-XFS link will not be prudent. Rather, a more fitting explanation is that the CDS variant risk reflects the interactions of these two sites with other proteins or other locations within LOXL1, which in turn, contain variants alternatively facilitating or obstructing, the process of XFM formation. This conjecture fits the facts a) that the Japanese and the Bantu, which end having no variant commonality for risk, happens to be the two populations at both ends of the human genomic clustering map (Fig. 4) and b) the uniqueness of the genomic variant complement in the Bantu is suggested by the fact that, intriguingly the Bantu is the only population in the world where the G/D ratio at aa 153 deviates from a universal 70/30 % ratio (Rautenbach et al., 2011; Williams et al., 2010). More recently, a functional variant in the *LOXL1* locus was identified immediately upstream of the long noncoding RNA *LOXL1-AS1,* that changes promotor activity and thus the level of its expression. Investigations into *LOXL1-AS1*'s function indicate involvement in the maintenance of the extracellular matrix and

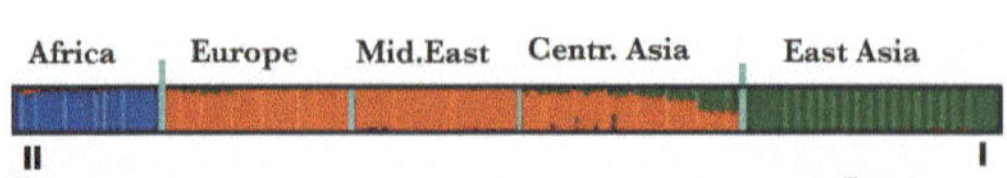

Fig. 4 ***Human genetic clustering.*** The South African Bantu and Japanese sit at the opposite sides of the spectrum of the differences in expression of genetic variants. One representative gene cluster, of several studied, is shown.

other pathways integral to the pathophysiology of XFG. (Hauser et al., 2015). Future studies will test the impact of *LOXL1-AS1* on the development of Exfoliation Glaucoma.

4. LOXL1 autophagy and LOXL1 in XFG patient cells

For studies based on actual disease, presently the only available option consists of Tenon's capsule tissue obtained from trabeculectomies (glaucoma surgery to relieve intraocular pressure) of XFG patients and the tenon fibroblasts (XFG-TF) that can be cultured out of these minute biopsies (Kottler et al., 2005; Zenkel et al., 2011). While the tenon tissue is not intraocular, the systemic nature of the syndrome and the fact that the samples derive from individuals with glaucoma and documentable XFM accumulations, make XFG-TFs a useful conduit to understanding the drivers of XFG. Studies were based on the comparison against TF derived from non-XFG patients, either primary open-angle glaucoma (POAG) or non-glaucoma (NG) donors (Aboobakar et al., 2016; Bernstein, Ritch, & Wolosin, 2018; Want et al., 2016; Wolosin, Ritch, & Bernstein, 2018). Our studies have shown that XFG-TFs are larger than POAG and NG-TFs, proliferate at a slower rate and fail to organize in parallel bundles typical of cultured fibroblasts. Furthermore, XFG-TFs display the multiple abnormalities and dysfunctions observed in cells from the main neurodegenerative diseases and other age-related aggregopathies, including, impeded autophagic flux, disturbed LC3 expression, engorged and mislocalized lysosomes and endosomes, and decaying mitochondria (Bernstein et al., 2018; Want et al., 2016; Wolosin et al., 2018). A model for the formation of toxic intracellular aggregations leading to larger extracellular aggregates comes from Alzheimer's. For a long time, it was assumed that neuronal death was caused by bulky extracellular amyloid aggregations. Nonetheless, it is now understood that the toxicity is caused by small intracellular aggregations (Iadanza, Jackson, Hewitt, Ranson, & Radford, 2018; Jeong, 2017; Tam & Pasternak, 2012). As graphically indicated in Fig. 5, Alzheimer's extracellular deposits seem to result from the export of intracellular aggregates as a cell survival mechanism. It has been speculated that the intracellular aggregate toxicity reflects hindrances to cellular autophagy (Cruz, Kumar, Yuan, Arikkath, & Batra, 2018; Olzscha et al., 2011). The same possibility exists for XFG. In terms of LOXL1, we have found that in the XFG-derived fibroblasts LOXL1 accumulates in particles of larger size compared to controls and it

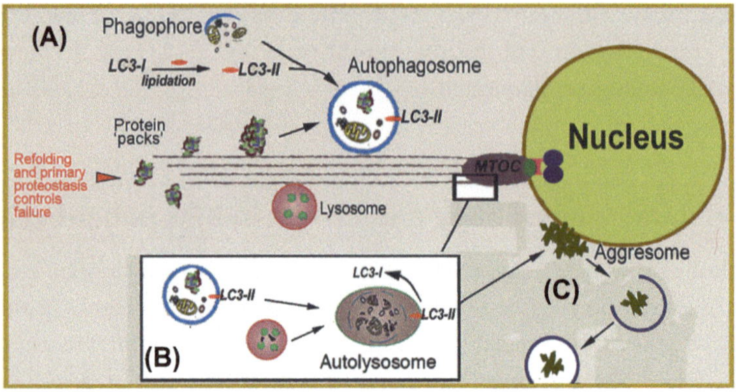

Fig. 5 ***Macroautophagy.*** (A) Biogenesis of autophagosomes from precursor phagophores within the cytosol is orchestrated by the lipidated form of LC3 (LC3-II). Autophagosomes engulf 'macroscopic' cellular detritus, naturally decaying organelles, in particular, mitochondria (i.e., mitophagy) and misfolded or denatured proteins that have resisted the primary protein folding quality control mechanisms. (B) Autophagosomes fuse with lysosomes at the microtubule organizing center (MTOC). (C) Detritus that resist the combined proteostatic processes may accumulate in body inclusions, (e.g., Lewy body in senile dementia) categorized alternatively as aggresomes. If as demonstrated for tau protein propagation in Alzheimer's and α-synuclein protein in Parkinson's, then the aggregates may propagate to adjacent cells by particulate phagocytosis or endocytosis.

is bound and exported by a chaperone that exports only misfolded proteins (*unpublished*). Additionally, when autophagy is blocked with the inhibitors Spautin-1 and Bafilomycin the amount of LOXL1 protein detected increases in XFG-TFs but not in POAG cells suggesting that LOXL1 is processed in the autophagy pathway that engulfs aggregates, in XFG-TFs only (Bernstein et al., 2018).

5. LOXL1 and intrinsic disorder

Given the observations described in the previous section on LOXL1 phenomena in XFG-TFs, it is plausible the LOXL1's role in XFG parallels the role played by proteins in other age-related aggregopathies, namely as a germinal factor for XFM genesis. Extensive studies across the spectrum of age-related aggregopathies have established that they are usually mediated by one or more proteins containing intrinsically disordered regions (IDR) s. The IDR aggregation propensity reflects their ability to transiently exist, or dynamically adapt, undesirable hydrophobic β-sheet conformations of low solubility that can 'stack'(De Simone et al., 2012; Martinelli, Lopes, John, Carlini, & Ligabue-Braun, 2019; Uversky, 2016) or even transfer the denatured conformation to an otherwise healthy protein, in a prion-like manner (Frost & Diamond, 2010; Kumar & Udgaonkar, 2019; Vaquer-Alicea & Diamond, 2019). This potential for unchecked propagation of hydrophobic denaturation represents a serious danger to the cell from a number of different perspectives, in particular for their interference with microtubule transport processes (Clark, Yeaman, Blizzard, Chuckowree, & Dickson, 2016; Cruz et al., 2018; Dubey, Ratnakaran, & Koushika, 2015; Olzscha et al., 2011). Incipient, small aggregates are the most toxic elements. The copious extracellular amyloid brain deposits found in cadaver brains in Alzheimer's and probably, the large inclusion bodies observed in other neurodegenerative diseases, result in part from aging of the protecting and lytic mechanisms that neutralize aggregation within a cell (Fig. 5).

The study of proteins with extended disordered regions is challenging, they easily aggregate and cannot be crystallized. As a result, a large number of prediction-based systems providing disorder prediction (probability) that use a variety of different computational methods have been established (Atkins, Boateng, Sorensen, & McGuffin, 2015). Utilization of several of these servers to analyze the LOXL1 sequence uniformly yielded a prediction of disorder for most of the LOXL1 N-terminus extending up to amino acid (aa) 324. The results generated by one of the most advanced predictors, PdDOS, is shown in Fig. 6A. The most notable feature is the presence of two peaks of near > 95 % probability, centered around aa 80 and 155, respectively. The latter peak encompasses the two CDS variants associated with XFG risk. The C-terminus, containing the enzymatic function of LOXL1 (DP < 0.5), is fully structured. Intriguingly, all other members of the LOXL family are fully structured proteins (Fig. 6B) and BlastP analysis the human 1–324 IDR sequence against the GeneBank proteomic collections (eukaryotic and prokaryotic) yields high conservation across the LOXL1 mammalian spectrum but no significant homology to any other protein. This feature suggests that LOXL1 may play be involved in unique complex cellular interactions (Uversky, 2016).

6. The role of the IDR in LOXL1 aggregation

To examine whether the predicted disorder in LOXL1 confers to this protein an aggregation propensity, we employed lentivirus transduction of

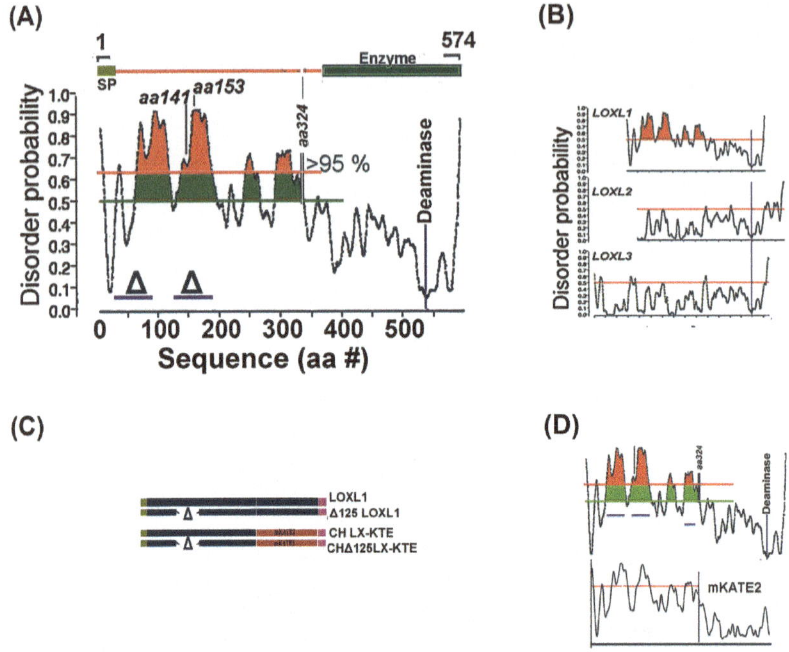

Fig. 6 ***IDR in LOXL proteins.*** (A) Order/disorder probability prediction in the LOXL1 sequence. The zones where prediction confidence is > 95% is highlighted in red and the location for the enzymatic activity is indicated. (B) Comparison of order/disorder in LOXL family. (C) Comparison of order/disorder pattern in parent LOXL1 sequence and its mKATE2 chimera. (D) Examples of transgene LOXL1 constructs and their equivalent chimeras used to test aggregation propensity in overexpression assays.

immortalized fibroblasts to generate, after antibiotic selection, cell sublines with high expression from a CMV promoter of either the whole LOXL1 sequence or sequences in which the aa stretches 28–95 (Δ28) or 125–185 (Δ125) were deleted (blue bars in Fig. 6A). Additionally, the structured enzymatic C-terminus of LOXL1, starting from aa 324, was replaced by the highly structured monomeric (m), red fluorescent protein mKATE2. The polypeptides were designed without tags or with either 6xHis (HIS) or FLAG C-terminus tags (Fig. 6C, schematic of Δ125). Flow cytometry demonstrated that all mKATE2 cells lines had a comparable mean level of fluorescence and Gaussian intensity distributions. Selected results from the over-expression studies are shown in Fig. 7. Over expression of whole LOXL1 creates small aggregates when grown in FBS-containing media (Fig. 7A) However, when cells are stressed with growth in supplemented-serum free media (SSFM, Fig. 7B) this phenotype is exacerbated, ring-like

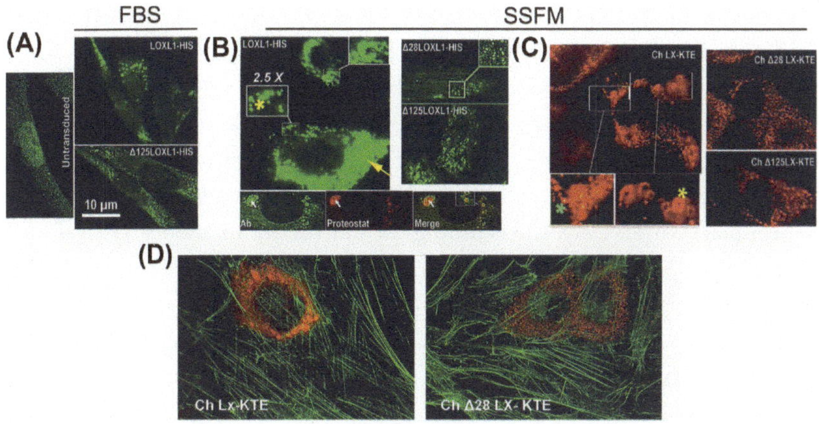

Fig. 7 Aggregation in overexpressed LOXL1 and homolog mKATE2 sequences. Representative images of BJ5a cells transduced with the indicated constructs. LOXL1 constructs were stained with an N-terminus anti-LOXL1 monoclonal Ab (H11 clone). Chimeras are red fluorescent. (A) Cultures in FBS-containing media. Left. Untransduced cells. Top. Cells transduced with LOXL1-HIS. Enlarged vesicles accumulated near the nuclei are the most common feature of cells expressing high levels of the transgene. Bottom. Typical stain profile for the Δ125-185 LOXL1 version. Vesicles are smaller and distribute throughout the cytosol. (B) Left top. The same transduction with LOXL1-HIS cDNA in supplemented-serum free media (SSFM). In addition to engorged vesicles (yellow arrow), very large particles and ring-shaped profiles (yellow asterisks) are frequent. Bottom 3-frame composite. Anti-LOXL1- Proteostat dual stain. Whereas the LOXL1 Ab stains the periphery of particles generating ring-shaped fluorescence, the aggregated-protein dye (Proteostat) labels the center of the same particles (white arrow). Right. LOXL1 with deleted domains lack this grossly vesicular phenotype. (C) The fluorescence patterns of the mKATE2 chimeras. Left. Ch LX-KTE fluorescence. Large vesicles and ring-like profiles (yellow asterisk) and engorged vesicles. There are also vesicles where the fluorescence intensity is higher at the center, rather than absent or lower (green asterisk). Right. The fluorescence of Δ versions. There are no aggregation profiles. (D) Micrographs of cells transduced with whole sequence (1–324) mKATE2 chimera and a chimera with a 28–95 aa deletion (Δ28) and counterstained with FITC-phalloidin.

vesicular structures are apparent. In both cases the Δ 28 and Δ 125 have reduced aggregate formation compared to the whole LOXL1. Interestingly, the center of the ring-like LOXL1-staining structures in Fig. 7B can be identified with a protein aggregation-specific dye (Proteostat, Fig. 7B, bottom). Like LOXL1-HIS, images of the chimera Ch LX-KTE, demonstrate the same ring-like structures (Fig. 7C). A plausible explanation for the LOXL1-Proteostat staining profiles (and similarly, the mKATE2 fluorescence patterns) is that the center of large vesicles is undergoing intense aggregation of LOXL1 or CH LX-KTE and, while initially reflected in increased fluorescence at the center (LOXL1-binding secondary Ab or mKATE fluorescence), eventually inhibits (quenches) fluorophore function. As in the case of the parent LOXL1 sequences the Δ versions of CH LX-KTE showed only enlarged vesicular structures distributed through the cytosol (Fig. 7C). In Fig. 7D, the LX-KTE fluorescent images are set in the cellular context by the staining of the actin filaments with FITC-phalloidin.

7. Summary

The aggregation of the LOXL1 protein coupled with age-related decline in cellular proteostasis capabilities may result in a build-up of LOXL1-containing protein aggregates. *In silico* and experimental work demonstrate LOXL1's propensity to aggregate and provide a road map to identify the loci that determine or facilitate the initiation of aggregation. This latter knowledge could be instrumental in the design of strategies to neutralize aggregation without the need to fully understand the process of fibrous growth.

Acknowledgments

This work was supported by The MYS Family US Charitable Foundation, Inc., Research to Prevent Blindness, The Bright Focus Foundation, The Glaucoma Foundation, and a donation from Barbie and Tony Mayer (AMB and JMW), NIH-NEI R01 EY024942 and Lions District 20-Y (AMB) and NIH-NEI R01 EY018748 (JMW).

References

Aboobakar, I. F., Johnson, W. M., Stamer, W. D., Hauser, M. A., & Allingham, R. R. (2016). Major review: Exfoliation syndrome; advances in disease genetics, molecular biology, and epidemiology. *Experimental Eye Research, 154*, 88–103. http://dx.doi.org/10.1016/j.exer.2016.11.011.

Abu-Amero, K. K., Osman, E. A., Dewedar, A. S., Schmidt, S., Allingham, R. R., & Al-Obeidan, S. A. (2010). Analysis of LOXL1 polymorphisms in a Saudi Arabian population with pseudoexfoliation glaucoma. *Molecular Vision, 16*, 2805–2810. Retrieved from https://www.ncbi.nlm.nih.gov/pubmed/21197115.

Anastasopoulos, E., Founti, P., & Topouzis, F. (2015). Update on pseudoexfoliation syndrome pathogenesis and associations with intraocular pressure, glaucoma and systemic diseases. *Current Opinion in Ophthalmology, 26*(2), 82–89. http://dx.doi.org/10.1097/ICU.0000000000000132.

Atkins, J. D., Boateng, S. Y., Sorensen, T., & McGuffin, L. J. (2015). Disorder prediction methods, their applicability to different protein targets and their usefulness for guiding experimental studies. *International Journal of Molecular Sciences, 16*(8), 19040–19054. http://dx.doi.org/10.3390/ijms160819040.

Bernstein, A. M., Ritch, R., & Wolosin, J. M. (2018). Exfoliation syndrome: A disease of autophagy and LOXL1 proteopathy. *Journal of Glaucoma*. http://dx.doi.org/10.1097/IJG.0000000000000919.

Bialasiewicz, A. A., Wali, U., Shenoy, R., & Al-Saeidi, R. (2005). Patients with secondary open-angle glaucoma in pseudoexfoliation (PEX) syndrome among a population with high prevalence of PEX. Clinical findings and morphological and surgical characteristics. *Ophthalmologe, Der, 102*(11), 1064–1068. http://dx.doi.org/10.1007/s00347-005-1226-2.

Chen, H., Chen, L. J., Zhang, M., Gong, W., Tam, P. O., Lam, D. S., et al. (2010). Ethnicity-based subgroup meta-analysis of the association of LOXL1 polymorphisms with glaucoma. *Molecular Vision, 16*, 167–177. Retrieved from https://www.ncbi.nlm.nih.gov/pubmed/20142848.

Clark, J. A., Yeaman, E. J., Blizzard, C. A., Chuckowree, J. A., & Dickson, T. C. (2016). A case for microtubule vulnerability in amyotrophic lateral sclerosis: Altered dynamics during disease. *Frontiers in Cellular Neuroscience, 10*, 204. http://dx.doi.org/10.3389/fncel.2016.00204.

Cruz, E., Kumar, S., Yuan, L., Arikkath, J., & Batra, S. K. (2018). Intracellular amyloid beta expression leads to dysregulation of the mitogen-activated protein kinase and bone morphogenetic protein-2 signaling axis. *PLoS One, 13*(2), e0191696. http://dx.doi.org/10.1371/journal.pone.0191696.

De Simone, A., Kitchen, C., Kwan, A. H., Sunde, M., Dobson, C. M., & Frenkel, D. (2012). Intrinsic disorder modulates protein self-assembly and aggregation. *Proceedings of the National Academy of Sciences of the United States of America, 109*(18), 6951–6956. http://dx.doi.org/10.1073/pnas.1118048109.

Dubey, J., Ratnakaran, N., & Koushika, S. P. (2015). Neurodegeneration and microtubule dynamics: Death by a thousand cuts. *Frontiers in Cellular Neuroscience, 9*, 343. http://dx.doi.org/10.3389/fncel.2015.00343.

Fan, B. J., Pasquale, L., Grosskreutz, C. L., Rhee, D., Chen, T., DeAngelis, M. M., ... Wiggs, J. L. (2008). DNA sequence variants in the LOXL1 gene are associated with pseudoexfoliation glaucoma in a U.S. clinic-based population with broad ethnic diversity. *BMC Medical Genetics, 9*, 5. http://dx.doi.org/10.1186/1471-2350-9-5.

Forsius, H. (1988). Exfoliation syndrome in various ethnic populations. *Acta Ophthalmologica*, (Suppl. 184), 71–85. Retrieved from https://www.ncbi.nlm.nih.gov/pubmed/2853925.

Frost, B., & Diamond, M. I. (2010). Prion-like mechanisms in neurodegenerative diseases. *Nature Reviews Neuroscience, 11*(3), 155–159. http://dx.doi.org/10.1038/nrn2786.

Hauser, M. A., Aboobakar, I. F., Liu, Y., Miura, S., Whigham, B. T., Challa, P., ... Allingham, R. R. (2015). Genetic variants and cellular stressors associated with exfoliation syndrome modulate promoter activity of a lncRNA within the LOXL1 locus. *Human Molecular Genetics, 24*(22), 6552–6563. http://dx.doi.org/10.1093/hmg/ddv347.

Hayashi, H., Gotoh, N., Ueda, Y., Nakanishi, H., & Yoshimura, N. (2008). Lysyl oxidase-like 1 polymorphisms and exfoliation syndrome in the Japanese population. *American Journal of Ophthalmology, 145*(3), 582–585. http://dx.doi.org/10.1016/j.ajo.2007.10.023.

Henry, J. C., Krupin, T., Schmitt, M., Lauffer, J., Miller, E., Ewing, M. Q., et al. (1987). Long-term follow-up of pseudoexfoliation and the development of elevated intraocular pressure. *Ophthalmology, 94*(5), 545–552. Retrieved from https://www.ncbi.nlm.nih.gov/pubmed/3601370.

Hewitt, A. W., Sharma, S., Burdon, K. P., Wang, J. J., Baird, P. N., Dimasi, D. P., ... Craig, J. E. (2008). Ancestral LOXL1 variants are associated with pseudoexfoliation in

Caucasian Australians but with markedly lower penetrance than in Nordic people. *Human Molecular Genetics, 17*(5), 710–716. http://dx.doi.org/10.1093/hmg/ddm342.

Iadanza, M. G., Jackson, M. P., Hewitt, E. W., Ranson, N. A., & Radford, S. E. (2018). A new era for understanding amyloid structures and disease. *Nature Reviews Molecular Cell Biology*. http://dx.doi.org/10.1038/s41580-018-0060-8.

Jeong, S. (2017). Molecular and cellular basis of neurodegeneration in Alzheimer's disease. *Molecules and Cells, 40*(9), 613–620. http://dx.doi.org/10.14348/molcells.2017.0096.

de Juan-Marcos, L., Escudero-Dominguez, F. A., Hernandez-Galilea, E., Cabrillo-Estevez, L., Cruz-Gonzalez, F., Cieza-Borrella, C., … Gonzalez-Sarmiento, R. (2016). Association of lysyl oxidase-like 1 gene polymorphisms in pseudoexfoliation syndrome and pseudoexfoliation glaucoma in a Spanish population. *Ophthalmic Genetics, 37*(1), 25–30. http://dx.doi.org/10.3109/13816810.2014.921316.

Kastner, C., Lobler, M., Sternberg, K., Reske, T., Stachs, O., Guthoff, R., et al. (2013). Permeability of the anterior lens capsule for large molecules and small drugs. *Current Eye Research, 38*(10), 1057–1063. http://dx.doi.org/10.3109/02713683.2013.803288.

Konstas, A. G. P., & Ringvold, A. (2018). Epidemiology of exfoliation syndrome. *Journal of Glaucoma, 27*(Suppl. 1), S4–S11. http://dx.doi.org/10.1097/IJG.0000000000000908.

Kottler, U. B., Junemann, A. G., Aigner, T., Zenkel, M., Rummelt, C., & Schlotzer-Schrehardt, U. (2005). Comparative effects of TGF-beta 1 and TGF-beta 2 on extracellular matrix production, proliferation, migration, and collagen contraction of human Tenon's capsule fibroblasts in pseudoexfoliation and primary open-angle glaucoma. *Experimental Eye Research, 80*(1), 121–134. http://dx.doi.org/10.1016/j.exer.2004.08.018.

Krause, U., Alanko, H. I., Karna, J., Miettinen, R., Larmi, T., Jaanio, E., … Takala, J. (1988). Prevalence of exfoliation syndrome in Finland. *Acta Ophthalmologica*, (Suppl. 184), 120–122. Retrieved from https://www.ncbi.nlm.nih.gov/pubmed/2853908.

Kumar, H., & Udgaonkar, J. B. (2019). Mechanistic approaches to understand the prion-like propagation of aggregates of the human tau protein. *Biochimica et Biophysica Acta - Proteins and Proteomics*. http://dx.doi.org/10.1016/j.bbapap.2019.04.004.

Lee, C. J., Vroom, J. A., Fishman, H. A., & Bent, S. F. (2006). Determination of human lens capsule permeability and its feasibility as a replacement for Bruch's membrane. *Biomaterials, 27*(8), 1670–1678. http://dx.doi.org/10.1016/j.biomaterials.2005.09.008.

Malukiewicz, G., Lesiewska-Junk, H., Linkowska, K., Mielnik, M., Grzybowski, T., & Sulima, N. (2011). Analysis of LOXL1 single nucleotide polymorphisms in Polish population with pseudoexfoliation syndrome. *Acta Ophthalmologica, 89*(1), e64–66. http://dx.doi.org/10.1111/j.1755-3768.2010.02083.x.

Martinelli, A. H. S., Lopes, F. C., John, E. B. O., Carlini, C. R., & Ligabue-Braun, R. (2019). Modulation of disordered proteins with a Focus on neurodegenerative diseases and other pathologies. *International Journal of Molecular Sciences, 20*(6). http://dx.doi.org/10.3390/ijms20061322.

Nazarali, S., Damji, F., & Damji, K. F. (2018). What have we learned about exfoliation syndrome since its discovery by John Lindberg 100 years ago? *British Journal of Ophthalmology, 102*(10), 1342–1350. http://dx.doi.org/10.1136/bjophthalmol-2017-311321.

Olzscha, H., Schermann, S. M., Woerner, A. C., Pinkert, S., Hecht, M. H., Tartaglia, G. G., … Vabulas, R. M. (2011). Amyloid-like aggregates sequester numerous metastable proteins with essential cellular functions. *Cell, 144*(1), 67–78. http://dx.doi.org/10.1016/j.cell.2010.11.050.

Padhy, B., Kapuganti, R. S., Hayat, B., Mohanty, P. P., & Alone, D. P. (2019). De novo variants in an extracellular matrix protein coding gene, fibulin-5 (FBLN5) are associated with pseudoexfoliation. *European Journal of Human Genetics*. http://dx.doi.org/10.1038/s41431-019-0482-6.

Pasquale, L. R., Jiwani, A. Z., Zehavi-Dorin, T., Majd, A., Rhee, D. J., Chen, T., ... Levkovitch-Verbin, H. (2014). Solar exposure and residential geographic history in relation to exfoliation syndrome in the United States and Israel. *JAMA Ophthalmology, 132*(12), 1439–1445. http://dx.doi.org/10.1001/jamaophthalmol.2014.3326.

Prince, A. M., & Ritch, R. (1986). Clinical signs of the pseudoexfoliation syndrome. *Ophthalmology, 93*(6), 803–807. Retrieved from https://www.ncbi.nlm.nih.gov/pubmed/3737125.

Rautenbach, R. M., Bardien, S., Harvey, J., & Ziskind, A. (2011). An investigation into LOXL1 variants in black South African individuals with exfoliation syndrome. *Archives of Ophthalmology, 129*(2), 206–210. http://dx.doi.org/10.1001/archophthalmol.2010.349.

Ringvold, A. (1996). Epidemiology of glaucoma in northern Europe. *European Journal of Ophthalmology, 6*(1), 26–29. Retrieved from https://www.ncbi.nlm.nih.gov/pubmed/8744847.

Ritch, R., & Schlotzer-Schrehardt, U. (2001). Exfoliation syndrome. *Survey of Ophthalmology, 45*(4), 265–315. pii:S0039-6257(00)00196-X.

Ritch, R., Schlotzer-Schrehardt, U., & Konstas, A. G. (2003). Why is glaucoma associated with exfoliation syndrome? *Progress in Retinal and Eye Research, 22*(3), 253–275. Retrieved from https://www.ncbi.nlm.nih.gov/pubmed/12852486.

Ronci, M., Sharma, S., Martin, S., Craig, J. E., & Voelcker, N. H. (2013). MALDI MS imaging analysis of apolipoprotein E and lysyl oxidase-like 1 in human lens capsules affected by pseudoexfoliation syndrome. *Journal of Proteomics, 82*, 27–34. http://dx.doi.org/10.1016/j.jprot.2013.01.008.

Schlotzer-Schrehardt, U. (2009). Molecular pathology of pseudoexfoliation syndrome/glaucoma–new insights from LOXL1 gene associations. *Experimental Eye Research, 88*(4), 776–785. http://dx.doi.org/10.1016/j.exer.2008.08.012.

Sharma, S., Chataway, T., Burdon, K. P., Jonavicius, L., Klebe, S., Hewitt, A. W., ... Craig, J. E. (2009). Identification of LOXL1 protein and Apolipoprotein E as components of surgically isolated pseudoexfoliation material by direct mass spectrometry. *Experimental Eye Research, 89*(4), 479–485. http://dx.doi.org/10.1016/j.exer.2009.05.001.

Sharma, S., Chataway, T., Klebe, S., Griggs, K., Martin, S., Chegeni, N., ... Craig, J. E. (2018). Novel protein constituents of pathological ocular pseudoexfoliation syndrome deposits identified with mass spectrometry. *Molecular Vision, 24*, 801–817. Retrieved from https://www.ncbi.nlm.nih.gov/pubmed/30713420.

Tam, J. H., & Pasternak, S. H. (2012). Amyloid and Alzheimer's disease: Inside and out. *The Canadian Journal of Neurological Sciences, 39*(3), 286–298. Retrieved from https://www.ncbi.nlm.nih.gov/pubmed/22547507.

Thorleifsson, G., Magnusson, K. P., Sulem, P., Walters, G. B., Gudbjartsson, D. F., Stefansson, H., ... Stefansson, K. (2007). Common sequence variants in the LOXL1 gene confer susceptibility to exfoliation glaucoma. *Science, 317*(5843), 1397–1400. http://dx.doi.org/10.1126/science.1146554.

Uversky, V. N. (2016). Paradoxes and wonders of intrinsic disorder: Complexity of simplicity. *Intrinsically Disordered Proteins, 4*(1), e1135015. http://dx.doi.org/10.1080/21690707.2015.1135015.

Vaquer-Alicea, J., & Diamond, M. I. (2019). Propagation of protein aggregation in neurodegenerative diseases. *Annual Review of Biochemistry, 88*, 785–810. http://dx.doi.org/10.1146/annurev-biochem-061516-045049.

Want, A., Gillespie, S. R., Wang, Z., Gordon, R., Iomini, C., Ritch, R., ... Bernstein, A. M. (2016). Autophagy and mitochondrial dysfunction in tenon fibroblasts from exfoliation glaucoma patients. *PLoS One, 11*(7), e0157404. http://dx.doi.org/10.1371/journal.pone.0157404.

Williams, S. E., Whigham, B. T., Liu, Y., Carmichael, T. R., Qin, X., Schmidt, S., ... Allingham, R. R. (2010). Major LOXL1 risk allele is reversed in exfoliation glaucoma in a black South African population. *Molecular Vision, 16*, 705–712. Retrieved from https://www.ncbi.nlm.nih.gov/pubmed/20431720.

Wolosin, J. M., Ritch, R., & Bernstein, A. M. (2018). Is autophagy dysfunction a key to exfoliation glaucoma? *Journal of Glaucoma, 27*(3), 197–201. http://dx.doi.org/10.1097/IJG.0000000000000606.

Zenkel, M., Krysta, A., Pasutto, F., Juenemann, A., Kruse, F. E., & Schlotzer-Schrehardt, U. (2011). Regulation of lysyl oxidase-like 1 (LOXL1) and elastin-related genes by pathogenic factors associated with pseudoexfoliation syndrome. *Investigative Ophthalmology & Visual Science, 52*(11), 8488–8495. http://dx.doi.org/10.1167/iovs.11-8361.

CHAPTER NINE

Liquid-liquid phase transitions and amyloid aggregation in proteins related to cancer and neurodegenerative diseases

Guilherme A.P. de Oliveira[a,b], Yraima Cordeiro[c], Jerson L. Silva[a,*] and Tuane C.R.G. Vieira[a,*]

[a]Programa de Biologia Estrutural, Instituto de Bioquímica Médica Leopoldo de Meis, Instituto Nacional de Biologia Estrutural e Bioimagem, Centro Nacional de Ressonância Magnética Nuclear Jiri Jonas, Universidade Federal do Rio de Janeiro, Rio de Janeiro, RJ, Brazil
[b]Department of Biochemistry and Molecular Genetics, University of Virginia, Charlottesville, VA, United States
[c]Faculty of Pharmacy, Federal University of Rio de Janeiro, Rio de Janeiro-RJ, Brazil
*Corresponding author: E-mails: jerson@bioqmed.ufrj.br; tuane@bioqmed.ufrj.br

Contents

Advances in Protein Chemistry and Structural Biology, Volume 118
ISSN 1876-1623
https://doi.org/10.1016/bs.apcsb.2019.08.002
© 2020 Elsevier Inc.
All rights reserved.

Abstract

Liquid-liquid phase separation (LLPS) and phase transition (LLPT) of proteins and nucleic acids have emerged as a new paradigm in cell biology. Here we will describe the recent findings about LLPS and LLPT, including the molecular and physical determinants leading to their formation, the resulting functions and their implications in cell physiology and disease. Amyloid aggregation is implicated in many neurodegenerative diseases and cancer, and LLPS of proteins involved in these diseases appear to be related to their function in different cell contexts. Amyloid formation would correspond to an irreversible liquid-to-solid transition, as clearly observed in the case of PrP, TDP43, FUS/TLS and tau protein in neurodegenerative pathologies as well as with the mutant tumor suppressor p53 in cancer. Nucleic acids play a modulatory effect on both LLPS and amyloid aggregation. Understanding the molecular events regulating how the demixing process advances to solid-like fibril materials is crucial for the development of novel therapeutic strategies against cancer and neurodegenerative maladies.

1. Introduction

In his 1944 book "What is Life? The Physical Aspects of Living Cells", Erwin Schrödinger warned us that "not on the ground that there is any 'new force' or what not, directing the behavior of the single atoms within a living organism, but because the construction is different from anything we have yet tested in the physical laboratory". After 75 years, Schrödinger's statement is still true, as revealed by recent findings on how biomolecular condensates are assembled and play an important role in physiology and disease.

From human sociology to organs to individual cells, compartmentalization is a prerequisite to pave the way for functional organization, avoiding chaos. As humans organize themselves in boundary countries, cities, and family communities based upon cultural commonalities, inner cell borders likely behave the same. The presence of lipid membranes surrounding cell compartments requires specialized hubs to mediate transport and communication among defined milieu, such as the nuclear pore complex (NPC) connecting the nucleus to the cytoplasmic fluid. The evolutionary achievement of compartmentalization guarantees, to some extent, the hierarchy and the spatiotemporal regulation of metabolic reactions, signaling pathways, energy production, and, ultimately, life maintenance. Findings from ancient organisms living at extremely harsh conditions showed us how compartments preserve life. For example, the hyperthermophilic archaea *Acidianus hospitalis* is infected by the filamentous virus AFV1, and during the budding process, the virus 'steals' part of the host's membrane to protect it from extreme surroundings and to help search for new cells to infect. Intriguingly, the *A. hospitalis* membrane is formed by a 2-nm thick monolayer, and the virion seems to selectively incorporate lipids from the organism, providing support to maintain the compartments for life preservation (Kasson et al., 2017).

Interestingly, recent reports have revealed a compartmentalization phenomenon that nullifies the need for a lipid layer membrane, the so-called membraneless organelles (Alberti & Carra, 2018; Alberti, Mateju, Mediani, & Carra, 2017; Toretsky & Wright, 2014). Although these compartments (e.g., the nucleolus, the stress granules, the P and Cajal bodies, the speckles, the paraspeckles, and the promyelocytic nuclear bodies) were discovered a long time ago, only now are the physical and molecular determinants for their formation and potential functionalities being exploited in physiology and disease (Boeynaems et al., 2018; Zhu & Brangwynne, 2015). Other important examples include inflammasomes (Lu et al., 2014), signaling complexes (Su et al., 2016), purinosomes (An, Kumar, Sheets, & Benkovic, 2008), centrosomes (Woodruff et al., 2017), postsynaptic densities (PSDs) (Baron et al., 2006), and inclusion bodies (De Smet et al., 2017; Ogrodnik et al., 2014).

The discovery of most of these membraneless organelles relies on the fact that they are formed by a solution of proteinaceous composition that mixes and 'demixes' dynamically, a condition that allows a physical state transition, i.e., a phase separation (PS) coexisting within the cellular environment. In this scenario, a protein-enriched dense/viscous state and a protein-depleted liquid state can reversibly form, or not, depending on ionic strength, concentration, temperature, pressure, multivalency components, and RNA or protein composition (Boeynaems et al., 2018). These dynamic condensates or biomolecule coacervates have instigated a new era of cells as colloidal systems, first proposed by Edmund Wilson in his book, "The Physical Basis of Life": "No conception of modern biology offers greater promise for the physico-chemical analysis of vital phenomena than that the cell is a colloidal system; and that what we call life is, a complex of innumerable chemical reactions in the substance of this system" (Wilson, 2018).

The process of PS in proteins was initially observed in crystallin, a lens protein solution purified from calf (Broide, Berland, Pande, Ogun, & Benedek, 1991). A later observation reported PS occurring reversibly in solutions of human hemoglobin at near-physiological temperatures (Galkin, Chen, Nagel, Hirsch, & Vekilov, 2002). Intriguingly, dense droplets facilitate the nucleation of hemoglobin polymers, an event occurring in sickle cell anemia (Galkin et al., 2002). For the potential physiological roles involving PS, an incredible example was observed with nucleoporin, one of the building blocks of the NPC. This protein reversibly crosslinks between the Phe and Gly (FG)-rich motifs, forming a meshwork with hydrogel-like properties that is necessary for yeast viability and to explain the selectivity of cytosol cargos (Frey, Richter, & Gorlich, 2006). These seminal examples paved the way for our discussion in this review. Hereafter, we will focus on the most recent discoveries about PS and phase transition, including the molecular and physical determinants driving their formation, their potential functionalities and their implications in cell physiology and disease with a focus on cancer and neurodegenerative disorders. Especially in the last five years, a connection has been established between PS and diseases related to protein misfolding and amyloid aggregation (Molliex et al., 2015; Patel et al., 2015). Amyloid aggregates have been implicated in many neurodegenerative diseases (Chiti & Dobson, 2006) as well in cancer, as in the case of mutant p53 aggregation (Costa et al., 2016; Silva, De Moura Gallo, Costa, & Rangel, 2014). Whereas PS is a dynamic reversible event, amyloid formation corresponds to an irreversible liquid-to-solid transition. Because the culprits of these diseases are small oligomers rather than large fibrils, the liquid-liquid phase separation (LLPS) and amyloid oligomerization are likely different facets of the same moiety. We and others have also found that nucleic acids (NAs) exert a modulatory effect on both LLPS and amyloid aggregation (Cordeiro, Vieira, Kovachev, Sanyal, & Silva, 2019; Silva & Cordeiro, 2016) that will be explored here.

2. Phase separation terminology in biomolecular systems

2.1 States of matter and properties of phase transition and separation

The recollection of the four states of matter, e.g., solid, liquid, gas, and plasma, and their properties is an important first step for a broader comprehension of phase transition in biomolecules. Phase transition simply refers to when a molecule or biomolecule changes from one state to the other. In this scenario, water is a good example. Water molecules as liquids have a volume but not a defined shape and tend to minimize the surface area to achieve reduced surface tension, which explains why water droplets are often spherical. These molecules also present dynamic H-bond breakage and formation within the picosecond (ps) timescale, but recent discoveries noted an ultrafast regime of ~100 fs and a H-bond network of tetrahedral coordination (Perakis et al., 2018).

Despite the complex phase diagram of ice containing eight distinct and stable ice polymorphs, ordinary ice is formed when water is cooled below 0°C at atmospheric pressure (Fletcher, 1971). Under these conditions, water assumes a crystalline long-range order periodicity and a hexagonal H-bond patterning resulting in lower density (mass/volume), the reason why ice floats in liquid water. When frozen at extremely high rates, water forms a non-crystalline glass-like state known as an amorphous state, which is a tremendous advantage for solving the structure of biomolecules in their native state by cryo-electron microscopy, as water crystals will no longer damage biological samples (Dubochet et al., 1988). On the other hand, when water molecules are heated above their boiling point, the molecules acquire enough kinetic energy to expand out of the system, losing their colligative properties. This phenomenon is observed when boiling water entirely 'disappears' after a few minutes. Finally, when gases such as hydrogen or oxygen are exposed to extremely high temperatures or voltage differences, they ionize, forming the fourth state of matter known as plasma. An example is an electric arcing obtained from a traditional Tesla coil apparatus.

Although more complex, the understanding of phase transition properties does not substantially change when transposing from the aforementioned scenario to heteropolymers such as polypeptide chains and RNA molecules. At this level, phase transition should not be confused with PS, the latter referring to a demixing process within the same state of matter. At least five properties should be taken into consideration to provide an accurate classification of PS and phase transitions in biology (Fig. 1): (i) the density of the molecules, meaning the concentration contained within the condensate; (ii) the long-range spatial arrangement and directional preferences to define the intermolecular organization; (iii) the level of crosslinking occurring within involved molecules; (iv) the interfacial tension occurring between the condensate and its surroundings; and (v) the kinetics of forming and breaking bonds, also called the aging or maturation of the coacervate.

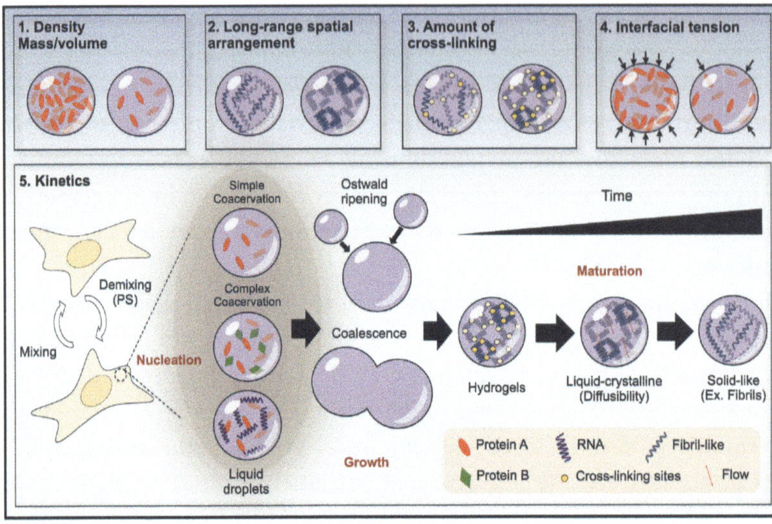

Fig. 1 ***Features of PS/transition events.*** Schematics illustrating five important parameters regulating PS/transition processes. Density, long-range order, cross-linking sites, superficial tension, and the kinetics itself are determinants of the formation of membraneless organelles. Kinetics involve nucleation, growth and droplet maturation. The maturation route depicted in the picture may vary according to the droplet composition.

Exploring each of these parameters in the context of living cells may add affluents to the PS/transition terminology, such as liquid-liquid droplets, hydrogels (liquid-gels, solid-gels), solid-crystalline, and liquid-crystalline.

2.2 Liquid-liquid droplets

Starting from a homogenized solution of solutes (exemplified as proteins) and solvents (exemplified as water and ions), the solution undergoes a demixing process that culminates in a protein-enriched dense partition and a protein-depleted liquid partition within the cell. Importantly, liquid droplets involve separation and spatial rearrangement of preferred protein interactions and do not initially evolve to a phase transition phenomenon. These segregated compartments, previously unknown to play a role in cell physiology and disease, are now appearing as important regulators of cell functionalities (Hyman, Weber, & Julicher, 2014; Lin, Protter, Rosen, & Parker, 2015; Xiang et al., 2015). *In vitro* assays indicate droplet formation in supersaturated solutions, which is a reversible process (Lin et al., 2015; Xiang et al., 2015). Underpinning the physics behind PS requires the principles of the Flory-Huggins solution theory (Brangwynne, Tompa, Pappu,

2015). If the protein-solvent system is considered a repertoire of molecular dipoles, three types of dipole competitions are possible: protein-solvent, protein-protein, and solvent-solvent. Based on this finding, the Flory coefficient quantifies the energetic costs for an occupancy of solvents surrounding the protein. The higher the coefficient, the larger the energetic cost is and the less favorable the protein-solvent proximity is, a condition that suits PS (Brangwynne et al., 2015). As noted, solubility represents an important parameter favoring PS. The molecular determinants underlying droplet formation will be further explored in later sections.

2.3 Hydrogels

Gelation is the process of gel formation and is characterized by a crosslinking network reaction of constituent biomolecules forming noncovalent interactions (physical gel) or covalent interactions (chemical gel). As an example of this process, gelatin or hydrolyzed collagen upon dissolution in warm water forms crosslinked molecules that, when cooled, produce hydrogels. The properties of gels vary according to the extent and disposition of the crosslinking reactions (Boeynaems et al., 2018). One of the driving forces related to gel formation underlies protein binding motifs organized in tandem. Hydrogels are important in biomedical applications (Burdick & Stevens, 2005), and multiple examples of hydrogel formation have been reported in physiology and disease (Frey et al., 2006; Kato et al., 2012; Kumar et al., 2018; Xiang et al., 2015). These processes will be further discussed in later sections.

2.4 Liquid-crystalline, semi-crystalline, and solid-crystalline

A liquid- or solid-crystalline relies on a phase transition state with shared features of liquids and crystals. The biomolecular composition within these states is oriented in a crystal pattern, i.e., in a repetitive and periodic manner; however, liquid properties such as flow are still present. The classification within crystalline phases (liquid-, semi-, and solid-) strongly depends on the long-range spatial and directional ordering of the material. As an illustration, cellular membranes containing mono- or bilayered lipids represent an important example of liquid-crystalline material because they are formed by a periodic array of phospholipids but show substantial flexibility, fluidity, and diffusibility. Furthermore, lipid membranes participate in important processes, including bending, fusion, fission, and lipid movements, through the process of transverse diffusion mediated by flippases, supporting their liquid-like properties. Polymeric protein assemblages, such as helical filaments, are important examples of solid-crystalline materials in biology. Notably, these kinds of helical polymers, such as actin, microtubules, viral proteins, membrane tubules, and amyloid filaments, represent a significant proportion of macromolecules participating in life and disease (Sachse, 2015). Their classification as solid-crystalline materials is general because an ideal solid-crystalline material would be better represented by protein crystals produced in crystallization trials. Indeed, the long-range order that is maintained in a crystal by the presence of a space group symmetry is lost within helical filaments due to the accumulation of thermal fluctuations and because local interactions are mostly predominant in helical assemblies (Egelman, 2007). However, most of these filaments are easily harvested from a solution applying moderate centrifugal force, supporting its solid-like properties. Therefore, even though helical filaments are interpreted as 2D crystals presenting defined periodicities, long-range order is hardly maintained within these materials. The polymorphic nature of several amyloid fibrils in terms of internal periodicity along the fibril axis and quaternary conformation (different structural polymorphs) (Annamalai et al., 2016; Guo et al., 2013; Li et al., 2018) supports their classification of nonclassical solid-crystalline or even semi-crystalline materials.

2.5 Maturation or molecular aging of coacervates

This process consists of the evolution of these condensates over time from a reversible and dynamic PS to an irreversible reaction involving phase transitions (liquids to hydrogels to solid-like materials). The beginning of a PS event requires a nucleation step and may involve one single protein (Fig. 1), in which simple coacervation is the adopted nomenclature. Conceivably, when multiple proteins or other biomolecules, such as RNA, are involved, the phenomenon is referred to as complex coacervation. The density parameter expressed by the concentration of biomolecules within a defined volume seems to play a pivotal role in the initial steps of PS processes (Banani et al., 2016). The critical concentration required to start a PS nucleation event depends on the polypeptide(s) sequence(s) involved. In complex coacervation (Fig. 1), one hypothesis states that the protein containing the lowest critical concentration initiates the partition process, recruiting additional proteins to the droplet (Boeynaems et al., 2018). Abrupt changes in protein concentration inside cells can easily occur, for example, in cases of transcription bursts or when scaffold proteins are released from clusters of vesicles, such as in the case of synaptic vesicles

(SVs). Synapsin, a protein involved in clustering SVs, has been shown to do so by forming liquid droplets (Milovanovic, Wu, Bian, & De Camilli, 2018). More importantly, dissolution of these droplets to trigger SV membrane fusion and neurotransmitter release occurs due to synapsin phosphorylation by Ca^{2+}/calmodulin-dependent protein kinase II (Milovanovic et al., 2018).

Proteins are not the only biomolecules involved in nucleation and regulation of PS/transition formation (Fig. 1). Other biomolecules, such as RNA and polyADP-ribose (PAR), also participate in this process (McGurk et al., 2018; Schwartz, Wang, Podell, & Cech, 2013). The RNA-binding protein FUS binds to RNA polymerase II in an RNA-dependent manner, affecting transcription. RNA molecules have been shown to seed the formation of higher-order FUS assemblies containing fibril-like properties that bind the RNA polymerase II (Schwartz et al., 2013). In the case of the TAR DNA-binding protein of 43 kDa (TDP-43) involved in amyotrophic lateral sclerosis (ALS) and frontotemporal degeneration (FTD), PAR binds to TDP-43 and stimulates liquid PS and stress granule localization (McGurk et al., 2018). In contrast, in *Caenorhabditis elegans,* the RNA helicase LAF-1 participates in P granule formation by driving liquid PS, but RNA molecules have been shown to decrease droplet viscosity and increase molecular dynamics (Elbaum-Garfinkle et al., 2015). RNA and RNA-like molecules such as PAR seem to regulate PS/transitions either by fluidization or maturation to solid-like materials.

As dynamic partitions, these droplets tend to grow (Fig. 1). Their mechanisms are still elusive in cells, but these mechanisms have been investigated to explain how the growth process is achieved. One mechanism is known as Ostwald ripening, in which small molecules move to find larger droplets (Li, Lubchenko, Vorontsova, Filobelo, & Vekilov, 2012). This is a spontaneous event, and because molecules located close to the surface of droplets are energetically less stable to those inside, the process is driven to minimize the surface free energy of these partitions (Li et al., 2012). A similar phenomenon is observed when oil is mixed in warm water. Another potential growth mechanism is by passive coalescence in which droplets move freely by Brownian motion (Boeynaems et al., 2018). This process occurs when two droplets similar in size touch each other and fuse into a larger droplet (Fig. 1). Active coalescence is another type of growth event and was shown to be mediated by microtubules and motor proteins in stress granules and processing bodies (P-bodies) (Lomakin & Nadezhdina, 2010; Nadezhdina, Lomakin, Shpilman, Chudinova, & Ivanov, 2010).

Growing evidence has shown that phase transitions occur during the maturation process of liquid droplets, especially for proteins containing RNA-binding motifs and those participating in neurodegenerative disorders. For example, the hnRNA1 protein forming RNP granules progresses from liquid droplets to fibrous assemblies over time (Lin et al., 2015; Molliex et al., 2015). In another scenario, the ALS-associated protein FUS can form liquid-PS either *in vitro* or *in vivo* after DNA damage (Patel et al., 2015). Over time, these liquid droplets also convert to solid-like fibrous material. Intriguingly, ALS-related FUS mutations accelerate this phase transition to fibrils, supporting a link between PS maturation and this kind of aging disorder (Patel et al., 2015). Furthermore, ALS/FTD mutations induce FUS irreversible hydrogels (Murakami et al., 2015). Controversially, in other ALS-related proteins, such as TDP-43, pathological mutations in this protein can disrupt PS, revealing the complexity of how partition formation is tuned during pathological processes (Conicella, Zerze, Mittal, & Fawzi, 2016).

3. Molecular determinants involved in PS/transition and regulation

3.1 Multivalency of flexible linkers and modular domains

The heteropolymeric nature of proteins opens up a multitude of molecular possibilities regulating PS (Fig. 2). The determinants participating in PS have just recently started to be dissected (Li et al., 2012; Molliex et al., 2015; Su et al., 2016). Notably, protein-protein and protein-RNA motifs were shown to play a role in PS processes (Kato et al., 2012; Li et al., 2012). The observation that several proteins involved in cell signaling are composed of multidomains spaced by flexible linkers suggested that multivalency would contribute to PS. Complex proteins, including kinases, undergo autoregulation via Src domains (SH3 and SH2) and linkers to control functionalities inside cells. For their autoregulatory role, these proteins possess a series of different binding sites organized in tandem so different levels of activity/inhibition can be achieved. The multivalency behavior of these complex hubs composed by flexible linkers and modular domains has been shown to participate in PS as the number of domains and interacting sites increases (Li et al., 2012). For illustration of the multivalency effect in PS, copies of the SH3 from the Nck adaptor and the proline region motif (PRM) from the c-Abl kinase were produced in tandem [$(SH3)_n$ and

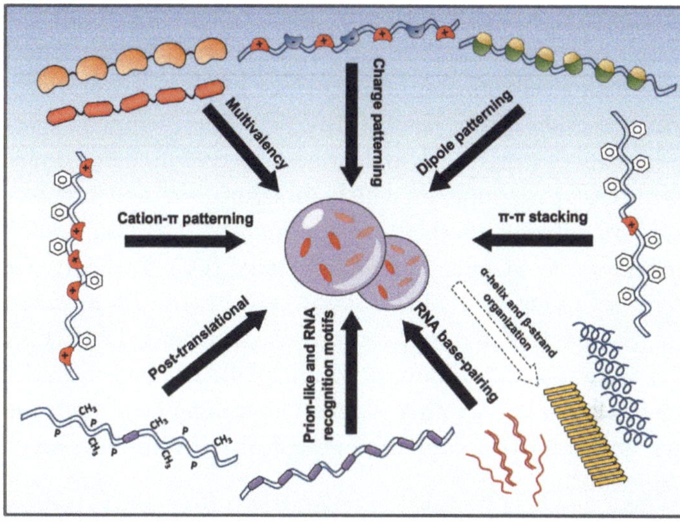

Fig. 2 ***Summary of the molecular determinants participating in PS.*** Multivalency of globular domains and IDRs, different motif patterning, post-translational modifications, prion-like motifs, RRMs, and nucleotide base-pairing include the major protagonists of PS events. During the maturation of liquid-like assemblies, phase transitions to solid-like materials containing helix or strand organization may occur, but these materials do not exclusively form from this route (dashed arrow).

$(PRM)_n$, n = 3, 4, or 5 domains] and mixed to each other at equimolar concentrations (Li et al., 2012). Intriguingly, mixing $(SH3)_3$ with $(PRM)_3$ did not induce PS, which was different from what was observed when the $(SH3)_5$ and $(PRM)_5$ domains were mixed. PS was induced in a concentration-dependent manner as the valency (number of domains) was increased (Li et al., 2012). Therefore, the valency of modular domains and flexible linkers represents one driving force of PS.

3.2 Intrinsically disordered regions and motif patterning

Intrinsically disordered proteins (IDPs) are those in which secondary and tertiary structures are not sampled during their lifetime. In contrast, structured proteins are those with a high proportion of folded elements. Notably, ~44% of the human protein-coding genes translate proteins containing both folded and unfolded segments, the latter called intrinsically disordered regions (IDRs) (van der Lee et al., 2014). The paradigm of sequence-structure-function is no longer an absolute idea, as the previously unappreciated IDRs have upended the paradigm of functionality. Several observations have shown the importance of these IDRs in cell physiology. To exemplify,

IDRs are subjected to post-translational modifications (PTMs) that affect the activity levels of kinases, and they also expose short linear motifs to binding partners regulating cell signaling, as in the case of the c-Abl nonreceptor kinase (Hantschel et al., 2003; Nagar et al., 2003). These segments do not carry a similar diversity of amino acids as the ones observed in folded domains, but they share motif patterning in which functionalities are just now being exploited. The different composition and frequency of amino acids within IDRs underlies the so-called low-complexity (LC) segments (Conicella et al., 2016; Monahan et al., 2017; Ryan et al., 2018). More interestingly, the LC segment seems to adopt patterns of short linear interaction motifs (SLiMs) (Tompa, Davey, Gibson, & Babu, 2014). The recent breakthrough in PS/transition events is another example illustrating how motif patterning of LC segments in IDRs occupies an unquestionable role in the repertoire of protein functionality.

3.2.1 Positively and negatively charged patterning

One class of LC segments is those in which highly positive or negatively charged residues are distributed along the amino acid sequence (Fig. 2). These opposing charges play an important role in long-range attraction and PS/transition. Sequences containing alternating charges [e.g., $(EK)_{25}$] produce a counterbalance of attractions and repulsions and lead to increased tendencies to form random coils (Das & Pappu, 2013). In contrast, segregation of opposing charges within linear motifs [e.g.,$(E)_{12}(K)_{12}$] favors long-range attractions and hairpin-like conformations (Das & Pappu, 2013). Motifs composed of RG and QN are important during the formation of RNA granules, stress granules, and P-bodies (Molliex et al., 2015; Xiang et al., 2015). In the case of Ddx4, a protein containing a central Dead-box RNA helicase domain and intrinsic N– and C-termini, its IDR at the N-domain is composed of alternating net charge repeats of FG, GF, RG, and GR residues (Nott et al., 2015). These LC sequences participate in Ddx4 PS and are mainly mediated by electrostatic interactions because changes in ionic strength dissolve the droplets (Nott et al., 2015). Likewise, the K-rich segments of the microtubule-binding Tau protein participate in liquid droplet formation and are driven by electrostatic interactions and the valency of Tau repeats (Ambadipudi, Biernat, Riedel, Mandelkow, & Zweckstetter, 2017). This demixing process eventually progresses to solid-like fibril materials related to Alzheimer's disease (AD) (Ambadipudi et al., 2017).

3.2.2 Dipole-dipole patterning

Another type of LC segment commonly found within IDRs that participates in PS/transition belongs to a group of amino acids containing polar side chains (Fig. 2). Dipole-dipole interactions occur very frequently during PS, and common motif patterning includes Q/N-rich segments. In the case of the Q-rich protein Whi3, PS to liquid droplets and maturation occurs, leading to gels containing fibril-like assemblies (Zhang et al., 2015). This phenomenon may be an important regulatory element in cell divisions in fungi (Zhang et al., 2015). A concentration-dependent PS behavior and maturation process can also be observed in other proteins containing enriched dipole-dipole patterning, such as FUS, hnRNPA1 and hnRNPA2 (Molliex et al., 2015; Murakami et al., 2015; Xiang et al., 2015). For the Huntingtin (Htt) protein, the presence of an N-terminal polyglutamine tail flanked by amphipathic residues leads to a lower saturation concentration of the Htt, culminating in a phase transition to solid-like fibrillar material (Crick, Ruff, Garai, Frieden, & Pappu, 2013). More importantly, this type of fibrillar material is typically observed in cases of neurodegenerative Huntington disorder (Crick et al., 2013).

3.2.3 Cation-π and π-π stacking patterning

The third patterning behavior occurring in LC segments that plays a role in PS events relates to weak interactions formed by aromatic-aromatic and aromatic-charged residues (Pak et al., 2016; Vernon et al., 2018) (Fig. 2). These interactions have been shown to contribute in a short-range and directional arrangement during PS (Brangwynne et al., 2015). A good example is the FG motifs occurring in proteins of the NPC. One of these proteins, the so-called nucleoporin (Nup98), is present from animals to protozoans and is composed of FG-rich motifs that show PS into a meshwork (Frey et al., 2006; Schmidt & Gorlich, 2015). This process may represent an important mechanism of 'molecular sieving' under which the NPC dictates the nucleus-to-cytoplasm transit of cargos (Schmidt & Gorlich, 2015).

3.3 Prion-like and RNA-recognition motifs

Prions are proteins that can adopt different conformations that can self-propagate, many through self-assembly and aggregation. Prion-like domains (PLDs) are predicted to be present in 70% of human proteins, and many of these domains are involved in RNA binding and metabolism (King, Gitler, & Shorter, 2012; March, King, & Shorter, 2016). Such domains are characterized by the similarity to sequences present in yeast prion. These sequences are at least 60 amino acids long, are enriched in glycine and uncharged polar amino acids (mostly asparagine, glutamine and tyrosine), and show low complexity and a high tendency to be intrinsically disordered (Alberti, Halfmann, King, Kapila, & Lindquist, 2009).

PLDs are sequences that in most proteins are involved in the formation of macromolecular complexes, which can be seen as an advantage by concentrating molecules that are important for performing a certain function in the cell, acting as a reservoir of mainly proteins and RNA molecules. However, PLDs have been identified as important regions for mediating the formation of macromolecular complexes in disease events. Paraspeckles are subnuclear bodies that are stress-responsive and important for the control of gene expression (Fox & Lamond, 2010). These bodies are abundant in different RNA-binding proteins interacting with nonprotein-coding RNA species. Hennig and collaborators showed that the PLDs of the RBM14 and FUS proteins are important for mediating protein-protein interactions and liquid-phase transitions, leading to paraspeckle targeting and formation (Hennig et al., 2015). Mutation of these proteins is associated with neurodegenerative disorders, and paraspeckles are found in ALS motor neurons. Perturbation of these structures must account for the origin and/or development of the disease.

RNA-recognition motifs (RRM) are present in almost 1% of human genes (Venter et al., 2001), and in addition to being involved in processes that involve direct interaction with RNA (stability and metabolism, for example), they also interact with DNA, with themselves and with other proteins (Clery, Blatter, & Allain, 2008). RRMs have a sequence of 80–90 amino acids and two conserved consensus sequences termed RNP1 ([RK]-G-[FY]-GA-[FY]-[ILV]-X-[FY]) and RNP2 ([ILV]-[FY]-[ILV]-X-N-L), where X is any amino acid residue. This protein motif has a secondary structure with four-stranded β-sheets and two α-helices with the β1α1β2β3α2β4 topology. Aromatic rings found in the motif interact with the bases and sugar rings of NAs (Clery et al., 2008).

RNA-binding proteins include RRMs, LCDs and nuclear import/export sequences and are responsible for interaction with partners leading to the formation of important structures inside the cell, such as stress granules, P-bodies, and nuclear granules. Changes in these proteins or in their interactions are associated with various pathological states. How much each of these regions contributes to PS as well as phase transition will be scrutinized in later sections for some diseases.

3.4 Post-translational modifications (PTMs)

The importance of PTMs in cell signaling is well established. Notwithstanding, the effects PTMs may have within SLiMs and ultimately PS/transition events are still growing (Fig. 2). Notably, common targets of PTMs include phosphorylation of serines, threonines, and tyrosines; citrullination and sumoylation of arginines; and methylation of arginines and lysines. Interestingly, several of these PTM targets are present in LC SLiMs, prompting researchers to investigate how they might regulate PS/transition events. Indeed, several reports have shown the effects of PTMs in different protein models and their impact on PS (Ambadipudi et al., 2017; Kwon et al., 2013; Milovanovic et al., 2018). Droplet disassembly to trigger neurotransmitter release from SVs has been shown due to synapsin phosphorylation by Ca^{2+}/calmodulin-dependent protein kinase II (Milovanovic et al., 2018). Controversially, in the microtubule-binding Tau protein involved in AD pathology, phosphorylation leads to liquid droplet formation (Ambadipudi et al., 2017), but for other amyloidogenic-related proteins, such as FUS, phosphorylation of its LC domain impairs PS (Monahan et al., 2017). Furthermore, arginine methylation of Ddx4 has been shown to dissolve germ granules (Nott et al., 2015). The foregoing examples show how PTMs affect droplet formation and further illustrate the repertoire cells have to use this driving force to modulate PS/transitions.

3.5 Helix-helix and β-strand organization

An important observation is that the maturation route from a liquid-like droplet to a semi-crystalline or solid-like fibrillar assembly may vary depending on the protein sequence composition. Some proteins, under defined circumstances, do not necessarily phase separate into liquid droplets prior to evolving to solid-like species but undergo phase transitions to fibrillar material directly. This direct route is frequently the result of misfolding processes that require exposure of hydrophobic motifs to the solvent with subsequent conversion of random coils to secondary structure elements such as β-strand or α-helices (Chiti & Dobson, 2017). In these examples, we might highlight the recent discovery of a peptide virulence factor secreted from *Staphylococcus aureus* that curls into nonclassical amyloid fibrils composed of α-helices (Tayeb-Fligelman et al., 2017). Several amyloidogenic proteins, including transthyretin (TTR), typically undergo misfolding prior to forming oligomers that subsequently grow by monomer incorporation to form cross-β amyloid materials (Ferrao-Gonzales, Souto, Silva, & Foguel, 2000). Classical

IDPs, such as α-synuclein (α-syn) and amyloid-β, likely form β-strand amyloid fibrils that do not require PS as an intermediate step (Cohen et al., 2013; Lorenzen et al., 2014). However, a recent study reported the presence of a gel-like α-syn state when α-syn monomers were mixed at high concentrations and physiological ionic strength (Kumar et al., 2018). More interestingly, this gel-like material is composed of cytotoxic oligomers and fibrils (Kumar et al., 2018). These cytotoxic α-syn assemblies raise the question of whether they might have any significance during Parkinson's disease (PD) development.

3.6 RNA intermolecular base-pairing

An unusual form of PS is that in which protein components do not participate. This process occurs and is driven exclusively by nucleotide repeat expansions of RNA molecules (Fig. 2). Nucleotide repeat expansions are commonly observed in several disorders and are typically characterized by codons carrying increased G/C content. This is the case in Huntington's disease (CAG codon expansion), myotonic dystrophy (CTG expansion) and familial ALS (GGGGCC hexanucleotide expansion) (La Spada & Taylor, 2010). Jain and Vale have reported that triplets and hexanucleotide repeats convert to hydrogels with a critical number of nucleotide repeats (Jain & Vale, 2017). The multivalency RNA base-pairing mechanistically explains the process of RNA gelation and sequestration within the nucleus. This situation might help to explain why disease-associated RNA repeat expansions should reach a threshold concentration to trigger disease occurrence (Jain & Vale, 2017).

4. The functionalities of PS/transitions in physiology and disease

From cells to humans, PS plays a fundamental role in everyday life. Demixing biomolecular components to liquid-like droplets or solid-like filaments has been shown to occur both at the cellular and systemic levels. At the systemic level, many examples of PS events have been observed. To illustrate, mucus is a viscous colloid produced in the respiratory, digestive and reproductive systems and contains inorganic salts and various antiseptic enzymes to protect against infectious agents and to facilitate movement (Bansil & Turner, 2018). Another example is the vitreous humor, a fluid that fills the space between the lens and the retina of the eyeball. This fluid is composed of water and a mixture of collagen, salts, and sugars that help maintain the eye shape and keep the retina in place (Reddy & Kinsey, 1960). Further examples include cartilage, a rubber-like material protecting the end of bones; glycogen granules, a multibranched polysaccharide of glucose serving as a short-term energy source (Prats, Graham, & Shearer, 2018); and fibrin gels, an essential meshwork responsible for blood clotting and wound healing (Burdick & Stevens, 2005).

Within the crowded environment of cells, various demixing events have been shown to participate in biological processes, such as biochemical reactions (Li et al., 2012; Su et al., 2016) and assembly of complex machineries such as the centrosomes (Woodruff et al., 2017), spliceosomes (Will & Luhrmann, 2011), ribosomes, purinosomes (An et al., 2008), and inflammasomes (Lu et al., 2014), among others. The colligative properties of the nucleoplasm and the cytoplasm synchronously organize the 'omes' to form droplet-like clusters, also known as membraneless organelles. In the next sections, we will describe, on a cell-based level, the physiological roles of membraneless organelles in cell homeostasis. Later, we will provide examples of how PS may be related to pathological conditions.

4.1 Membraneless compartments and their physiological role in cell biology

Among most of the membraneless organelles, the nucleolus is of pivotal importance in ribosome biogenesis. This organelle is composed of proteins, DNA, and RNA molecules and has been shown to have a complex subcompartmentalization architecture (Feric et al., 2016). Three well-defined subcompartments are appreciable: the fibrillar center in which RNA polymerase I works to transcribe rDNA, the dense fibrillar component containing fibrillarin that participates in the initial steps of rRNA processing, and the granular subcompartment enriched in nucleophosmin that aids in ribosome biogenesis (Feric et al., 2016). Incredibly, these nucleolar subcompartments behave as immiscible PSs both *in vivo* and *in vitro,* supporting a layered droplet organization. This multilayered arrangement is triggered mostly by the sequence-specific composition of subcompartment constituents, a phenomenon that may help RNA processing reactions and ultimately ribosome assembly (Feric et al., 2016).

Signaling complexes composed of multiple protein assemblies likely phase separate into liquid-like droplets and have been shown to be an important event for biochemical reactions in cells. For example, reconstitution of T cell receptor signaling resulted in a downstream cascade that phase separates, culminating in actin polymerization (Su et al., 2016). Other examples of PS participating in cell signaling and genetic regulation include the dissolution of liquid-like clusters of SVs for neurotransmitter release (Milovanovic et al., 2018), the molecular sieving mechanism of nucleoporins within NPCs (Frey et al., 2006; Schmidt & Gorlich, 2015), and heterochromatin formation (Strom et al., 2017). In the particular case of the crowded environment of centrosomes, the main partition of tubulin molecules, the scaffold SPD-5 protein favors PS, a condition that recruits other proteins to participate in microtubule nucleation (Woodruff et al., 2017). In addition to the role such proteins have during PS formation, small molecules such as adenosine triphosphate (ATP), a well-known energy source in cell metabolism, may also behave as a hydrotrope participating in the dissolution of liquid-like droplets (Patel et al., 2017).

4.2 The pathological role of PS/transitions in the context of cancer

Despite increasing evidence that PS of amyloidogenic proteins and their progress to solid-like fibrils are linked to aging disorders, the participation of phase transitions in the context of cancer is receiving attention. In that sense, the tumor suppressor p53 is one of the most studied proteins, as the mutation frequency within the TP53 gene nears 50% in human cancers. Our group and others have identified p53 as an amyloidogenic protein and elucidated the implications of p53 aggregation in cancer progression and oncogenic transformation (Ano Bom et al., 2012; Ghosh et al., 2017; Ishimaru et al., 2003) (Fig. 3). The first *in vitro* evidence of p53 forming solid-like fibrillar materials was reported in 2003 by our group (Ishimaru et al., 2003). This finding was not surprising at that time because p53 is nearly unstable at physiological temperature (Bullock et al., 1997), a condition that favors phase transitions to amorphous-like or ordered-like assemblies. Later, in 2011 and 2012, we detected p53 aggregation in breast cancer cells and breast tumor biopsies (Ano Bom et al., 2012; Levy et al., 2011). Using a plethora of biochemical and cell-based approaches, we noticed that aggregation of mutant p53 occurs faster than that of wild-type p53 and that the aggregates share some commonalities with amyloid-like materials, such as thioflavin T binding and β-strand conversion (Ano Bom et al., 2012). Notably, the p53 phase transition has been attributed to a misfolding phenomenon, as solvent exposure of hydrophobic motifs is the driving force of p53 collapse and ultimately aggregation (Ghosh et al., 2014; Pedrote et al., 2018; Silva, Cino, Soares, Ferreira, & G, 2018). Likewise, the kinetics of p53 aggregation do not proceed as in

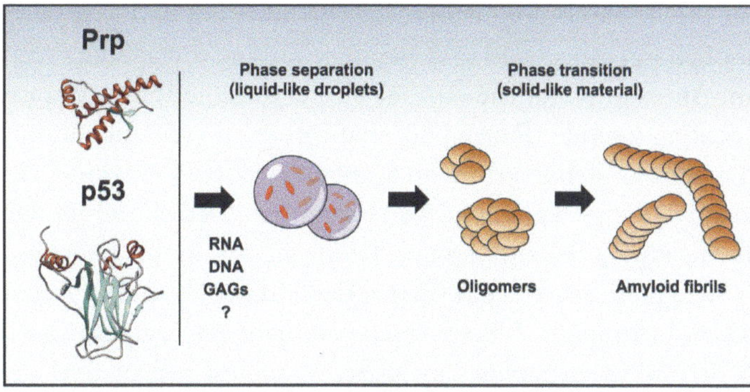

Fig. 3 ***Phase separation and transition events in cancer-associated and aging-related proteins.*** Scheme of potential factors participating during liquid- and solid-like transitions of the prion (PrP) and the tumor suppressor p53 protein. These conversions do not exclusively progress from the illustrated route.

classical amyloidogenic proteins (Wang & Fersht, 2015; Wilcken, Wang, Boeckler, & Fersht, 2012), in which nucleation and monomer incorporation favor the growth of semi-crystalline amyloid filaments. In the case of p53, a mixture of amyloid-like and amorphous precipitates is present (Ano Bom et al., 2012).

Recently, we investigated the molecular determinants involved in p53 misfolding and aggregation. The p53 structural integrity is compromised by its high degree of exposed backbone hydrogen bonds (BHBs) not evident in the related p53-family members p63 and p73 (Cino, Soares, Pedrote, de Oliveira, & Silva, 2016). These exposed BHBs are susceptible sites for solvent attack and structural instability that may locally contribute to nucleate aggregation. Subsequently, we further showed that modulating the average-weighted distribution of p53 native states results in globule-like conformers (Pedrote et al., 2018). These states show increased hydrophobic patches and expanded volumes but preserve secondary and tertiary contacts. Most importantly, they are pre-amyloidogenic because kinetic experiments evolved rapidly compared to those of non-perturbed p53 native states (Pedrote et al., 2018). These seminal studies resulted in many investigations of p53 aggregation, and increasing evidence now exists to support p53 aggregates as an essential factor during oncogenic transformation (De Smet et al., 2017; Ghosh et al., 2014, 2017; Kluth et al., 2014; Lasagna-Reeves et al., 2013; Sengupta, Maji, & Ghosh, 2017; Silva, De Moura Gallo et al., 2014; Wang & Fersht, 2017; Xu et al., 2011; Yang-Hartwich et al., 2015). These exquisite insights on p53 aggregation now help researchers to exploit the negative-dominance, loss-of-function, and gain-of-function behavior of p53 in cancer and to develop promising therapeutics to avert aggregation. An example of the current efforts to target p53 aggregation is a cell-penetrating peptide (ReACp53) that can rescue p53 function in cancer cells and organoids derived from an aggressive cancer (Soragni et al., 2016). Concordantly, the polyphenolic compound resveratrol and its derivatives pterostilbene and piceatannol have been shown to be effective against mutant p53 aggregation in cells (Ferraz da Costa et al., 2018). Accordingly, the PRIMA-1 compound likely rescues mutant p53 aggregation (Rangel et al., 2019). Further strategies include the exploration of new druggable sites within pre-amyloidogenic p53 states (Pedrote et al., 2018; Silva, Cino, Soares, Ferreira, & de Oliveira, 2018).

The p53 tumor suppressor has been shown to participate as a transcription factor upon genotoxic stress. This molecule binds to responsive elements in the DNA and regulates the expression of apoptotic-related genes and those involved in cell cycle arrest and DNA repair. Although p53 bears a DNA-binding domain, it lacks RRMs. Intriguingly, a recent investigation induced the modulation of p53 aggregation by RNA molecules (Kovachev et al., 2017). Despite this breakthrough, interpretation of how RNA affects p53 aggregation is mechanistically challenging (Fig. 3). Hypotheses include RNA acting as scaffold architectures and occluding aggregation-prone segments to RNA inducing structural changes and favoring nucleation. Notwithstanding, RNA seems to modulate p53 aggregation and may trigger PS processes such as liquid-like droplets. At this level, RNA participates in phase transitions to solid-like amorphous and amyloid-like materials (Cordeiro et al., 2019; Kovachev et al., 2017; Silva & Cordeiro, 2016). The significance of RNA-bound p53 assemblies and their potential roles in either cell physiology or pathological conversion requires further elucidation (Fig. 3). Although PS to liquid droplets has not yet been reported for p53, a recent study reported an anomalous liquid condensate of wild-type p53 (Safari et al., 2019). These structures differ from classical droplets in several ways, such as their nanometric size, their independence of p53 concentration, and the lack of constant solubility. In contrast to classical droplets, however, their maturation facilitates fibril assemblies (Safari et al., 2019).

Tumor suppressors usually exist in the cells as isoforms because of different mechanisms, such as alternative splicing, alternative promoter usage or alternative codon usage. The human *TP53* gene expresses approximately twelve p53 isoforms. Recently, we found that the Δ40p53 isoform has a high propensity to form aggregates in the cytoplasm of endometrial carcinoma cells, and biophysical studies confirmed that loss of the p53 transactivation domain results in high amyloid aggregation (Melo Dos Santos et al., 2019). Thus, not only mutations but also truncated variants may lead to phase transition resulting in amyloid structures.

The speckle-type POZ (pox virus and zinc finger protein) protein (SPOP) is involved in ubiquitination of target proteins to their proteasomal degradation as a component of the E3 ubiquitin-protein ligase complex (Cullin-RING E3 family). This protein also regulates the transcriptional activation of ETS-1 target genes and changes protein subcellular localization. SPOP is mutated in solid tumors, such as breast, gastric, and renal tumors, gliomas and especially prostate cancer, and is associated with early events in tumorigenesis (Mani, 2014). SPOP interacts with many oncogenic proteins regulating their activity. For example, when mutated, this protein does not interact with SRC-3, failing to regulate androgen receptor (AR) (Geng et al., 2013). Low SPOP expression and loss of SPOP genomic DNA are also associated with cancer (Li et al., 2017).

SPOP was named for its localization in nuclear speckles (Nagai et al., 1997). Nuclear speckles (NS) are subcompartments involved in RNA metabolism, splicing regulation, and gene expression. NS is structured by the protein-protein and protein-RNA interactions, many of these proteins have very flexible LC regions, and these proteins tend to form liquid droplets at high concentrations (Molliex et al., 2015). The multivalence of LC regions enables the formation of protein oligomeric species. Oligomeric forms of SPOP show increased ubiquitination efficiency, and mutations that disrupt self-association are found in cancers (Marzahn et al., 2016). NS localization was shown to be dependent on SPOP oligomerization, but oligomerization is insufficient to drive LLPS (Marzahn et al., 2016). Binding of this protein to its substrate DAXX or AR triggers a demixing process, forming liquid droplets (Bouchard et al., 2018). SPOP oligomerization forms filamentous assemblies that evolve to droplets in the presence of its substrate. In this context, disruption of SPOP PS may be related to oncogenesis and may be an interesting target (Bouchard et al., 2018).

4.3 The pathological role of PS/transitions in the context of neurodegenerative disorders

Neurodegenerative diseases are classically characterized by the presence of stable protein aggregates, such as Lewy bodies in PD, neurofibrillary tangles (NFTs) and amyloid plaques in AD, and amyloid plaques in transmissible spongiform encephalopathy (TSE). Each of these aggregates involves a major protein connected to the process of neurodegeneration. Although these aggregates have been characterized over several decades, the molecular details that lead to the aggregation of these proteins are not yet fully understood. As Dobson notes, most proteins have the potential to aggregate and form amyloid fibers, depending on the environment and the partners it encounters (Vendruscolo, Knowles, & Dobson, 2011). For several of the amyloidogenic proteins, molecules such as polyanions [NAs and glycosaminoglycans (GAGs), among others] are capable of modulating the folding and aggregation routes.

LLPS has emerged as a mechanism that may also be common for several proteins and has been shown both *in vitro* and in cells for proteins related to neurodegenerative disorders, such as FUS and TDP-43 protein in ALS and frontotemporal lobar dementia (FTLD), Tau in AD and several tauopathies, and PrP^{Sc} in TSEs.

Although PS may be a functional mechanism (Zeng et al., 2016), for many proteins, this process has been observed in events associated with its dysfunction, such as protein accumulation in regions of the cell where it does not perform its regular function or alteration in the interaction with ligands, leading to a loss of function and/or gain of toxic function. These changes would lead to a PS that would be an important step in a phase transition.

4.3.1 FUS

FUS/TLS (fused in sarcoma/translocated in liposarcoma) is a protein first described by its association with cancer and is involved in transcription, DNA repair, and RNA biogenesis. FUS is part of a family of heterogeneous nuclear ribonucleoprotein particle (hnRNP) proteins called FET, together with EWSR1 and TAF15. These proteins have 4 types of domains: an N-terminal LC domain, RGG (arginine/glycine)-rich domains, a zinc finger (ZnF) domain, and an RRM. RGG, ZnF and RRM are domains involved in NA interaction, binding various RNA molecules. The RGG and LC domains are important for forming intermolecular assemblies, including self-assembly. Their amino-terminal LC domain is fused to the DNA-binding domain of transcription factors through chromosomal translocation, deregulating gene expression in cancer (Kato & McKnight, 2018).

FUS mutations have been associated with neurodegenerative disorders, such as ALS and FTLD. Mutations in the C-terminal nuclear localization sequence lead to protein accumulation in the cytoplasm, which is associated with aggregation (Vance et al., 2009). The LC domain has a prion-like motif, with many glycine, serine, glutamine, and tyrosine residues, showing a high tendency to form stacked β-amyloid assemblies and forming fibers that were shown to regulate transcription *in vivo* (Kwon et al., 2013). Interaction with RNA was shown to trigger FUS aggregation, leading to interaction with RNA Pol II, stimulating its elongation activity. FUS phosphorylation prevents aggregation, consequently affecting transcriptional regulation (Kwon et al., 2013; Schwartz et al., 2013).

FUS is found in neuronal granules of dendritic spines and interacts with RNAs stored for local translation at the synapse (Kato & McKnight, 2018). This protein is important for proper neuronal function, as FUS-knockout mice show neurons with altered dendritic morphology, which was also observed in disease mutant animal models (Fujii et al., 2005; Huang et al., 2012). Because it has an RRM domain and the self-association LC domain, FUS may be important for RNA granule assembly. Recombinant FUS was able to form amyloid-like cross beta filaments that behave as hydrogels *in vitro* at high concentrations. This capacity was found to be associated with the LC domain, as truncation of the LC sequence abrogates this behavior (Kato et al., 2012).

FUS localizes in unstressed cells mainly in the nucleus and in the nucleoplasm. This protein can also be found in the cytoplasm in stress granules after heat treatment (Patel et al., 2015). Regardless of the condition, these spherical granules showed a liquid state, which was 10–100 times more viscous than water, with FUS localization fluctuating in a dynamic way, forming a liquid droplet. *In vitro*, at concentrations near the physiologic and in the presence of molecular crowding agents, FUS containing the LC domain was also able to phase separate (Patel et al., 2015). These liquid droplets fused over time, increased in size, and then aggregated into fibers, and this process was accelerated when the LC domain contained disease-related mutations (Patel et al., 2015). This finding is characteristic of a liquid-solid state transition, although these aggregates are not as stable as those observed for classical amyloid proteins (Kato et al., 2012). These findings suggest that PS is an early event in the FUS route of aggregation. Although FUS fibers showed cross beta assemblies (Kwon et al., 2013), FUS LC was shown to have a disordered structure at the LLPS state in the monomeric form but had restricted mobility (Patel et al., 2015).

In cells, FUS is phosphorylated at its LC domain during the DNA damage response, resulting in translocation of the protein from the nucleus to the cytoplasm (Deng, Gao, & Jankovic, 2014). *In vitro* FUS phosphorylation at the LC domain unleashed the dissolution of liquid droplets (Murray et al.,

2017), suggesting that this PTM may modulate PS and fibril assembly *in vivo*. Full-length FUS also forms the LLPS state, and RNA was shown to concentration-dependently modulate this process and was inversely proportional *in vitro* (low RNA concentrations increased PS) (Burke, Janke, Rhine, & Fawzi, 2015). This finding suggests that the RMM domain may also be important for modulation of LLPS *in vivo* because it is an RNA-binding protein.

4.3.2 TDP-43

TDP-43 is the main protein found accumulated in ALS and FLTD. It is an hnRNP protein containing a folded N-terminal domain with two RRMs and a disordered C-terminal domain (CTD). The CTD is a PLD due to its low complexity and enrichment in glutamine and asparagine residues and has a glycine-rich domain. It has a transcriptional repressor role, first described for the transactivation response element of human immunodeficiency virus. This protein is also involved in RNA metabolism, and similar to FUS, it is found in stress granules. TDP-43 cytoplasmic aggregation is related to a decrease in TDP-43 in the nucleus and downregulation of its RNA targets, which contribute to pathogenesis (Ling, 2018).

In ALS and FLTD, this nuclear protein shows increased cytoplasmic localization and forms ubiquitinated inclusions. The recombinant form of TDP-43 readily aggregates *in vitro,* forming oligomeric species, which are neurotoxic in cell and *in vivo* experiments, in addition to being present in the brains of patients (Fang et al., 2014). The CTD was shown to aggregate as the full-length protein, demonstrating the importance of this region for aggregation (Johnson et al., 2009).

The CTD domain is the site of many ALS-linked mutations. It undergoes LLPS *in vitro* with increasing salt concentration (still physiological) and in the presence of RNA. As observed for FUS, NMR data suggested that TDP-43 shows an intrinsically disordered structure, resembling the monomeric form, when forming liquid droplets (Conicella et al., 2016). Enhancing the protein concentration from 20 μM to more than 100 μM leads to solid/hydrogel formation. Most ALS mutations decreased or abrogated LLPS, increasing the transition to solid aggregates (Conicella et al., 2016). Thus, mutations in the CTD region may disrupt TDP-43 function and enhance a toxic aggregation pathway, gaining toxic function and losing normal function.

In the nucleus, TDP-43 exists as dimer, trimer, tetramer and higher oligomers via interaction of the N-terminal domain. These oligomers were shown to be important for TDP-43 function, which is a physiological oligomerization (Afroz et al., 2017). Upon stress, this molecule migrates to stress granules and forms higher molecular weight complexes in the cytoplasm via the CTD; these complexes are phosphorylated, ubiquitinated, acetylated, and sumoylated, leading to irreversible aggregation. These aggregates are formed by the full protein, and the CTD originates from proteolytic cleavage (Afroz et al., 2017). Whether PS of this domain is involved with the assembly of stress granules and causes toxicity of these aggregates should be elucidated.

4.3.3 Tau

The microtubule-associated protein Tau is important for microtubule polymerization and stabilization and axonal transport, among other functions. Tau is expressed by alternative splicing of the MAPT gene (of exons 2 and 3), forming six different isoforms (Nizynski, Dzwolak, & Nieznanski, 2017). These isoforms can be differentiated by the presence of 29 or 58 amino acid inserts in the N-terminal region and of three (3R) or four (4R) tandem repeat amino acids. This repeat region is an important site for microtubule binding (microtubule binding domain, MBD) and is the site for many disease mutations. Tau undergoes different PTMs, mainly phosphorylation, and hyperphosphorylation has an important role in triggering pathology (Nizynski et al., 2017). This protein can be found in stress granules and speckles in cells (Lu et al., 2014; Vanderweyde et al., 2012).

Tau is an IDP with an amino acid sequence of low complexity that is rich in proline and positively charged residues, with a fraction of hydrophobic amino acid side-chains. Misfolding and aggregation of this protein, forming the so-called NFTs, is associated with different types of neurodegenerative disorders known collectively as tauopathies. AD is the most prevalent type of tauopathy (Nizynski et al., 2017).

In AD, Tau aggregation is connected to Aβ pathology. Aβ aggregation stimulates Tau hyperphosphorylation, leading to fibrillization. Tau can aggregate *in vitro* and *in vivo* with protein monomers interacting to form β-sheet-rich amyloid fibrils, ribbon-like fibrils, and/or straight filaments (SFs) (Goedert, Crowther, & Spillantini, 1998) via template-based conversion (Nizynski et al., 2017). Polyanions, such as RNAs and GAGs, are associated with Tau in the NFTs of patient brains. These molecules also stimulate Tau aggregation. All these inducers appear to cause neutralization of the positive charges of the amino acid residues, contributing to aggregation.

Another AD-related PTM of Tau is truncation. Tau can be cleaved by caspases or calpains, generating fragments found in AD brains that lack both N− and C-termini, and as in phosphorylation, oligomers of Aβ stimulate Tau cleavage (Hanger & Wray, 2010). The main fragment found in many tauopathies harbors four MBDs, the 4R-protein. The 4R-protein was shown to form liquid droplets at 37 °C with increasing protein concentrations and at a pH close to the pI. This process was also dependent on ionic strength, and phosphorylated 4R-protein undergoes LLPS. The secondary structure of the LLPS 4R-protein remains largely disordered but apparently has a tendency to form a β-structure.

Heparin is a GAG known to induce Tau fibrilization (Goedert et al., 1996), and its addition to 4R-protein preformed liquid droplets induced the formation of amyloid structures. The 4R-protein LLPS was not observed at low and high temperatures, as well as heparin-induced fibrillation, suggesting a dependence between fibrillation and demixing conditions (Ambadipudi et al., 2017).

4.3.4 PrP

The cellular prion protein (PrP^C) is a membrane protein attached by a glycosyl phosphatidylinositol (GPI) anchor that can interact with a wide range of partners, such as proteins, NAs, metal ions, lipids, and carbohydrates. This ability to interact with different molecules is related to the wide variety of functions postulated for this protein, such as regulation of myelin maintenance, cellular differentiation, iron uptake, and circadian modulation (Wulf, Senatore, & Aguzzi, 2017). Although described as a membrane protein, it can be found in the cytosol and nucleus under stress conditions (Godsave, Peters, & Wille, 2015). Mature PrP is a protein of 209 amino acids with an intrinsically disordered N-terminal domain and a folded CTD with three α-helices and one small antiparallel β-sheet. This protein may also show glycosylation as a PTM.

PrP^{Sc} is the main causative agent of TSEs. This PrP has the same amino acid sequence as PrP^C, except in hereditary cases of the disease, but has a very different secondary structure. Its secondary structure is enriched in β-sheets, which consequently lead to changes in its physicochemical properties and modes of interaction, resulting in a strong tendency to aggregate (both amorphous and fibrillary aggregates), leading to the acquisition of a toxic function to the cell. These aggregates are self-replicative (a shared characteristic between different amyloid proteins), and acting as a template can convert PrP^C into new PrP^{Sc}. PrP^{Sc} is able to propagate between cells and individuals and is an infectious disease (Scheckel & Aguzzi, 2018).

The term "prion" is also used to designate other proteins capable of self-propagating and thus transmit a phenotype (advantageous or not) independent of changes in the coding gene sequence, as it was used to name yeast prions. The term "prionoid" or "prion like" has been used for proteins related to diseases that share the same characteristics of PrP^{Sc} except for the infectious character (Scheckel & Aguzzi, 2018).

The PrP N-terminal domain is a PLD identified as the site of interaction for several molecules. Its intrinsically unfolded nature allows it to adapt to molecules with different structures with diverse consequences. Although this domain has been related to PrP signaling activity and was shown to modulate PrP conversion efficiency, it is a dispensable domain for the formation of infectious PrP^{Sc} because the PrP protease K-resistant core lacks amino acids 23−90 (McKinley, Bolton, & Prusiner, 1983). This domain is also important for interaction with Aβ oligomers, mediating neurotoxic signaling in AD (Ulbrich et al., 2018).

Kostylev and collaborators, investigating PrP and Aβ oligomer (Aβo) interactions, suggested that PrP^C underwent LLPS at physiologic pH and salinity at concentrations higher than 30 μM and dissolved upon dilution (Kostylev et al., 2018). Interaction with Aβo demixed protein and formed a dynamic hydrogel in a stoichiometric-dependent manner. This transition appeared to be achieved by stimulating the coalescence of the liquid droplets, and the N-terminal adopted an α-helical conformation together with unfolding in helix-α2 (Kostylev et al., 2018).

Polyanions are the most relevant molecules as cofactors for prion conversion and aggregation, such as NAs and sulfated polysaccharides (SPs) (Cordeiro et al., 2019). Many of these molecules were shown to induce PrP aggregation and are highlighted as adjuvant factors important for conversion and propagation. Our group showed in 2001 that DNA molecules are capable of inducing recombinant PrP aggregation (Cordeiro et al., 2001). Since then, several researchers have corroborated the idea that NA and other cofactors are important in this process (Deleault et al., 2012).

The interaction of PrP with NAs, both DNA and RNA, is quite varied in terms of affinity, polynucleotide sequence and structure important for interaction, as well as effect on the protein itself (Silva & Cordeiro, 2016). Although some NA sequences are capable of inducing PrP aggregation, others are capable of reducing the formation of PrP^{Sc}. RNA was shown to bind primarily to the N-terminal domain, while DNA bound the

N-terminal and the CTD (Gomes et al., 2008; Lima et al., 2006; Silva & Cordeiro, 2016). However, the exact mechanism of how the interaction occurs still needs to be elucidated. Once PrP is an RNA-binding protein, interaction with NA may lead to less favorable molecule-solvent interactions, creating demixing conditions (Fig. 3).

In a recent publication, Matos and collaborators showed that two identified 25-mer single-stranded DNA aptamers bind PrP globular domain with a high affinity (Matos et al., 2019) (Fig. 3). PrP^C underwent LLPS (at pH 7.4, 100 mM NaCl, and w/wo 10% PEG), and 1,6-hexanodiol abolished droplet formation, suggesting that hydrophobic contacts must be important for PrP PS. PrP droplets suffered phase transition after time into fiber-like structures. Aptamers could increase the number of droplets, but with a reduced size, in a protein:NA proportion-dependent manner and accelerated phase transition.

SPs are also known to induce the aggregation of PrP under some conditions and inhibit PrP aggregation in other conditions. Our group showed that low molecular weight heparin, a sulfated GAG, induced rapid aggregation of PrP *in vitro*, followed by a slow process of disaggregation (Vieira, et al., 2011). This interaction did not cause major changes in the protein folding; however, it increased the protein stability and protected PrP from aggregation mediated by different factors (temperature, RNA, PrP^{Sc} seeded conversion) (Vieira, Cordeiro, Caughey, & Silva, 2014). As previously reported for Tau, induction of aggregation by heparin may involve reliance on the prior formation of LLPS. In the case of PrP, we still do not know whether heparin could induce PS and/or interfere with the aggregation of PrP in liquid droplets.

5. Methodological approaches to detect PS

Based on the dynamic properties of membraneless organelles formed by LLPS and the multitude of components and interactions they can harbor, assessment of PS *in vivo* and even in *in vitro* assays is difficult. Bioinformatic tools are valuable in predicting IDRs based on the protein sequence and potential regions responsible for driving LLPS (Alberti, Gladfelter, & Mittag, 2019). Significant efforts have been made to understand the 'molecular grammar' behind LLPS, which was firstly conceived as a property of LCR (Wang et al., 2018). Even though LLPS determinants are complex (Boeynaems et al., 2018), certain properties stand out: multivalency (multiple binding domains) and the presence of **S**hort **Li**near amino acids **M**otifs (SLiMs) (Harmon, Holehouse, Rosen, & Pappu, 2017). These SLiMs are usually composed by polar amino acids interleaved by aromatic residues or glycine-rich domains, such as RGG (Arg, Gly, Gly) in LC regions that indicates that the investigated protein can bind RNA; RNA-binding proteins are important scaffolds in PS. *In vitro* evaluation of PS is based on the use of highly purified recombinant proteins usually produced in bacteria. PTMs, such as methylation, acetylation, phosphorylation and ubiquitination, can be implicated in the modulation of phase transition (Hofweber & Dormann, 2019); thus, protein expression in eukaryotic systems might be considered. For detailed information on how to produce high-quality proteins—including PTMs, if required for PS—consult (Alberti et al., 2018). Approaches to estimate the intracellular concentration of a target protein, such as by immunofluorescence, are relevant, especially if the goal is investigation of PS in a particular compartment, such as cell nuclei (Maharana et al., 2018). *In vitro* PS assessment starts with optical microscopy and sample preparation mimicking physiological conditions. Differential interference contrast (DIC) is the microscopy method of choice for the initial observation of micrometer-sized spherical structures. DIC images show *pseudo three-dimensional relief shading, which* provides optimal contrast for glassy protein-rich condensate observation, in contrast with phase contrast microscopy. Liquid droplet characteristic fusion and fission events usually occur in a timescale of seconds. These events can be time lapse-recorded and provide a hint of the condensate dynamics, i.e., liquid-like condensates rapidly fuse in contrast to more ordered gel-like condensates whose morphology does not change over time. Microscopy investigation of protein samples includes the addition of polymers that resemble the crowded intracellular *milieu,* such as polyethylene glycol (PEG), and known protein interactors, such as NAs or client proteins. Some proteins phase separate even in the absence of PEG (or other crowding compounds), and increasing protein concentrations to determine the critical concentration for PS is useful. In addition, fine control of external factors, such as temperature and incubation time prior to imaging, is pivotal when comparing different conditions, which must be analyzed with the same parameters. Additionally, meticulous sample preparation is essential, as air bubbles and scratches when pipetting protein samples to microscopy chambers have been shown to misguide PS observations (Alberti et al., 2018). After protein purification, the sample must be stored in a non-demixed state, for example, in the presence of high salt concentration and kept at -80 °C. This enables

PS induction by dilution in low salt buffer, for example, and can be controlled in a timely manner to account for different microrheology properties of "fresh" and "old" droplets. Optical tweezer assays and fluorescence recovery after photobleaching (FRAP) are useful in analyzing the dynamic physical properties. A screening of PS at different temperatures, pH and ionic strength is ideal, as PS is highly susceptible to these physicochemical properties. LLPS reversibility upon temperature screening is in agreement with IDPs known resistance to thermal fluctuations; thus, LLPS is better investigated using a microscope equipped with thermal stage (Nott et al., 2015). Extrinsic compounds can also be used to indicate that the droplets observed by DIC, for instance, are not an artifact. This process includes the addition of 1,6-hexanediol, which disrupts protein droplets while lipid vesicles are resistant, and SYPRO orange (Matos et al., 2019) to confirm the proteinaceous nature of liquid droplets. Fluorescent labeling of proteins (e.g., fusion to green fluorescent protein, GFP) and/or ligands can probe the colocalization of molecules within the biomolecular condensates (Nott et al., 2015). Furthermore, for NA-binding proteins, the addition of RNA or DNA at different protein-NA molar ratios will demonstrate the dependence on the binding stoichiometry for efficient PS. This is a hallmark for RNA-binding proteins that phase separate in the presence of the cognate NA (Maharana et al., 2018; Matos et al., 2019). Fluorescence correlation spectroscopy (FCS) allows the identification of labeled protein interactions with RNA within living cells and the characterization of the size/mobility of the condensates (Maharana et al., 2018). Provided that PS in the membrane has been demonstrated to organize receptors in micrometer-scale signal transduction hubs (Banjade & Rosen, 2014), *in vitro* observation can be performed by lipid coating of glass slides.

The main challenge in characterization of cell-LLPS is to differentiate liquid-like condensates from membrane-bound compartments. This procedure can be performed by exclusion by immunostaining membrane-delimited organelles, such as endosomes, endoplasmic reticulum, and mitochondria, to name a few desirable controls in mammalian cells. In addition, optogenetic tools were developed to specifically manipulate the cell formation of biomolecular condensates, such as a combination of protein IDRs with light-dependent oligomerization domains of other proteins that, upon illumination, can hetero-oligomerize and form condensates (Brangwynne, 2013; Shin et al., 2017). Electron spectroscopy imaging (ESI) can map elements such as phosphorous and nitrogen, showing the elementary composition of membraneless organelles in cells. Live-cell single particle tracking (SPT) together with FRAP is a valuable tool to determine diffusion coefficients of photostable fluorescently labeled proteins in cells and to corroborate *in vitro* data. If the intracellular concentration of IDPs is significantly lower than that of ordered proteins, perhaps as a means to regulate liquid droplet assembly, careful evaluation of cellular systems that overexpress the target protein must be taken into account. Despite sophisticated methodological approaches, one can argue that any evidence about PS lies in the "eye of the beholder". For example, highly concentrated protein samples, such as those produced for nuclear magnetic resonance studies, can exhibit visual clouding that is reversible upon temperature changes. A simple verification of the remaining soluble protein after centrifugation, i.e., the supernatants of PS screening, can be loaded on an SDS-PAGE gel, and the presence of soluble protein can be easily assessed. In addition, turbidity measurements (wavelength range of 350–600 nm) have been extensively used for PS screening. However, it is necessary to differentiate turbidity due to irreversible aggregation or to LLPS events. This can be assessed in parallel by other methods, such as microscopy, lack of thioflavin T fluorescence, and by reversibility of structural changes on temperature-based PS screening by spectroscopy, such as NMR and circular dichroism (CD).

In summary, both *in vitro* and *in vivo* (in-cell) evaluations of PS have its drawbacks, and the detection of PS in living tissues/animals is most challenging of all.

6. Perspectives

A new concept in biology has emerged in recent years, in which membraneless organelles formed from the process of liquid–liquid PS are highly crucial to health and disease. These biomolecular condensates act by speeding up reactions among their macromolecular components (proteins and NAs). The homeostasis of the membraneless condensates is controlled by the delicate balance of the component concentration, their dynamics and spatial distribution. Phase transitions in these structures occur with changes in these parameters or by mutations, with the potential appearance of pathologies, such as neurodegenerative diseases and cancer. We appear to be at a turning point of putting together a substantial amount of information on amyloid diseases accumulated in the last two decades. Perhaps LLPS and phase transitions were the missing data that prevented a full understanding of these diseases. For example, we have known for many years that fibrils are likely not pathogenic factors in most maladies. Instead, small amyloid oligomers are the culprits that results in toxic gain of function in neurodegenerative diseases or oncogenic gain of function in cancer.

Amyloid oligomers may be side products of a transition from liquid droplets to solid fibrils, which are mediated by the cellular machinery (Fig. 3). Only the future will say how fast we will be able to use this new knowledge to understand these transitions, such as LLPS→LLPT (fibrils and oligomers) → aggresomes, as well as to apply them as therapeutic targets. With few exceptions, only a few inhibitors of amyloid aggregation have progressed into clinical trials. In the case of mutant p53 aggregation, researchers hope that it can be used as a therapeutic target (Ferraz da Costa et al., 2018; Rangel et al., 2019; Silva et al., 2018; Soragni et al., 2016). However, more basic studies on the biology of the formation of these structures in cells are needed. As noted recently, therapeutic targets in oncology might be easier to use than those in other diseases because the main goal is to kill cancer cells (Mullard, 2019).

Undoubtedly, these new discoveries about biomolecular condensates represent a change in the biological paradigm of how proteins and other molecules interact in the cell. In 1975, Gregorio Weber published an essay in *Advances in Protein Chemistry* on the interactions of proteins and small ligands and the energetics of the biological specificity (Weber, 1975). Some years later, Weber extended his thermodynamic analysis to protein-protein interactions and to protein condensation reactions (Silva, Weber, 1993; Weber, 1992). Accordingly, condensation reactions would be driven by an increase in volume and changes in entropy and enthalpy (positive or negative). Several amyloids are assembled with positive changes in volume and can be dissociated by pressure (Ferrao-Gonzales, Souto, Silva, & Foguel, 2000; Foguel et al., 2003; Silva, Oliveira, et al., 2014). Recently, Winter and coworkers published the striking result that the LLPS of γ-crystallin is much more sensitive to pressure than folded and associated states of globular proteins, indicating that phase separation of γ-crystallin occurs with volume increase (Cinar, Cinar, Chan, & Winter 2019).

We anticipate that, in the next years, biological science will unveil more exciting discoveries on the mechanisms of assembly of biomolecular condensates and their roles in cell function and disease.

References

Afroz, T., Hock, E. M., Ernst, P., Foglieni, C., Jambeau, M., Gilhespy, L. A. B., ... Polymenidou, M. (2017). Functional and dynamic polymerization of the ALS-linked protein TDP-43 antagonizes its pathologic aggregation. *Nature Communications, 8*(1), 45. http://dx.doi.org/10.1038/s41467-017-00062-0.

Alberti, S., & Carra, S. (2018). Quality control of membraneless organelles. *Journal of Molecular Biology, 430*(23), 4711–4729. http://dx.doi.org/10.1016/j.jmb.2018.05.013.

Alberti, S., Gladfelter, A., & Mittag, T. (2019). Considerations and challenges in studying liquid-liquid phase separation and biomolecular condensates. *Cell, 176*(3), 419–434. http://dx.doi.org/10.1016/j.cell.2018.12.035.

Alberti, S., Halfmann, R., King, O., Kapila, A., & Lindquist, S. (2009). A systematic survey identifies prions and illuminates sequence features of prionogenic proteins. *Cell, 137*(1), 146–158. http://dx.doi.org/10.1016/j.cell.2009.02.044.

Alberti, S., Mateju, D., Mediani, L., & Carra, S. (2017). Granulostasis: Protein quality control of RNP granules. *Frontiers in Molecular Neuroscience, 10*, 84. http://dx.doi.org/10.3389/fnmol.2017.00084.

Alberti, S., Saha, S., Woodruff, J. B., Franzmann, T. M., Wang, J., & Hyman, A. A. (2018). A user's guide for phase separation assays with purified proteins. *Journal of Molecular Biology, 430*(23), 4806–4820. http://dx.doi.org/10.1016/j.jmb.2018.06.038.

Ambadipudi, S., Biernat, J., Riedel, D., Mandelkow, E., & Zweckstetter, M. (2017). Liquid-liquid phase separation of the microtubule-binding repeats of the Alzheimer-related protein Tau. *Nature Communications, 8*(1), 275. http://dx.doi.org/10.1038/s41467-017-00480-0.

An, S., Kumar, R., Sheets, E. D., & Benkovic, S. J. (2008). Reversible compartmentalization of de novo purine biosynthetic complexes in living cells. *Science, 320*(5872), 103–106. http://dx.doi.org/10.1126/science.1152241.

Annamalai, K., Guhrs, K. H., Koehler, R., Schmidt, M., Michel, H., Loos, C., ... Fandrich, M. (2016). Polymorphism of amyloid fibrils in vivo. *Angewandte Chemie International Edition in English, 55*(15), 4822–4825. http://dx.doi.org/10.1002/anie.201511524.

Ano Bom, A. P., Rangel, L. P., Costa, D. C., de Oliveira, G. A., Sanches, D., Braga, C. A., ... Silva, J. L. (2012). Mutant p53 aggregates into prion-like amyloid oligomers and fibrils: Implications for cancer. *Journal of Biological Chemistry, 287*(33), 28152–28162. http://dx.doi.org/10.1074/jbc.M112.340638.

Banani, S. F., Rice, A. M., Peeples, W. B., Lin, Y., Jain, S., Parker, R., et al. (2016). Compositional control of phase-separated cellular bodies. *Cell, 166*(3), 651–663. http://dx.doi.org/10.1016/j.cell.2016.06.010.

Banjade, S., & Rosen, M. K. (2014). Phase transitions of multivalent proteins can promote clustering of membrane receptors. *Elife, 3*. http://dx.doi.org/10.7554/eLife.04123.

Bansil, R., & Turner, B. S. (2018). The biology of mucus: Composition, synthesis and organization. *Advanced Drug Delivery Reviews, 124*, 3–15. http://dx.doi.org/10.1016/j.addr.2017.09.023.

Baron, M. K., Boeckers, T. M., Vaida, B., Faham, S., Gingery, M., Sawaya, M. R., ... Bowie, J. U. (2006). An architectural framework that may lie at the core of the postsynaptic density. *Science, 311*(5760), 531–535. http://dx.doi.org/10.1126/science.1118995.

Boeynaems, S., Alberti, S., Fawzi, N. L., Mittag, T., Polymenidou, M., Rousseau, F., ... Fuxreiter, M. (2018). Protein phase separation: A new phase in cell biology. *Trends in Cell Biology, 28*(6), 420–435. http://dx.doi.org/10.1016/j.tcb.2018.02.004.

Bouchard, J. J., Otero, J. H., Scott, D. C., Szulc, E., Martin, E. W., Sabri, N., ... Mittag, T. (2018). Cancer mutations of the tumor suppressor SPOP disrupt the formation of active, phase-separated compartments. *Molecular Cell, 72*(1), 19–36 e18. http://dx.doi.org/10.1016/j.molcel.2018.08.027.

Brangwynne, C. P. (2013). Phase transitions and size scaling of membrane-less organelles. *The Journal of Cell Biology, 203*(6), 875–881. http://dx.doi.org/10.1083/jcb.201308087.

Brangwynne, C. P., Tompa, P., & Pappu, R. V. (2015). Polymer physics of intracellular phase transitions. *Nature Physics, 11*, 899–904.

Broide, M. L., Berland, C. R., Pande, J., Ogun, O. O., & Benedek, G. B. (1991). Binary-liquid phase separation of lens protein solutions. *Proceedings of the National Academy of Sciences of the U S A, 88*(13), 5660–5664.

Bullock, A. N., Henckel, J., DeDecker, B. S., Johnson, C. M., Nikolova, P. V., Proctor, M. R., … Fersht, A. R. (1997). Thermodynamic stability of wild-type and mutant p53 core domain. *Proceedings of the National Academy of Sciences of the U S A, 94*(26), 14338–14342.

Burdick, J. A., & Stevens, M. M. (2005). Biomedical hydrogels. Biomaterials. *Artificial Organs and Tissue Engineering*, 107–115. http://dx.doi.org/10.1533/9781845690861.2.107.

Burke, K. A., Janke, A. M., Rhine, C. L., & Fawzi, N. L. (2015). Residue-by-Residue view of in vitro FUS granules that bind the C-terminal domain of RNA polymerase II. *Molecular Cell, 60*(2), 231–241. http://dx.doi.org/10.1016/j.molcel.2015.09.006.

Chiti, F., & Dobson, C. M. (2006). Protein misfolding, functional amyloid, and human disease. *Annual Review of Biochemistry, 75*, 333–366. http://dx.doi.org/10.1146/annurev.biochem.75.101304.123901.

Chiti, F., & Dobson, C. M. (2017). Protein misfolding, amyloid formation, and human disease: A summary of progress over the last decade. *Annual Review of Biochemistry, 86*, 27–68. http://dx.doi.org/10.1146/annurev-biochem-061516-045115.

Cinar, S., Cinar, H., Chan, H. S., & Winter, R. (2019). Pressure-sensitive and osmolyte-modulated liquid-liquid phase separation of eye-lens gamma-crystallins. *Journal of the American Chemical Society, 141*(18), 7347–7354. http://dx.doi.org/10.1021/jacs.8b13636.

Cino, E. A., Soares, I. N., Pedrote, M. M., de Oliveira, G. A., & Silva, J. L. (2016). Aggregation tendencies in the p53 family are modulated by backbone hydrogen bonds. *Scientific Reports, 6*, 32535. http://dx.doi.org/10.1038/srep32535.

Clery, A., Blatter, M., & Allain, F. H. (2008). RNA recognition motifs: Boring? Not quite. *Current Opinion in Structural Biology, 18*(3), 290–298. http://dx.doi.org/10.1016/j.sbi.2008.04.002.

Cohen, S. I., Linse, S., Luheshi, L. M., Hellstrand, E., White, D. A., Rajah, L., … Knowles, T. P. (2013). Proliferation of amyloid-beta42 aggregates occurs through a secondary nucleation mechanism. *Proceedings of the National Academy of Sciences of the U S A, 110*(24), 9758–9763. http://dx.doi.org/10.1073/pnas.1218402110.

Conicella, A. E., Zerze, G. H., Mittal, J., & Fawzi, N. L. (2016). ALS mutations disrupt phase separation mediated by alpha-helical structure in the TDP-43 low-complexity C-terminal domain. *Structure, 24*(9), 1537–1549. http://dx.doi.org/10.1016/j.str.2016.07.007.

Cordeiro, Y., Machado, F., Juliano, L., Juliano, M. A., Brentani, R. R., Foguel, D., et al. (2001). DNA converts cellular prion protein into the beta-sheet conformation and inhibits prion peptide aggregation. *Journal of Biological Chemistry, 276*(52), 49400–49409. http://dx.doi.org/10.1074/jbc.M106707200.

Cordeiro, Y., Vieira, T., Kovachev, P. S., Sanyal, S., & Silva, J. L. (2019). Modulation of p53 and prion protein aggregation by RNA. *Biochimica et Biophysica Acta (BBA) - Proteins & Proteomics*. http://dx.doi.org/10.1016/j.bbapap.2019.02.006.

Costa, D. C., de Oliveira, G. A., Cino, E. A., Soares, I. N., Rangel, L. P., & Silva, J. L. (2016). Aggregation and prion-like properties of misfolded tumor suppressors: Is cancer a prion disease? *Cold Spring Harb Perspect Biol, 8*(10). http://dx.doi.org/10.1101/cshperspect.a023614.

Crick, S. L., Ruff, K. M., Garai, K., Frieden, C., & Pappu, R. V. (2013). Unmasking the roles of N- and C-terminal flanking sequences from exon 1 of huntingtin as modulators of polyglutamine aggregation. *Proceedings of the National Academy of Sciences of the U S A, 110*(50), 20075–20080. http://dx.doi.org/10.1073/pnas.1320626110.
Das, R. K., & Pappu, R. V. (2013). Conformations of intrinsically disordered proteins are influenced by linear sequence distributions of oppositely charged residues. *Proceedings of the National Academy of Sciences of the U S A, 110*(33), 13392–13397. http://dx.doi.org/10.1073/pnas.1304749110.
De Smet, F., Saiz Rubio, M., Hompes, D., Naus, E., De Baets, G., Langenberg, T., ... Rousseau, F. (2017). Nuclear inclusion bodies of mutant and wild-type p53 in cancer: A hallmark of p53 inactivation and proteostasis remodelling by p53 aggregation. *The Journal of Pathology, 242*(1), 24–38. http://dx.doi.org/10.1002/path.4872.
Deleault, N. R., Walsh, D. J., Piro, J. R., Wang, F., Wang, X., Ma, J., ... Supattapone, S. (2012). Cofactor molecules maintain infectious conformation and restrict strain properties in purified prions. *Proceedings of the National Academy of Sciences of the U S A, 109*(28), E1938–E1946. http://dx.doi.org/10.1073/pnas.1206999109.
Deng, H., Gao, K., & Jankovic, J. (2014). The role of FUS gene variants in neurodegenerative diseases. *Nature Reviews Neurology, 10*(6), 337–348. http://dx.doi.org/10.1038/nrneurol.2014.78.
Dubochet, J., Adrian, M., Chang, J. J., Homo, J. C., Lepault, J., McDowall, A. W., et al. (1988). Cryo-electron microscopy of vitrified specimens. *Quarterly Reviews of Biophysics, 21*(2), 129–228.
Egelman, E. H. (2007). The iterative helical real space reconstruction method: Surmounting the problems posed by real polymers. *Journal of Structural Biology, 157*(1), 83–94. http://dx.doi.org/10.1016/j.jsb.2006.05.015.
Elbaum-Garfinkle, S., Kim, Y., Szczepaniak, K., Chen, C. C., Eckmann, C. R., Myong, S., et al. (2015). The disordered P granule protein LAF-1 drives phase separation into droplets with tunable viscosity and dynamics. *Proceedings of the National Academy of Sciences of the U S A, 112*(23), 7189–7194. http://dx.doi.org/10.1073/pnas.1504822112.
Fang, Y. S., Tsai, K. J., Chang, Y. J., Kao, P., Woods, R., Kuo, P. H., ... Chen, Y. R. (2014). Full-length TDP-43 forms toxic amyloid oligomers that are present in frontotemporal lobar dementia-TDP patients. *Nature Communications, 5*, 4824. http://dx.doi.org/10.1038/ncomms5824.
Feric, M., Vaidya, N., Harmon, T. S., Mitrea, D. M., Zhu, L., Richardson, T. M., ... Brangwynne, C. P. (2016). Coexisting liquid phases underlie nucleolar subcompartments. *Cell, 165*(7), 1686–1697. http://dx.doi.org/10.1016/j.cell.2016.04.047.
Ferrao-Gonzales, A. D., Souto, S. O., Silva, J. L., & Foguel, D. (2000). The preaggregated state of an amyloidogenic protein: Hydrostatic pressure converts native transthyretin into the amyloidogenic state. *Proceedings of the National Academy of Sciences of the U S A, 97*(12), 6445–6450.
Ferraz da Costa, D. C., Campos, N. P. C., Santos, R. A., Guedes-da-Silva, F. H., Martins-Dinis, M., Zanphorlin, L., ... Silva, J. L. (2018). Resveratrol prevents p53 aggregation in vitro and in breast cancer cells. *Oncotarget, 9*(49), 29112–29122. http://dx.doi.org/10.18632/oncotarget.25631.
Fletcher, N. H. (1971). Structural aspects of the ice-water system. *Reports on Progress in Physics, 34*, 913–994.
Foguel, D., Suarez, M. C., Ferrao-Gonzales, A. D., Porto, T. C., Palmieri, L., Einsiedler, C. M., ... Silva, J. L. (2003). Dissociation of amyloid fibrils of alpha-synuclein and transthyretin by pressure reveals their reversible nature and the formation of water-excluded cavities. *Proceedings of the National Academy of Sciences of the U S A, 100*(17), 9831–9836. http://dx.doi.org/10.1073/pnas.1734009100.

Fox, A. H., & Lamond, A. I. (2010). Paraspeckles. *Cold Spring Harb Perspect Biol, 2*(7), a000687. http://dx.doi.org/10.1101/cshperspect.a000687.

Frey, S., Richter, R. P., & Gorlich, D. (2006). FG-rich repeats of nuclear pore proteins form a three-dimensional meshwork with hydrogel-like properties. *Science, 314*(5800), 815–817. http://dx.doi.org/10.1126/science.1132516.

Fujii, R., Okabe, S., Urushido, T., Inoue, K., Yoshimura, A., Tachibana, T., … Takumi, T. (2005). The RNA binding protein TLS is translocated to dendritic spines by mGluR5 activation and regulates spine morphology. *Current Biology, 15*(6), 587–593. http://dx.doi.org/10.1016/j.cub.2005.01.058.

Galkin, O., Chen, K., Nagel, R. L., Hirsch, R. E., & Vekilov, P. G. (2002). Liquid-liquid separation in solutions of normal and sickle cell hemoglobin. *Proceedings of the National Academy of Sciences of the U S A, 99*(13), 8479–8483. http://dx.doi.org/10.1073/pnas.122055299.

Geng, C., He, B., Xu, L., Barbieri, C. E., Eedunuri, V. K., Chew, S. A., … Mitsiades, N. (2013). Prostate cancer-associated mutations in speckle-type POZ protein (SPOP) regulate steroid receptor coactivator 3 protein turnover. *Proceedings of the National Academy of Sciences of the U S A, 110*(17), 6997–7002. http://dx.doi.org/10.1073/pnas.1304502110.

Ghosh, S., Ghosh, D., Ranganathan, S., Anoop, A., P, S. K., Jha, N. N., … Maji, S. K. (2014). Investigating the intrinsic aggregation potential of evolutionarily conserved segments in p53. *Biochemistry, 53*(38), 5995–6010. http://dx.doi.org/10.1021/bi500825d.

Ghosh, S., Salot, S., Sengupta, S., Navalkar, A., Ghosh, D., Jacob, R., … Maji, S. K. (2017). p53 amyloid formation leading to its loss of function: implications in cancer pathogenesis. *Cell Death & Differentiation, 24*(10), 1784–1798. http://dx.doi.org/10.1038/cdd.2017.105.

Godsave, S. F., Peters, P. J., & Wille, H. (2015). Subcellular distribution of the prion protein in sickness and in health. *Virus Research, 207*, 136–145. http://dx.doi.org/10.1016/j.virusres.2015.02.004.

Goedert, M., Crowther, R. A., & Spillantini, M. G. (1998). Tau mutations cause frontotemporal dementias. *Neuron, 21*(5), 955–958.

Goedert, M., Jakes, R., Spillantini, M. G., Hasegawa, M., Smith, M. J., & Crowther, R. A. (1996). Assembly of microtubule-associated protein tau into Alzheimer-like filaments induced by sulphated glycosaminoglycans. *Nature, 383*(6600), 550–553. http://dx.doi.org/10.1038/383550a0.

Gomes, M. P., Millen, T. A., Ferreira, P. S., e Silva, N. L., Vieira, T. C., Almeida, M. S., … Cordeiro, Y. (2008). Prion protein complexed to N2a cellular RNAs through its N-terminal domain forms aggregates and is toxic to murine neuroblastoma cells. *Journal of Biological Chemistry, 283*(28), 19616–19625. http://dx.doi.org/10.1074/jbc.M802102200.

Guo, J. L., Covell, D. J., Daniels, J. P., Iba, M., Stieber, A., Zhang, B., … Lee, V. M. (2013). Distinct alpha-synuclein strains differentially promote tau inclusions in neurons. *Cell, 154*(1), 103–117. http://dx.doi.org/10.1016/j.cell.2013.05.057.

Hanger, D. P., & Wray, S. (2010). Tau cleavage and tau aggregation in neurodegenerative disease. *Biochemical Society Transactions, 38*(4), 1016–1020. http://dx.doi.org/10.1042/BST0381016.

Hantschel, O., Nagar, B., Guettler, S., Kretzschmar, J., Dorey, K., Kuriyan, J., et al. (2003). A myristoyl/phosphotyrosine switch regulates c-Abl. *Cell, 112*(6), 845–857.

Harmon, T. S., Holehouse, A. S., Rosen, M. K., & Pappu, R. V. (2017). Intrinsically disordered linkers determine the interplay between phase separation and gelation in multivalent proteins. *Elife, 6*. http://dx.doi.org/10.7554/eLife.30294.

Hennig, S., Kong, G., Mannen, T., Sadowska, A., Kobelke, S., Blythe, A., … Fox, A. H. (2015). Prion-like domains in RNA binding proteins are essential for building subnuclear

paraspeckles. *The Journal of Cell Biology, 210*(4), 529–539. http://dx.doi.org/10.1083/jcb.201504117.
Hofweber, M., & Dormann, D. (2019). Friend or foe-Post-translational modifications as regulators of phase separation and RNP granule dynamics. *Journal of Biological Chemistry, 294*(18), 7137–7150. http://dx.doi.org/10.1074/jbc.TM118.001189.
Huang, C., Tong, J., Bi, F., Wu, Q., Huang, B., Zhou, H., et al. (2012). Entorhinal cortical neurons are the primary targets of FUS mislocalization and ubiquitin aggregation in FUS transgenic rats. *Human Molecular Genetics, 21*(21), 4602–4614. http://dx.doi.org/10.1093/hmg/dds299.
Hyman, A. A., Weber, C. A., & Julicher, F. (2014). Liquid-liquid phase separation in biology. *Annual Review of Cell and Developmental Biology, 30*, 39–58. http://dx.doi.org/10.1146/annurev-cellbio-100913-013325.
Ishimaru, D., Andrade, L. R., Teixeira, L. S., Quesado, P. A., Maiolino, L. M., Lopez, P. M., ... Silva, J. L. (2003). Fibrillar aggregates of the tumor suppressor p53 core domain. *Biochemistry, 42*(30), 9022–9027. http://dx.doi.org/10.1021/bi034218k.
Jain, A., & Vale, R. D. (2017). RNA phase transitions in repeat expansion disorders. *Nature, 546*(7657), 243–247. http://dx.doi.org/10.1038/nature22386.
Johnson, B. S., Snead, D., Lee, J. J., McCaffery, J. M., Shorter, J., & Gitler, A. D. (2009). TDP-43 is intrinsically aggregation-prone, and amyotrophic lateral sclerosis-linked mutations accelerate aggregation and increase toxicity. *Journal of Biological Chemistry, 284*(30), 20329–20339. http://dx.doi.org/10.1074/jbc.M109.010264.
Kasson, P., DiMaio, F., Yu, X., Lucas-Staat, S., Krupovic, M., Schouten, S., ... Egelman, E. H. (2017). Model for a novel membrane envelope in a filamentous hyperthermophilic virus. *Elife, 6*. http://dx.doi.org/10.7554/eLife.26268.
Kato, M., Han, T. W., Xie, S., Shi, K., Du, X., Wu, L. C., ... McKnight, S. L. (2012). Cell-free formation of RNA granules: Low complexity sequence domains form dynamic fibers within hydrogels. *Cell, 149*(4), 753–767. http://dx.doi.org/10.1016/j.cell.2012.04.017.
Kato, M., & McKnight, S. L. (2018). A solid-state conceptualization of information transfer from gene to message to protein. *Annual Review of Biochemistry, 87*, 351–390. http://dx.doi.org/10.1146/annurev-biochem-061516-044700.
King, O. D., Gitler, A. D., & Shorter, J. (2012). The tip of the iceberg: RNA-binding proteins with prion-like domains in neurodegenerative disease. *Brain Research, 1462*, 61–80. http://dx.doi.org/10.1016/j.brainres.2012.01.016.
Kluth, M., Harasimowicz, S., Burkhardt, L., Grupp, K., Krohn, A., Prien, K., ... Schlomm, T. (2014). Clinical significance of different types of p53 gene alteration in surgically treated prostate cancer. *International Journal of Cancer, 135*(6), 1369–1380. http://dx.doi.org/10.1002/ijc.28784.
Kostylev, M. A., Tuttle, M. D., Lee, S., Klein, L. E., Takahashi, H., Cox, T. O., ... Strittmatter, S. M. (2018). Liquid and hydrogel phases of PrP(C) linked to conformation shifts and triggered by alzheimer's amyloid-beta oligomers. *Molecular Cell, 72*(3), 426–443 e412. http://dx.doi.org/10.1016/j.molcel.2018.10.009.
Kovachev, P. S., Banerjee, D., Rangel, L. P., Eriksson, J., Pedrote, M. M., Martins-Dinis, M., ... Sanyal, S. (2017). Distinct modulatory role of RNA in the aggregation of the tumor suppressor protein p53 core domain. *Journal of Biological Chemistry, 292*(22), 9345–9357. http://dx.doi.org/10.1074/jbc.M116.762096.
Kumar, R., Das, S., Mohite, G. M., Rout, S. K., Halder, S., Jha, N. N., ... Maji, S. K. (2018). Cytotoxic oligomers and fibrils trapped in a gel-like state of alpha-synuclein assemblies. *Angewandte Chemie International Edition in English*. http://dx.doi.org/10.1002/anie.201711854.
Kwon, I., Kato, M., Xiang, S., Wu, L., Theodoropoulos, P., Mirzaei, H., ... McKnight, S. L. (2013). Phosphorylation-regulated binding of RNA polymerase II to fibrous polymers of

low-complexity domains. *Cell, 155*(5), 1049–1060. http://dx.doi.org/10.1016/j.cell.2013.10.033.
La Spada, A. R., & Taylor, J. P. (2010). Repeat expansion disease: Progress and puzzles in disease pathogenesis. *Nature Reviews Genetics, 11*(4), 247–258. http://dx.doi.org/10.1038/nrg2748.
Lasagna-Reeves, C. A., Clos, A. L., Castillo-Carranza, D., Sengupta, U., Guerrero-Munoz, M., Kelly, B., … Kayed, R. (2013). Dual role of p53 amyloid formation in cancer; loss of function and gain of toxicity. *Biochemical and Biophysical Research Communications, 430*(3), 963–968. http://dx.doi.org/10.1016/j.bbrc.2012.11.130.
van der Lee, R., Buljan, M., Lang, B., Weatheritt, R. J., Daughdrill, G. W., Dunker, A. K., … Babu, M. M. (2014). Classification of intrinsically disordered regions and proteins. *Chemistry Review, 114*(13), 6589–6631. http://dx.doi.org/10.1021/cr400525m.
Levy, C. B., Stumbo, A. C., Ano Bom, A. P., Portari, E. A., Cordeiro, Y., Silva, J. L., et al. (2011). Co-localization of mutant p53 and amyloid-like protein aggregates in breast tumors. *The International Journal of Biochemistry & Cell Biology, 43*(1), 60–64. http://dx.doi.org/10.1016/j.biocel.2010.10.017.
Li, P., Banjade, S., Cheng, H. C., Kim, S., Chen, B., Guo, L., … Rosen, M. K. (2012). Phase transitions in the assembly of multivalent signalling proteins. *Nature, 483*(7389), 336–340. http://dx.doi.org/10.1038/nature10879.
Li, B., Ge, P., Murray, K. A., Sheth, P., Zhang, M., Nair, G., … Jiang, L. (2018). Cryo-EM of full-length alpha-synuclein reveals fibril polymorphs with a common structural kernel. *Nature Communications, 9*(1), 3609. http://dx.doi.org/10.1038/s41467-018-05971-2.
Li, Y., Lubchenko, V., Vorontsova, M. A., Filobelo, L., & Vekilov, P. G. (2012). Ostwald-like ripening of the anomalous mesoscopic clusters in protein solutions. *Journal of Physical Chemistry B, 116*(35), 10657–10664. http://dx.doi.org/10.1021/jp303316s.
Lima, L. M., Cordeiro, Y., Tinoco, L. W., Marques, A. F., Oliveira, C. L., Sampath, S., … Silva, J. L. (2006). Structural insights into the interaction between prion protein and nucleic acid. *Biochemistry, 45*(30), 9180–9187. http://dx.doi.org/10.1021/bi060532d.
Ling, S. C. (2018). Synaptic paths to neurodegeneration: The emerging role of TDP-43 and FUS in synaptic functions. *Neural Plasticity, 2018*, 8413496. http://dx.doi.org/10.1155/2018/8413496.
Lin, Y., Protter, D. S., Rosen, M. K., & Parker, R. (2015). formation and maturation of phase-separated liquid droplets by RNA-binding proteins. *Molecular Cell, 60*(2), 208–219. http://dx.doi.org/10.1016/j.molcel.2015.08.018.
Li, J. J., Zhang, J. F., Yao, S. M., Huang, H., Zhang, S., Zhao, M., et al. (2017). Decreased expression of speckle-type POZ protein for the prediction of poor prognosis in patients with non-small cell lung cancer. *Oncol Lett, 14*(3), 2743–2748. http://dx.doi.org/10.3892/ol.2017.6567.
Lomakin, A. Y., & Nadezhdina, E. S. (2010). Dynamics of nonmembranous cell components: Role of active transport along microtubules. *Biochemistry, 75*(1), 7–18.
Lorenzen, N., Nielsen, S. B., Buell, A. K., Kaspersen, J. D., Arosio, P., Vad, B. S., … Otzen, D. E. (2014). The role of stable alpha-synuclein oligomers in the molecular events underlying amyloid formation. *Journal of the American Chemical Society, 136*(10), 3859–3868. http://dx.doi.org/10.1021/ja411577t.
Lu, A., Magupalli, V. G., Ruan, J., Yin, Q., Atianand, M. K., Vos, M. R., … Egelman, E. H. (2014). Unified polymerization mechanism for the assembly of ASC-dependent inflammasomes. *Cell, 156*(6), 1193–1206. http://dx.doi.org/10.1016/j.cell.2014.02.008.
Maharana, S., Wang, J., Papadopoulos, D. K., Richter, D., Pozniakovsky, A., Poser, I., … Alberti, S. (2018). RNA buffers the phase separation behavior of prion-like RNA binding proteins. *Science, 360*(6391), 918–921. http://dx.doi.org/10.1126/science.aar7366.

Mani, R. S. (2014). The emerging role of speckle-type POZ protein (SPOP) in cancer development. *Drug Discovery Today, 19*(9), 1498–1502. http://dx.doi.org/10.1016/j.drudis.2014.07.009.
March, Z. M., King, O. D., & Shorter, J. (2016). Prion-like domains as epigenetic regulators, scaffolds for subcellular organization, and drivers of neurodegenerative disease. *Brain Research, 1647*, 9–18. http://dx.doi.org/10.1016/j.brainres.2016.02.037.
Marzahn, M. R., Marada, S., Lee, J., Nourse, A., Kenrick, S., Zhao, H., … Mittag, T. (2016). Higher-order oligomerization promotes localization of SPOP to liquid nuclear speckles. *The EMBO Journal, 35*(12), 1254–1275. http://dx.doi.org/10.15252/embj.201593169.
Matos, C. O., Passos, Y. M., do Amaral, M. J., Macedo, B., Tempone, M., Bezerra, O. C. L., et al. (2019). Liquid-liquid phase separation and aggregation of the prion protein globular domain modulated by a high-affinity DNA aptamer. *bioRxiv*. http://dx.doi.org/10.1101/659037.
McGurk, L., Gomes, E., Guo, L., Mojsilovic-Petrovic, J., Tran, V., Kalb, R. G., … Bonini, N. M. (2018). Poly(ADP-Ribose) prevents pathological phase separation of TDP-43 by promoting liquid demixing and stress granule localization. *Molecular Cell, 71*(5), 703–717 e709. http://dx.doi.org/10.1016/j.molcel.2018.07.002.
McKinley, M. P., Bolton, D. C., & Prusiner, S. B. (1983). A protease-resistant protein is a structural component of the scrapie prion. *Cell, 35*(1), 57–62.
Melo Dos Santos, N., de Oliveira, G. A. P., Ramos Rocha, M., Pedrote, M. M., Diniz da Silva Ferretti, G., Pereira Rangel, L., … Rodrigues Pereira Gimba, E. (2019). Loss of the p53 transactivation domain results in high amyloid aggregation of the Delta40p53 isoform in endometrial carcinoma cells. *Journal of Biological Chemistry, 294*(24), 9430–9439. http://dx.doi.org/10.1074/jbc.RA119.007566.
Milovanovic, D., Wu, Y., Bian, X., & De Camilli, P. (2018). A liquid phase of synapsin and lipid vesicles. *Science, 361*(6402), 604–607. http://dx.doi.org/10.1126/science.aat5671.
Molliex, A., Temirov, J., Lee, J., Coughlin, M., Kanagaraj, A. P., Kim, H. J., … Taylor, J. P. (2015). Phase separation by low complexity domains promotes stress granule assembly and drives pathological fibrillization. *Cell, 163*(1), 123–133. http://dx.doi.org/10.1016/j.cell.2015.09.015.
Monahan, Z., Ryan, V. H., Janke, A. M., Burke, K. A., Rhoads, S. N., Zerze, G. H., … Fawzi, N. L. (2017). Phosphorylation of the FUS low-complexity domain disrupts phase separation, aggregation, and toxicity. *The EMBO Journal, 36*(20), 2951–2967. http://dx.doi.org/10.15252/embj.201696394.
Mullard, A. (2019). Biomolecular condensates pique drug discovery curiosity. *Nature Reviews Drug Discovery*. http://dx.doi.org/10.1038/d41573-019-00069-w.
Murakami, T., Qamar, S., Lin, J. Q., Schierle, G. S., Rees, E., Miyashita, A., … St George-Hyslop, P. (2015). ALS/FTD mutation-induced phase transition of FUS liquid droplets and reversible hydrogels into irreversible hydrogels impairs RNP granule function. *Neuron, 88*(4), 678–690. http://dx.doi.org/10.1016/j.neuron.2015.10.030.
Murray, D. T., Kato, M., Lin, Y., Thurber, K. R., Hung, I., McKnight, S. L., et al. (2017). Structure of FUS protein fibrils and its relevance to self-assembly and phase separation of low-complexity domains. *Cell, 171*(3), 615–627 e616. http://dx.doi.org/10.1016/j.cell.2017.08.048.
Nadezhdina, E. S., Lomakin, A. J., Shpilman, A. A., Chudinova, E. M., & Ivanov, P. A. (2010). Microtubules govern stress granule mobility and dynamics. *Biochimica et Biophysica Acta, 1803*(3), 361–371. http://dx.doi.org/10.1016/j.bbamcr.2009.12.004.
Nagai, Y., Kojima, T., Muro, Y., Hachiya, T., Nishizawa, Y., Wakabayashi, T., et al. (1997). Identification of a novel nuclear speckle-type protein, SPOP. *FEBS Letters, 418*(1–2), 23–26. http://dx.doi.org/10.1016/s0014-5793(97)01340-9.

Nagar, B., Hantschel, O., Young, M. A., Scheffzek, K., Veach, D., Bornmann, W., ... Kuriyan, J. (2003). Structural basis for the autoinhibition of c-Abl tyrosine kinase. *Cell, 112*(6), 859–871.

Nizynski, B., Dzwolak, W., & Nieznanski, K. (2017). Amyloidogenesis of tau protein. *Protein Science, 26*(11), 2126–2150. http://dx.doi.org/10.1002/pro.3275.

Nott, T. J., Petsalaki, E., Farber, P., Jervis, D., Fussner, E., Plochowietz, A., ... Baldwin, A. J. (2015). Phase transition of a disordered nuage protein generates environmentally responsive membraneless organelles. *Molecular Cell, 57*(5), 936–947. http://dx.doi.org/10.1016/j.molcel.2015.01.013.

Ogrodnik, M., Salmonowicz, H., Brown, R., Turkowska, J., Sredniawa, W., Pattabiraman, S., ... Kaganovich, D. (2014). Dynamic JUNQ inclusion bodies are asymmetrically inherited in mammalian cell lines through the asymmetric partitioning of vimentin. *Proceedings of the National Academy of Sciences of the U S A, 111*(22), 8049–8054. http://dx.doi.org/10.1073/pnas.1324035111.

Pak, C. W., Kosno, M., Holehouse, A. S., Padrick, S. B., Mittal, A., Ali, R., ... Rosen, M. K. (2016). Sequence determinants of intracellular phase separation by complex coacervation of a disordered protein. *Molecular Cell, 63*(1), 72–85. http://dx.doi.org/10.1016/j.molcel.2016.05.042.

Patel, A., Lee, H. O., Jawerth, L., Maharana, S., Jahnel, M., Hein, M. Y., ... Alberti, S. (2015). A liquid-to-solid phase transition of the ALS protein FUS accelerated by disease mutation. *Cell, 162*(5), 1066–1077. http://dx.doi.org/10.1016/j.cell.2015.07.047.

Patel, A., Malinovska, L., Saha, S., Wang, J., Alberti, S., Krishnan, Y., et al. (2017). ATP as a biological hydrotrope. *Science, 356*(6339), 753–756. http://dx.doi.org/10.1126/science.aaf6846.

Pedrote, M. M., de Oliveira, G. A. P., Felix, A. L., Mota, M. F., Marques, M. A., Soares, I. N., ... Silva, J. L. (2018). Aggregation-primed molten globule conformers of the p53 core domain provide potential tools for studying p53C aggregation in cancer. *Journal of Biological Chemistry, 293*(29), 11374–11387. http://dx.doi.org/10.1074/jbc.RA118.003285.

Perakis, F., Camisasca, G., Lane, T. J., Spah, A., Wikfeldt, K. T., Sellberg, J. A., ... Nilsson, A. (2018). Coherent X-rays reveal the influence of cage effects on ultrafast water dynamics. *Nature Communications, 9*(1), 1917. http://dx.doi.org/10.1038/s41467-018-04330-5.

Prats, C., Graham, T. E., & Shearer, J. (2018). The dynamic life of the glycogen granule. *Journal of Biological Chemistry, 293*(19), 7089–7098. http://dx.doi.org/10.1074/jbc.R117.802843.

Rangel, L. P., Ferretti, G. D. S., Costa, C. L., Andrade, S., Carvalho, R. S., Costa, D. C. F., et al. (2019). p53 reactivation with induction of massive apoptosis-1 (PRIMA-1) inhibits amyloid aggregation of mutant p53 in cancer cells. *Journal of Biological Chemistry*. http://dx.doi.org/10.1074/jbc.RA118.004671.

Reddy, D. V., & Kinsey, V. E. (1960). Composition of the vitreous humor in relation to that of plasma and aqueous humors. *Archives of Ophthalmology, 63*, 715–720.

Ryan, V. H., Dignon, G. L., Zerze, G. H., Chabata, C. V., Silva, R., Conicella, A. E., ... Fawzi, N. L. (2018). Mechanistic view of hnRNPA2 low-complexity domain structure, interactions, and phase separation altered by mutation and arginine methylation. *Molecular Cell, 69*(3), 465–479 e467. http://dx.doi.org/10.1016/j.molcel.2017.12.022.

Sachse, C. (2015). Single-particle based helical reconstruction—how to make the most of real and Fourier space. *AIMS Biophysics, 2*(2), 219–244. http://dx.doi.org/10.3934/biophy.2015.2.219.

Safari, M. S., Wang, Z., Tailor, K., Kolomeisky, A. B., Conrad, J. C., & Vekilov, P. G. (2019). Anomalous dense liquid condensates host the nucleation of tumor suppressor p53 fibrils. *iScience, 12*, 342–355. http://dx.doi.org/10.1016/j.isci.2019.01.027.

Scheckel, C., & Aguzzi, A. (2018). Prions, prionoids and protein misfolding disorders. *Nature Reviews Genetics, 19*(7), 405–418. http://dx.doi.org/10.1038/s41576-018-0011-4.

Schmidt, H. B., & Gorlich, D. (2015). Nup98 FG domains from diverse species spontaneously phase-separate into particles with nuclear pore-like permselectivity. *Elife, 4*. http://dx.doi.org/10.7554/eLife.04251.

Schwartz, J. C., Wang, X., Podell, E. R., & Cech, T. R. (2013). RNA seeds higher-order assembly of FUS protein. *Cell Reports, 5*(4), 918–925. http://dx.doi.org/10.1016/j.celrep.2013.11.017.

Sengupta, S., Maji, S. K., & Ghosh, S. K. (2017). Evidence of a prion-like transmission of p53 amyloid in *Saccharomyces cerevisiae*. *Molecular and Cellular Biology*. http://dx.doi.org/10.1128/MCB.00118-17.

Shin, Y., Berry, J., Pannucci, N., Haataja, M. P., Toettcher, J. E., & Brangwynne, C. P. (2017). Spatiotemporal control of intracellular phase transitions using light-activated optoDroplets. *Cell, 168*(1–2), 159–171 e114. http://dx.doi.org/10.1016/j.cell.2016.11.054.

Silva, J. L., Cino, E. A., Soares, I. N., Ferreira, V. F., & A P de Oliveira, G. (2018). Targeting the prion-like aggregation of mutant p53 to combat cancer. *Accounts of Chemical Research, 51*(1), 181–190. http://dx.doi.org/10.1021/acs.accounts.7b00473.

Silva, J. L., & Cordeiro, Y. (2016). The "Jekyll and Hyde" actions of nucleic acids on the prion-like aggregation of proteins. *Journal of Biological Chemistry, 291*(30), 15482–15490. http://dx.doi.org/10.1074/jbc.R116.733428.

Silva, J. L., De Moura Gallo, C. V., Costa, D. C., & Rangel, L. P. (2014). Prion-like aggregation of mutant p53 in cancer. *Trends in Biochemical Sciences, 39*(6), 260–267. http://dx.doi.org/10.1016/j.tibs.2014.04.001.

Silva, J. L., Oliveira, A. C., Vieira, T. C., de Oliveira, G. A., Suarez, M. C., & Foguel, D. (2014). High-pressure chemical biology and biotechnology. *Chemistry Review, 114*(14), 7239–7267. http://dx.doi.org/10.1021/cr400204z.

Silva, J. L., & Weber, G. (1993). Pressure stability of proteins. *Annual Review of Physical Chemistry, 44*, 89–113.

Soragni, A., Janzen, D. M., Johnson, L. M., Lindgren, A. G., Thai-Quynh Nguyen, A., Tiourin, E., … Eisenberg, D. S. (2016). A designed inhibitor of p53 aggregation rescues p53 tumor suppression in ovarian carcinomas. *Cancer Cell, 29*(1), 90–103. http://dx.doi.org/10.1016/j.ccell.2015.12.002.

Strom, A. R., Emelyanov, A. V., Mir, M., Fyodorov, D. V., Darzacq, X., & Karpen, G. H. (2017). Phase separation drives heterochromatin domain formation. *Nature, 547*(7662), 241–245. http://dx.doi.org/10.1038/nature22989.

Su, X., Ditlev, J. A., Hui, E., Xing, W., Banjade, S., Okrut, J., … Vale, R. D. (2016). Phase separation of signaling molecules promotes T cell receptor signal transduction. *Science, 352*(6285), 595–599. http://dx.doi.org/10.1126/science.aad9964.

Tayeb-Fligelman, E., Tabachnikov, O., Moshe, A., Goldshmidt-Tran, O., Sawaya, M. R., Coquelle, N., … Landau, M. (2017). The cytotoxic *Staphylococcus aureus* PSMalpha3 reveals a cross-alpha amyloid-like fibril. *Science, 355*(6327), 831–833. http://dx.doi.org/10.1126/science.aaf4901.

Tompa, P., Davey, N. E., Gibson, T. J., & Babu, M. M. (2014). A million peptide motifs for the molecular biologist. *Molecular Cell, 55*(2), 161–169. http://dx.doi.org/10.1016/j.molcel.2014.05.032.

Toretsky, J. A., & Wright, P. E. (2014). Assemblages: Functional units formed by cellular phase separation. *The Journal of Cell Biology, 206*(5), 579–588. http://dx.doi.org/10.1083/jcb.201404124.

Ulbrich, S., Janning, P., Seidel, R., Matschke, J., Gonsberg, A., Jung, S., … Tatzelt, J. (2018). Alterations in the brain interactome of the intrinsically disordered N-terminal domain of the cellular prion protein (PrPC) in Alzheimer's disease. *PLoS One, 13*(5), e0197659. http://dx.doi.org/10.1371/journal.pone.0197659.

Vance, C., Rogelj, B., Hortobagyi, T., De Vos, K. J., Nishimura, A. L., Sreedharan, J., … Shaw, C. E. (2009). Mutations in FUS, an RNA processing protein, cause familial amyotrophic lateral sclerosis type 6. *Science, 323*(5918), 1208–1211. http://dx.doi.org/10.1126/science.1165942.

Vanderweyde, T., Yu, H., Varnum, M., Liu-Yesucevitz, L., Citro, A., Ikezu, T., … Wolozin, B. (2012). Contrasting pathology of the stress granule proteins TIA-1 and G3BP in tauopathies. *Journal of Neuroscience, 32*(24), 8270–8283. http://dx.doi.org/10.1523/JNEUROSCI.1592-12.2012.

Vendruscolo, M., Knowles, T. P., & Dobson, C. M. (2011). Protein solubility and protein homeostasis: A generic view of protein misfolding disorders. *Cold Spring Harb Perspect Biol, 3*(12). http://dx.doi.org/10.1101/cshperspect.a010454.

Venter, J. C., Adams, M. D., Myers, E. W., Li, P. W., Mural, R. J., Sutton, G. G., … Zhu, X. (2001). The sequence of the human genome. *Science, 291*(5507), 1304–1351. http://dx.doi.org/10.1126/science.1058040.

Vernon, R. M., Chong, P. A., Tsang, B., Kim, T. H., Bah, A., Farber, P., … Forman-Kay, J. D. (2018). Pi-Pi contacts are an overlooked protein feature relevant to phase separation. *Elife, 7*. http://dx.doi.org/10.7554/eLife.31486.

Vieira, T. C., Cordeiro, Y., Caughey, B., & Silva, J. L. (2014). Heparin binding confers prion stability and impairs its aggregation. *The FASEB Journal, 28*(6), 2667–2676. http://dx.doi.org/10.1096/fj.13-246777.

Vieira, T. C. R. G., Reynaldo, D. P., Gomes, M. P. B., Almeida, M. S., Cordeiro, Y., & Silva, J. L. (2011). Heparin binding by murine recombinant prion protein leads to transient aggregation and formation of RNA-resistant species. *Journal of the American Chemical Society, 133*, 334–344.

Wang, J., Choi, J. M., Holehouse, A. S., Lee, H. O., Zhang, X., Jahnel, M., … Hyman, A. A. (2018). A molecular grammar governing the driving forces for phase separation of prion-like RNA binding proteins. *Cell, 174*(3), 688–699 e616. http://dx.doi.org/10.1016/j.cell.2018.06.006.

Wang, G., & Fersht, A. R. (2015). Propagation of aggregated p53: Cross-reaction and coaggregation vs. seeding. *Proceedings of the National Academy of Sciences of the U S A, 112*(8), 2443–2448. http://dx.doi.org/10.1073/pnas.1500262112.

Wang, G., & Fersht, A. R. (2017). Multisite aggregation of p53 and implications for drug rescue. *Proceedings of the National Academy of Sciences of the U S A, 114*(13), E2634–E2643. http://dx.doi.org/10.1073/pnas.1700308114.

Weber, G. (1975). Energetics of ligand binding to proteins. *Advances in Protein Chemistry, 29*, 1–83.

Weber, G. (1992). *Protein interactions*. New York: Chapman and Hall.

Wilcken, R., Wang, G., Boeckler, F. M., & Fersht, A. R. (2012). Kinetic mechanism of p53 oncogenic mutant aggregation and its inhibition. *Proceedings of the National Academy of Sciences of the U S A, 109*(34), 13584–13589. http://dx.doi.org/10.1073/pnas.1211550109.

Will, C. L., & Luhrmann, R. (2011). Spliceosome structure and function. *Cold Spring Harb Perspect Biol, 3*(7). http://dx.doi.org/10.1101/cshperspect.a003707.

Wilson, E. B. (2018). *The physical basis of life*. Forgotten Books.

Woodruff, J. B., Ferreira Gomes, B., Widlund, P. O., Mahamid, J., Honigmann, A., & Hyman, A. A. (2017). The centrosome is a selective condensate that nucleates microtubules by concentrating tubulin. *Cell, 169*(6), 1066–1077 e1010. http://dx.doi.org/10.1016/j.cell.2017.05.028.

Wulf, M. A., Senatore, A., & Aguzzi, A. (2017). The biological function of the cellular prion protein: An update. *BMC Biology, 15*(1), 34. http://dx.doi.org/10.1186/s12915-017-0375-5.

Xiang, S., Kato, M., Wu, L. C., Lin, Y., Ding, M., Zhang, Y., ... McKnight, S. L. (2015). The LC domain of hnRNPA2 adopts similar conformations in hydrogel polymers, liquid-like droplets, and nuclei. *Cell, 163*(4), 829–839. http://dx.doi.org/10.1016/j.cell.2015.10.040.

Xu, J., Reumers, J., Couceiro, J. R., De Smet, F., Gallardo, R., Rudyak, S., ... Schymkowitz, J. (2011). Gain of function of mutant p53 by coaggregation with multiple tumor suppressors. *Nature Chemical Biology,* 7(5), 285–295. http://dx.doi.org/10.1038/nchembio.546.

Yang-Hartwich, Y., Soteras, M. G., Lin, Z. P., Holmberg, J., Sumi, N., Craveiro, V., ... Mor, G. (2015). p53 protein aggregation promotes platinum resistance in ovarian cancer. *Oncogene, 34*(27), 3605–3616. http://dx.doi.org/10.1038/onc.2014.296.

Zeng, M., Shang, Y., Araki, Y., Guo, T., Huganir, R. L., & Zhang, M. (2016). Phase transition in postsynaptic densities underlies formation of synaptic complexes and synaptic plasticity. *Cell, 166*(5), 1163–1175 e1112. http://dx.doi.org/10.1016/j.cell.2016.07.008.

Zhang, H., Elbaum-Garfinkle, S., Langdon, E. M., Taylor, N., Occhipinti, P., Bridges, A. A., ... Gladfelter, A. S. (2015). RNA controls PolyQ protein phase transitions. *Molecular Cell, 60*(2), 220–230. http://dx.doi.org/10.1016/j.molcel.2015.09.017.

Zhu, L., & Brangwynne, C. P. (2015). Nuclear bodies: The emerging biophysics of nucleoplasmic phases. *Current Opinion in Cell Biology, 34,* 23–30. http://dx.doi.org/10.1016/j.ceb.2015.04.003.

CHAPTER TEN

Cytotoxic species in amyloid-associated diseases: Oligomers or mature fibrils

Mohammad Khursheed Siddiqi, Sadia Malik, Nabeela Majid, Parvez Alam and Rizwan Hasan Khan*

Interdisciplinary Biotechnology Unit, Aligarh Muslim University, Aligarh, India

*Corresponding author: E-mail: rizwanhkhan1@gmail.com

Contents

Advances in Protein Chemistry and Structural Biology, Volume 118
ISSN 1876-1623
https://doi.org/10.1016/bs.apcsb.2019.06.001

© 2020 Elsevier Inc.
All rights reserved.

Abstract

Amyloid diseases especially, Alzheimer's disease (AD), is characterized by an imbalance between the production and clearance of amyloid-β (Aβ) species. Amyloidogenic proteins or peptides can transform structurally from monomers into β-stranded fibrils via multiple oligomeric states. Among various amyloid species, structured oligomers are proposed to be more toxic than fibrils; however, the identification of amyloid oligomers has been challenging due to their heterogeneous and metastable nature. Multiple techniques have recently helped in better understanding of oligomer's assembly details and structural properties. Moreover, some progress on elucidating the mechanisms of oligomer-triggered toxicity has been made. Based on the collection of current findings, there is growing consensus that control of toxic amyloid oligomers could be a valid approach to regulate amyloid-associated toxicity, which could advance development of new diagnostics and therapeutics for amyloid-related diseases. In this review, we have described the recent scenario of amyloid diseases with a great deal of information about the recent understanding of oligomers' assembly, structural properties, and toxicity. Also comprehensive details have been provided to differentiate the degree of toxicity associated with prefibrillar aggregates.

1. Introduction

Protein misfolding is a deleterious process for living organisms due to its involvement in the prevention of protein from adopting its functional state. The misfolded protein aberrantly self-assemble in an infinitely propagating fashion and form large molecular weight aggregates that acts as a key pathognomonic entity of various neurodegenerative diseases affecting different tissues with diverse symptoms (Alam, Siddiqi, Chturvedi, & Khan, 2017; Chaturvedi, Siddiqi, Alam, & Khan, 2016; Dobson, 2003; Siddiqi, Alam, Chaturvedi, Shahein, & Khan, 2017) (Table 1). Protein aggregates can be classified into amyloid fibrils, amorphous aggregate and amyloid oligomers. The core of systemic diseases such as diabetes type II and immunoglobulin light-chain amyloidosis as well as localized amyloid diseases particularly neurodegenerative disorders such as Alzheimer's, Parkinson's, Huntington's, motor neuron diseases, different ataxias and prion-related diseases, is the ordered aggregation of proteins to amyloid fibrils (Alam et al., 2019; Almeida, Takahashi, & Gouras, 2006; Morris, Watzky, & Finke, 2009; Ross & Poirier, 2004; Siddiqi, Chaturvedi et al., 2018; Uversky, 2011). Each disease is characterized by the amyloid deposition of different protein precursors. The lethal consequences of amyloid deposition is widely denoted by the term 'amyloidosis' and the process of self-association of peptide or protein molecules into dimers, oligomers and

Table 1 Amyloid disease and proteins involved.

Disease	Protein
Alzheimer's disease	Amyloid beta peptide/tau (Dobson, 2001)
Parkinson's disease	Alpha-Synuclein (Dobson, 2001)
Familial amyloidosis	Transthyretin/lysozyme (Dobson, 2001)
Scrapie/Creutzfeldt–Jakob disease	Prion protein (Dobson, 2001)
Cataracts	Crystallins (Dobson, 2001)
Cancer	p53 (Dobson, 2001)
Amyotrophic lateral sclerosis	Superoxide dismutase 1 (Andersen, 2006)
Finnish hereditary amyloidosis	Fragments of gelsolin (Ardalan, Shoja, & Kiuru-Enari, 2007) mutants
Injection-localized amyloidosis	Insulin (Shikama et al., 2010)
Cutaneous lichen amyloidosis	Keratins (Oiso, Yudate, Kawara, & Kawada, 2009)
Corneal amylodosis associated with trichiasis	Lactoferrin (Araki-Sasaki et al., 2005)
Pituitary prolactinoma	Prolactin (Fleseriu et al., 2006)
Medullary carcinoma of the thyroid	Calcitonin (Costante et al., 2007)
Type II diabetes	Amylin, also called islet amyloid polypeptide (IAPP) (Konarkowska, Aitken, Kistler, Zhang, & Cooper, 2006)
Icelandic hereditary cerebral amyloid angiopathy	Mutant of cystatin C (Ao, 2006)

subsequently to mature fibrillar amyloid aggregates is known as Amyloidogenesis (Ngoungoure, Schluesener, Moundipa, & Schluesener, 2015). During amyloidogenesis, toxic protein fibrils that arise from misfolded monomers get deposited in tissue or cells and leads to serious pathological consequences. The deposition in neuronal cells leads to neuron degeneration which manifests symptoms such as memory loss and dementia in Neurodegenerative diseases. In fact, Human amyloid diseases constitutes a group of pathologies derived from 27 different proteins such as Amyloid β, Tau, Huntingtin, α-Synuclein and prion protein etc., identified by the International Nomenclature Committee (Sipe et al., 2010) which are involved in neurodegenerative diseases, like Alzheimer's disease (AD), Parkinson's disease, Huntington disease, Cruetzfelt-Jakob disease that have been reported to have a characteristic deposition of abnormal protein aggregates in the brain **TABLE**. Some of these amyloid diseases are rare, whereas others, such as Alzheimer's disease, are very common within the aging population and are considered a serious health care concern. A lot of research is under progress to develop therapeutic strategies for slowing down the progression of neurodegenerative diseases since only symptomatic treatments are available at this moment (Chiti & Dobson, 2017; Dobson, 2017; Usmani et al., 2017; Usmani, Kumar, Bhalla, Kumar, & Raghava, 2018).

The toxicity of amyloids is usually categorized as either loss of function or gain of function (Winklhofer, Tatzelt, & Haass, 2008). Protein which are important for the normal function of the cell/tissue/organ, if aggregated, then the critical functions associated with these proteins get hampered (e.g., p53); this is termed as loss of function toxicity whereas in case of gain of function toxicity, aggregate itself shows toxicity and leads to the cell or tissue degeneration (e.g., beta amyloid). Initially, it was hypothesized that fibrillar species, reported to be present in the region of brain of the patient suffering from amyloid diseases, were the key toxic agents responsible for the pathogenicity associated with the amyloid diseases. But it was true only for some systemic amyloidosis conditions, in which abundant fibrils are deposited in the vital organs. However, in other amyloid disorders including most of the neurodegenerative diseases, there is a lack of direct relation between the extent of protein fibrillation and the severity of the diseases. This suggested that some other form is responsible for the pathogenesis of neurodegenerative diseases (disease condition associated with degeneration of neurons arises due to the deposition of fibrils). It becomes increasingly evident from the analysis of human patients that amyloid intermediate species, particularly oligomeric species, play an important role in the progression and severity of these disease (Haass & Selkoe, 2007; Kirkitadze, Bitan, & Teplow, 2002). This conclusion is also consistent with the hypothesis of the effect of pre-fibrillar aggregates, even when the aggregates are not related with any known disorders. This chapter will focus on the importance of the biochemical and biophysical features of oligomers and mature fibrils.

2. Protein folding misfolding and aggregation

Proteins, carbohydrates, nucleic acids and lipids are the four key macromolecules that compose the living system. All these macromolecules perform diverse and vital roles within the body of the living organisms. The life of every organism depends upon thousands of proteins that are large complex biomolecules synthesized on ribosomes within the cells of the living organisms and are required for the structure, function, and regulation of the body's tissues and organ (Dobson, 2003). Being very diverse macromolecule, proteins vary in size from small peptides to large multimers composed of amino acids that are building blocks of protein and can be distinguished from each other based on their constituent polymeric sequence of amino acids (Chothia & Lesk, 1986). However despite their molecular diversity, protein possess a common property, that is to fold correctly into native conformation by rapid, spontaneous and complex folding pathways in order to acquire specific function. The order of the amino acids create a relatively flexible unique polypeptide chain and determines the primary structure of the protein. It has been reported that all the necessary information to specify protein's three-dimensional structure to attain functional active state is carried by the linear sequence of the polypeptide chain (Anfinsen, 1973). The process by which a specific peptide or protein assumes its functional shape or conformation is known as Protein Folding (Fig. 1). In some cases, protein folding starts immediately in a co-translational manner, where the nascent polypeptide chain folds while still attached to the ribosomes in contrast other proteins follow a folding pathway in the endoplasmic reticulum after the translation process. Folding

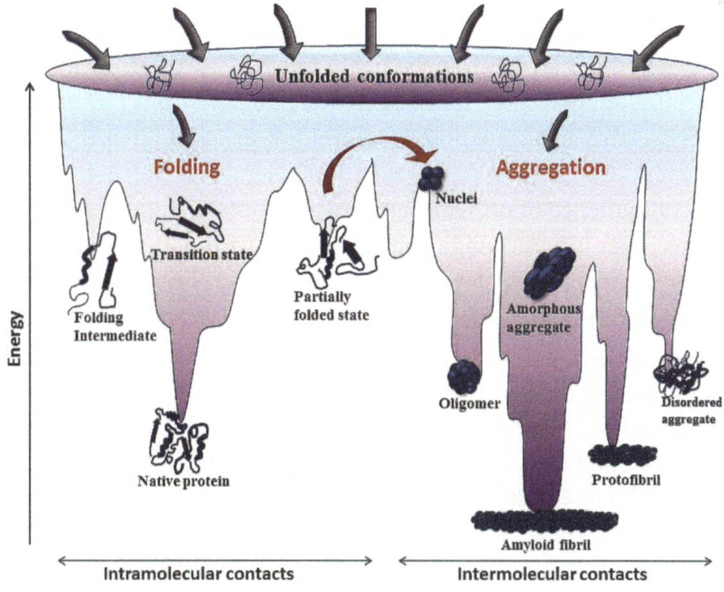

Fig. 1 The energy landscape of protein folding and protein aggregation phenomenon.

involves a complex phenomenon of molecular recognition that depends on the cooperation of a large number of relatively weak, noncovalent interactions involving thousands of atoms. The major forces driving protein folding are hydrophobic (Dill, 1990; Kauzmann, 1959), electrostatic interaction (Avbelj & Moult, 1995; Brant & Flory, 1965; Perutz, 1978), van der Waals interactions (Klapper, 1971; Richards, 1977); peptide hydrogen bonds (Schellman, 1954; Sticke, Presta, Dill, & Rose, 1992) and peptide solvation (Makhatadze & Privalov, 1993; Wolfenden, 1978).

Thermodynamically, native state of a protein tends to be the most stabilized state and the propensity to undergo aberrant aggregation is minimal because most of its hydrophobic moieties and a large portion of the backbone are sequestered inside the protein (Fig. 1). Also the undesired intermolecular association between folded protein molecules, such as the peripheral strands of β-sheets, are protected by structural adaptations (Richardson & Richardson, 2002). But various triggering factors such as mutation, aging, glycation and environmental factors like increased temperature, low and high pH, local increase of protein concentration, agitation and oxidative stress may perturb the protein to attain correctly folded conformation and result in unfolding or partial misfolding that is associated with the tendency to aggregate (Uversky, 2014). It occurs because most of the hydrophobic groups and backbone amide and carbonyl groups, normally buried in the interior of the protein, become solvent exposed. Also the intermediates formed during the folding process contain surface exposed hydrophobic amino acids which are more prone to interact with other such molecules to form protein aggregate. Thus, protein aggregation is a process resulting from nonspecific amalgamation of misfolded proteins (Jucker & Walker, 2013).

As a consequence of protein aggregation processes highly heterogeneous species are observed. Although the conversion of monomeric soluble peptides or proteins into an array of aggregate structures is the overall reaction that occurs during aggregation process, but standard terminology to describe the specific aggregated species has not yet developed. Protein aggregates are usually either classified as amyloid fibrils that are the ordered linear aggregates with repetitive cross-β structure of hundreds to thousands of monomers (a single protein or polypeptide chain) that can be observed both in vivo and in vitro, amorphous aggregates that are disordered aggregates, or prefibrillar aggregates. Through recent in vivo and in vitro discoveries smaller, prefibrillar aggregates have now emerged as the primary toxic species in several of these diseases (Bucciantini et al., 2002; Cleary et al., 2005;

Glabe, 2008; Kayed et al., 2003). Prefibrillar aggregates include the prefibrillar oligomers (globular aggregates lacking the ordered cross-stacked β-sheet structure) and fibrillar oligomers/protofibrils (aggregates with the cross-stacked β-sheet structure). The term oligomer refers to aggregates up to a certain size, e.g. dimer, dodecamer or even higher. However, several possible definitions are given in literature by which oligomers can be described as small aggregates exhibiting different structure and lower growth rate with respect to fibrils. They appear as intermediates stage and some time they directly participate in the in the fibrillation process known as on-pathway oligomers and some time they don't known as off-pathway oligomers. Even some off-pathway oligomers inhibits the mature fibril formation.

In all amyloid diseases, the basic mechanism of fibril formation is same. According to the fibril formation model, native proteins undergo conformational change and misfolds. After reaching a critical concentration, these misfolded proteins forms oligomers that further result in the formation of protofibrils and finally culminate in mature fibrils. Based on the kinetic relationship between oligomers and amyloid fibrils, oligomeric intermediates are referred to as prefibrillar or on-pathway oligomers and off-pathway oligomers (Ehrnhoefer et al., 2008; Wu et al., 2010). The on-pathway oligomers are capable of conversion into mature fibrils without being dissociating into monomers whereas off-pathway oligomers are incapable of such conversion and end up in forming fibrils via partially unfolded monomers. The on-pathway and off-pathway oligomers can be distinguished according to their ability to accelerate or delay amyloid fibril formation upon addition of monomers respectively (Bemporad & Chiti, 2012; Gosal et al., 2005; Souillac, Uversky, & Fink, 2003). In general, amyloid oligomers are non-monomeric aggregates that are soluble in nature and are inherently toxic (Chiti & Dobson, 2006; Walsh & Selkoe, 2007). Many of them are highly cytotoxic and are believed to play an important role in various protein aggregation-related diseases. In general, Cytotoxicity of amyloid oligomers is correlated with their size and hydrophobic surface exposure. Similar in structure to amyloid oligomers several examples of functional protein oligomers including milk colloids and oligomers of small heat shock proteins are also reported (Haslbeck & Vierling, 2019; Redwan, Xue, Almehdar, & Uversky, 2015; Sudnitsyna, Mymrikov, Seit-Nebi, & Gusev, 2012). In the process of aggregation these oligomeric species appear as intermediate or final product that lacks one or more of the hallmarks of fibrillar structure (Glabe, 2006; Jucker & Walker, 2013; Lesne, 2013; Verma, Vats, & Taneja, 2015). They are highly heterogeneous in size, stability and morphology and tend to be highly structurally variable. Over the past 10–15 years, study on protein oligomers have acquired increasing importance since it has been pointed out that the soluble oligomers and fibrillar aggregates are cytotoxic and appears to be responsible for the onset of neurodegeneration, in contrast fibrils seems to be relatively harmless until and unless a very high level of assembly is formed to cause organ malfunction and subsequent failure (Pepys et al., 1993). The major properties of amyloid oligomers will be discussed in this review.

3. Structural features of amyloid fibrils and oligomers

Amyloid intermediates have been reported to be present in the aggregation path of many disease associated proteins (in *in vitro* and/or *in vivo*) such as beta amyloid, amylin, prion, alpha synuclein, transthyretin protein, and others (Stefani, 2004). Similar intermediate species have also been found in aggregation path of protein which is not associated with amyloid disease. However these entities finally result in the formation of mature amyloid fibrils (Stefani, 2012).

Much information has recently been gained on the hierarchical formation of amyloid fibrils from structurally more simple precursors through a number of steps (Fig. 2). Amyloid fibrils are the consequences of ordered self-assembly of protein molecules, characterized by the presence of cross beta sheet rich structure in which beta sheet runs perpendicular to the fibril axis (Eisenberg & Jucker, 2012; Tycko, 2004). Morphologically, amyloids are unbranched and threadlike structures, just a few nanometers in diameter, and are composed of several protofilaments that wrap around each other (Otzen, 2013).

In addition to mature fibrils, understanding the structural information of oligomers is also significant. These species are frequently observed to accumulate during the process of fibrils formation. Although trapping these intermediate species during the path of amyloid fibrillation is challenging, but study of oligomeric forms has opened up the possibility of gaining insights into the nature and structure of these species. Various experimental procedures have been used to isolate these species (Bitan, Lomakin, & Teplow, 2001).

The very early species in the path of amyloid formation are found to be dimers, trimers, tetramers (collectively termed as oligomers) (Podlisny et al., 1995). Oligomers may have variable sizes, with an average of approximately

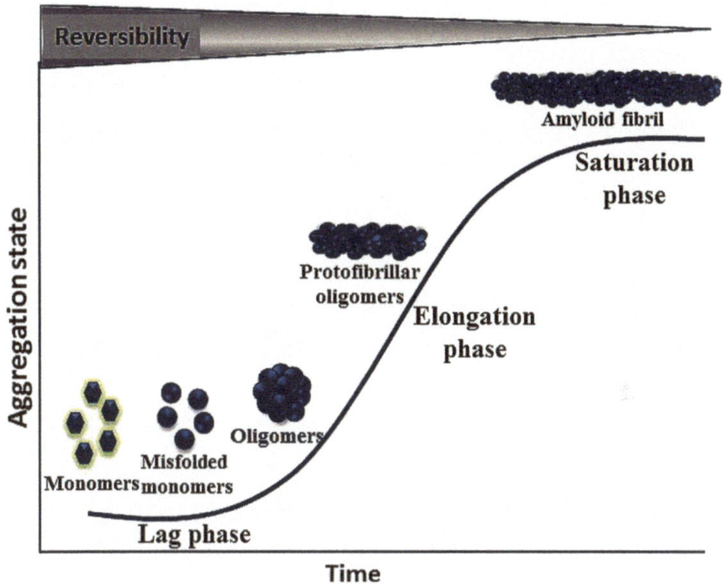

Fig. 2 Schematic representation of protein fibrillation process involving different intermediate species.

25–30 monomers molecules. These oligomers are able to interact with membrane, causes cell death by permeabilizing the membrane. The oligomeric form of αS possesses beta sheet structure formed in between the monomeric and mature fibrils of αS. Oligomers range in size from 20 to >50 kDa. Further addition of monomeric unit to oligomers results in the formation of bead-like structures up to 200 nm in length (4–11 nm in diameter) called protofibrils (Wong, Sheehan, & Lieber, 1997). Study on insulin and α-synuclein fibril suggests that three intertwining protofibrils constitute mature insulin fibril whereas one for α-synuclein fibril respectively. Each protofibril is assumed to consists of two intertwining proto-filaments (Vestergaard et al., 2007).

Another striking difference between the αS oligomer and mature fibrils is that mature fibrils consist of parallel beta sheet arrangement whereas anti-parallel beta sheet dominates in oligomeric form, however, the core residues of both forms are almost similar (Kim et al., 2009). Importantly, further study also suggested that information obtained for αS oligomer such as pathogenicity, physico-chemical properties possess high degree of similarity with oligomeric species obtained from different protein and peptides associated with amyloid diseases. Additionally, variable parallel and anti-parallel beta sheets are also observed for other amyloidogenic proteins such as Aβ-peptide, lysozyme, β2-microglobulin and a prion-related peptide (Burdick et al., 1992; Forloni et al., 1993; Wadai et al., 2005; Westermark, 2005). Further study shows that oligomer which differ in size also differ in their beta sheet content, larger size possess more beta sheet structure (Glabe, 2008). The similarities in the architecture of the fibrillar and the oligomeric forms, despite the differences in the organization of their β-sheet structure, suggest, that similar types of interactions to those that stabilize amyloid fibrils are likely to be responsible for the initial acquisition of cross-β structure in the oligomeric species.

4. Techniques used to characterize oligomer and mature fibrils

Multiple traditional in vitro techniques have allowed a deep analysis of the structural properties of different types of protein aggregates.

4.1 Electrophoresis

In electrophoresis, the separation of molecules take place on the basis of size, shape, and charge hence electrophoresis can be used for characterizing protein oligomers provided that they are soluble and stable to the separation conditions and are analyzed in pure form or detectable when present in complex mixtures such as biofluids. For example, Aß monomers and oligomers are readily separated after covalent cross-linking via polyacrylamide gel electrophoresis (PAGE) (Levine Iii, 2004). SDS-PAGE (Sodium Dodecyl Sulfate-PAGE) can't be used straightforward for the same because in some studies it has been shown to be responsible for the induction of oligomerization (Bitan, Fradinger, Spring, & Teplow, 2005). Capillary electrophoresis (CE) can also be used to monitor the changes in the concentration of both monomers and early, still soluble oligomers over time during kinetic analysis as demonstrated for both Aß (Pedersen, Ostergaard, Rozlosnik, Gammelgaard, & Heegaard, 2011) and insulin oligomers (Pryor, Kotarek, Moss, & Hestekin, 2011). Alternatively, if CE is coupled with specialized detectors, e.g., laser-induced fluorescence anisotropy (LIFA) detectors, an online assessment of the size/shape of the separated Aß species can be done simultaneously (Picou et al., 2012). Although CE is convenient for monitoring the contents of small biological samples (including micro dialysis fluids) over time, due to its very low sample volume consumption but generally electrophoretic methods are not able to provide much structural information about separated analytes. In native gel electrophoresis (having no anionic detergent), aggregates are more likely to be stable than in denaturing electrophoresis and this idea has been exploited to analyze Aß, yeast prion aggregates (Bagriantsev, Kushnirov, & Liebman, 2006) and oligomer formation of neuroserpin variants (Miranda et al., 2008). Native gel electrophoresis is not reliable because in the absence of anionic detergent the separation is dependent on a combination of analyte charge, shape and mass, and it cannot separate oligomers with an identical size-to charge ratio and thus for strongly complexed aggregates denaturing gel electrophoresis found to be useful (Bullock, 1993; Lawrence & Payne, 1983). For example, soluble Aß oligomers and prion protein aggregates (SDS-resistant) are characterized using sodium dodecyl sulfate SDS-PAGE but not the non-SDS resistant oligomers (Coalier, Paranjape, Karki, & Nichols, 2013; Mc Donald et al., 2010).

4.2 Analytical ultracentrifugation

Analytical ultracentrifugation (AUC) is a versatile and powerful hydrodynamic technique used for the analysis of macromolecular properties (Cantor & Schimmel, 1981). Sedimentation coefficient is responsible here for the separation of macromolecules which directly correlates with the molecular mass and inversely with the frictional coefficient (Schuck, 2000). Since such an experiment covers a size range in molar mass of 3 orders of magnitude thus making sedimentation velocity AUC suitable for characterizing the molecular weight distribution of proteins in relevant solutions where the quantity of biological material is not the limiting factor (Brown, Balbo, & Schuck, 2008; Gabrielson et al. 2010). Moreover, AUC can also be used for evaluating noncovalent interactions in a 2 (maximum 3) component protein solution under equilibrium conditions, if provided with enough material and less complex mixtures. Sedimentation equilibrium AUC has been reported for providing the data suggesting tetramer assembly of native α-synuclein which contrasts with the prevailing view of this Parkinson's disease related aggregating protein which is normally found as an unstructured monomer. But this technique has a drawback associated with it that it may require several days to perform.

4.3 Size exclusion chromatography

Size exclusion chromatography (SEC) is a form of partition chromatography where the hydrodynamic volume (i.e., analyte size and shape) determines analyte path length and the time to emerge at the detector gives the information about the molecular mass of the analyte. Smaller the analyte, the greater the time it will take to emerge at the detector and vice versa as the small analyte will pass through the bead because of large mobile phase distribution while the larger analytes will pass through the void volume. SEC has been reported for the analysis of Aß oligomers in several studies. Analytical SEC can also be used for the separation of Aß monomers from dimers and higher order oligomers and hence may prove useful for the preparative isolation of well-defined monomers or oligomers for further studies (Welzel et al., 2012). Conflicting results were always obtained while characterizing α-synuclein from various sources using SEC (Fauvet et al., 2012; Podlisny et al., 1998). SEC columns are still being designed for different molecular size ranges although, the approach will normally not be useful in the case of higher order oligomers, polymers, and fibrillar species which will either not enter the bead of the column at all or elute with the void volume without size separation.

4.4 Mass spectrometry and ion mobility mass spectrometry

Mass spectrometry (MS) separates the molecules on the basis of their mass to charge ratio and the samples are either ionized and volatilized by matrix assisted laser desorption ionization of sample crystallized with matrix molecules or by electrospray ionization of liquid samples from direct injection. MS provides data on aggregate stoichiometry and mass distribution hence it is widely used to characterize and monitor the formation of ionizable and non-covalent oligomers and aggregates in vitro. MS offers certain advantages such as very high selectivity, requires little material, can be used for mapping post-translational modifications and may also be performed directly on tissues including neuropathological samples containing Aß deposits (Caprioli, Farmer, & Gile, 1997; Luxembourg, Mize, McDonnell, & Heeren, 2004; Stoeckli et al., 2006). Electrospray ionization (ESI) and desorption-ESI techniques require molecules to be in solution which may pose a problem since amyloidogenic proteins aggregate and precipitate out of solution over time however by using MALDI-MS (Matrix-assisted laser desorption/ionization-MS) insoluble species can also be studied. But sample preparation in MALDI-MS causes disruption of non-covalent complexes and hence provide us with the imprecise results as compared to ESI.

MS in combination with gas-phase electrophoretic molecular analysis allows two dimensional separation approaches where ESI-generated ions are charge-reduced and separated according to their size dependent electrophoretic mobility in air or a gas such as helium in the first dimension. The second dimension is the time-of-flight MS analysis (Karasek, 1974; Woods, Radford, & Ashcroft, 2013). Thus, conformers with the same masses but different size (e.g., different conformations of the same species) can be separated. Different oligomeric species with identical m/z (mass/charge) value but different absolute size and charge can also be separated hence it is very useful for characterizing very early unfolding events and thus the prefibrillar aggregates too, formed during aggregation. It is also useful for studying the interaction of small molecules with Aß isoforms.

Hydrogen-deuterium (H/D) exchange monitored by MS is a very sensitive approach used to measure the dynamics of conformational fluctuations in proteins which exploits the fact that the hydrogens that are shielded by structure or are present in core, e.g., involved in hydrogen bonding, exchange much more slowly than the accessible free hydrogens of backbone amides. When any conformational change takes place the hydrogens participating in hydrogen bonding are set free which were previously inaccessible and are exposed to solvent and readily exchange with deuterium in the solvent. This approach thereby helps to differentiate between states with or without protecting structures and thus has been reported for investigating the unstructured/structured core state of soluble and fibrillar α-synuclein (Paslawski, Mysling, Thomsen, Jergensen, & Otzen, 2014; Vilar et al., 2008) and it was observed that fibrils are more dynamic structures than previously assumed.

4.5 Turbidity and light scattering analysis

Turbidity is an optical kinetic method that indicates the presence of suspended particles and can thus be used to investigate the formation of protein aggregates (with variable size) with a hydrodynamic radius (R_H) greater than the wavelength of the incident light (λ_i) (Mahler, Friess, Grauschopf, & Kiese, 2009). A decrease in the transmitted light/an increase in absorbance or turbidity is observed at wavelength of around 350 nm (since soluble monomeric proteins do not absorb at this wavelength) when prefibrillar aggregates and aggregates that are large enough to scatter light are present in the solution. Turbidity measurement offers a single advantage that the protein need not to be labeled but it is not reliable as it cannot differentiate between different types of aggregates, can not even detect the oligomeric intermediates and hence it must be supported with some other biophysical technique.

Dynamic light scattering (DLS) is a sensitive and label-free method where the sample is irradiated by a light source and the fluctuations of the scattered light intensity is being recorded, to determine the size distribution of formed oligomers, protofibrils and aggregates as well as aggregation kinetics (Siddiqi, Alam, Iqbal et al., 2018). The size of aggregates is measured in terms of hydrodynamic radius (R_H) ranging from 1 nm to 4 μm. The fluctuation in the scattering pattern is caused by the Brownian motions of the molecules which are a measure of the molecular diffusion coefficient as it is dependent upon the size and shape of the molecule and thus the small molecules diffuse faster than larger molecules so the small molecules will cause less fluctuation in the intensity of the scattered light (Hassan, Rana, & Verma, 2014). This important information can be exploited to study the evolution of oligomeric species with time; to determine the elongation rate and the fibril length as has been reported for Aß and Huntingtin protein.

4.6 Dye binding assays

To examine protein aggregation the extrinsic fluorescence dyes like Thioflavin T (ThT), Congo Red (CR) and 1-anilino-8-naphthalenesulfonate (ANS) are widely used due to ease of performing the dye-binding assay as compared to techniques requiring extensive sample preparation. In addition, the assay are usually performed using fluorescence spectroscopic technique, hence are highly sensitive when used for detecting variety of amyloidogenic protein fibrils. Not only the presence of aggregate is detected but the degree of exposed hydrophobic patches as well as the characterization of folding intermediates can be done using dye binding methods (Cardamone & Puri, 1992; Goto & Fink, 1989).

4.6.1 1-ThioflavinT binding assay

ThT is a small molecule cationic dye, earlier used for staining of ex-vivo extract (Kelienyi, 1967; Vassar & Culling, 1959) however later Thioflavin T (ThT) assay became the most widely used technique to identify the presence of Amyloid fibril. ThT binds to the β-sheet groove structure of fibrillar protein aggregates and displays a fluorescence emission maximum at around the wavelength of 485 nm when exited at around 450 nm (Reinke & Gestwicki, 2007; Siddiqi, Ajmal et al., 2017; Siddiqi, Shahein et al., 2017). Detection is based on the enhancement of intensity of fluorescence spectra upon binding of ThT with amyloid fibril that can be observed by fluorescence microscopy or by fluorescent spectroscopy (LeVine, 1995). A Characteristic sigmoidal increase in fluorescence is observed that occurs between monomers and end fibril state (Bhak, Lee, Hahn, & Paik, 2009; Ladiwala, Dordick, & Tessier, 2011; Lee, Bhak, Lee, & Paik, 2008; Wong

& Kwon, 2011). Usually ThT spectroscopic assay is performed to monitor fibrillation kinetics. Several mechanism of interaction of ThT with amyloid fibril was proposed. Previously it was believed that ThT binds strictly to β-sheet structure since these are predominantly present in amyloid fibrils but later the idea was given up because it was discovered that non-β sheet cavities can also induce fluorescence (eg; Acetylcholinesterase or γ-cyclodextrin) and not all structures rich in β-sheet can (eg; Transthyretin) (Vestergaard et al., 2007). Prefibrillar aggregates containing β-sheet groove binding site (toxic protofibrils and fibrillar oligomers) have been found to bind ThT whereas prefibrillar oligomer lacking defined β-sheet structure cannot be detected by ThT binding assay (Ahmed et al., 2010; Ladiwala et al., 2010). Evidences have shown that most probably ThT molecules intercalate between the grooves that are present between side chains of amyloid exposed to the solvent. Lastly, ThT should be used cautiously when monitoring Inhibition of Amyloid by exogenous compounds due to possibility of potential spectral interference with ThT fluorescence (Hudson, Ecroyd, Kee, & Carver, 2009; Wong & Kwon, 2011). The most probable reasons for this to happen could be that either the drug is spectroscopically active within the wavelength range of ThT or it competes with ThT for binding to protein aggregates or it may directly interact with ThT molecules.

4.6.2 Congo red binding assay

Congo red is another fluorescent dye similar to ThT which is traditionally been used for identifying amyloid fibrils. It's chemical name is 3, 3-[(1,1-biphenyl)-4,4-diylbis (azo)] bis-(4-amino-1-naphthalene acid) disodium salt and has symmetrical linear structure. In 1884, German chemist Paul Böttiger invented CR (Bottiger, 1884). Earlier it was used for the examination of in situ and ex vivo amyloids, later use was extended to in-vitro amyloid detection (Klunk, Jacob, & Mason, 1999; Maezawa et al., 2008; Nilsson, 2004; Reinke & Gestwicki, 2011). The basis of detection is that when the protein-CR complex is formed, a characteristic apple green birefringence is observed under cross polarized light. Functional groups of CR such as the primary amino groups and sulfonate could be involved in interacting with amyloid fibrils by Hydrogen bonding and ionic interaction respectively. Hydrophobic interactions could be between the aromatic rings and the fibril or the dye may also intercalate between β-sheets of the fibrils. Since CR partially inhibits the formation of fibrils that is why in-situ assay should not be performed using CR binding assay. For that reason single time-point experiments that employ recording of CR absorption band immediately after CR addition to amyloid forming polypeptide has been well established (Klunk et al., 1999; Lorenzo & Yankner, 1994). A characteristic increase in absorption and red shift from 490 to 540 nm in the CR absorption band is observed when CR binds to the β-sheet rich structures of amyloids (Hudson et al., 2009). As the absorption spectra is obtained at a longer wavelength compared to ThT fluorescence assay, So the amyloid fibrils can be quantitated in the presence of colored compounds that generally interfere with ThT fluorescence (Klunk et al., 1999; Nilsson, 2004). However as with the case of ThT, CR has been demonstrated to be applicable in characterizing protofibrils and fibrillar oligomers, but not prefibrillar oligomers that lack defined stacked β-sheet structure because upon CR binding to prefibrillar oligomers no shift is observed in absorbance maximum.

4.6.3 ANS fluorescence assay

1-anilinonaphthalene 8-sulfonate (ANS) and 4,4′-bis-1-anilinonaphthalene 8-sulfonate i.e. Bis-ANS (dimer of ANS) are widely used dyes for characterizing array of proteins on the basis of surface hydrophobicity (Bhak et al., 2009; Hawe, Sutter, & Jiskoot, 2008; Lindgren & Hammarstrom, 2010; Siddiqi, Alam, Chaturvedi, & Khan, 2016). The microenvironment of the dye that is solvent polarity, viscosity and temperature effects the wavelength shifts and intensity of the ANS and Bis-ANS fluorescence. An increase in fluorescence intensity and a blue shift is observed in fluorescence maxima when exposed to hydrophobic surfaces on the proteins and also with decrease in the dielectric constant of the solvent (Hawe et al., 2008; Stryer, 1965) whereas the free ANS in aqueous environment has a weak fluorescence peak along with a decrease in emission wavelength. A part from detection of amyloids, ANS can be used for monitoring the presence of oligomeric intermediates during the process of folding and refolding and to investigate the protein ligand interactions as well as the conformational change in protein upon chemical treatment. Recent studies have reported a correlation between increased ANS fluorescence and toxicity. Ladiwala et al. characterized Aβ-42 fibrils, soluble prefibrillar oligomers and freshly disaggregated having oligomers/fibrils low molecular weight using ANS and it was shown that prefibrillar oligomers of Aβ42 peptide exhibited largest increase in fluorescence and maximum blue shift compared to fibrils. It is due to the fact that as the aggregation proceeds, more ordered fibrillar structures appear which have reduced surface to volume ratio, hence the ANS fluorescence decreases (Bolognesi et al., 2010; Campioni et al.,

2010; Ladiwala et al., 2011; Ladiwala et al., 2012). Another property of ANS is the ability to bind to cluster of hydrophobic patches rather than the individual hydrophobic residues. In the aggregate either the individual hydrophobic molecule or the neighboring hydrophobic molecules comprises the cluster of hydrophobic residues. Also ANS does not binds to the tertiary structure of aggregate because hydrophobic patches unable to form clusters and bury themselves from solvent via intra-aggregate interactions (Kundu & Guptasarma, 2002).

4.7 Imaging methods

Light microscopy readily reveals the CR-stained amyloid deposits but the visualization of morphology of insoluble protein deposits need high resolution that can be obtained using ultra structural imaging techniques including electron microscopy (EM) and Atomic Force Microscopy (AFM) (Shirahama & Cohen, 1967; Stine et al., 1996). Electron microscope was used for the first imaging of amyloid deposits indicating the presence of long, unbranched fibrils. Two types of EM are being used i.e., transmission electron microscope (TEM) and scanning electron microscope (SEM) which are useful for examining the morphology of aggregates especially ordered aggregates such as oligomer, protofibrils and mature fibrillar species (Goldsbury et al., 2011; Shirahama & Cohen, 1967).

Electron microscopy especially TEM, confirms the fibrillar morphology of amyloid deposits. The TEM analysis of amyloids shows that fibrils irrespective of different nature of proteins have similar unbranched fibrillar morphology (Tosoni, Barbiano di Belgiojoso, & Nebuloni, 2011). In-vitro synthesized aggregates as well as amyloid aggregates extracted from tissue sections can be visualized using TEM. Under TEM observation amyloid aggregates are found to consist of 5–6 protofilaments having a diameter of 2–5 nm which are parallel to fibril axis and seem to have wavy, twisted or rod like structure with 5–25 nm diameter (Makin, Atkins, Sikorski, Johansson, & Serpell, 2005).

In classical TEM, a focused electron beams of high-voltage is passed through the specimen in vacuum resulting in the formation of an image that depends upon the scattering of the electrons by the specimen. Negative staining is sometimes done to enhance the imaging but it gives limited resolution. Cryo-EM can also be utilized which allows the examination of unstained and unfixed specimens rapidly frozen hydrated, preserved in vitrified (noncrystalline) water at -180 °C using low-to-moderate radiation dose EM and it has the potential of near atomic resolution, and has led 3D-imaging of samples possible. Cryo-EM has been reported for characterization of amyloid fibrils of Aß, α-synuclein, yeast prion, and the protein polymers of the serpinopathies.

In SEM, the sample is irradiated to a subnanometer 100 kV electron beam that scans the specimen in a systematic fashion which provides scanning EM with analytical capabilities. Thus, the scattered electrons are recorded to get intensities that are directly proportional to the mass of the irradiated region. In such a manner protein masses and mass-per-length data can be determined.

Direct visualization of the 3D-structure (topography) of a solid sample is achieved by AFM. A cantilever with a tip of radius in nanometer under piezoelectrical control scans the surface of a sample either continuously in contact (static mode) or as an oscillating probe (noncontact or tapping mode). The deflections of cantilever are recorded by optical interferometry, laser reflection, or other means and combined into a 3D-image of the specimen (Goldsbury et al., 2011; Gosal, Myers, Radford, & Thomson, 2006). AFM provides resolution at subnanometer level and time-lapse studies offer us to study morphology and growth of aggregates, nanostructures, oligomers, protofibrils, and fibrils of, e.g., Aß, α-synuclein, etc. The main disadvantage of AFM is the limited scan area compared to TEM and SEM and slow scanning speed (several minutes vs. seconds).

4.8 Circular dichroism measurements

CD is a reliable and sensitive technique to determine the conformational changes in the structure of protein and it can give the idea about the intermediates that are formed during the aggregation process in terms of increase and decrease in the ellipticity. To monitor the secondary and tertiary structural change of protein during aggregation, CD spectra is being recorded in far- UV (200—250 nm) and near -UV (250—300 nm) range respectively (Kelly & Price, 2000; Siddiqi, Ajmal et al., 2017; Siddiqi, Shahein et al., 2017). Usually amyloids are characterized by the presence of beta sheet rich structure (Sabate & Ventura, 2012).

In the far-UV region, the secondary structural changes are indicated in the form of peaks e.g., the alpha helical proteins give rise to two negative minima around 208 and 222 nm while the beta sheet rich proteins give a single minima around 218 nm (Siddiqi, Nusrat et al., 2018).

In the near-UV region the two minima are found around 262 and 268 nm in case of alpha helical protein whereas the ellipticity of these minima got decreased in case of beta sheet rich proteins indicating the tertiary structural changes (Siddiqi, Alam, Malik et al., 2018).

4.9 Nuclear magnetic resonance spectroscopy

Nuclear magnetic resonance (NMR) spectroscopy is a versatile and powerful spectroscopic technique which exploits the fact that many atomic nuclei behave like small electromagnets when placed in a strong external magnetic field for probing structures of isolated homogeneous species in protein aggregation pathways. NMR may provide structural information about monomeric, secondary (e.g., identification of ß-strand segments and quantification of backbone torsion angles), tertiary (e.g., alignment of ß-strands in parallel or antiparallel ß-sheets), and quaternary (e.g., relative orientation of ß-sheets) structure, dynamics, folding kinetics of purified proteins and the residue-specific structural changes during self-assembly can also be obtained by using different and often complex radiofrequency pulse sequences (Chen, 2015).

NMR relaxation is an important solution NMR approach which may provide information on the dynamics and kinetics of the aggregation events, the exchange kinetics between monomeric Aß and oligomeric Aß (protofibrils) but it is associated with certain limitations including the size limit which is around 40 kDa, time consuming for large proteins (Fawzi, Ying, Torchia, & Clore, 2010).

Solid state NMR (ssNMR) provides good quality data and is primarily used for the samples that cannot be analyzed by X-ray crystallography and cannot be crystallized e.g, for detailed structural determination of insoluble homogenous amyloid fibrils including Aß, α-Synuclein, and prion protein (Tycko, 2006).

4.10 Immunochemical methods

Antibodies have been widely used to quantitate specific proteins in complex biological sample via ELISA (enzyme-linked immunosorbent assay), RIA (radioimmunoassay), Western blot etc., and the tunable properties of antibodies have also been useful for specific detection of protein oligomers, protofibrils, fibrils and conformers associated with neurodegenerative diseases (MacBeath, 2002). As we know that monoclonal antibodies bind to a single epitope on the target thus they can be used for the detection of specific type of conformation (e.g., oligomers). In this way only dimeric or higher order oligomers will bind to the antibody recognition region unless the self-assembly leads to occlusion of the antibody epitope. Monoclonal antibodies can also be designed for protofibrils and amyloid fibrils e.g., mAb158 for protofibrils and MABN687 for amyloid-β fibrils that has already been reported (Haass & Selkoe, 2007; Tucker et al., 2015). Moreover, this method provides us with the advantage that it requires low sample volume and has high specificity.

5. Cytotoxicity of Oligomer and mature fibril

Protein aggregation is a dynamic process which involves the formation of transient intermediate structures as well as dissociation of mature fibrils making it challenging to define the real culprit leading to toxicity (Verma et al., 2015). The close link of protein aggregation with neurodegeneration was supported by the presence of amyloid plaques rich in Aß fibrils in the brains of AD patients (Guivernau et al., 2016; Irvine, El-Agnaf, Shankar, & Walsh, 2008). According to initial studies amyloid fibrils were supposed to be the primary cause of cell death and disease pathogenesis (Pimplikar, 2009). This was demonstrated by some experimental approaches, involving the application of extracellular Aß fibrils which induced AD-like changes in cultured neurons including the increased frequency of action potential, membrane depolarization and reduced cell viability (Kress & Mennerick, 2009). Cognitive decline and the neuronal cell death was also observed as a result of injecting Aß fibrils in primary hippocampal neurons in rat-impaired synaptic transmission (Stephan, Laroche, & Davis, 2001). However, there has always been a controversy regarding the correlation between amyloid plaque burden and severity of neuronal loss and other AD-symptoms (Morris, Clark, & Vissel, 2014). Increasing number of recent studies suggests that prefibrillar species (especially oligomer) formed during fibrillation are more toxic species than the mature fibrils reported by Hey et al. in 2012, they demonstrated the significant impairment of learning and memory functions on infusing oligomeric Aß oligomers into left ventricles of brain while no significant effect was observed due to fibrils (Dasari et al., 2011; Kitazawa, Medeiros, & LaFerla, 2012).

Moreover, Aß oligomers were shown to suppress synaptic plasticity by specifically inhibiting presynaptic P/Q calcium currents unlike protein monomers and fibrils (Catterall, Leal, & Nanou, 2013; Mezler, Barghorn, Schoemaker, Gross, & Nimmrich, 2012). Instead in some instances, formation of mature fibrils is believed to be protective by acting as a harmless reservoir of toxic oligomers but it can also result in the release of off pathway oligomers which are quite toxic (Hubin, Van Nuland, Broersen, & Pauwels, 2014). Hence, the original amyloid cascade hypothesis has been modified to include the intermediary species formed during the aggregation process and is named as oligomer hypothesis (Musiek & Holtzman, 2015). Thus we can say that prefibrillar species (such oligomer and protofibrils) are relatively more toxic than mature fibrils.

Progressive research on amyloid oligomers have revealed some of the structural features of these aggregates that account for their cytotoxicity e.g., small size and high surface hydrophobicity (Mannini et al., 2014). Oligomer cytotoxicity has been shown by multiple studies to depend on their size (Table 2 and Table 3). In case of Oligomers of small Aβ (Tian et al.,

Table 2 Summary of the technique used to characterize different protein species.

Techniques of study	Prefibrillar structures/mature amyloid fibrils
Spectroscopy: UV, Fluoro, CD, FTIR Dyes: ThT, ANS, CongoRed Solution NMR, Mass spectroscopy, X ray crystallography, Light Scattering/Turbidity, Electrophoresis, Size exclusion chromatography Analytical ultracentrifugation, Immunoassay	Native Monomer Or Partially Unfolded monomer ⇅
Spectroscopy: UV, Fluoro, CD, FTIR Dyes: ThT, ANS, CongoRed Solution NMR, Mass Spectroscopy, Light Scattering/Turbidity, Electrophoresis Size exclusion chromatography, Analytical ultracentrifugation, Immunoassay	Critical nuclei ⇅
Spectroscopy: UV, Fluoro, CD, FTIR Dyes: ThT, ANS, CongoRed Solution NMR, Mass Spectroscopy, Light Scattering/Turbidity, Electrophoresis, Size exclusion chromatography Analytical ultracentrifugation, Immunoassay	Soluble higher order Oligomers ⇅
Spectroscopy: UV, Fluoro, CD, FTIR Dyes: ThT, ANS, CongoRed Solid state NMR, Mass Spectroscopy, Light Scattering/Turbidity Ultrastructural imaging (AFM, EM), Fiber Diffraction, Analytical ultracentrifugation, immunoassay	Amyloid Fibrils/Amorphous aggregates

Table 3 Depending upon the abundance of membrane GM1, oligomer shows variable cytotoxicity depending.

		Membrane GM1		
		Low	**Basal**	**High**
Oligomer Type	Loose oligomers, high hydrophobic exposure	Non toxic	Toxic	Highly Toxic
	Compact oligomers, low hydrophobic exposure	Non toxic	Non toxic	Toxic

2013) and tau (Bemporad & Chiti, 2012), cytotoxicity increases with oligomer size whereas larger Aβ oligomer (12-mers or higher) show decreased cytotoxicity with size (Ono, Condron, & Teplow, 2009). This suggest that cytotoxicity may be maximal for the oligomers of intermediate size (Table 4).

Additional to the size of oligomers, structural features also have significant effect on their toxicity. Less well-packed hydrophobic core in toxic oligomers results in solvent exposure of the most hydrophobic regions of the protein sequence (Campioni et al., 2010). Differences of the exposed hydrophobic surfaces among oligomers may also make them toxic and non-toxic.

In addition to size and hydrophobic exposure, another proposed determinant of oligomer toxicity is the shape of the aggregates. It has been proposed that either the pore-like oligomers are formed by the association of monomers that binds to the membrane or alternatively the direct self-assembly of monomers into pores at the membrane interface takes place

Table 4 Depending upon the abundance of membrane cholesterol, oligomer shows variable cytotoxicity depending.

		Membrane cholesterol		
		Low	**Basal**	**High**
Oligomer Type	Loose oligomers, high hydrophobic exposure	Highly Toxic	Toxic	Non toxic
	Compact oligomers, low hydrophobic exposure	Toxic	Non toxic	Non toxic

(Giehm, Svergun, Otzen, & Vestergaard, 2011; Lashuel, Hartley, Petre, Walz, & Lansbury, 2002; Lashuel, Petre et al., 2002; Last, Rhoades, & Miranker, 2011).

Although a well-defined set of parameters such as size and hydrophobic exposure of the protein oligomer determine the oligomeric toxicity, in addition it should be emphasized that toxicity does not reside in one or limited number of oligomeric forms of a given protein. Different oligomers from the same protein affect cell viability to different degree. Moreover, observation that protein oligomers formed by different proteins share similar levels of toxicity suggest that toxicity is a shared property of protein misfolded oligomers.

Thus, the differential toxicity of amyloid oligomers and fibrils can be the result of their structural arrangements:

1. Hydrophobic surfaces are exposed in oligomers comprising of ß-sheets while they are hidden inside the interacting stacks in fibrils (Stefani, 2010).
2. Because of their small size oligomers can easily diffuse in tissues as compared to longer fibrils (Sengupta, Nilson, & Kayed, 2016).
3. Improved interaction of oligomers with cellular target is attributed to the presence of more number of open active ends compared to the fibrils (Stefani, 2012).
4. Oligomers are rich in disordered structures hence unstable whereas fibrils are stable organized molecules (Chen et al., 2015).

From the piece of knowledge that has been reviewed here, it can be concluded that there is a link between protein aggregation and disease pathogenesis (Spires-Jones, Attems, & Thal, 2017; Valastyan & Lindquist, 2014). However, to get a clear and detailed insight of the protein aggregation process and to provide the definite clinical correlation of disease stage or severity associated with oligomeric intermediates and mature fibrils there is an urgent need of developing of in vivo imaging techniques.

6. Mechanism of cytotoxicity

Since a common structure is shared by different amyloid oligomers that are generically toxic to the cells, predicts that they have the same primary mechanism of toxicity in degenerative diseases and the common mechanism of toxicity would predict their action on the same primary target. As some amyloids are derived from cytosolic proteins or may arise from secretary or extracellular proteins, suggests that the primary target of oligomers must be accessible to both the cytosolic and extracellular compartments. Therefore, the most probable target accessible to both the cytosolic and extracellular compartments is the interface between the two compartments that is the plasma membrane (Fig. 3).

6.1 Membrane interaction and perturbation of calcium homeostasis

It has been reported that membrane permeabilization is the primary mechanism of pathogenesis of amyloid related diseases (Fig. 3). Amyloid toxicity is associated with an increase in membrane permeability and intracellular calcium concentration (Mattson, 1994; Mattson et al., 1992). Aβ, α-synuclein, polyglutamine and IAPP are examples of some amyloidogenic proteins and peptides that have been widely reported to form discrete pores or single channels in membranes (Arispe, Pollard, & Rojas, 1994; Arispe, Rojas, & Pollard, 1993; Hirakura, Azimov, Azimova, & Kagan, 2000; Mirzabekov, Lin, & Kagan, 1996) and it is due to the amyloid oligomers that permeabilize the cell membrane (Bucciantini et al., 2004; Demuro et al., 2005; Mina

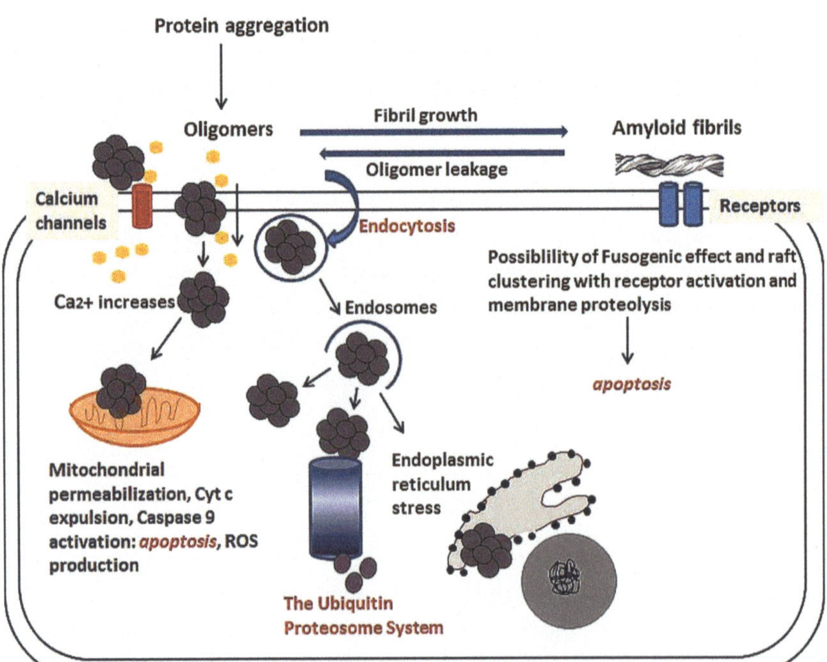

Fig. 3 Schematic illustration of oligomer and mature fibril mediated cell toxicity.

et al., 2004). It is reported that external application of oligomers leads intracellular stores of Ca^{2+} to liberate. This report is in accordance with the fact that penetration of oligomers into the cells causes disruption of intracellular membranes leading to leakage of sequestered Ca^{2+} and elevated Ca^{2+} level has been proposed as the central mechanism of toxicity (Bucciantini et al., 2004).

In all the amyloid-related degenerative diseases, membrane permeabilization may constitute a core group of common pathological events that ultimately result in cell dysfunction and death since the membrane perturbation by soluble oligomers could have direct effect on a wide variety of trans membrane signaling processes and the production of reactive oxygen species (Mattson, 1995; Saitoh, Horsburgh, & Masliah, 1993). Many pathogenic pathways like reactive oxygen species (ROSs) production (Schubert et al., 1995), altered signaling pathways (Mattson, 1995) and mitochondrial dysfunction (Shoffner, 1997) are believed to be initiated with the permeabilization of membrane by amyloid oligomers and with the concomitant increase in intracellular calcium levels. Therefore, Membrane permeabilization and Calcium dyshomeostasis has been found to be the most ubiquitous toxicity mechanism of amyloid oligomer.

6.2 Intermediary oligomers as potential biomarker

Biomarkers that can determine brain pathologies underlying the symptoms of an individual patient, will be very helpful when selecting patients with early symptoms for new clinical trials to evaluate new disease-modifying therapies (Hansson et al., 2014) Findings suggested that α-synuclein oligomers and Aβ oligomers count in CSF (cerebrospinal fluid) could be potential diagnostic biomarkers for the diagnosis and early detection of PD and AD respectively (Park, Cheon, Bae, Kim, & Kim, 2011; Tokuda et al., 2010) In these studies patients' body fluid samples were found to have elevated oligomers as compared to controls (Hansson et al., 2014; Holtta et al., 2013). Thus, it is desirable to have in place a panel of appropriate sensitive diagnostic technologies and specific assays that can detect oligomeric amyloid species. Aβ-PMCA (protein misfolding cyclic amplification assay) and monoclonal single antibody sandwich ELISA assay for Aβ and time-resolved Förster resonance energy transfer (TRFRET)-based immunoassays for α-synuclein oligomer detection are some of the most sensitive assays developed so far (Bidinosti et al., 2012; Esparza et al., 2013; Herskovits, Locascio, Peskind, Li, & Hyman, 2013; Salvadores, Shahnawaz, Scarpini, Tagliavini, & Soto, 2014). These oligomeric assays do not recognize APP (amyloid precursor protein), monomeric Aβ and other non-Aβ-peptide oligomers and are specific and sensitive to Aβ oligomers. Thus, these studies suggest Aβ oligomer as potential biomarker but further research is needed for considering the analysis of oligomers as a clinically routine diagnostic assay. Also, similar oligomeric-based assays for other amyloid proteins need to be developed.

6.3 Amyloid fibrils can be toxic either directly or as a source of toxic species in tissue

Early oligomers, prefibrillar aggregates are presently reported as the most toxic species, in contrast amyloid fibrils are considered far less cytotoxic and inert reservoir of toxic species possessing protective significance for the cell. In fact, some cases of cytotoxicity associated with fibril assembly and growth on lipid membrane (Bucciantini et al., 2012; Engel et al., 2008) or direct fibril cytotoxicity have been reported over the years (Gharibyan et al., 2007; Novitskaya, Bocharova, Bronstein, & Baskakov, 2006). The idea that mature fibrils are not as stable as usually believed and disassembled under suitable conditions has been shown to be correct by various evidences. Also, the data obtained from direct and indirect toxicity indicates that the amyloid plaques should not always be considered as protective recruiters of toxic assemblies in tissues arising from protein misfolding rather they could also act as potential sources of toxicity. In addition, recent data suggest that fibrils of the same protein/peptide that are grown from structurally different oligomers or deposited under differing conditions, can display different cytotoxicities that might be a result of their differing abilities to leak toxic oligomers. Thus, fibril decomposition or fragmentation in tissue spontaneously can be seen as a possible important determinant of amyloid cytotoxicity.

7. Conclusion

Amyloid diseases are characterized by the accumulation of aggregated proteins or peptides. During aggregation heterogeneous population of transient on-pathway amyloid oligomers are generated that are eventually transform to larger protofibrils and fibrils. The structured oligomeric intermediates are suggested to be the primary toxic agents while fibrils are indicated to be relatively less toxic, but can still act as a reservoir of toxic oligomers by fragmentation and secondary nucleation.

Along with advances in Aβ oligomers' characterization, several toxicity mechanisms mediated by structured amyloid oligomers have also been proposed. Oligomers can exert toxic effects through extracellular interactions with membranes, accumulation at intracellular organelles, and cell-to-cell transmission. Until now, a clear relationship between oligomers' structural properties and toxicity has not been fully established; however, the involvement of Aβ oligomers in AD pathogenesis has been recognized. This has initiated a search for strategies to manage amyloid oligomers to alleviate toxicity. Therefore, the development of compounds to inhibit Aβ aggregation—and thus to avoid the formation of toxic oligomeric intermediates—is urgently needed as they could be potential drugs to treat AD.

Acknowledgments

Facilities provided by Interdisciplinary Biotechnology Unit, Aligarh Muslim University, Aligarh are gratefully acknowledged. For providing financial assistance, M.K. Siddiqi (Senior author) is thankful to Department of Biotechnology (DBT), New Delhi, India. S. Malik is thankful to Indian Council of Medical Research (ICMR), New Delhi, India. P. Alam and N. Majid are thankful to Council of Scientific and Industrial Research (CSIR), New Delhi, India, R.H.K. is thankful to CSIR and UGC for project referenced as 37(1676)/17/EMR – II and F. 19–219/2018, respectively.

Conflict of interest

The authors declare that they have no conflicts of interest with the contents of this article.

References

Ahmed, M., Davis, J., Aucoin, D., Sato, T., Ahuja, S., & Aimoto, S. (2010). Structural conversion of neurotoxic amyloid-[beta] 1-42 oligomers to fibrils. *Nature Structural & Molecular Biology, 17*(5), 561–567.

Alam, P., Siddiqi, M. K., Chturvedi, S. K., & Khan, R. H. (2017). Protein aggregation: From background to inhibition strategies. *International Journal of Biological Macromolecules, 109*, 208–219.

Alam, P., Siddiqi, M. K., Malik, S., Chaturvedi, S. K., Uddin, M., & Khan, R. H. (2019). Elucidating the inhibitory potential of Vitamin A against fibrillation and amyloid associated cytotoxicity. *International Journal of Biological Macromolecules, 129*, 333–338.

Almeida, C. G., Takahashi, R. H., & Gouras, G. K. (2006). β-Amyloid accumulation impairs multivesicular body sorting by inhibiting the ubiquitin-proteasome system. *Journal of Neuroscience, 26*(16), 4277–4288.

Andersen, P. M. (2006). Amyotrophic lateral sclerosis associated with mutations in the CuZn superoxide dismutase gene. *Current Neurology and Neuroscience Reports, 6*(1), 37–46.

Anfinsen, C. B. (1973). Principles that govern the folding of protein chains. *Science, 181*(4096), 223–230.

Ao, S. (2006). Hereditary cystatin C amyloid angiopathy: Genetic, clinical, and pathological aspects. *Brain Pathology, 16*(1), 55–59.

Araki-Sasaki, K., Ando, Y., Nakamura, M., Kitagawa, K., Ikemizu, S., & Kawaji, T. (2005). Lactoferrin Glu561Asp facilitates secondary amyloidosis in the cornea. *British Journal of Ophthalmology, 89*(6), 684–688.

Ardalan, M. R., Shoja, M. M., & Kiuru-Enari, S. (2007). Amyloidosis-related nephrotic syndrome due to a G654A gelsolin mutation: The first report from the Middle East. *Nephrology Dialysis Transplantation, 22*(1), 272–275.

Arispe, N., Pollard, H. B., & Rojas, E. (1994). β-Amyloid $Ca2^{+}$-channel hypothesis for neuronal death in Alzheimer disease. *Molecular and Cellular Biochemistry, 140*(2), 119–125.

Arispe, N., Rojas, E., & Pollard, H. B. (1993). Alzheimer disease amyloid beta protein forms calcium channels in bilayer membranes: Blockade by tromethamine and aluminum. *Proceedings of the National Academy of Sciences, 90*(2), 567–571.

Avbelj, F., & Moult, J. (1995). Role of electrostatic screening in determining protein main chain conformational preferences. *Biochemistry, 34*(3), 755–764.

Bagriantsev, S. N., Kushnirov, V. V., & Liebman, S. W. (2006). Analysis of amyloid aggregates using agarose gel electrophoresis. *Methods in Enzymology, 412*, 33–48.

Bemporad, F., & Chiti, F. (2012). Protein misfolded oligomers: Experimental approaches, mechanism of formation, and structure-toxicity relationships. *Chemistry & Biology, 19*(3), 315–327.

Bhak, G., Lee, J.-H., Hahn, J. S., & Paik, S. R. (2009). Granular assembly of α-synuclein leading to the accelerated amyloid fibril formation with shear stress. *PLoS One, 4*(1), e4177.

Bidinosti, M., Shimshek, D. R., Mollenhauer, B., Marcellin, D., Schweizer, T., & Lotz, G. P. (2012). Novel One-step Immunoassays to Quantify α-Synuclein applications for biomarker development and high throughput screening. *Journal of Biological Chemistry, 287*(40), 33691–33705.

Bitan, G., Fradinger, E. A., Spring, S. M., & Teplow, D. B. (2005). Neurotoxic protein oligomers - what you see is not always what you get. *Amyloid: International Journal of Experimental & Clinical Investigation, 12*(2), 88–95.

Bitan, G., Lomakin, A., & Teplow, D. B. (2001). Amyloid β-protein oligomerization prenucleation interactions revealed by photo-induced cross-linking of unmodified proteins. *Journal of Biological Chemistry, 276*(37), 35176–35184.

Bolognesi, B., Kumita, J. R., Barros, T. P., Esbjorner, E. K., Luheshi, L. M., & Crowther, D. C. (2010). ANS binding reveals common features of cytotoxic amyloid species. *ACS Chemical Biology, 5*(8), 735–740.

Bottiger, P. (1884). *Deutsches reichs patent* [German Imperial Patent] 28753. In: August.

Brant, D. A., & Flory, P. J. (1965). The role of dipole interactions in determining polypeptide configurations. *Journal of the American Chemical Society, 87*(3), 663–664.

Brown, P. H., Balbo, A., & Schuck, P. (2008). Characterizing protein - protein interactions by sedimentation velocity analytical ultracentrifugation. *Current Protocols in Immunology, 81*(1).

Bucciantini, M., Calloni, G., Chiti, F., Formigli, L., Nosi, D., & Dobson, C. M. (2004). Prefibrillar amyloid protein aggregates share common features of cytotoxicity. *Journal of Biological Chemistry, 279*(30), 31374–31382.

Bucciantini, M., Giannoni, E., Chiti, F., Baroni, F., Formigli, L., & Zurdo, J. s. (2002). Inherent toxicity of aggregates implies a common mechanism for protein misfolding diseases. *Nature, 416*(6880), 507–511.

Bucciantini, M., Nosi, D., Forzan, M., Russo, E., Calamai, M., & Pieri, L. (2012). Toxic effects of amyloid fibrils on cell membranes: The importance of ganglioside GM1. *The FASEB Journal, 26*(2), 818–831.

Bullock, J. (1993). Application of capillary electrophoresis to the analysis of the oligomeric distribution of polydisperse polymers. *Journal of Chromatography A, 645*(1), 169–177.

Burdick, D., Soreghan, B., Kwon, M., Kosmoski, J., Knauer, M., & Henschen, A. (1992). Assembly and aggregation properties of synthetic Alzheimer's A4/beta amyloid peptide analogs. *Journal of Biological Chemistry, 267*(1), 546–554.

Campioni, S., Mannini, B., Zampagni, M., Pensalfini, A., Parrini, C., & Evangelisti, E. (2010). A causative link between the structure of aberrant protein oligomers and their toxicity. *Nature Chemical Biology, 6*(2), 140–147.
Cantor, C. R., & Schimmel, P. R. (1981). Biophysical chemistry: Part II - techniques for the study of biological structure and function. *Biochemical Education, 1*, 157.
Caprioli, R. M., Farmer, T. B., & Gile, J. (1997). Molecular imaging of biological samples: Localization of peptides and proteins using MALDI-TOF MS. *Analytical Chemistry, 69*(23), 4751–4760.
Cardamone, M., & Puri, N. K. (1992). Spectrofluorimetric assessment of the surface hydrophobicity of proteins. *Biochemical Journal, 282*(2), 589–593.
Catterall, W. A., Leal, K., & Nanou, E. (2013). Calcium channels and short-term synaptic plasticity. *Journal of Biological Chemistry, 288*(15), 10742–10749.
Chaturvedi, S. K., Siddiqi, M. K., Alam, P., & Khan, R. H. (2016). Protein misfolding and aggregation: Mechanism, factors and detection. *Process Biochemistry, 51*(9), 1183–1192.
Chen, L. (2015). *De novo protein structure modeling and energy function design*. Old Dominion University.
Chen, S. W., Drakulic, S., Deas, E., Ouberai, M., Aprile, F. A., & Arranz, R. (2015). Structural characterization of toxic oligomers that are kinetically trapped during α-synuclein fibril formation. *Proceedings of the National Academy of Sciences, 112*(16), E1994–E2003.
Chiti, F., & Dobson, C. M. (2006). Protein misfolding, functional amyloid, and human disease. *Annual Review of Biochemistry, 75*, 333–366.
Chiti, F., & Dobson, C. M. (2017). Protein misfolding, amyloid formation, and human disease: A summary of progress over the last decade. *Annual Review of Biochemistry, 86*, 27–68.
Chothia, C., & Lesk, A. M. (1986). The relation between the divergence of sequence and structure in proteins. *The EMBO Journal, 5*(4), 823–826.
Cleary, J. P., Walsh, D. M., Hofmeister, J. J., Shankar, G. M., Kuskowski, M. A., & Selkoe, D. J. (2005). Natural oligomers of the amyloid-β protein specifically disrupt cognitive function. *Nature Neuroscience, 8*(1), 79–84.
Coalier, K. A., Paranjape, G. S., Karki, S., & Nichols, M. R. (2013). Stability of early-stage amyloid-β (1-42) aggregation species. *Biochimica et Biophysica Acta (BBA) - Proteins & Proteomics, 1834*(1), 65–70.
Costante, G., Meringolo, D., Durante, C., Bianchi, D., Nocera, M., & Tumino, S. (2007). Predictive value of serum calcitonin levels for preoperative diagnosis of medullary thyroid carcinoma in a cohort of 5817 consecutive patients with thyroid nodules. *The Journal of Clinical Endocrinology & Metabolism, 92*(2), 450–455.
Dasari, M., Espargaro, A., Sabate, R., Lopez del Amo, J. M., Fink, U., & Grelle, G. (2011). Bacterial inclusion bodies of Alzheimer's disease β-amyloid peptides can Be employed to study native - Like aggregation intermediate states. *ChemBioChem, 12*(3), 407–423.
Demuro, A., Mina, E., Kayed, R., Milton, S. C., Parker, I., & Glabe, C. G. (2005). Calcium dysregulation and membrane disruption as a ubiquitous neurotoxic mechanism of soluble amyloid oligomers. *Journal of Biological Chemistry, 280*(17), 17294–17300.
Dill, K. A. (1990). Dominant forces in protein folding. *Biochemistry, 29*(31), 7133–7155.
Dobson, C. M. (2001). The structural basis of protein folding and its links with human disease. *Philosophical Transactions of the Royal Society of London B: Biological Sciences, 356*(1406), 133–145.
Dobson, C. M. (2003). Protein folding and misfolding. *Nature, 426*(6968), 884–890.
Dobson, C. M. (2017). The amyloid phenomenon and its links with human disease. *Cold Spring Harbor Perspectives in Biology, 9*(6), 1–12.
Ehrnhoefer, D. E., Bieschke, J., Boeddrich, A., Herbst, M., Masino, L., & Lurz, R. (2008). EGCG redirects amyloidogenic polypeptides into unstructured, off-pathway oligomers. *Nature Structural and Molecular Biology, 15*(6), 558–566.

Eisenberg, D., & Jucker, M. (2012). The amyloid state of proteins in human diseases. *Cell, 148*(6), 1188–1203.

Engel, M. F. M., Khemtemourian, L., Kleijer, C. C., Meeldijk, H. J. D., Jacobs, J., & Verkleij, A. J. (2008). Membrane damage by human islet amyloid polypeptide through fibril growth at the membrane. *Proceedings of the National Academy of Sciences, 105*(16), 6033–6038.

Esparza, T. J., Zhao, H., Cirrito, J. R., Cairns, N. J., Bateman, R. J., & Holtzman, D. M. (2013). Amyloid-beta oligomerization in Alzheimer dementia versus high pathology controls. *Annals of Neurology, 73*(1), 104–119.

Fauvet, B., Kamdem, M. M., Fares, M.-B., Desobry, C., Michael, S., & Ardah, M. T. (2012). Alpha-synuclein in the central nervous system and from erythrocytes, mammalian cells and *E. coli* exists predominantly as a disordered monomer. *Journal of Biological Chemistry*. http://dx.doi.org/10.1074/jbc. M111. 318949.

Fawzi, N. L., Ying, J., Torchia, D. A., & Clore, G. M. (2010). Kinetics of amyloid β monomer-to-oligomer exchange by NMR relaxation. *Journal of the American Chemical Society, 132*(29), 9948–9951.

Fleseriu, M., Lee, M., Pineyro, M. M., Skugor, M., Reddy, S. K., & Siraj, E. S. (2006). Giant invasive pituitary prolactinoma with falsely low serum prolactin: The significance of 'hook effect. *Journal of Neuro-oncology, 79*(1), 41–43.

Forloni, G., Angeretti, N., Chiesa, R., Monzani, E., Salmona, M., & Bugiani, O. (1993). Neurotoxicity of a prion protein fragment. *Nature, 362*(6420), 543–546.

Gabrielson, J. P., Arthur, K. K., Stoner, M. R., Winn, B. C., Kendrick, B. S., & Razinkov, V. (2010). Precision of protein aggregation measurements by sedimentation velocity analytical ultracentrifugation in biopharmaceutical applications. *Analytical Biochemistry, 396*(2), 231–241.

Gharibyan, A. L., Zamotin, V., Yanamandra, K., Moskaleva, O. S., Margulis, B. A., & Kostanyan, I. A. (2007). Lysozyme amyloid oligomers and fibrils induce cellular death via different apoptotic/necrotic pathways. *Journal of Molecular Biology, 365*(5), 1337–1349.

Giehm, L., Svergun, D. I., Otzen, D. E., & Vestergaard, B. (2011). Low-resolution structure of a vesicle disrupting α-synuclein oligomer that accumulates during fibrillation. *Proceedings of the National Academy of Sciences, 108*(8), 3246–3251.

Glabe, C. G. (2006). Common mechanisms of amyloid oligomer pathogenesis in degenerative disease. *Neurobiology of Aging, 27*(4), 570–575.

Glabe, C. G. (2008). Structural classification of toxic amyloid oligomers. *Journal of Biological Chemistry, 283*(44), 29639–29643.

Goldsbury, C., Baxa, U., Simon, M. N., Steven, A. C., Engel, A., & Wall, J. S. (2011). Amyloid structure and assembly: Insights from scanning transmission electron microscopy. *Journal of Structural Biology, 173*(1), 1–13.

Gosal, W. S., Morten, I. J., Hewitt, E. W., Smith, D. A., Thomson, N. H., & Radford, S. E. (2005). Competing pathways determine fibril morphology in the self-assembly of β2-microglobulin into amyloid. *Journal of Molecular Biology, 351*(4), 850–864.

Gosal, W. S., Myers, S. L., Radford, S. E., & Thomson, N. H. (2006). Amyloid under the atomic force microscope. *Protein and Peptide Letters, 13*(3), 261–270.

Goto, Y., & Fink, A. L. (1989). Conformational states in. beta.-lactamase: Molten-globule states at acidic and alkaline pH with high salt. *Biochemistry, 28*(3), 945–952.

Guivernau, B., Bonet, J., Valls-Comamala, V., Bosch-Morate, M., Godoy, J. A., & Inestrosa, N. C. (2016). Amyloid-β peptide nitrotyrosination stabilizes oligomers and enhances nmdar-mediated toxicity. *Journal of Neuroscience, 36*(46), 11693–11703.

Haass, C., & Selkoe, D. J. (2007). Soluble protein oligomers in neurodegeneration: Lessons from the Alzheimer's amyloid β-peptide. *Nature Reviews Molecular Cell Biology, 8*(2), 101–112.

Hansson, O., Hall, S., Ohrfelt, A., Zetterberg, H., Blennow, K., & Minthon, L. (2014). Levels of cerebrospinal fluid α-synuclein oligomers are increased in Parkinson's disease with dementia and dementia with Lewy bodies compared to Alzheimer's disease. *Alzheimer's Research & Therapy, 6*(3), 1–6.

Haslbeck, M., & Vierling, E. (2019). A first line of stress defense: Small heat shock proteins and their function in protein homeostasis. *Journal of Molecular Biology, 427*(7), 1537–1548.

Hassan, P. A., Rana, S., & Verma, G. (2014). Making sense of Brownian motion: Colloid characterization by dynamic light scattering. *Langmuir, 31*(1), 3–12.

Hawe, A., Sutter, M., & Jiskoot, W. (2008). Extrinsic fluorescent dyes as tools for protein characterization. *Pharmaceutical Research, 25*(7), 1487–1499.

Herskovits, A. Z., Locascio, J. J., Peskind, E. R., Li, G., & Hyman, B. T. (2013). A Luminex assay detects amyloid β oligomers in Alzheimer's disease cerebrospinal fluid. *PLoS One, 8*(7). e67898.

Hirakura, Y., Azimov, R., Azimova, R., & Kagan, B. L. (2000). Polyglutamine - induced ion channels: A possible mechanism for the neurotoxicity of Huntington and other CAG repeat diseases. *Journal of Neuroscience Research, 60*(4), 490–494.

Holtta, M., Hansson, O., Andreasson, U., Hertze, J., Minthon, L., & Nagga, K. (2013). Evaluating amyloid-β oligomers in cerebrospinal fluid as a biomarker for Alzheimer's disease. *PLoS One, 8*(6). e66381.

Hubin, E., Van Nuland, N. A. J., Broersen, K., & Pauwels, K. (2014). Transient dynamics of Aβ contribute to toxicity in Alzheimer's disease. *Cellular and Molecular Life Sciences, 71*(18), 3507–3521.

Hudson, S. A., Ecroyd, H., Kee, T. W., & Carver, J. A. (2009). The thioflavin T fluorescence assay for amyloid fibril detection can be biased by the presence of exogenous compounds. *FEBS Journal, 276*(20), 5960–5972.

Irvine, G. B., El-Agnaf, O. M., Shankar, G. M., & Walsh, D. M. (2008). Protein aggregation in the brain: The molecular basis for Alzheimer's and Parkinson's diseases. *Molecular Medicine, 14*(7–8), 451–464.

Jucker, M., & Walker, L. C. (2013). Self-propagation of pathogenic protein aggregates in neurodegenerative diseases. *Nature, 501*(7465), 45–51.

Karasek, F. W. (1974). Plasma chromatography. *Analytical Chemistry, 46*(8), 710A–720A.

Kauzmann, W. (1959). Some factors in the interpretation of protein denaturation. *Advances in Protein Chemistry, 14*, 1–63.

Kayed, R., Head, E., Thompson, J. L., McIntire, T. M., Milton, S. C., & Cotman, C. W. (2003). Common structure of soluble amyloid oligomers implies common mechanism of pathogenesis. *Science, 300*(5618), 486–489.

Kelienyi, G. (1967). On the histochemistry of azo group-free thiazole dyes. *Journal of Histochemistry & Cytochemistry, 15*(3), 172–180.

Kelly, S. M., & Price, N. C. (2000). The use of circular dichroism in the investigation of protein structure and function. *Current Protein and Peptide Science, 1*(4), 349–384.

Kim, H. Y., Cho, M. K., Kumar, A., Maier, E., Siebenhaar, C., & Becker, S. (2009). Structural properties of pore-forming oligomers of α-synuclein. *Journal of the American Chemical Society, 131*(47), 17482–17489.

Kirkitadze, M. D., Bitan, G., & Teplow, D. B. (2002). Paradigm shifts in Alzheimer's disease and other neurodegenerative disorders: The emerging role of oligomeric assemblies. *Journal of Neuroscience Research, 69*(5), 567–577.

Kitazawa, M., Medeiros, R., & LaFerla, F. M. (2012). Transgenic mouse models of Alzheimer disease: Developing a better model as a tool for therapeutic interventions. *Current Pharmaceutical Design, 18*(8), 1131–1147.

Klapper, M. H. (1971). On the nature of the protein interior. *Biochimica et Biophysica Acta (BBA) - Protein Structure, 229*(3), 557–566.

Klunk, W. E., Jacob, R. F., & Mason, R. P. (1999). [19] Quantifying amyloid by Congo red spectral shift assay. *Methods in Enzymology, 309*, 285–305.

Konarkowska, B., Aitken, J. F., Kistler, J., Zhang, S., & Cooper, G. J. S. (2006). The aggregation potential of human amylin determines its cytotoxicity towards islet β cells. *FEBS Journal, 273*(15), 3614–3624.

Kress, G. J., & Mennerick, S. (2009). Action potential initiation and propagation: Upstream influences on neurotransmission. *Neuroscience, 158*(1), 211–222.

Kundu, B., & Guptasarma, P. (2002). Use of a hydrophobic dye to indirectly probe the structural organization and conformational plasticity of molecules in amorphous aggregates of carbonic anhydrase. *Biochemical and Biophysical Research Communications, 293*(1), 572–577.

Ladiwala, A. R. A., Dordick, J. S., & Tessier, P. M. (2011). Aromatic small molecules remodel toxic soluble oligomers of amyloid β through three independent pathways. *Journal of Biological Chemistry, 286*(5), 3209–3218.

Ladiwala, A. R. A., Lin, J. C., Bale, S. S., Marcelino-Cruz, A. M., Bhattacharya, M., & Dordick, J. S. (2010). Resveratrol selectively remodels soluble oligomers and fibrils of amyloid Aβ into off-pathway conformers. *Journal of Biological Chemistry, 285*(31), 24228–24237.

Ladiwala, A. R. A., Perchiacca, J. M., Fishman, Z. S., Bhattacharya, M., Hickey, A. M., & Domigan, B. G. (2012). Polyphenolic disaccharides endow proteins with unusual resistance to aggregation. *Biotechnology and Bioengineering, 109*(7), 1869–1874.

Lashuel, H. A., Hartley, D., Petre, B. M., Walz, T., & Lansbury, P. T., Jr. (2002). Neurodegenerative disease: Amyloid pores from pathogenic mutations. *Nature, 418*(6895), 291–298.

Lashuel, H. A., Petre, B. M., Wall, J., Simon, M., Nowak, R. J., & Walz, T. (2002). α-Synuclein, especially the Parkinson's disease-associated mutants, forms pore-like annular and tubular protofibrils. *Journal of Molecular Biology, 322*(5), 1089–1102.

Last, N. B., Rhoades, E., & Miranker, A. D. (2011). Islet amyloid polypeptide demonstrates a persistent capacity to disrupt membrane integrity. *Proceedings of the National Academy of Sciences, 108*(23), 9460–9465.

Lawrence, G. J., & Payne, P. I. (1983). Detection by gel electrophoresis of oligomers formed by the association of high-molecular-weight glutenin protein subunits of wheat endosperm. *Journal of Experimental Botany, 34*(3), 254–267.

Lee, J. H., Bhak, G., Lee, S.-G., & Paik, S. R. (2008). Instantaneous amyloid fibril formation of α-synuclein from the oligomeric granular structures in the presence of hexane. *Biophysical Journal, 95*(2), L16–L18.

Lesne, S. E. (2013). Breaking the code of amyloid-oligomers. *International Journal of Cell Biology, 2013*, 1–7.

LeVine, H. (1995). Thioflavine T interaction with amyloid β-sheet structures. *Amyloid: International Journal of Experimental & Clinical Investigation, 2*(1), 1–6.

Levine, H., Iii (2004). Alzheimer's β-peptide oligomer formation at physiologic concentrations. *Analytical Biochemistry, 335*(1), 81–90.

Lindgren, M., & Hammarstrom, P. (2010). Amyloid oligomers: Spectroscopic characterization of amyloidogenic protein states. *FEBS Journal, 277*(6), 1380–1388.

Lorenzo, A., & Yankner, B. A. (1994). Beta-amyloid neurotoxicity requires fibril formation and is inhibited by Congo red. *Proceedings of the National Academy of Sciences, 91*(25), 12243–12247.

Luxembourg, S. L., Mize, T. H., McDonnell, L. A., & Heeren, R. M. A. (2004). High-spatial resolution mass spectrometric imaging of peptide and protein distributions on a surface. *Analytical Chemistry, 76*(18), 5339–5344.

MacBeath, G. (2002). Protein microarrays and proteomics. *Nature Genetics, 32*(4s), 526–532.

Maezawa, I., Hong, H. S., Liu, R., Wu, C. Y., Cheng, R. H., & Kung, M. P. (2008). Congo red and thioflavin - T analogs detect Aβ oligomers. *Journal of Neurochemistry, 104*(2), 457–468.

Mahler, H.-C., Friess, W., Grauschopf, U., & Kiese, S. (2009). Protein aggregation: Pathways, induction factors and analysis. *Journal of Pharmaceutical Sciences, 98*(9), 2909–2934.

Makhatadze, G. I., & Privalov, P. L. (1993). Contribution of hydration to protein folding thermodynamics: I. The enthalpy of hydration. *Journal of Molecular Biology, 232*(2), 639–659.

Makin, O. S., Atkins, E., Sikorski, P., Johansson, J., & Serpell, L. C. (2005). Molecular basis for amyloid fibril formation and stability. *Proceedings of the National Academy of Sciences of the United States of America, 102*(2), 315–320.

Mannini, B., Mulvihill, E., Sgromo, C., Cascella, R., Khodarahmi, R., & Ramazzotti, M. (2014). Toxicity of protein oligomers is rationalized by a function combining size and surface hydrophobicity. *ACS Chemical Biology, 9*(10), 2309–2317.

Mattson, M. P. (1994). Calcium and neuronal injury in Alzheimer's disease. *Annals of the New York Academy of Sciences, 747*(1), 50–76.

Mattson, M. P. (1995). Degenerative and protective signaling mechanisms in the neurofibrillary pathology of AD. *Neurobiology of Aging, 16*(3), 447–457.

Mattson, M. P., Cheng, B., Davis, D., Bryant, K., Lieberburg, I., & Rydel, R. E. (1992). beta-Amyloid peptides destabilize calcium homeostasis and render human cortical neurons vulnerable to excitotoxicity. *Journal of Neuroscience, 12*(2), 376–389.

Mc Donald, J. M., Savva, G. M., Brayne, C., Welzel, A. T., Forster, G., & Shankar, G. M. (2010). The presence of sodium dodecyl sulphate-stable Aβ dimers is strongly associated with Alzheimer-type dementia. *Brain, 133*(5), 1328–1341.

Mezler, M., Barghorn, S., Schoemaker, H., Gross, G., & Nimmrich, V. (2012). A β-amyloid oligomer directly modulates P/Q - type calcium currents in Xenopus oocytes. *British Journal of Pharmacology, 165*(5), 1572–1583.

Mina, E. W., Demuro, A., Kayed, R., Milton, S., Parker, I., & Glabe, C. G. (2004). Membrane disruption and elevated intracellular calcium as a common mechanism of amyloid oligomer-induced neurodegeneration. *Neuroscience Abstracts, 449*, 20.

Miranda, E., MacLeod, I., Davies, M. J., Perez, J., Romisch, K., & Crowther, D. C. (2008). The intracellular accumulation of polymeric neuroserpin explains the severity of the dementia FENIB. *Human Molecular Genetics, 17*(11), 1527–1539.

Mirzabekov, T. A., Lin, M.-c., & Kagan, B. L. (1996). Pore formation by the cytotoxic islet amyloid peptide amylin. *Journal of Biological Chemistry, 271*(4), 1988–1992.

Morris, A. M., Watzky, M. A., & Finke, R. G. (2009). Protein aggregation kinetics, mechanism, and curve-fitting: A review of the literature. *Biochimica et Biophysica Acta (BBA) - Proteins & Proteomics, 1794*(3), 375–397.

Morris, G. P., Clark, I. A., & Vissel, B. (2014). Inconsistencies and controversies surrounding the amyloid hypothesis of Alzheimer's disease. *Acta Neuropathologica Communications, 2*(1), 1–21.

Musiek, E. S., & Holtzman, D. M. (2015). Three dimensions of the amyloid hypothesis: Time, space and 'wingmen'. *Nature Neuroscience, 18*(6), 800–806.

Ngoungoure, V. L. N., Schluesener, J., Moundipa, P. F., & Schluesener, H. (2015). Natural polyphenols binding to amyloid: A broad class of compounds to treat different human amyloid diseases. *Molecular Nutrition & Food Research, 59*(1), 8–20.

Nilsson, M. R. (2004). Techniques to study amyloid fibril formation in vitro. *Methods, 34*(1), 151–160.

Novitskaya, V., Bocharova, O. V., Bronstein, I., & Baskakov, I. V. (2006). Amyloid fibrils of mammalian prion protein are highly toxic to cultured cells and primary neurons. *Journal of Biological Chemistry, 281*(19), 13828–13836.

Oiso, N., Yudate, T., Kawara, S., & Kawada, A. (2009). Successful treatment of lichen amyloidosus associated with atopic dermatitis using a combination of narrowband ultraviolet B phototherapy, topical corticosteroids and an antihistamine. *Clinical and Experimental Dermatology, 34*(8), e833–e836.

Ono, K., Condron, M. M., & Teplow, D. B. (2009). Structure - neurotoxicity relationships of amyloid β-protein oligomers. *Proceedings of the National Academy of Sciences, 106*(35), 14745–14750.

Otzen, D. E. (2013). *Amyloid fibrils and prefibrillar aggregates: Molecular and biological properties.* John Wiley & Sons.

Park, M. J., Cheon, S. M., Bae, H. R., Kim, S. H., & Kim, J. W. (2011). Elevated levels of α-synuclein oligomer in the cerebrospinal fluid of drug-naive patients with Parkinson's disease. *Journal of Clinical Neurology, 7*(4), 215–222.

Paslawski, W., Mysling, S., Thomsen, K., Jergensen, T. J. D., & Otzen, D. E. (2014). Co - existence of two different α-synuclein oligomers with different core structures determined by hydrogen/deuterium exchange mass spectrometry. *Angewandte Chemie International Edition, 53*(29), 7560–7563.

Pedersen, J. T., Ostergaard, J., Rozlosnik, N., Gammelgaard, B., & Heegaard, N. H. H. (2011). Cu (II) mediates kinetically distinct, non-amyloidogenic aggregation of amyloid-β peptides. *Journal of Biological Chemistry, 286*(30), 26952–26963.

Pepys, M. B., Hawkins, P. N., Booth, D. R., Vigushin, D. M., Tennent, G. A., & Soutar, A. K. (1993). Human lysozyme gene mutations cause hereditary systemic amyloidosis. *Nature, 362*(6420), 553–557.

Perutz, M. F. (1978). Electrostatic effects in proteins. *Science, 201*(4362), 1187–1191.

Picou, R. A., Schrum, D. P., Ku, G., Cerqua, R. A., Kheterpal, I., & Gilman, S. D. (2012). Separation and detection of individual Aβ aggregates by capillary electrophoresis with laser-induced fluorescence detection. *Analytical Biochemistry, 425*(2), 104–112.

Pimplikar, S. W. (2009). Reassessing the amyloid cascade hypothesis of Alzheimer's disease. *The International Journal of Biochemistry & Cell Biology, 41*(6), 1261–1268.

Podlisny, M. B., Ostaszewski, B. L., Squazzo, S. L., Koo, E. H., Rydell, R. E., & Teplow, D. B. (1995). Aggregation of secreted amyloid-protein into sodium dodecyl sulfate-stable oligomers in cell culture. *Journal of Biological Chemistry, 270*(16), 9564–9570.

Podlisny, M. B., Walsh, D. M., Amarante, P., Ostaszewski, B. L., Stimson, E. R., & Maggio, J. E. (1998). Oligomerization of endogenous and synthetic amyloid β-protein at nanomolar levels in cell culture and stabilization of monomer by Congo red. *Biochemistry, 37*(11), 3602–3611.

Pryor, E., Kotarek, J. A., Moss, M. A., & Hestekin, C. N. (2011). Monitoring insulin aggregation via capillary electrophoresis. *International Journal of Molecular Sciences, 12*(12), 9369–9388.

Redwan, E. M., Xue, B., Almehdar, H. A., & Uversky, V. N. (2015). Disorder in milk proteins: Caseins, intrinsically disordered colloids. *Current Protein and Peptide Science, 16*(3), 228–242.

Reinke, A. A., & Gestwicki, J. E. (2007). Structure - activity relationships of amyloid beta - aggregation inhibitors based on curcumin: Influence of linker length and flexibility. *Chemical Biology & Drug Design, 70*(3), 206–215.

Reinke, A. A., & Gestwicki, J. E. (2011). Insight into amyloid structure using chemical probes. *Chemical Biology & Drug Design, 77*(6), 399–411.

Richards, F. M. (1977). Areas, volumes, packing, and protein structure. *Annual Review of Biophysics and Bioengineering, 6*(1), 151–176.

Richardson, J. S., & Richardson, D. C. (2002). Natural β-sheet proteins use negative design to avoid edge-to-edge aggregation. *Proceedings of the National Academy of Sciences, 99*(5), 2754–2759.

Ross, C. A., & Poirier, M. A. (2004). Protein aggregation and neurodegenerative disease. *Nature Medicine, 10*(7), 553–557.

Sabate, R., & Ventura, S. (2012). Cross-β-sheet supersecondary structure in amyloid folds: Techniques for detection and characterization. *Protein Supersecondary Structures, 932*, 237–257.

Saitoh, T., Horsburgh, K., & Masliah, E. (1993). Hyperactivation of signal transduction systems in Alzheimer's disease. *Annals of the New York Academy of Sciences, 695*(1), 34–41.

Salvadores, N., Shahnawaz, M., Scarpini, E., Tagliavini, F., & Soto, C. (2014). Detection of misfolded Aβ oligomers for sensitive biochemical diagnosis of Alzheimer's disease. *Cell Reports, 7*(1), 261–268.

Schellman, J. A. (1954). The stability of hydrogen-bonded peptide structures in aqueous solution. *Comptes Rendus des Travaux du Laboratoire Carlsberg Serie Chimique, 29*(14–15), 230–259.

Schubert, D., Behl, C., Lesley, R., Brack, A., Dargusch, R., & Sagara, Y. (1995). Amyloid peptides are toxic via a common oxidative mechanism. *Proceedings of the National Academy of Sciences, 92*(6), 1989–1993.

Schuck, P. (2000). Size-distribution analysis of macromolecules by sedimentation velocity ultracentrifugation and lamm equation modeling. *Biophysical Journal, 78*(3), 1606–1619.

Sengupta, U., Nilson, A. N., & Kayed, R. (2016). The role of amyloid-β oligomers in toxicity, propagation, and immunotherapy. *EBioMedicine, 6*, 42–49.

Shikama, Y., Kitazawa, J.i., Yagihashi, N., Uehara, O., Murata, Y., & Yajima, N. (2010). Localized amyloidosis at the site of repeated insulin injection in a diabetic patient. *Internal Medicine, 49*(5), 397–401.

Shirahama, T., & Cohen, A. S. (1967). High-resolution electron microscopic analysis of the amyloid fibril. *The Journal of Cell Biology, 33*(3), 679–708.

Shoffner, J. M. (1997). Oxidative phosphorylation defects and Alzheimer's disease. *Neurogenetics, 1*(1), 13–19.

Siddiqi, M. K., Alam, P., Chaturvedi, S. K., Khan, M. V., Nusrat, S., & Malik, S. (2018). Capreomycin inhibits the initiation of amyloid fibrillation and suppresses amyloid induced cell toxicity. *Biochimica et Biophysica Acta (BBA) - Proteins & Proteomics, 1866*(4), 549–557.

Siddiqi, M. K., Alam, P., Chaturvedi, S. K., & Khan, R. H. (2016). Anti-amyloidogenic behavior and interaction of Diallylsulfide with human serum albumin. *International Journal of Biological Macromolecules, 92*, 1220–1228.

Siddiqi, M. K., Alam, P., Chaturvedi, S. K., Nusrat, S., Ajmal, M. R., & Abdelhameed, A. S. (2017). Probing the interaction of cephalosporin antibiotic - ceftazidime with human serum albumin: A biophysical investigation. *International Journal of Biological Macromolecules, 105*, 292–299.

Siddiqi, M. K., Alam, P., Chaturvedi, S. K., Nusrat, S., Shahein, Y. E., & Khan, R. H. (2017). Attenuation of amyloid fibrillation in presence of Warfarin: A biophysical investigation. *International Journal of Biological Macromolecules, 95*, 713–718.

Siddiqi, M. K., Alam, P., Chaturvedi, S. K., Shahein, Y. E., & Khan, R. H. (2017). Mechanisms of protein aggregation and inhibition. *Frontiers in Bioscience, 9*(1–20).

Siddiqi, M. K., Alam, P., Iqbal, T., Majid, N., Malik, S., & Alam, A. (2018). Elucidating the inhibitory potential of designed peptides against amyloid fibrillation and amyloid associated cytotoxicity. *Frontiers in Chemistry, 6*, 311.

Siddiqi, M. K., Alam, P., Malik, S., Majid, N., Chaturvedi, S. K., & Rajan, S. (2018). Stabilizing proteins to prevent conformational changes required for amyloid fibril formation. *Journal of Cellular Biochemistry, 120*(2), 2642–2656.

Siddiqi, M., Nusrat, S., Alam, P., Malik, S., Chaturvedi, S. K., & Ajmal, M. R. (2018). Investigating the site selective binding of busulfan to human serum albumin: Biophysical and

molecular docking approaches. *International Journal of Biological Macromolecules, 107*, 1414–1421.
Sipe, J. D., Benson, M. D., Buxbaum, J. N., Ikeda, S.-I., Merlini, G., & Saraiva, M. J. M. (2010). Amyloid fibril protein nomenclature: 2010 recommendations from the nomenclature committee of the international society of amyloidosis. *Amyloid: International Journal of Experimental & Clinical Investigation, 17*(3–4), 101–104.
Souillac, P. O., Uversky, V. N., & Fink, A. L. (2003). Structural transformations of oligomeric intermediates in the fibrillation of the immunoglobulin light chain LEN. *Biochemistry, 42*(26), 8094–8104.
Spires-Jones, T. L., Attems, J., & Thal, D. R. (2017). Interactions of pathological proteins in neurodegenerative diseases. *Acta Neuropathologica, 134*(2), 187–205.
Stefani, M. (2004). Protein misfolding and aggregation: New examples in medicine and biology of the dark side of the protein world. *Biochimica et Biophysica Acta - Molecular Basis of Disease, 1739*(1), 5–25.
Stefani, M. (2010). Structural polymorphism of amyloid oligomers and fibrils underlies different fibrillization pathways: Immunogenicity and cytotoxicity. *Current Protein and Peptide Science, 11*(5), 343–354.
Stefani, M. (2012). Structural features and cytotoxicity of amyloid oligomers: Implications in Alzheimer's disease and other diseases with amyloid deposits. *Progress in Neurobiology, 99*(3), 226–245.
Stephan, A., Laroche, S., & Davis, S. (2001). Generation of aggregated β-amyloid in the rat hippocampus impairs synaptic transmission and plasticity and causes memory deficits. *Journal of Neuroscience, 21*(15), 5703–5714.
Sticke, D. F., Presta, L. G., Dill, K. A., & Rose, G. D. (1992). Hydrogen bonding in globular proteins. *Journal of Molecular Biology, 226*(4), 1143–1159.
Stine, W. B., Snyder, S. W., Ladror, U. S., Wade, W. S., Miller, M. F., & Perun, T. J. (1996). The nanometer-scale structure of amyloid-β visualized by atomic force microscopy. *Journal of Protein Chemistry, 15*(2), 193–203.
Stoeckli, M., Knochenmuss, R., McCombie, G., Mueller, D., Rohner, T., & Staab, D. (2006). MALDI MS imaging of amyloid. *Methods in Enzymology, 412*, 94–106.
Stryer, L. (1965). The interaction of a naphthalene dye with apomyoglobin and apohemoglobin: A fluorescent probe of non-polar binding sites. *Journal of Molecular Biology, 13*(2), 482–495.
Sudnitsyna, M. V., Mymrikov, E. V., Seit-Nebi, A. S., & Gusev, N. B. (2012). The role of intrinsically disordered regions in the structure and functioning of small heat shock proteins. *Current Protein and Peptide Science, 13*(1), 76–85.
Tian, H., Davidowitz, E., Lopez, P., Emadi, S., Moe, J., & Sierks, M. (2013). Trimeric tau is toxic to human neuronal cells at low nanomolar concentrations. *International Journal of Cell Biology, 2013*, 1–10.
Tokuda, T., Qureshi, M. M., Ardah, M. T., Varghese, S., Shehab, S. A. S., & Kasai, T. (2010). Detection of elevated levels of α-synuclein oligomers in CSF from patients with Parkinson disease. *Neurology, 75*(20), 1766–1770.
Tosoni, A., Barbiano di Belgiojoso, G., & Nebuloni, M. (2011). Electron microscopy in the diagnosis of amyloidosis. In *Amyloidosis-mechanisms and prospects for therapy*. InTech.
Tucker, S., Muller, C., Tegerstedt, K., Lord, A., Laudon, H., & Sjudahl, J. (2015). The murine version of BAN2401 (mAb158) selectively reduces amyloid-β protofibrils in brain and cerebrospinal fluid of tg-ArcSwe mice. *Journal of Alzheimer's Disease, 43*(2), 575–588.
Tycko, R. (2004). Progress towards a molecular-level structural understanding of amyloid fibrils. *Current Opinion in Structural Biology, 14*(1), 96–103.
Tycko, R. (2006). Molecular structure of amyloid fibrils: Insights from solid-state NMR. *Quarterly Reviews of Biophysics, 39*(1), 1–55.

Usmani, S. S., Bedi, G., Samuel, J. S., Singh, S., Kalra, S., & Kumar, P. (2017). THPdb: Database of FDA-approved peptide and protein therapeutics. *PLoS One, 12*(7). e0181748.
Usmani, S. S., Kumar, R., Bhalla, S., Kumar, V., & Raghava, G. P. S. (2018). In silico tools and databases for designing peptide-based vaccine and drugs. *Advances in Protein Chemistry and Structural Biology, 112*, 221–263.
Uversky, V. N. (2011). Mysterious oligomerization of the amyloidogenic proteins. *FEBS Journal, 277*(14), 2940–2953.
Uversky, V. N. (2014). The triple power of D (3): Protein intrinsic disorder in degenerative diseases. *Frontiers in Bioscience, 19*, 181–258.
Valastyan, J. S., & Lindquist, S. (2014). Mechanisms of protein-folding diseases at a glance. *Disease Models & Mechanisms,* 7(1), 9–14.
Vassar, P. S., & Culling, C. F. (1959). Fluorescent stains, with special reference to amyloid and connective tissues. *Archives of Pathology, 68*, 487–498.
Verma, M., Vats, A., & Taneja, V. (2015). Toxic species in amyloid disorders: Oligomers or mature fibrils. *Annals of Indian Academy of Neurology, 18*(2), 138–145.
Vestergaard, B., Groenning, M., Roessle, M., Kastrup, J. S., Van De Weert, M., & Flink, J. M. (2007). A helical structural nucleus is the primary elongating unit of insulin amyloid fibrils. *PLoS Biology, 5*(5), e134.
Vilar, M., Chou, H.-T., Lohrs, T., Maji, S. K., Riek-Loher, D., & Verel, R. (2008). The fold of α-synuclein fibrils. *Proceedings of the National Academy of Sciences, 105*(25), 8637–8642.
Wadai, H., Yamaguchi, K.-i., Takahashi, S., Kanno, T., Kawai, T., & Naiki, H. (2005). Stereospecific amyloid-like fibril formation by a peptide fragment of β2-microglobulin. *Biochemistry, 44*(1), 157–164.
Walsh, D. M., & Selkoe, D. J. (2007). Aβ oligomers - a decade of discovery. *Journal of Neurochemistry, 101*(5), 1172–1184.
Welzel, A. T., Williams, A. D., McWilliams-Koeppen, H. P., Acero, L., Weber, A., & Blinder, V. (2012). Human anti-Aβ IgGs target conformational epitopes on synthetic dimer assemblies and the AD brain-derived peptide. *PLoS One,* 7(11). e50317.
Westermark, P. (2005). Aspects on human amyloid forms and their fibril polypeptides. *FEBS Journal, 272*(23), 5942–5949.
Winklhofer, K. F., Tatzelt, J., & Haass, C. (2008). The two faces of protein misfolding: Gain - and loss - of function in neurodegenerative diseases. *The EMBO Journal, 27*(2), 336–349.
Wolfenden, R. (1978). Interaction of the peptide bond with solvent water: A vapor phase analysis. *Biochemistry, 17*(1), 201–204.
Wong, H. E., & Kwon, I. (2011). Xanthene food dye, as a modulator of Alzheimer's disease amyloid-beta peptide aggregation and the associated impaired neuronal cell function. *PLoS One, 6*(10). e25752.
Wong, E. W., Sheehan, P. E., & Lieber, C. M. (1997). Observation of metastable abeta amyloid protofibrils by atomic force microscopy. *Science, 277*, 1971–1975.
Woods, L. A., Radford, S. E., & Ashcroft, A. E. (2013). Advances in ion mobility spectrometry - mass spectrometry reveal key insights into amyloid assembly. *Biochimica et Biophysica Acta (BBA) - Proteins & Proteomics, 1834*(6), 1257–1268.
Wu, J. W., Breydo, L., Isas, J. M., Lee, J., Kuznetsov, Y. G., & Langen, R. (2010). Fibrillar oligomers nucleate the oligomerization of monomeric amyloid β but do not seed fibril formation. *Journal of Biological Chemistry, 285*(9), 6071–6079.

CHAPTER ELEVEN

Misfolded proteins as a therapeutic target in Alzheimer's disease

S. Imindu Liyanage[a] and Donald F. Weaver[a,b,*]

[a]Krembil Research Institute, University Health Network, Toronto, ON, Canada
[b]Departments of Medicine (Neurology), Chemistry and Pharmaceutical Sciences, University of Toronto, Toronto, ON, Canada
*Corresponding author: E-mail: donald.weaver@utoronto.ca

Contents

Advances in Protein Chemistry and Structural Biology, Volume 118
ISSN 1876-1623
https://doi.org/10.1016/bs.apcsb.2019.08.003

© 2020 Elsevier Inc.
All rights reserved.

Abstract

For decades, Alzheimer's Disease (AD) was defined as a disorder of protein misfolding and aggregation. In particular, the extracellular peptide fragment: amyloid-β (Aβ), and the intracellular microtubule-associated protein: tau, were thought to initiate a neurodegenerative cascade which culminated in AD's progressive loss of memory and executive function. As such, both proteins became the focus of intense scrutiny, and served as the principal pathogenic target for hundreds of clinical trials. However, with varying efficacy, none of these investigations produced a disease-modifying therapy — offering patients with AD little recourse aside from transient, symptomatic medications. The near universal failure of clinical trials is unprecedented for a major research discipline. In part, this has motivated an increasing skepticism of the relevance of protein misfolding to AD's etiology. Several recent observations, principally the presence of significant protein pathologies in non-demented seniors, have lent credence to an apparent cursory role for Aβ and tau. Herein, we review both Aβ and tau, examining the processes from their biosynthesis to their pathogenesis and evaluate their vulnerability to medicinal intervention. We further attempt to reconcile the apparent failure of trials with the potential these targets hold. Ultimately, we seek to answer if protein misfolding is a viable platform in the pursuit of a disease-arresting strategy for AD.

1. Alzheimer's disease

In the span of a few decades, Alzheimer's Disease (AD) emerged from an obscure disorder of the elderly to a leading cause of worldwide mortality (Bondi, Edmonds, & Salmon, 2017; Dementia Collaborators, 2019). Presently, it is classified as the most prevalent neurodegenerative disorder, the most common cause of dementia, and the sixth leading cause of death (Alzheimer's Association, 2016; Dementia Collaborators, 2019; Hickman, Faustin, & Wisniewski, 2016). Both industry and academia were initially slow to respond; though considerable resources have now been invested, neither a definitive etiology nor a viable therapy have been discovered. Rather, our understanding of AD's pathogenesis is rapidly evolving into a quagmire of differing yet interconnected neurochemical processes; and clinical trials continue to fail universally.

In seeking to explain these failures, blame is often ascribed to an inadequate characterization of AD at a molecular level. By convention, AD was conceptualized as a disorder of protein misfolding and aggregation. Though the events initiating and propagating this misfolding and aggregation were ambiguous, AD's pathologies were thought to originate from the dyshomeostases of a few well-defined proteins. Principal among these are the extracellular peptide fragment β-amyloid (Aβ), and the intracellular microtubule associated protein tau. Both are prone to self-propagative misfolding (such that one misfolded peptide can induce another to misfold) and virulent aggregation – creating a neurotoxic oligomer.

Together, the misfolding and aggregation of these proteins, in tandem with uncontrolled chronic neuroinflammation, and a series of associated pathologies can perpetuate neural membrane damage (Askarova, Yang, & Lee, 2011; Donev, Kolev, Millet, & Thome, 2009), synaptotoxicity (Marttinen et al., 2018), mitochondrial damage (Swerdlow, Burns, & Khan, 2010), oxidative stress and homeostatic dysregulation (Greenough, Camakaris, & Bush, 2013) leading to neuronal death and brain atrophy (Donev et al., 2009; Pini et al., 2016). In patients, this manifests as progressive deterioration in memory, executive function, cognition and information processing domains – eventually resulting in a terminal disorder, typically within 3–10 years of diagnosis (Ganguli, Dodge, Shen, Pandav, & DeKosky, 2005; Helzner et al., 2008; Zanetti, Solerte, & Cantoni, 2009).

AD is currently managed with several medications offering transient symptomatic relief – often by manipulating cholinergic neurotransmitter perturbations (Yiannopoulou & Papageorgiou, 2013). These do not address the underlying pathologies of the brain, thus AD's progression is unimpeded. A disease-modifying intervention capable of halting, perhaps reversing AD's neural damage has been the long sought-after aim for therapeutic inquiry. However, notwithstanding vigorous investigation, the failures of clinical trials have meant that no such treatment exists. In a review of study outcomes, Cummings et al. report that of 413 AD trials initiated in the decade preceding 2012, all but one failed to reach final approval (Cummings, Morstorf, & Zhong, 2014). This failure rate (99.6%) exceeds all other major disciplines of comparable scope.

Many of these trials focused on abating AD's well established proteopathies. Aβ, in particular, dominated research and became increasingly viewed as AD's key pathognomonic element (Kametani & Hasegawa, 2018; Morris, Clark, & Vissel, 2014). Thus, trials frequently sought to arrest AD by attenuating β-amyloid aggregation, or promoting its clearance. Yet, the overwhelming failure of this strategy in human trials raised doubts as to the relevance of Aβ, and the other proteopathies as therapeutic targets. Some initial correlations are even suggesting that inhibition of Aβ may yield deleterious outcomes (Espay et al., 2019; Kametani & Hasegawa, 2018) – implying it may serve an as yet undiscerned, physiological role within the neurochemistry of the brain. Studies have also proposed that Aβ and tau may be downstream co-pathologies, or reactionary elements to an underlying (again undiscerned) pathogenic stimulus (Espay et al., 2019; Kametani & Hasegawa, 2018). The conceptualization of AD's proteopathies have therefore emerged as increasingly nuanced, with both their viability as a therapeutic target and their pathogenicity growing progressively unclear.

Herein, we explore AD's principal protein misfolding pathologies from a molecular to a physiologic level. We further seek to define their role within the pathogenic paradigm of AD and identify synergies with associated pathologies. Finally, we seek to evaluate the susceptibility of these systems to a medicinal intervention, and their potential utility as therapeutic targets.

2. Forms of Alzheimer's dementia

Dementia was initially conceptualized as a singular disorder of memory; it has since evolved into multiple subtypes, which share a broad symptomology (the progressive deterioration of memory and executive function) but differ significantly in etiology (Boller & Forbes, 1998; Bondi et al., 2017). These include vascular dementia, in which pathology arises from perturbed circulation to the brain (Venkat, Chopp, & Chen, 2015); and protein associated dementias, which are diagnosed by the presence of misfolded protein species (Weller & Budson, 2018). These dementias include frontotemporal dementia (defined by tau pathologies), Lewy Body dementia (defined by α-synuclein pathologies) and AD (defined by both Aβ and tau pathologies). In the case of AD, it can be subcategorized further still into early onset AD (EOAD) and late onset AD (LOAD) (Fig. 1). EOAD is relatively rare, representing only 5% of AD cases, but progresses aggressively upon the onset of symptoms (Masellis et al., 2013; Mendez, 2017). EOAD is also referred to a familial AD, reflecting a strong genetic component, often in genes responsible for Aβ synthesis or processing (Lanoiselee et al., 2017; Masellis et al., 2013; Murrell, Hake, Quaid, Farlow, & Ghetti, 2000). LOAD is, to the best of current knowledge, idiopathic but may derive from a cumulative accumulation of incremental risk factors, including (but not limited to) diet, brain trauma, mental stimulation and environmental factors such as metal exposure (Ramos-Cejudo et al., 2018; Schultz et al., 2015; Swaminathan & Jicha, 2014; Wainaina, Chen, & Zhong, 2014). The progression of LOAD is considerably more protracted than EOAD, though it too culminates in a fatal disorder. There remains considerable variability, both in prognosis and pathology, even among the subtypes of AD. This has led to the supposition that AD may be more of a syndrome than a disease — being

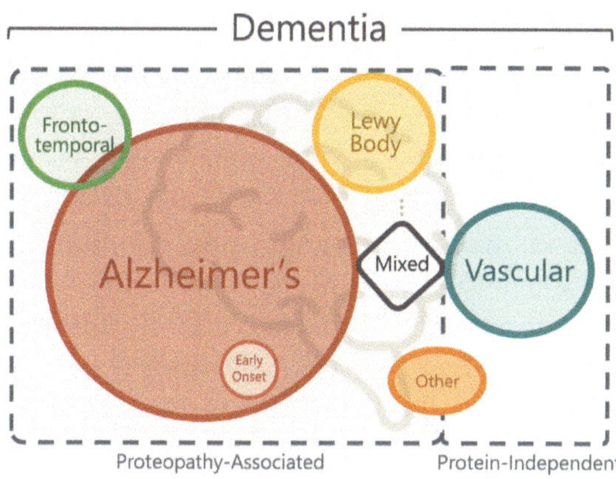

Fig. 1 ***Forms of dementia, by relative prevalence.*** Dementia describes a series of disorders which are symptomatically similar, but differ in etiology and cause. A fundamental distinction can be made among dementias which have a vascular origin (in which memory impairment arises from compromised cerebral blood flow), and those with major associated protein pathologies (in which the dementia presents with significant brain proteopathies). The latter includes Alzheimer's disease, in which Aβ and tau are found in aberrant states; frontotemporal dementia, which is mainly associated with tauopathies; and Lewy Body dementia, which are associated with pathologies of α-synuclein. Mixed dementias, demonstrating both vascular and protein pathologies, and other idiopathic dementias can also occur.

composed of multiple disease states that we are currently unable to differentiate.

3. Amyloid-β

For much of the past three decades, research into both forms of Alzheimer's dementia was driven by a quest to manage Aβ. While its association to AD was initially thought cursory (Morris et al., 2014), the peptide's prominence within AD plaques, as well as numerous pathogenic capacities lead to an ever more prominent role within the presumed etiology of AD. These were loosely organized into an abstract paradigm, termed the amyloid hypothesis, which posits (amidst considerable variation) that the synthesis, misfolding, aggregation, propagation and signaling of Aβ are the principle causative elements in the initiation of AD (Du, Wang, & Geng, 2018). Moreover, it implies that each of these steps may provide a possible therapeutic target against the progression or onset of AD. Herein we review each of these pathological steps as potential druggable targets in the development of disease modifying therapies for AD.

3.1 Synthesis of Aβ

Aβ is a peptide fragment, typically comprising 39–43 amino acids, or approximately 4 kDa (D. L. Price, Sisodia, & Gandy, 1995). It is derived from the proteolytic cleavage of a larger intramembranous macromolecule, called Amyloid Precursor Protein (APP) (O'Brien & Wong, 2011). Transcripts of APP are known to exist in up to eight isoforms, and APP itself is associated with a gene family consisting of two other functionally related proteins (amyloid precursor-like proteins, APLP 1 and 2) (Bayer, Cappai, Masters, Beyreuther, & Multhaup, 1999). Whereas these and other expression characteristics are now well-defined, neither the function of APP nor its protein family has yet been discerned. Studies have demonstrated that multiple knockouts of APP and related genes are lethal (van der Kant & Goldstein, 2015), and that knockouts of APP alone, though still viable, exhibit significant behavioral and cellular anomalies, particularly relevant to the operation of axons (Gunawardena & Goldstein, 2001; Marik, Olsen, Tessier-Lavigne, & Gilbert, 2016). Overexpression models have similarly reported phenotypic changes, typically associated with beneficial or gain-in-function, characteristics (O'Brien & Wong, 2011). Yet, beyond vague formulations for a role in development, and axonal activity, the reason for APP's regular expression in the adult brain remains equivocal.

To derive Aβ, APP must be sequentially cleaved by two membrane-associated proteases. First, β-secretase (beta-site amyloid precursor protein cleaving enzyme 1), liberates a large extracellular fragment (APPsβ), exposing the N-terminus of Aβ. Next, γ-secretase, cleaves the C-terminus – freeing Aβ from the rest of the membrane bound fragment (Amyloid Precursor Protein Intracellular Domain, AICD), and releasing it into the extracellular space. The mechanism of γ-secretase is irregular; it can cleave Aβ at several sites – yielding a peptide with considerable heterogeneity (O'Brien & Wong, 2011). Longer fragments (particularly $A\beta_{42}$) have greater terminal hydrophobicity, and are thus generally more fibrillogenic (Lambermon, Rappaport, & McLaurin, 2005). In fact, dysfunction of γ-secretase leading to the production of altered fragments of Aβ is thought to be a major predisposition toward early-onset AD (Wolfe, 2019). APP can also be cleaved by the non-amyloidogenic enzyme: α-secretase. As with β-secretase, it cleaves the ectodomain of APP, however it does so at a site closer to the plasma membrane, severing the aggregation-prone core of Aβ. An estimated 90% of APP is constitutively processed via α-secretase's non-amyloidogenic pathway, β-secretase is thus the rate-limiting step in the production of Aβ (Murphy & LeVine, 2010; Stockley & O'Neill, 2007) (Fig. 2).

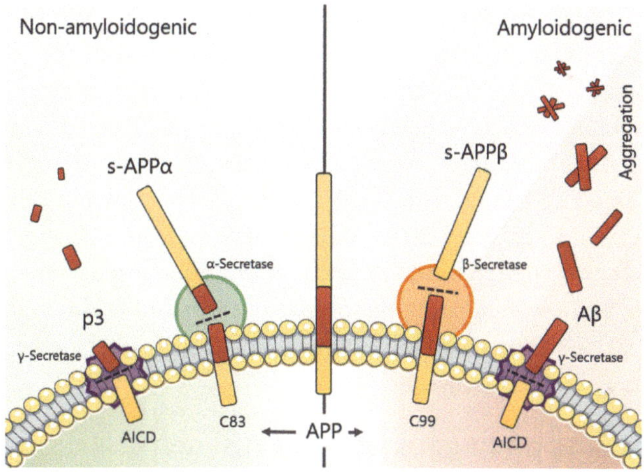

Fig. 2 ***Synthesis of Aβ.*** The synthesis of Aβ occurs when intramembranous Amyloid Precursor Protein (APP) is consecutively cleaved by the enzymes β-secretase, and γ-secretase. Cleavage by β-secretase preserves the aggregation-prone core of Aβ. Once synthesized, Aβ can interact with other Aβ peptides and aggregate, forming cytotoxic species. Alternatively, Aβ can be cleaved by α-secretase, in which the critical aggregation domain is severed. Subsequent cleavage by γ-secretase generates an unaggregative fragment termed P3. In both cases a small remnant of APP is left within the membrane (the amyloid precursor protein intracellular domain or AICD). This may have critical functions as a mediator of gene expression and cell cycling.

Predictably, the inhibition of β-secretase and the shuttling of Aβ toward non-amyloidogenic processing was a field of intense research. Unlike targeting subsequent downstream processes, such as aggregation, this strategy offers an effective prophylactic course against β-amyloid misfolding and toxicity, before the onset of significant pathology (Kamenetz et al., 2003). An array of β-secretase inhibitors were thus proposed, several of which showed significant efficacy in transgenic murine models, and reached clinical trials (Ghosh & Osswald, 2014). Yet, with varying efficacy, these compounds were unable to yield significant gains in clinical outcomes; the viability of this strategy has now largely been brought into contention – though multiple β-secretase inhibitors remain under investigation in phase III clinical trials (Cummings, Lee, Ritter, & Zhong, 2018; Moussa, 2017).

An alternate strategy was the inhibition of γ-secretase. Though this would inhibit both the amyloidogenic and non-amyloidogenic mechanisms of APP processing, it was theorized that leaving APP anchored to the cell membrane would alleviate subsequent pathology. Moreover, mutations in γ-secretases key functional domains, Presenilin-1 & -2 (PS1 & PS2), were associated with an elevated risk of familial or early-onset AD — attributable to the generation of longer, and more aggregation-prone Aβ species (Sherrington et al., 1995). However, in clinical trials this strategy proved not only ineffective, but deleterious (Doody et al., 2013; Penninkilampi, Brothers, & Eslick, 2016). It is now known that in addition to the synthesis of Aβ, PS1 and PS2 have critical roles in developmental and cell proliferation pathways (Ho & Shen, 2011; Veeraraghavalu, Choi, Zhang, & Sisodia, 2010; Xia et al., 2015). Murine knockouts and knockins of presenilin genes also exhibit visible neural and phenotypic abnormalities (Ho & Shen, 2011; Xia et al., 2015). Thus, the inhibition of γ-secretase is likely to be associated with broader, harmful consequences.

3.2 Misfolding and aggregation of Aβ

Contrary to its popular conception, Aβ is not universally toxic. Rather, factors such as the peptide length (Lin, Bowman, Beauchamp, & Pande, 2012), 3D conformation (Lal, Lin, & Quist, 2007), the aqueous microenvironment and the degree of aggregation can all influence final toxicity of the peptide (Jamasbi, Wade, Separovic, & Hossain, 2016). The result is that a spectrum of Aβ species exist in an AD brain, which vary from highly pathogenic, to physiologically beneficial. Discerning which Aβ species to target as proteotoxins, and which to leave alone remains highly controversial, and yet another evolving field of continued study.

At present, the accepted narrative of Aβ′s pathogenesis begins after its synthesis. It is believed that upon release from the plasma membrane, Aβ adopts a benign conformation dominated by two α-helices (Soto, Castano, Frangione, & Inestrosa, 1995). This purported native conformation does not readily aggregate nor induce broad pathologies (Nerelius et al., 2009). However, if this α-helix is unwound, the polarity of the amino acid side-chains can drive a pathogenic misfolding, in which the α-helices transmute to β-sheets (Soto et al., 1995). This process is slow, energetically costly, and may require stabilizing interactions from associated Aβ molecules to proceed. It is thus referred to as the lag phase of aggregation.

Misfolding of Aβ can be significantly expediated with the addition of pre-formed, higher order species. This nucleation principle suggests that misfolded Aβ can induce the misfolding of other native Aβ peptides, instigating a neurotoxic chain reaction. The mechanism again has not been definitively established, though it is likely that exposed hydrophobic residues may interact with sufficient strength to overcome the significant activation energy required to unfold and then refold a peptide (Flock, Colacino,

Colombo, & Di Nola, 2006; Kim & Hecht, 2006; Urbanc et al., 2004; Vugmeyster et al., 2016). Ultimately, these misfolded peptides can interact in a mutually stabilizing association, driving aggregation – first into small aggregates, then proto-fibrils, and finally into large, insoluble fibrils and plaques. The process of systematically larger aggregation is referred to as primary nucleation. A secondary nucleation process has also been devised, stemming from the observation that Aβ forms oligomers only upon the addition of monomers and fibrils (Cohen et al., 2013). In this model, fibrils can nucleate the aggregation of monomers to oligomers at multiple sites along its length (Cohen et al., 2013). It is likely that upon reaching a sufficient concentration, both mechanisms function concurrently to drive Aβ aggregation (Fig. 3).

In early studies, the dominance of amyloid plaques in the AD brain led to the correlative assumption that plaques and similarly large species ought to be the pathogenic element of AD, and thus should be the focus of

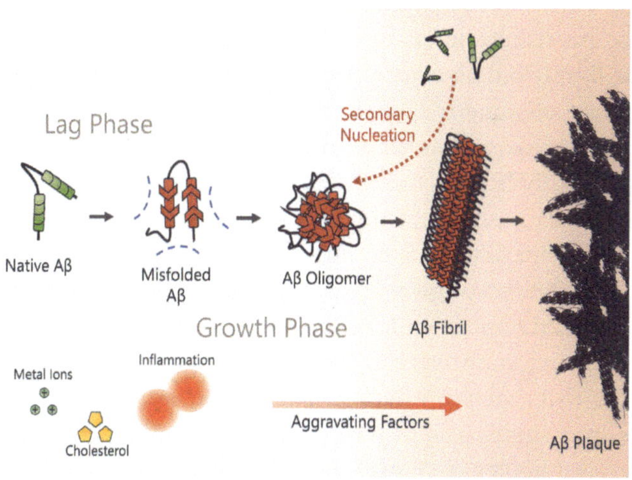

Fig. 3 ***Aggregation of Aβ.*** In its native state, Aβ exists as an α-helical dominated peptide. Misfolding to achieve the aggregative β-pleated sheet arrangement is energetically costly to initiate, and may require stabilizing interactions from other Aβ molecules, cellular membranes and/or cholesterol to proceed. Once misfolded, Aβ can associate with other peptides to form a highly toxic oligomer. This can aggregate further to form a larger fibril-like species (a proto-fibgril), in which Aβ stacks upon itself in an ordered manner. Proto-fibrils can go on to aggregate in to larger fibrils, and finally into large extracellular plaques. Fibrils may also be able to nucleate the misfolding and aggregation of Aβ independently (secondary nucleation). External factors such as cholesterol, inflammation and metal ions may expedite aggregation, and worsen amyloid toxicity.

therapeutic targeting. This hypothesis, however, has been reversed in recent years and it is widely accepted that smaller species (Cizas et al., 2010; Verma, Vats, & Taneja, 2015; Zhao, Long, Mu, & Chew, 2012), from dimers to dodecamers (a species of 12 monomers) are the principal disease-causing elements. Beyond this size, aggregates become insoluble, and thus have weaker interactions with the cellular elements necessary to effectuate toxicity. This led to the assumption that the size of the Aβ aggregate is inversely correlated with toxicity.

The association, however, is more complicated than solubility alone. Smaller amyloid species are more mobile, yet lacking the size of larger aggregates, their disruption to cellular systems is less. Taking, for example, Aβ′s membrane insertion into membranes, *in silico* studies from our group demonstrated that smaller peptidic species can readily attack and permeate the membrane — yet they do not form extensive pores, and thus the insult to membrane integrity is minimal. Conversely, larger oligomeric species can form extensive and destructive pores, yet the energy expenditure required to permeate the membrane is high, and correspondingly fewer of these species manage to successfully insert into the membrane leaflets. Therefore, the toxicity of β-amyloid aggregates exhibits a skewed quadratic association, with moderate oligomeric species (about the size of trimers or tetramers) being the most toxic, and either extremes being less relevant to AD pathologies (Sengupta, Nilson, & Kayed, 2016).

Unlike other parameters within AD's pathogenesis, the management of β-amyloid oligomerization is comparatively easy to assay (Esparza et al., 2016; LeVine, 2006; Savage et al., 2014). It has therefore garnered intense scrutiny as an AD therapeutic strategy. Yet, the ideal management of oligomers remains controversial. If midsize oligomers are most toxic, it is conceivable that agents which expedite the formation of insoluble aggregates (pro-aggregants) as well as those that inhibit the formation of oligomers (anti-aggregants) may both be viable therapeutic platforms (Giorgetti, Greco, Tortora, & Aprile, 2018). Indeed, both strategies are currently under investigation, though the inhibition of oligomers is more often examined. In either case, agents are typically designed to bind to the aggregative domains of Aβ. They then disrupt interaction with additional Aβ moieties, preventing oligomerization, or in the case of pro-aggregants, facilitate them to accelerate aggregation to benign larger aggregate fibrils. This is often done with small molecules, derived from synthetic chemistry, though antibodies, peptides and other synthetic species like nanoparticles (ameliorating blood-brain-barrier transport) have all been assayed as potential anti-amyloid therapies (Aisen, 2005; Cummings et al., 2014; Giorgetti et al., 2018).

3.3 Aβ clearance and degradation

The accumulation of a substance in any system is dependent on both a steady inflow, and an inadequate outflow. AD's amyloidosis is similarly attributable to elevated production and impeded removal. Targeting and restoring Aβ's clearance mechanisms may thus offer yet another therapeutic approach against AD. This strategy may also allow for the removal of existing pathologies, offering the prospect of not only disease-arresting, but a curative therapy.

Physiologic Aβ clearance involves both proteosomic and lysosomic processes (Baranello et al., 2015; Yoon & Jo, 2012). The principal proteases thus far identified include neprilysin (Hersh & Rodgers, 2008; Marr et al., 2004), endothelin converting enzyme (ECE) (Eckman, Reed, & Eckman, 2001), insulin degrading enzyme (IDE) (Farris et al., 2003), plasmin (Tucker et al., 2000), and cathepsin-B (CatB) (Wang, Sun, Zhou, Grubb, & Gan, 2012; Yoon & Jo, 2012). All have been associated with significant Aβ perturbations among knockouts/overexpression models. Moreover, neprilysin has demonstrated an ability to degrade oligomers and fibrils of Aβ – making them especially relevant toward an Aβ therapeutic (Baranello et al., 2015; Kanemitsu, Tomiyama, & Mori, 2003). Lysosomic elimination of Aβ relies on the operation of glial cells, namely astrocytes and microglia. These cells, also sources of Aβ degrading enzymes, can internalize Aβ via pinocytosis (via receptors such as P2Y4) or phagocytosis (facilitated by the complement system) and expose them to lysosomal peptidases (Lee & Landreth, 2010; Ries & Sastre, 2016; Yoon & Jo, 2012).

Aβ can also be cleared from the brain by receptor-mediated transcytosis across the blood-brain-barrier (BBB). This typically involves the formation of complexes with chaperones, and subsequent transport across the BBB via the corresponding receptor. Apolipoprotein E (ApoE) has been the best studied of these chaperones, and though the mechanism of clearance has emerged as somewhat controversial, it is known that enhancing ApoE expression can elevate Aβ clearance, likely mediated by the *low density lipoprotein receptor-related protein 1* (LRP1) (Bachmeier et al., 2013; Baranello et al., 2015; Donahue & Johanson, 2008). Notably, mutations in lipoprotein genes were among the first genetic predispositions identified in AD, and they continue to share robust associations with AD onset and prognosis. The mediation of Aβ clearance may be one mechanism by which this

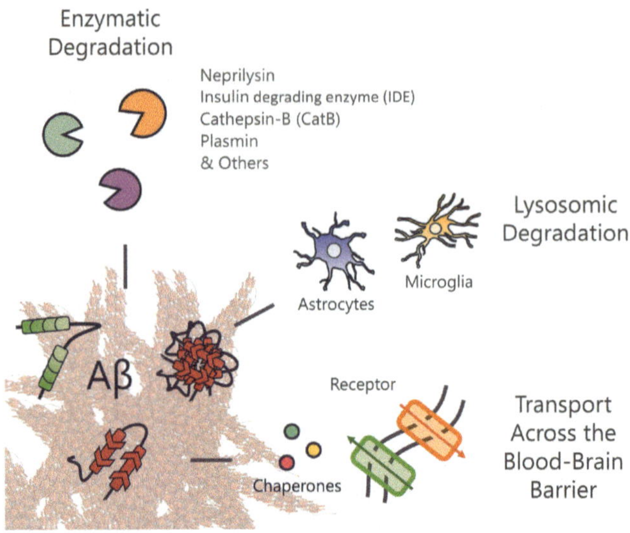

Fig. 4 ***Clearance and degradation of Aβ.*** Aβ can be degraded either by one of several enzymes (above) or via exposure to the lysosomes of neural glial cells (astrocytes and microglia). Aβ can also be cleared from the brain via chaperone and receptor mediated transport across the blood brain barrier. Notably, these systems are paralleled by Aβ import mechanisms into the brain, suggesting the potential for a dynamic homeostasis of cerebral amyloid.

genetic association occurs (Bachmeier et al., 2013), and underscores the possible relevance of β-amyloid clearance to AD (Fig. 4).

The brain also possesses a potent Aβ influx mechanism via the *receptor for advanced glycation endproducts* (RAGE) (Deane, Bell, Sagare, & Zlokovic, 2009). This directly countermands the action of LRP1 and transports Aβ from the blood into the brain via a complex signal transduction pathway — also associated with the inflammasome (Bierhaus, Stern, & Nawroth, 2006; Kierdorf & Fritz, 2013). As yet, the consequences of these two opposing systems for AD's amyloid burden have not been clarified; though the presence of both an influx and efflux mechanism suggests a relative homeostatic counterbalance for Aβ in the brain. The accumulation of Aβ may then represent a homeostatic dysregulation of native excretory/inhibitory signaling — suggesting the possibility for an endogenous therapeutic strategy. To date, the viability of these targets has not been systematically studied, though several agents seeking to inhibit β-amyloid influx and promote β-amyloid efflux are being explored. One such agent, PF-04494700 — a potent RAGE inhibitor, was trialed in a cohort of nearly 400 subjects with AD; however high-doses of the intervention were associated with deleterious outcomes and low-doses proved inconclusive (Galasko et al., 2014). Another well-studied agent is Somatostatin; AD brains are known to lack somatostatin in regions critical to memory formation, and its supplementation has shown some promise as a multifunctional Aβ therapeutic (Burgos-Ramos et al., 2008; Solarski, Wang, Wille, & Schmitt-Ulms, 2018). One potential mechanism of operation may be upregulating the activity of neprilysin and IDE, which suggests the possibility of artificially manipulating Aβ degradation (Hama & Saido, 2005; Tundo et al., 2012). These correlations have yet to yield a viable clinical therapy, though several avenues in the clearance and degradation of Aβ remain under investigation.

3.4 Indirect Aβ therapeutics

Failures in the direct manipulation of Aβ has driven some attention away from the peptide itself, and toward the management of physiologic factors known to influence Aβ dynamics. While it may seem an unlikely therapeutic course, the management of brain physiology may be more easily achieved than the management of a singular peptide's operation. In AD, three characteristics appear critical to Aβ′s behavior: metal ion homeostasis, cholesterol levels and inflammatory stress. All have come under investigation as potential therapeutic avenues.

3.4.1 Metal ions

The role of metal ions in AD is multifaceted and increasingly complex (Maynard, Bush, Masters, Cappai, & Li, 2005). It is well established that AD brains exhibit significant metal perturbations, but the precise direction of these changes has proved difficult to identify. Broadly, it appears that there is a decrease of serum ions such as Zn^{2+} and Fe^{3+} (Li, Zhang, Wang, & Zhao, 2017; Paglia et al., 2016), accompanied by possible elevation of these ions in the brain (locally, if not systemically) (Nuttall & Oteiza, 2014; van Duijn et al., 2017). The phenomenon has been termed 'flooding' or 'pooling' and may represent a physiologic mechanism to counteract AD's neural damage. Once in the brain, however, the concentration of these ions are known to be elevated in and around amyloid plaques, and they can promote the interaction and aggregation of Aβ (Hane & Leonenko, 2014; Lovell, Robertson, Teesdale, Campbell, & Markesbery, 1998; Smith, Harris, Sayre, & Perry, 1997). Data also suggests that metal ions may stabilize oligomeric species, such that their aggregation to benign higher order species is perturbed (Lee, Kim, Na, & Eom, 2018). In this way, metal ions may preserve Aβ in its most neurotoxic state, and contribute to the pathogenesis of AD.

Work in our group has also identified that Zn^{2+} can also perpetuate the association of Aβ to cellular membranes, via interaction with anionic cell surface macromolecules, like glycosaminoglycans. As such, the chelation of metals from the brain may offer multiple benefits toward ameliorating Aβ pathologies. However, the role of these metals within the dynamic state of the brain is unclear. Furthermore, while the levels of ions in the brain may be elevated, serum levels are reduced, thus any chelation strategy must be restricted to the brain, lest broader dyshomeostases be worsened. Indeed, trials supplementing metal ions have also been proposed — specifically seeking to alleviate the serum deficits of critical metals (Adlard & Bush, 2018). Presently, there is ample evidence suggesting some critical role for metals in AD, and Aβ processing/toxicity, however translating these to a clear therapeutic strategy has not been achieved.

3.4.2 Cholesterol

The association with cholesterol has proved somewhat easier to discern. While no clear therapy has again been identified, cholesterol has been well-established as an aggravator of Aβ pathology. This correlation arose from genetic associations between cholesterol processing mechanisms and AD (Wollmer, 2010), and subsequent epidemiological data (upheld in meta-analyses) correlating high cholesterol consumption to the risk of AD (Anstey, Ashby-Mitchell, & Peters, 2017). Further studies have identified cholesterol may also play critical roles in the regulation of the proteolytic enzymes involved in APP processing (Beel, Sakakura, Barrett, & Sanders, 2010; Xiong et al., 2008) and may even have a direct role in the misfolding and aggregation of Aβ — possibly by acting as a nucleating factor in tandem with membranes (Habchi et al., 2018). Cholesterol also likely has broader synergies with vascular health and other organ systems which may further contribute toward a systemic impact on AD. These multiple associations have led to the testing of several cholesterol inhibitory agents, such as statins and other anti-biosynthetic agents. While multiple studies, including meta-analyses have now suggested a possible protective association against the risk of AD onset (Chu et al., 2018), no cholesterol therapy has yet to be shown effective as a disease modifying therapy, after disease onset.

3.4.3 Inflammation

The role of neuroinflammation on Aβ dynamics is appreciably nuanced and complex. The traditional conception of AD suggests that oligomers of Aβ trigger the initiation and propagation of inflammatory signaling, either directly is through danger/microbe-associated molecular patterns (DAMPS and MAMPS), indirectly via the activation of glial cells, or by inducing cellular damage — whose products go on to elicit an immune cascade. In this conceptualization, Aβ pathology precedes and triggers AD's inflammatory pathology. However, studies have reported that inflammatory pathologies exist prior to the onset of other Aβ mediated pathologies (Kinney et al., 2018). Likewise, AD is significantly more prevalent among cohorts with elevated inflammation, as well as those with repeated episodes of head trauma and/or infection (Fleminger, Oliver, Lovestone, Rabe-Hesketh, & Giora, 2003; Sochocka, Zwolinska, & Leszek, 2017). The current state of study suggests that Aβ and inflammation operate in a functional synergism in which inflammation can drive Aβ release, and Aβ can in turn elevate inflammatory stress. The presence of these mutually amplifying elements could generate a feedforward-loop, accounting for the unrelenting progression of AD symptoms after onset.

The abatement of inflammation is correspondingly a major target in the pursuit of a disease modifying therapy for AD. The outcomes of trials, however, parallel those of cholesterol. Prophylactic anti-inflammatory use (excluding corticosteroids) (J. Wang et al., 2015), as well as anti-inflammatory dietary nutrients (Gardener, Rainey-Smith, & Martins, 2016) are associated with a reduced risk of AD, however there has been no definitive evidence of any anti-inflammatory course, either alone or in combination, yielding significant clinical benefits after disease onset.

3.4.4 Failures of therapies

The repeated frustration of these physiologic strategies lends credence to the conceptualization of AD as a vicious feed-forward cycle, in that once AD's pathological machinery is active, it may not be feasible to arrest it without significant intervention. Alternatively, it may suggest that these targets may not be therapeutic, even upon successful inhibition. Under this paradigm, either we have yet to identify a viable modulator of Aβ, or perhaps manipulation of Aβ itself is irrelevant to disease course and onset. Current study is thus approaching a critical furcation, in which the amyloid hypothesis itself is in contention.

4. The amyloid hypothesis: fact or folly

4.1 Amyloid toxicity and therapeutic promise

Despite the ongoing failure of clinical trials, the amyloid hypothesis has been sustained for decades by the continuing discovery of novel and

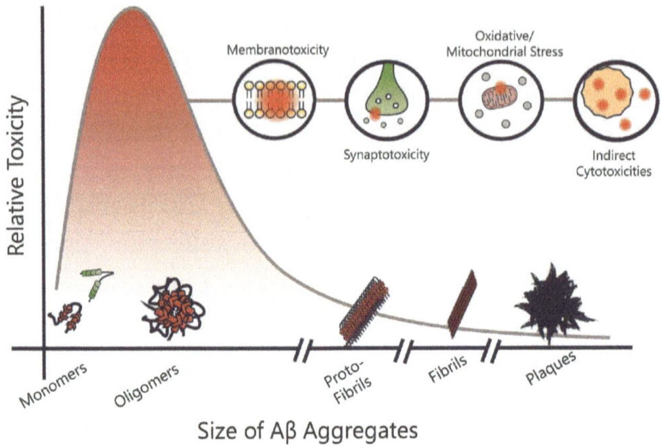

Fig. 5 ***Aβ toxicity.*** The toxicity of Aβ varies with the degree of aggregation. Monomers are mobile and soluble but tend to be relatively nontoxic. Conversely oligomers, though less soluble, are highly toxic – inflicting damage to membranes, synapses, mitochondria (inducing oxidative stress), as well as associating with concurrent pathologies such as tau and inflammation to induce broader cytotoxicities. As aggregates continue to grow, becoming protofibrils, fibrils and eventually plaques, they become increasingly insoluble and less harmful to cells.

increasingly diverse pathological roles for Aβ. These subsumed an array of cellular processes including membrane integrity, oxidative/mitochondrial stress, synapse regulation and activation of cell death mechanisms – which cumulatively left little doubt as to the critical role of Aβ within the pathogenesis of AD. Even as the etiology of AD grew more complex, and Aβ seemed but one of many concurrent proteopathic and immunopathic functionaries, its cytotoxicity justified its continued scrutiny as a therapeutic target. Aβ′s cytotoxicity remains among the strongest justifications for the amyloid hypothesis and suggests that though the etiology of AD is among the most complex disorders known, Aβ does play a role in its onset and propagation (Fig. 5).

4.1.1 Membranotoxicity

Deriving from a transmembrane peptide, it is unsurprising that the neuronal membrane would be a major site of β-amyloid toxicity. However, the mechanism and the consequences of this toxicity has proved challenging to substantiate. In the traditional conceptualization of the amyloid hypothesis, it was assumed that Aβ′s sequence of nonpolar residues, in regular patterns, would facilitate a simple insertion into the lipophilic environments between the leaflets of a cell membrane. However, quantum mechanical and other molecular modeling calculations from our own group and others have revealed that the sequence of residues alone is insufficient to account for the membrane penetration, and that not all species of Aβ can penetrate membranes (Bode, Baker, & Viles, 2017; Tsai, Lee, Tseng, Pan, & Shih, 2010). The process of insertion is in fact energetically unfavorable among model phospholipid membranes, and therefore should not readily occur.

We could only model insertion upon anchoring of the peptide to the surface of the membrane, which occurred with the introduction of metallic cations (principally Zn^{2+} and Cu^{2+}) and anionic macromolecules to the surface. Membranal cholesterol also facilitated insertion and lowered the activation energy required for penetration to occur. Despite this, our calculations observed insertion in only 5% of attempts, suggesting penetration of the membrane remains unfavorable, though conceivable with sufficient concentrations and exposure. Other groups overcame the energy barrier by sampling different species of Aβ in their molecular modeling computations (Bode et al., 2017; Tsai et al., 2010).

These calculations were also scalable to oligomers. If preformed prior to insertion, oligomers could insert into the membrane, and substantially disrupt a localized area of lipids. This does not form a traditional, ordered pore — however the disruption is sufficient to induce leakage of ions such as Ca^{2+}. Recent atomic force and electron microscopy data have further suggested that inserted oligomers may yield extensive discontinuities beyond the initial insertion site, paralleling the activity of detergents (Bode, Freeley, Nield, Palma, & Viles, 2019). Alternatively, Aβ may decrease membrane fluidity, in complexes with GM1 (Peters et al., 2009). Studies also suggest that insertion may initiate a neuroinflammatory response (Fernandez-Perez, Peters, & Aguayo, 2016). The impact of Aβ on membrane integrity may therefore be more pronounced that previously conceived. However, these disruptive effects were observed almost entirely for oligomers. Insertion of monomers appears benign, if not negligible in terms of cytotoxic effects (Bode et al., 2017; Poojari, Kukol, & Strodel, 2013; Rajasekhar, Chakrabarti, & Govindaraju, 2015). Likewise, proto-fibrils or larger species do not readily insert into membranes — rather they sequester the hydrophobic domains necessary for membrane solubilization within the core of the oligomers. Thus the elimination of oligomers, either by the inhibition of aggregation or the acceleration of aggregation past toxic thresholds may be prudent strategies to abate this toxicity.

4.1.2 Oxidative and mitochondrial stress

As with cellular membranes, mitochondria are similarly susceptible to penetration and poration by Aβ. The electrophysiology of these membranes, hypothesized to derive from an ancient bacterial symbiosis, tend to be more charged than typical somatic membranes – rendering them especially vulnerable to amyloid attack. Unlike cellular membranes, however, mitochondria house potent reactive radicals within their inner and outer membranes (Cadenas & Davies, 2000). These species are byproducts of the electron transport chain, which relies on the formation of reactive radicals to drive the synthesis of adenosine triphosphate (ATP); but if uncontrolled they can react rapidly with oxygen and nitrogen to form reactive oxygen species (ROS) and reactive nitrogen species (RNS) respectively (Di Meo, Reed, Venditti, & Victor, 2016). ROS and RNS may also be generated by cells and mitochondria enzymatically, as a marker of stress and damage. After generation, these free radical species can initiate propagative tissue damage, including lipid and protein peroxidation – and if unquenched, can lead to neuronal death (Ademowo, Dias, Burton, & Griffiths, 2017; Bergamini, Gambetti, Dondi, & Cervellati, 2004; Di Meo et al., 2016; Valencia & Moran, 2004).

A lesser known, but perhaps equally destructive mode of free radical generation involves the direct reduction of metals. This explanation arose from the observation of redox-active copper ions within plaques of Aβ (Rajasekhar et al., 2015). Subsequent studies revealed that these ions were not simply accumulating amidst the plaques, but were actively being generated by an undefined catalytic component within Aβ itself – likely involving one or more amino acid interactions (Opazo, Ruiz, & Inestrosa, 2000). The reduction of metals necessarily liberates an electron, if this is not incorporated in a corresponding oxidation reaction, an oxidative radical would be created. It is noteworthy that no singular residue has yet shown an ability to independently reduce and radicalize a metal ion; it is therefore likely that complex secondary and tertiary structural synergies drive the reduction of metals.

This observation confounds the already complex relationship between Aβ and metal ions. It suggests that irrespective of any existing cellular reserves of oxidative stress, Aβ acting with metal ions could be an independent source of reactive oxidative species. Pending further study, the elimination of Aβ species capable of producing reactive ions may therefore contribute toward a multifactorial therapeutic strategy.

4.1.3 Synaptotoxicity

Synaptic dysfunction was thought to be a direct consequence of β-amyloid accumulation. However, the association has recently faced significant contradiction, most prominently with the observation of a significant temporal discontinuity between the loss of synapses and the appearance of amyloid pathology (Sheng, Sabatini, & Sudhof, 2012). The loss of synapse activity is a major cause of AD symptoms; thus if symptoms manifest before the presence of plaques, then they are unlikely to be of relevance to a disease-modifying therapy. Yet, as with membrane poration, the pathological elements appear not to be large species, but rather smaller, toxic oligomers. Mechanistic studies have clarified that these oligomers can bind (if not strongly interact) with critical synaptic receptors, mediating neural functions (Mucke & Selkoe, 2012). These include: glutamate and N-Methyl-D-Aspartate (NMDA) receptors (Texido, Martin-Satue, Alberdi, Solsona, & Matute, 2011) (modulating cAMP response element-binding protein [CREB] activity) involved in learning and memory retention; nicotinic receptors involved in transmembrane importing and internalization of Aβ (Jurgensen & Ferreira, 2010); and Small Ubiquitin-like Modifier (SUMO) molecules (Lee, Sakurai, Matsuzaki, Arancio, & Fraser, 2013) which mediate long-term potentiation and hippocampal-dependent learning. Broader interactions with other synaptic elements are emerging, suggesting Aβ may act as a promiscuous agonist/antagonist in the region. It remains unclear why, or which Aβ elements offer this broad interaction, though the multiple associations suggest a possible functional role in synapse modulation (discussed below).

It is notable that while multiple receptor interactions have been shown, there are little data showing the *in vivo* effects of these interactions. Reports have identified that intraneuronally injected Aβ can instigate AD's characteristic neurotoxic perturbations (Nomura, Takechi, & Kato, 2012). Other models overexpressing Aβ or receiving similar injections into critical brain regions report comparable AD-like dysfunctions (Balducci et al., 2010; Faucher, Mons, Micheau, Louis, & Beracochea, 2015; Higgins, Holtzman, Rabin, Mobley, & Cordell, 1994; H. Y. Kim, Lee, Chung, Kim, & Kim, 2016). However, it is possible that the biological insult of the injection, in synergy with Aβ, may confound the onset of symptoms, in the same way that brain trauma is independently associated with dementia (Fleminger et al., 2003; Ramos-Cejudo et al., 2018). Thus, the association of Aβ and the synapse is strong from a molecular level, though yet to be demonstrated biologically.

4.2 The failure of amyloid therapies

The cumulative burden of these above-described pathologies would suggest Aβ would be an ideal therapeutic target. Indeed, these pathologies justified hundreds of clinical trials, thousands of therapeutic agents, and many more resource-intensive basic science, epidemiological, and pharmacological studies. Nonetheless, the overwhelming failure of these trials to yield a viable therapy has been well documented. Presently, fewer and fewer trials target Aβ on its own, and the amyloid hypothesis is increasingly rejected, even ridiculed as a failed endeavor. Several critical findings have dealt what some have considered the final, conclusive evidence on the failure of amyloid hypothesis. These include the observation of significant plaque pathology in nondemented or asymptomatic seniors (Erten-Lyons et al., 2009; Price et al., 2009; Zhan, Veerhuis, Kamphorst, & Eikelenboom, 1995), and a growing trend identifying deleterious outcomes in trials of apparently successful anti-amyloid therapies (Espay et al., 2019; Morris et al., 2014). The conclusion from these observation suggests either that β-amyloid is irrelevant to AD, and in the case of the latter, that Aβ may be protective or serve some beneficial biological function, thereby being simultaneously both neuroprotective and neurotoxic.

It is challenging to reconcile these two apparently contradictory observations, indeed investigations and critical appraisals tend to avoid addressing the paradox. Most either accept or reject the amyloid hypothesis and present their findings within their accepted perspective. Our best interpretation of current study is that targeting misfolded protein as a therapy for AD is both sound and flawed at the same time. First, it is clear that Aβ, in the appropriate form of oligomers, is toxic and that the administration of Aβ to neurons, be it *in vitro* or *in vivo*, is lethal (beyond a critical concentration). Yet, the failure of the hypothesis to yield a viable therapy, despite vigorous and diverse investigation is unequivocal. Herein, we propose three possible explanations to account for the failure; (1) Aβ and patients' response to pathologies are fundamentally heterogeneous; (2) traditional *in vivo* models and clinical trials may be inadequate to assess AD; (3) Aβ may not be entirely pathogenic.

4.2.1 The heterogeneity of Aβ & patient response to pathology

The presence of amyloid plaques in non-demented seniors is arguably the most challenging observation to reconcile with a pathologic role for Aβ. However, the manifestation of symptoms in most forms of brain injury is fundamentally heterogeneous; indeed patients with similar injuries can exhibit grossly different symptoms (Lingsma, Roozenbeek, Steyerberg, Murray, & Maas, 2010; Millis et al., 2001; Steward et al., 2018). The concept of reserve has been proposed to bridge the undefined gap between brain pathology and symptoms. It posits the presence of a series of factors, including plasticity, neurotransmitter regulation and synapse activity, which may offer a protective mechanism to the brain, such that symptoms may be delayed or lessened (Esiri & Chance, 2012). Though reserve remains a theoretical proposition, and its underlying operations have not been defined, factors known to increase reserve, such as multilingualism (Klein, Christie, & Parkvall, 2016) and participation in leisure activities (Akbaraly et al., 2009) can all significantly lower the risk of AD. Further evaluation of the functional elements responsible for reserve and its cognitive protection may yield, yet unexplored therapeutic avenues.

Another critically heterogeneous, but frequently uniformized element is Aβ itself. Commencing from its synthesis, it can exist as multiple different lengths, the longer of which are usually more aggregation prone (Lambermon et al., 2005). The process of misfolding too is varied, and may originate by interaction with multiple species, including membranes, cholesterol and preexisting misfolded proteins. Finally, the degree of aggregation amongst Aβ peptides can also be varied, namely a peptide can exist as monomers, oligomers, protofibrils, fibrils and plaques in a continuous spectrum. Plaques are large and easily visualized (*in vivo* and *in vitro*); thus inhibition of plaque formation drove much of anti-amyloid research. Only recently has the observation that oligomers, not plaques, are responsible for much of amyloid toxicity, been widely accepted. While it may seem logical that an agent capable of preventing plaque formation would also inhibit the formation of smaller species, it need not be so. The molecular forces and interactions governing the formation of plaques are substantially different than those responsible for the creation of oligomers. Thus agents capable of altering the formation of plaques may or may not be viable anti-oligomeric agents. Here, we present the case of two molecules synthesized by our group, one of which proved a potent anti-fibrillogenic agent, but was ineffective in preventing the formation of oligomers, and vice versa (Fig. 6).

The failure to assay the inhibition of oligomers, as opposed to easily visualized species like plaques and fibrils may account for the ineffective outcomes of Aβ therapeutics. Studies may also gain clarity by assaying their agent's efficacy against multiple sub-types of Aβ ($A\beta_{1-40}$, $A\beta_{1-42}$, etc). Pending further study to resolve which amyloid species are most pathogenic,

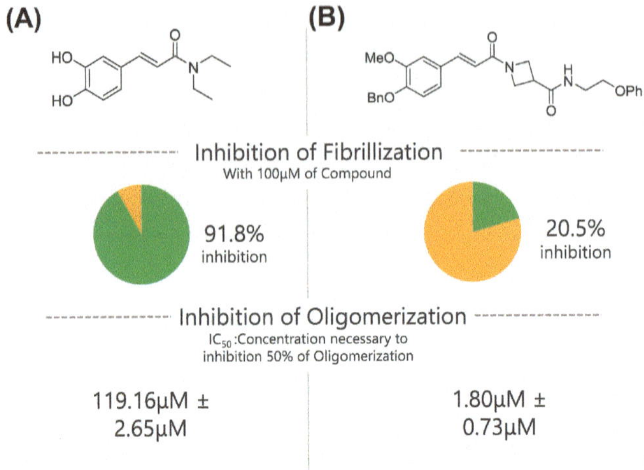

Fig. 6 ***Incongruity between therapeutic anti-aggregation strategies.*** Agents which inhibit the fibrillization of Aβ into large plaques need not be inhibitors of Aβ oligomerization. Two exemplary molecules, synthesized by or group, are presented here. Molecule A successfully inhibits Aβ fibrillization (91.8% inhibition, as assayed by Thioflavin T fluorescence), but requires inordinately high concentrations (119.16 μM) to inhibit oligomerization (assayed by ELISA). Conversely molecule B is a poor anti-fibrillogenic agent (20.5% inhibition), though is a potent anti-oligomeric agent (1.80 μM). Failing to assay oligomerization directly may thus leave the most toxic amyloid species unaffected.

noting that it may be different by patient, efficacy against a parameter of interest should be considered across multiple species of Aβ.

4.2.2 Trialing therapeutics

Therapeutics against AD, as with any drug development pathway, pass through a series of increasingly rigorous investigations, culminating in human clinical trials. In a typical therapeutic course, a perspective agent is first tested *in vitro* against a discrete parameter (such as amyloid aggregation). Successful agents then progress to *in vivo* testing, typically in mice and/or rats, in which efficacy as well as other fundamental drug criteria, such as toxicity, metabolism, absorption and excretion are assayed. Finally, agents showing both efficacy and acceptable drug characteristics are given to humans, in progressively larger trials, culminating in Phase III trials, in which thousands of subjects receive the agent, and their outcomes are followed for extended periods of time.

This methodology has been employed successfully for countless therapeutics, in varied disease states. However, AD trials undergoing this same paradigm have failed at an unprecedented rate. While this is often attributed to an incomplete understanding of AD, it is also conceivable that methodological inadequacies in the testing process may also be contributive. Specifically, we suggest current AD *in vivo* models and traditional clinical trial designs may not be sufficient to capture true disease state and response to investigational agents.

4.2.2.1 In vivo models of AD

Modeling AD *in vivo* is a near impossible challenge as the neurotoxic synergy between multiple proteopathies and immunopathies in the brain cannot yet be reliably recreated. Instead, most AD models rely on transgenic overexpression of mutated human genes associated with Aβ synthesis and processing (typically APP and/or Presenilin) (Malm, Magga, & Koistinaho, 2012). They offer a practical model of protein misfolding in a laboratory setting, and may possibly be a viable model of EOAD; however elevating Aβ production alone does not model the complex etiology of LOAD. Studies have attempted to assuage the shortcomings of the traditional models by testing therapeutics in other models which over express tau or neuroinflammatory elements (Kitazawa, Medeiros, & Laferla, 2012). These models are useful for assaying efficacy against non-amyloid pathologies, yet, these fail to capture the synergy between pathologies, which is critical to the manifestation of AD. Recently, some mouse models have been developed which simultaneously upregulate Aβ and tau (Kitazawa et al., 2012), but as with AD in humans, plaque development proved more variable, particularly with age.

Currently, it seems that the closer a model gets to recreating the complex etiology of human AD, the less reproducible and reliable it will be as a laboratory assay. It is therefore paradoxical to develop more representative models, though using current models risks advancement of ineffective agents into clinical trials. Consideration should also be given to the temporal manifestation of pathologies. As the complexity of the model grows, the time to manifest pathologies may differ, and may not reach a therapeutic threshold until later in life. It may therefore be unadvisable to assign treatment based solely on time from birth; considering the rate and extent of disease progression, perhaps assayed by outward memory or learning, may be advantageous.

4.2.2.2 AD clinical trials

The same complexities of patient heterogeneity, reserve and uncertain pathology discussed above inevitably interfere with AD clinical trials. Trials in AD are further encumbered by the fact that they must measure abstract concepts such as cognition and memory. Therefore elements not routinely considered in neurological trials, such as diet (Swaminathan & Jicha, 2014), exercise (Morris et al., 2017), trauma (Fleminger et al., 2003) and sleep (Brzecka et al., 2018) may be highly relevant to a patient's transient symptomology, and a corresponding assay of cognition. Factors relevant to cognitive reserve (mental exertion, leisure participation, multilingualism) may also be critical confounding/interfering variables, which may obscure the clarity of an outcome measure, if unaccounted or unbalanced.

In spite of their overwhelming failure, clinical trials tend to follow the traditional paradigm, controlling only for common parameters like age, gender and comorbidities. While these are undoubtedly critical parameters, trials in AD may benefit from including additional AD-specific factors in the consideration of their study design. We have previously outlined a more comprehensive list of variables, and strategies to abate their influence (Liyanage, Santos, & Weaver, 2018). The current clinical trial paradigm may be simply inadequate to afford meaningful results in trials of experimental drugs addressing AD.

4.2.3 Physiologic role for amyloid-β

A highly-conserved genetic signature of amyloid proteins can be found across numerous orders of life — spanning species from Hymenoptera to the Nematodes (Tharp & Sarkar, 2013). It is evolutionarily illogical for a purely self-destructive peptide to be so widespread and conserved, even if its production occurred after reproductive prime. Some role for Aβ within normal physiology is thus likely, though what this role is remains speculative.

At present, a role as an antimicrobial peptide (AMPs) is a leading candidate for a potential physiologic role (Gosztyla, Brothers, & Robinson, 2018; Kumar et al., 2016; Soscia et al., 2010). Human AMPs are simple innate immune effectors which work by compromising foreign organisms' membranes as well as mediating other cytotoxic and signaling effects (G. Wang, 2014). They are produced by the body in response to noxious stimuli, including infection and trauma, and are known to mediate broader immune responses. All these characteristics are shared by Aβ. Studies further report Aβ is a potent *in vitro* and *in vivo* antimicrobial — all of which lends credence to the notion that Aβ expression may arise from aberrant AMP signaling (Gosztyla et al., 2018). However, it remains unclear as to why a physiologic AMP would attack a 'self' entity, nor why Aβ expression could proceed from a transient AMP signal to chronic overexpression. One potential explanation is that neural membranes exhibit transmembrane potential charge gradients beyond typical somatic cells; therefore, Aβ may not be able to distinguish between neuronal and bacterial membranes.

Other roles for Aβ have been ascribed in neural development and differentiation, as well as synaptic plasticity and memory formation (Rajasekhar et al., 2015). Notably, small concentrations (picomolar range) of Aβ were shown to elevate neurogenesis and memory formation (Garcia-Osta & Alberini, 2009). It is then conceivable that AD and associated disorders may arise from dysregulation of an essential peptide, rather than the production of a pathological species. Moreover, these physiological roles may account for the worsening of symptoms observed in some anti-amyloid trials. The implications for a therapeutic outlook are unclear at this point, though considering Aβ as a nuanced peptide with both a pathological and physiological role may yield more fruitful therapeutic design strategies and outcomes.

4.3 Conclusions on Aβ

With the near universal failure of amyloid-based therapies, it is easy, even rational, to reject the amyloid hypothesis and consider new therapeutic avenues. However, the weight of current study is unable to substantiate its complete rejection; rather both the significant mechanistic understanding of Aβ′s pathological and physiological role, as well as the misfocus of prior study on benign elements such as plaques justify further inquiry. We therefore conclude that it is premature to reject the amyloid hypothesis, in its entirety. Yet, it is clear now that therapies which singularly target Aβ are likely to remain ineffective, particularly if administered late in the course of the disease when the patient is symptomatic and Aβ has been misfolding an accumulating for years. AD, aside from a small fraction of early onset cases arises from a complex interlay of pathogenic factors. This includes Aβ, but also other proteopathies and associated immunopathies. These pathologies appear to operate in synergy, amplifying each other in a vicious feed-forward cycle. Targeting Aβ alone seems insufficient to arrest this cycle, as such inquiry into multifactorial therapies inhibiting multiple elements within this combined proteopathic-immunopathic paradigm may be a fruitful course of investigation.

5. Tau protein

Were it not for its role as a key pathogenic hallmark of AD, tau would have remained one of several obscure microtubule associated proteins that mediate the structure and stability of tubulin filaments. It was characterized in 1975, not as a major neurodegenerative agent, but as a factor able to promote the self-assembly of tubulin (Weingarten, Lockwood, Hwo, & Kirschner, 1975). It took approximately another decade before tau was definitively identified as the core protein in intracellular neurofibrillary tangles, and began receiving attention as a potential therapeutic target in AD (Mandelkow & Mandelkow, 2012).

Tau's role in AD has been the subject of intense controversy, nearly since its characterization. Shortly after its discovery, β-amyloid processing was implicated in early onset AD; thus tau seemed to be a secondary pathology, and became somewhat sidelined as the brunt of AD research turned to the amyloid hypothesis. However, as research progressed, tau was identified as not only an independent protein pathology, but the likely causative agent in frontotemporal dementia (Gasparini, Terni, & Spillantini, 2007), Pick's disease (Probst, Tolnay, Langui, Goedert, & Spillantini, 1996) and other proteopathic disorders. Critical reviews soon observed that AD was one of few diseases in which tau co-existed with another proteopathy, and that no disorder yet found showed Aβ as the singular pathology (Rosenmann, Blum, Kayed, & Ittner, 2012). This led to yet another protein misfolding hypothesis — the tau hypothesis — which posits that tau, not Aβ is the instigating pathology of AD. The dynamic between tau, Aβ and AD is not clear, though tau has emerged as a potential druggable target, be it independently or in combination with other pathologies.

5.1 Tau protein structure

Human tau occurs largely in neural axons and can take the form of 6—7 isoforms, ranging from 352 to 441 amino acids (a 695 amino acid variant is also known to exist) (Goedert & Jakes, 1990; Goedert, Spillantini, & Crowther, 1992). They differ mainly by the inclusion or exclusion of two exons (exons 2 & 3) and various minor modifications thereafter (Mandelkow & Mandelkow, 2012). The peptide sequence of tau is notable for its preponderance of hydrophilic residues (Mandelkow & Mandelkow, 2012). The lack of hydrophobicity affords the final peptide an unusually loose folded structure, with minimal secondary features (such as β-pleated sheets or α-helices) (Jeganathan, von Bergen, Mandelkow, & Mandelkow, 2008). It is thought that this fluidity allows tau to preserve the integrity of microtubules, even under relatively extreme circumstances. The other fundamental structural feature of tau is a series of functional domains which perform tasks such as binding to the microtubule, and projecting into the cytosol (Trinczek, Biernat, Baumann, Mandelkow, & Mandelkow, 1995).

5.2 Tau pathology

The mechanism of tau pathology is considerably different from that of Aβ. Upon synthesis, an Aβ peptide is effectively pathogenic, however wild-type tau is a functional cellular protein. In order to become pathological, it must first undergo a series of critical modifications. Discerning what these changes are, and how to reverse them continues as a focus of significant research interest. Correspondingly, a number of post-translational modifications, including phosphorylation (Johnson & Stoothoff, 2004), ubiquitination (Oddo, 2008), acetylation (Lucke-Wold et al., 2017), nitration (Horiguchi et al., 2003), and other biochemical processes were all identified as possible modulators between the physiologic and pathologic states (Mandelkow & Mandelkow, 2012). Among these, phosphorylation proved especially relevant. Phosphorylation induces a structural change in tau, such that it loses affinity for the microtubule and is liberated into the cytosol. Once detached, the microtubule destabilizes, and is able to remodel and reform — a critical cellular process. Therefore, it is unsurprising that tau possesses numerous phosphorylation sites, and cells were shown to encode several tau kinases and phosphatases (Wang, Grundke-Iqbal, & Iqbal, 2007). However, in AD, tau becomes hyperphosphorylated, due either to overactivity of the kinases, or inhibition of the phosphatases (possibly both). The result is systemic tau detachment, and broad microtubule instability. Moreover, once free, the formerly loosely folded tau can associate into a toxic oligomer, and ultimately into sizable cellular neurofibrillary tangles.

As with Aβ, the large tangles were initially ascribed the primary pathogenic roles, however the locational association between tangles and neuronal loss was tenuous. This shifted focus to soluble oligomers, which proved the most cytotoxic assembly of tau (Kopeikina, Hyman, & Spires-Jones, 2012; Spires-Jones, Kopeikina, Koffie, de Calignon, & Hyman, 2011). Current data ascribes pathological roles for oligomers of tau in axonal transport (Terwel, Dewachter, & Van Leuven, 2002), mitochondrial damage (Cheng & Bai, 2018), synaptic dysregulation (Tracy & Gan, 2018) and a series of other less defined mechanisms (Fig. 7).

5.3 Tau propagation

As an intracellular pathology, tau should not propagate beyond the initial cell within which it presents. The supposed self-containment was one reason

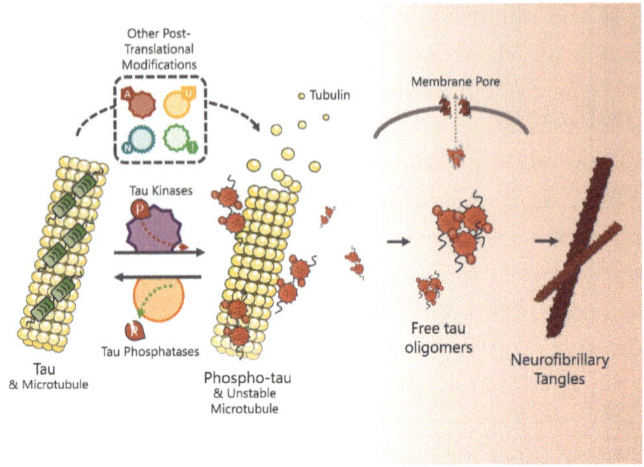

Fig. 7 ***Tau pathology.*** Tau is a multi-domain protein which stabilizes cytoskeletal microtubules. When phosphorylated - possibly accompanied by other post-translational modifications including acetylation (A), nitration (N), ubiquitination (U) and others (!), tau undergoes a conformational change, such that it no longer binds to the microtubule. This reduces the stability of the tubulin filaments, and free tau can aggregate to from neurotoxic oligomers and large neurofibrillary tangles. Tau may also escape to surrounding cells by poration of the plasma membrane, and subsequent uptake by an adjacent cell.

Aβ received greater interest, as it was thought that some external pathology was required to drive a reemergence of tau pathology in adjacent cells. However, injection of tau fibrils was able to produce endogenous tau pathologies in widespread brain regions (Peeraer et al., 2015). Upon further investigation, tau oligomers were revealed to translocate between synaptically affiliated neurons (DeVos et al., 2018). The mechanism of transport may be diverse, and involve receptor mediated endocytosis, proteoglycan-mediated micropinocytosis; even APP may serve as a receptor, pointing to yet another avenue of potential synergy between AD's two pathologies (Gerson & Kayed, 2013; Mudher et al., 2017). Tau may also pass through cells by forming membrane pores using its own oligomers, effectively encoding a self-propagation mechanism (Gerson & Kayed, 2013).

A critical observation — gaining credence among recent reports — identified that the manner of tau's propagation parallels the development of AD's clinical symptomology and neural pathology better than that of Aβ. This is unsurprising as a defined transmission between connected neurons is a more sensible disease course that Aβ′s effectively random diffusion in the extracellular space. These initial suppositions, as well as tau's apparent independent mechanism of initiation has renewed interest in tau as AD's central pathology.

Current study cannot definitively attribute either tau or Aβ as the absolute instigating pathology. It is conceivable that both may have an initiatory role, or that AD in different patients may commence differently. It is also possible that an upstream stimulus instigates both pathologies simultaneously. Until further mechanisms of AD pathogenesis are defined, discerning between the relative importance of tau and Aβ is speculative.

5.4 Therapeutic targets for tau

There are several steps in the pathogenesis of tau that are amenable to therapeutic intervention. In direct contrast to Aβ, tau cannot be wholly suppressed as this would deprive the brain of an essential microtubule stabilizing protein. Rather, interventions must be selective toward pathogenic processes only. Former studies (limited relative to amyloid targets) sought mainly to reduce the phosphorylation of tau (inhibit tau kinases) or reduce tau aggregation (Congdon & Sigurdsson, 2018). These proved ineffective, and focus has now shifted somewhat toward immunotherapies, and indirect strategies such as elevating baseline microtubule stability — as a means to alleviate tau's loss of function (Ballatore et al., 2012). None of these strategies have yet yielded a viable clinical agent, though the popularity of anti-tau agents has grown substantially.

6. Conclusions & future outlook

Evaluating misfolded proteins as a therapeutic target in AD is to confront a field of intense academic and industrial controversy. Both Aβ and tau have been the subject of vigorous investigation. At a mechanistic and basic science level, they have largely been fruitful. Data now support a clear pathogenic role for both in cellular operations as diverse as membrane integrity, mitochondrial operations, homeostasis, transport, cytoskeletal dynamics, cell cycle signaling and others. They have further begun to identify the dynamic synergies between cellular and other pathological systems (such as chronic inflammation) responsible for the unrelenting propagation of the pathologies throughout the brain. Yet, this immense body of basic and mechanistic study has been unable to yield a viable clinical therapy. Instead, the past few decades have seen hundreds of therapies fail in trial after trial. It is therefore entirely sensible to question the viability of protein misfolding as a therapeutic target.

While the failure of so many trials is compelling, the total rejection of the protein misfolding hypotheses seems premature. Many of these trials sought the singular inhibition of the wrong form of the proteopathy (plaques and tangles, not oligomers); they relied on models of AD which are poorly representative of the human disease; they employed potentially inadequate clinical trial design; and/or they were derived from an unclear etiology of AD. Notably, some of these deficiencies have not been completely addressed, despite the emergence of additional therapeutics against effectively the same targets.

Perhaps the only definitive conclusion that can be drawn, both from the failure of trials and the conclusive mechanistic data, is that AD is a multifactorial disorder of numerous, synergistic, concurrent pathologies. Protein misfolding is among these, but to discover meaningful therapies addressing both symptoms and disease progression, will likely require targeting and inhibiting multiple mechanistic points within this pathogenic cascade simultaneously. Arterial hypertension, a disorder that is pathogenically trivial compared to AD, is often treated with multiple agents addressing complementary targets; accordingly, it may be naïve to anticipate one single "magic bullet" therapeutic agent for AD. The manner by which to design or discover such a comprehensive agent, as well as how best to translate it into a clinical setting should be fundamental considerations moving forward.

Acknowledgments

DFW acknowledges support from a Tier 1, Canada Research Chair. The authors sincerely thank the Krembil Foundation for their ongoing support.

References

Ademowo, O. S., Dias, H. K. I., Burton, D. G. A., & Griffiths, H. R. (2017). Lipid (per) oxidation in mitochondria: An emerging target in the ageing process? *Biogerontology, 18*(6), 859–879. http://dx.doi.org/10.1007/s10522-017-9710-z.

Adlard, P. A., & Bush, A. I. (2018). Metals and Alzheimer's disease: How far have we come in the clinic? *Journal of Alzheimer's Disease, 62*(3), 1369–1379. http://dx.doi.org/10.3233/JAD-170662.

Aisen, P. S. (2005). The development of anti-amyloid therapy for Alzheimer's disease : From secretase modulators to polymerisation inhibitors. *CNS Drugs, 19*(12), 989–996. http://dx.doi.org/10.2165/00023210-200519120-00002.

Akbaraly, T. N., Portet, F., Fustinoni, S., Dartigues, J. F., Artero, S., Rouaud, O., … Berr, C. (2009). Leisure activities and the risk of dementia in the elderly: Results from the three-city study. *Neurology, 73*(11), 854–861. http://dx.doi.org/10.1212/WNL.0b013e3181b7849b.

Alzheimer's Association. (2016). 2016 Alzheimer's disease facts and figures. *Alzheimers Dement, 12*(4), 459–509.
Anstey, K. J., Ashby-Mitchell, K., & Peters, R. (2017). Updating the evidence on the association between serum cholesterol and risk of late-life dementia: Review and meta-analysis. *Journal of Alzheimer's Disease, 56*(1), 215–228. http://dx.doi.org/10.3233/JAD-160826.
Askarova, S., Yang, X., & Lee, J. C. (2011). Impacts of membrane biophysics in Alzheimer's disease: From amyloid precursor protein processing to abeta peptide-induced membrane changes. *International Journal of Alzheimer's Disease, 2011*, 134971. http://dx.doi.org/10.4061/2011/134971.
Bachmeier, C., Paris, D., Beaulieu-Abdelahad, D., Mouzon, B., Mullan, M., & Crawford, F. (2013). A multifaceted role for apoE in the clearance of beta-amyloid across the blood-brain barrier. *Neurodegenerative Diseases, 11*(1), 13–21. http://dx.doi.org/10.1159/000337231.
Balducci, C., Beeg, M., Stravalaci, M., Bastone, A., Sclip, A., Biasini, E., … Forloni, G. (2010). Synthetic amyloid-beta oligomers impair long-term memory independently of cellular prion protein. *Proceedings of the National Academy of Sciences of the U S A, 107*(5), 2295–2300. http://dx.doi.org/10.1073/pnas.0911829107.
Ballatore, C., Brunden, K. R., Huryn, D. M., Trojanowski, J. Q., Lee, V. M., & Smith, A. B., 3rd (2012). Microtubule stabilizing agents as potential treatment for Alzheimer's disease and related neurodegenerative tauopathies. *Journal of Medicinal Chemistry, 55*(21), 8979–8996. http://dx.doi.org/10.1021/jm301079z.
Baranello, R. J., Bharani, K. L., Padmaraju, V., Chopra, N., Lahiri, D. K., Greig, N. H., … Sambamurti, K. (2015). Amyloid-beta protein clearance and degradation (ABCD) pathways and their role in Alzheimer's disease. *Current Alzheimer Research, 12*(1), 32–46.
Bayer, T. A., Cappai, R., Masters, C. L., Beyreuther, K., & Multhaup, G. (1999). It all sticks together–the APP-related family of proteins and Alzheimer's disease. *Molecular Psychiatry, 4*(6), 524–528.
Beel, A. J., Sakakura, M., Barrett, P. J., & Sanders, C. R. (2010). Direct binding of cholesterol to the amyloid precursor protein: An important interaction in lipid-Alzheimer's disease relationships? *Biochimica et Biophysica Acta, 1801*(8), 975–982. http://dx.doi.org/10.1016/j.bbalip.2010.03.008.
Bergamini, C. M., Gambetti, S., Dondi, A., & Cervellati, C. (2004). Oxygen, reactive oxygen species and tissue damage. *Current Pharmaceutical Design, 10*(14), 1611–1626.
Bierhaus, A., Stern, D. M., & Nawroth, P. P. (2006). RAGE in inflammation: A new therapeutic target? *Current Opinion in Investigational Drugs,* 7(11), 985–991.
Bode, D. C., Baker, M. D., & Viles, J. H. (2017). Ion channel formation by amyloid-beta42 oligomers but not amyloid-beta40 in cellular membranes. *Journal of Biological Chemistry, 292*(4), 1404–1413. http://dx.doi.org/10.1074/jbc.M116.762526.
Bode, D. C., Freeley, M., Nield, J., Palma, M., & Viles, J. H. (2019). Amyloid-beta oligomers have a profound detergent-like effect on lipid membrane bilayers, imaged by atomic force and electron microscopy. *Journal of Biological Chemistry, 294*(19), 7566–7572. http://dx.doi.org/10.1074/jbc.AC118.007195.
Boller, F., & Forbes, M. M. (1998). History of dementia and dementia in history: An overview. *Journal of Neurological Sciences, 158*(2), 125–133.
Bondi, M. W., Edmonds, E. C., & Salmon, D. P. (2017). Alzheimer's disease: Past, present, and future. *Journal of the International Neuropsychological Society, 23*(9–10), 818–831. http://dx.doi.org/10.1017/S135561771700100X.
Brzecka, A., Leszek, J., Ashraf, G. M., Ejma, M., Avila-Rodriguez, M. F., Yarla, N. S., … Aliev, G. (2018). Sleep disorders associated with Alzheimer's disease:

A perspective. *Frontiers in Neuroscience, 12*, 330. http://dx.doi.org/10.3389/fnins.2018.00330.

Burgos-Ramos, E., Hervas-Aguilar, A., Aguado-Llera, D., Puebla-Jimenez, L., Hernandez-Pinto, A. M., Barrios, V., et al. (2008). Somatostatin and Alzheimer's disease. *Molecular and Cellular Endocrinology, 286*(1–2), 104–111. http://dx.doi.org/10.1016/j.mce.2008.01.014.

Cadenas, E., & Davies, K. J. (2000). Mitochondrial free radical generation, oxidative stress, and aging. *Free Radical Biology and Medicine, 29*(3–4), 222–230.

Cheng, Y., & Bai, F. (2018). The association of tau with mitochondrial dysfunction in Alzheimer's disease. *Frontiers in Neuroscience, 12*, 163. http://dx.doi.org/10.3389/fnins.2018.00163.

Chu, C. S., Tseng, P. T., Stubbs, B., Chen, T. Y., Tang, C. H., Li, D. J., … Lin, P. Y. (2018). Use of statins and the risk of dementia and mild cognitive impairment: A systematic review and meta-analysis. *Scientific Reports, 8*(1), 5804. http://dx.doi.org/10.1038/s41598-018-24248-8.

Cizas, P., Budvytyte, R., Morkuniene, R., Moldovan, R., Broccio, M., Losche, M., … Borutaite, V. (2010). Size-dependent neurotoxicity of beta-amyloid oligomers. *Archives of Biochemistry and Biophysics, 496*(2), 84–92. http://dx.doi.org/10.1016/j.abb.2010.02.001.

Cohen, S. I., Linse, S., Luheshi, L. M., Hellstrand, E., White, D. A., Rajah, L., … Knowles, T. P. (2013). Proliferation of amyloid-beta42 aggregates occurs through a secondary nucleation mechanism. *Proceedings of the National Academy of Sciences of the U S A, 110*(24), 9758–9763. http://dx.doi.org/10.1073/pnas.1218402110.

Congdon, E. E., & Sigurdsson, E. M. (2018). Tau-targeting therapies for Alzheimer disease. *Nature Reviews Neurology, 14*(7), 399–415. http://dx.doi.org/10.1038/s41582-018-0013-z.

Cummings, J., Lee, G., Ritter, A., & Zhong, K. (2018). Alzheimer's disease drug development pipeline: 2018. *Alzheimers Dement (N Y), 4*, 195–214. http://dx.doi.org/10.1016/j.trci.2018.03.009.

Cummings, J. L., Morstorf, T., & Zhong, K. (2014). Alzheimer's disease drug-development pipeline: Few candidates, frequent failures. *Alzheimer's Research & Therapy, 6*(4), 37. http://dx.doi.org/10.1186/alzrt269.

Deane, R., Bell, R. D., Sagare, A., & Zlokovic, B. V. (2009). Clearance of amyloid-beta peptide across the blood-brain barrier: Implication for therapies in Alzheimer's disease. *CNS & Neurological Disorders – Drug Targets, 8*(1), 16–30.

Dementia Collaborators, G. B. D. (2019). Global, regional, and national burden of Alzheimer's disease and other dementias, 1990–2016: A systematic analysis for the global burden of disease study 2016. *The Lancet Neurology, 18*(1), 88–106. http://dx.doi.org/10.1016/S1474-4422 (18)30403-4.

DeVos, S. L., Corjuc, B. T., Oakley, D. H., Nobuhara, C. K., Bannon, R. N., Chase, A., … Hyman, B. T. (2018). Synaptic tau seeding precedes tau pathology in human Alzheimer's disease brain. *Frontiers in Neuroscience, 12*, 267. http://dx.doi.org/10.3389/fnins.2018.00267.

Di Meo, S., Reed, T. T., Venditti, P., & Victor, V. M. (2016). Role of ROS and RNS sources in physiological and pathological conditions. *Oxid Med Cell Longev, 2016*, 1245049. http://dx.doi.org/10.1155/2016/1245049.

Donahue, J. E., & Johanson, C. E. (2008). Apolipoprotein E, amyloid-beta, and blood-brain barrier permeability in Alzheimer disease. *Journal of Neuropathology & Experimental Neurology, 67*(4), 261–270. http://dx.doi.org/10.1097/NEN.0b013e31816a0dc8.

Donev, R., Kolev, M., Millet, B., & Thome, J. (2009). Neuronal death in Alzheimer's disease and therapeutic opportunities. *Journal of Cellular and Molecular Medicine, 13*(11–12), 4329–4348. http://dx.doi.org/10.1111/j.1582-4934.2009.00889.x.

Doody, R. S., Raman, R., Farlow, M., Iwatsubo, T., Vellas, B., Joffe, S., ... Semagacestat Study, G. (2013). A phase 3 trial of semagacestat for treatment of Alzheimer's disease. *New England Journal of Medicine, 369*(4), 341–350. http://dx.doi.org/10.1056/NEJMoa1210951.

van Duijn, S., Bulk, M., van Duinen, S. G., Nabuurs, R. J. A., van Buchem, M. A., van der Weerd, L., et al. (2017). Cortical iron reflects severity of Alzheimer's disease. *Journal of Alzheimer's Disease, 60*(4), 1533–1545. http://dx.doi.org/10.3233/JAD-161143.

Du, X., Wang, X., & Geng, M. (2018). Alzheimer's disease hypothesis and related therapies. *Translational Neurodegeneration, 7*, 2. http://dx.doi.org/10.1186/s40035-018-0107-y.

Eckman, E. A., Reed, D. K., & Eckman, C. B. (2001). Degradation of the Alzheimer's amyloid beta peptide by endothelin-converting enzyme. *Journal of Biological Chemistry, 276*(27), 24540–24548. http://dx.doi.org/10.1074/jbc.M007579200.

Erten-Lyons, D., Woltjer, R. L., Dodge, H., Nixon, R., Vorobik, R., Calvert, J. F., ... Kaye, J. (2009). Factors associated with resistance to dementia despite high Alzheimer disease pathology. *Neurology, 72*(4), 354–360. http://dx.doi.org/10.1212/01.wnl.0000341273.18141.64.

Esiri, M. M., & Chance, S. A. (2012). Cognitive reserve, cortical plasticity and resistance to Alzheimer's disease. *Alzheimer's Research & Therapy, 4*(2), 7. http://dx.doi.org/10.1186/alzrt105.

Esparza, T. J., Wildburger, N. C., Jiang, H., Gangolli, M., Cairns, N. J., Bateman, R. J., et al. (2016). Soluble amyloid-beta aggregates from human Alzheimer's disease brains. *Scientific Reports, 6*, 38187. http://dx.doi.org/10.1038/srep38187.

Espay, A. J., Vizcarra, J. A., Marsili, L., Lang, A. E., Simon, D. K., Merola, A., ... Leverenz, J. B. (2019). Revisiting protein aggregation as pathogenic in sporadic Parkinson and Alzheimer diseases. *Neurology, 92*(7), 329–337. http://dx.doi.org/10.1212/WNL.0000000000006926.

Farris, W., Mansourian, S., Chang, Y., Lindsley, L., Eckman, E. A., Frosch, M. P., ... Guenette, S. (2003). Insulin-degrading enzyme regulates the levels of insulin, amyloid beta-protein, and the beta-amyloid precursor protein intracellular domain in vivo. *Proceedings of the National Academy of Sciences of the U S A, 100*(7), 4162–4167. http://dx.doi.org/10.1073/pnas.0230450100.

Faucher, P., Mons, N., Micheau, J., Louis, C., & Beracochea, D. J. (2015). Hippocampal injections of oligomeric amyloid beta-peptide (1-42) induce selective working memory deficits and long-lasting alterations of ERK signaling pathway. *Frontiers in Aging Neuroscience, 7*, 245. http://dx.doi.org/10.3389/fnagi.2015.00245.

Fernandez-Perez, E. J., Peters, C., & Aguayo, L. G. (2016). Membrane damage induced by amyloid beta and a potential link with neuroinflammation. *Current Pharmaceutical Design, 22*(10), 1295–1304.

Fleminger, S., Oliver, D. L., Lovestone, S., Rabe-Hesketh, S., & Giora, A. (2003). Head injury as a risk factor for Alzheimer's disease: The evidence 10 years on; a partial replication. *Journal of Neurology Neurosurgery and Psychiatry, 74*(7), 857–862. http://dx.doi.org/10.1136/jnnp.74.7.857.

Flock, D., Colacino, S., Colombo, G., & Di Nola, A. (2006). Misfolding of the amyloid beta-protein: A molecular dynamics study. *Proteins, 62*(1), 183–192. http://dx.doi.org/10.1002/prot.20683.

Galasko, D., Bell, J., Mancuso, J. Y., Kupiec, J. W., Sabbagh, M. N., van Dyck, C., ... Alzheimer's Disease Cooperative, S. (2014). Clinical trial of an inhibitor of RAGE-Abeta interactions in Alzheimer disease. *Neurology, 82*(17), 1536–1542. http://dx.doi.org/10.1212/WNL.0000000000000364.

Ganguli, M., Dodge, H. H., Shen, C., Pandav, R. S., & DeKosky, S. T. (2005). Alzheimer disease and mortality: A 15-year epidemiological study. *Archives of Neurology, 62*(5), 779–784. http://dx.doi.org/10.1001/archneur.62.5.779.

Garcia-Osta, A., & Alberini, C. M. (2009). Amyloid beta mediates memory formation. *Learning & Memory, 16*(4), 267–272. http://dx.doi.org/10.1101/lm.1310209.

Gardener, S. L., Rainey-Smith, S. R., & Martins, R. N. (2016). Diet and inflammation in Alzheimer's disease and related chronic diseases: A review. *Journal of Alzheimer's Disease, 50*(2), 301–334. http://dx.doi.org/10.3233/JAD-150765.

Gasparini, L., Terni, B., & Spillantini, M. G. (2007). Frontotemporal dementia with tau pathology. *Neurodegenerative Diseases, 4*(2–3), 236–253. http://dx.doi.org/10.1159/000101848.

Gerson, J. E., & Kayed, R. (2013). Formation and propagation of tau oligomeric seeds. *Frontiers in Neurology, 4*, 93. http://dx.doi.org/10.3389/fneur.2013.00093.

Ghosh, A. K., & Osswald, H. L. (2014). BACE1 (beta-secretase) inhibitors for the treatment of Alzheimer's disease. *Chemical Society Reviews, 43*(19), 6765–6813. http://dx.doi.org/10.1039/c3cs60460h.

Giorgetti, S., Greco, C., Tortora, P., & Aprile, F. A. (2018). Targeting amyloid aggregation: An overview of strategies and mechanisms. *International Journal of Molecular Sciences, 19*(9). http://dx.doi.org/10.3390/ijms19092677.

Goedert, M., & Jakes, R. (1990). Expression of separate isoforms of human tau protein: Correlation with the tau pattern in brain and effects on tubulin polymerization. *The EMBO Journal, 9*(13), 4225–4230.

Goedert, M., Spillantini, M. G., & Crowther, R. A. (1992). Cloning of a big tau microtubule-associated protein characteristic of the peripheral nervous system. *Proceedings of the National Academy of Sciences of the U S A, 89*(5), 1983–1987. http://dx.doi.org/10.1073/pnas.89.5.1983.

Gosztyla, M. L., Brothers, H. M., & Robinson, S. R. (2018). Alzheimer's amyloid-beta is an antimicrobial peptide: A review of the evidence. *Journal of Alzheimer's Disease, 62*(4), 1495–1506. http://dx.doi.org/10.3233/JAD-171133.

Greenough, M. A., Camakaris, J., & Bush, A. I. (2013). Metal dyshomeostasis and oxidative stress in Alzheimer's disease. *Neurochemistry International, 62*(5), 540–555. http://dx.doi.org/10.1016/j.neuint.2012.08.014.

Gunawardena, S., & Goldstein, L. S. (2001). Disruption of axonal transport and neuronal viability by amyloid precursor protein mutations in Drosophila. *Neuron, 32*(3), 389–401.

Habchi, J., Chia, S., Galvagnion, C., Michaels, T. C. T., Bellaiche, M. M. J., Ruggeri, F. S., … Vendruscolo, M. (2018). Cholesterol catalyses Abeta42 aggregation through a heterogeneous nucleation pathway in the presence of lipid membranes. *Nature Chemistry, 10*(6), 673–683. http://dx.doi.org/10.1038/s41557-018-0031-x.

Hama, E., & Saido, T. C. (2005). Etiology of sporadic Alzheimer's disease: Somatostatin, neprilysin, and amyloid beta peptide. *Medical Hypotheses, 65*(3), 498–500. http://dx.doi.org/10.1016/j.mehy.2005.02.045.

Hane, F., & Leonenko, Z. (2014). Effect of metals on kinetic pathways of amyloid-beta aggregation. *Biomolecules, 4*(1), 101–116. http://dx.doi.org/10.3390/biom4010101.

Helzner, E. P., Scarmeas, N., Cosentino, S., Tang, M. X., Schupf, N., & Stern, Y. (2008). Survival in Alzheimer disease: A multiethnic, population-based study of incident cases. *Neurology, 71*(19), 1489–1495. http://dx.doi.org/10.1212/01.wnl.0000334278.11022.42.

Hersh, L. B., & Rodgers, D. W. (2008). Neprilysin and amyloid beta peptide degradation. *Current Alzheimer Research, 5*(2), 225–231.

Hickman, R. A., Faustin, A., & Wisniewski, T. (2016). Alzheimer disease and its growing epidemic: Risk factors, biomarkers, and the urgent need for therapeutics. *Neurologic Clinics, 34*(4), 941–953. http://dx.doi.org/10.1016/j.ncl.2016.06.009.

Higgins, L. S., Holtzman, D. M., Rabin, J., Mobley, W. C., & Cordell, B. (1994). Transgenic mouse brain histopathology resembles early Alzheimer's disease. *Annals of Neurology, 35*(5), 598–607. http://dx.doi.org/10.1002/ana.410350514.

Horiguchi, T., Uryu, K., Giasson, B. I., Ischiropoulos, H., LightFoot, R., Bellmann, C., ... Trojanowski, J. Q. (2003). Nitration of tau protein is linked to neurodegeneration in tauopathies. *American Journal Of Pathology, 163*(3), 1021–1031. http://dx.doi.org/10.1016/S0002-9440 (10)63462-1.

Ho, A., & Shen, J. (2011). Presenilins in synaptic function and disease. *Trends in Molecular Medicine, 17*(11), 617–624. http://dx.doi.org/10.1016/j.molmed.2011.06.002.

Jamasbi, E., Wade, J. D., Separovic, F., & Hossain, M. A. (2016). Amyloid beta (abeta) peptide and factors that play important roles in Alzheimer's disease. *Current Medicinal Chemistry, 23*(9), 884–892.

Jeganathan, S., von Bergen, M., Mandelkow, E. M., & Mandelkow, E. (2008). The natively unfolded character of tau and its aggregation to Alzheimer-like paired helical filaments. *Biochemistry, 47*(40), 10526–10539. http://dx.doi.org/10.1021/bi800783d.

Johnson, G. V., & Stoothoff, W. H. (2004). Tau phosphorylation in neuronal cell function and dysfunction. *Journal of Cell Science, 117*(Pt 24), 5721–5729. http://dx.doi.org/10.1242/jcs.01558.

Jurgensen, S., & Ferreira, S. T. (2010). Nicotinic receptors, amyloid-beta, and synaptic failure in Alzheimer's disease. *Journal of Molecular Neuroscience, 40*(1–2), 221–229. http://dx.doi.org/10.1007/s12031-009-9237-0.

Kamenetz, F., Tomita, T., Hsieh, H., Seabrook, G., Borchelt, D., Iwatsubo, T., ... Malinow, R. (2003). APP processing and synaptic function. *Neuron, 37*(6), 925–937.

Kametani, F., & Hasegawa, M. (2018). Reconsideration of amyloid hypothesis and tau hypothesis in Alzheimer's disease. *Frontiers in Neuroscience, 12*, 25. http://dx.doi.org/10.3389/fnins.2018.00025.

Kanemitsu, H., Tomiyama, T., & Mori, H. (2003). Human neprilysin is capable of degrading amyloid beta peptide not only in the monomeric form but also the pathological oligomeric form. *Neuroscience Letters, 350*(2), 113–116.

van der Kant, R., & Goldstein, L. S. (2015). Cellular functions of the amyloid precursor protein from development to dementia. *Developmental Cell, 32*(4), 502–515. http://dx.doi.org/10.1016/j.devcel.2015.01.022.

Kierdorf, K., & Fritz, G. (2013). RAGE regulation and signaling in inflammation and beyond. *Journal of Leukocyte Biology, 94*(1), 55–68. http://dx.doi.org/10.1189/jlb.1012519.

Kim, W., & Hecht, M. H. (2006). Generic hydrophobic residues are sufficient to promote aggregation of the Alzheimer's Abeta42 peptide. *Proceedings of the National Academy of Sciences of the U S A, 103*(43), 15824–15829. http://dx.doi.org/10.1073/pnas.0605629103.

Kim, H. Y., Lee, D. K., Chung, B. R., Kim, H. V., & Kim, Y. (2016). Intracerebroventricular injection of amyloid-beta peptides in normal mice to acutely induce Alzheimer-like cognitive deficits. *Journal of Visualized Experiments*, (109)http://dx.doi.org/10.3791/53308.

Kinney, J. W., Bemiller, S. M., Murtishaw, A. S., Leisgang, A. M., Salazar, A. M., & Lamb, B. T. (2018). Inflammation as a central mechanism in Alzheimer's disease. *Alzheimers Dement (N Y), 4*, 575–590. http://dx.doi.org/10.1016/j.trci.2018.06.014.

Kitazawa, M., Medeiros, R., & Laferla, F. M. (2012). Transgenic mouse models of Alzheimer disease: Developing a better model as a tool for therapeutic interventions. *Current Pharmaceutical Design, 18*(8), 1131–1147.

Klein, R. M., Christie, J., & Parkvall, M. (2016). Does multilingualism affect the incidence of Alzheimer's disease?: A worldwide analysis by country. *SSM Popul Health, 2*, 463–467. http://dx.doi.org/10.1016/j.ssmph.2016.06.002.

Kopeikina, K. J., Hyman, B. T., & Spires-Jones, T. L. (2012). Soluble forms of tau are toxic in Alzheimer's disease. *Translational Neuroscience, 3*(3), 223–233. http://dx.doi.org/10.2478/s13380-012-0032-y.

Kumar, D. K., Choi, S. H., Washicosky, K. J., Eimer, W. A., Tucker, S., Ghofrani, J., … Moir, R. D. (2016). Amyloid-beta peptide protects against microbial infection in mouse and worm models of Alzheimer's disease. *Science Translational Medicine, 8*(340), 340ra372. http://dx.doi.org/10.1126/scitranslmed.aaf1059.

Lal, R., Lin, H., & Quist, A. P. (2007). Amyloid beta ion channel: 3D structure and relevance to amyloid channel paradigm. *Biochimica et Biophysica Acta, 1768*(8), 1966–1975. http://dx.doi.org/10.1016/j.bbamem.2007.04.021.

Lambermon, M. H., Rappaport, R. V., & McLaurin, J. (2005). Biophysical characterization of longer forms of amyloid beta peptides: Possible contribution to flocculent plaque formation. *Journal of Neurochemistry, 95*(6), 1667–1676. http://dx.doi.org/10.1111/j.1471-4159.2005.03497.x.

Lanoiselee, H. M., Nicolas, G., Wallon, D., Rovelet-Lecrux, A., Lacour, M., Rousseau, S., … collaborators of the, C. N. R. M. A. J. p.. (2017). APP, PSEN1, and PSEN2 mutations in early-onset Alzheimer disease: A genetic screening study of familial and sporadic cases. *PLoS Medicine, 14*(3), e1002270. http://dx.doi.org/10.1371/journal.pmed.1002270.

Lee, M., Kim, J. I., Na, S., & Eom, K. (2018). Metal ions affect the formation and stability of amyloid beta aggregates at multiple length scales. *Physical Chemistry Chemical Physics, 20*(13), 8951–8961. http://dx.doi.org/10.1039/C7CP05072K.

Lee, C. Y., & Landreth, G. E. (2010). The role of microglia in amyloid clearance from the AD brain. *Journal of Neural Transmission, 117*(8), 949–960. http://dx.doi.org/10.1007/s00702-010-0433-4.

Lee, L., Sakurai, M., Matsuzaki, S., Arancio, O., & Fraser, P. (2013). SUMO and Alzheimer's disease. *NeuroMolecular Medicine, 15*(4), 720–736. http://dx.doi.org/10.1007/s12017-013-8257-7.

LeVine, H., 3rd (2006). Biotin-avidin interaction-based screening assay for Alzheimer's beta-peptide oligomer inhibitors. *Analytical Biochemistry, 356*(2), 265–272. http://dx.doi.org/10.1016/j.ab.2006.04.036.

Lin, Y. S., Bowman, G. R., Beauchamp, K. A., & Pande, V. S. (2012). Investigating how peptide length and a pathogenic mutation modify the structural ensemble of amyloid beta monomer. *Biophysical Journal, 102*(2), 315–324. http://dx.doi.org/10.1016/j.bpj.2011.12.002.

Lingsma, H. F., Roozenbeek, B., Steyerberg, E. W., Murray, G. D., & Maas, A. I. (2010). Early prognosis in traumatic brain injury: From prophecies to predictions. *The Lancet Neurology, 9*(5), 543–554. http://dx.doi.org/10.1016/S1474-4422 (10)70065-X.

Liyanage, S. I., Santos, C., & Weaver, D. F. (2018). The hidden variables problem in Alzheimer's disease clinical trial design. *Alzheimers Dement (N Y), 4*, 628–635. http://dx.doi.org/10.1016/j.trci.2018.09.003.

Li, D. D., Zhang, W., Wang, Z. Y., & Zhao, P. (2017). Serum copper, zinc, and iron levels in patients with Alzheimer's disease: A meta-analysis of case-control studies. *Frontiers in Aging Neuroscience, 9*, 300. http://dx.doi.org/10.3389/fnagi.2017.00300.

Lovell, M. A., Robertson, J. D., Teesdale, W. J., Campbell, J. L., & Markesbery, W. R. (1998). Copper, iron and zinc in Alzheimer's disease senile plaques. *Journal of Neurological Sciences, 158*(1), 47–52.

Lucke-Wold, B., Seidel, K., Udo, R., Omalu, B., Ornstein, M., Nolan, R., … Ross, J. (2017). Role of tau acetylation in Alzheimer's disease and chronic traumatic encephalopathy: The way forward for successful treatment. *Journal of Neurology, Neurosurgery, 4*(2).

Malm, T., Magga, J., & Koistinaho, J. (2012). Animal models of Alzheimer's disease: Utilization of transgenic Alzheimer's disease models in studies of amyloid beta clearance.

Current Translational Geriatrics and Gerontology Reports, 1(1), 11–20. http://dx.doi.org/10.1007/s13670-011-0004-z.
Mandelkow, E. M., & Mandelkow, E. (2012). Biochemistry and cell biology of tau protein in neurofibrillary degeneration. *Cold Spring Harbor Perspectives in Medicine, 2*(7), a006247. http://dx.doi.org/10.1101/cshperspect.a006247.
Marik, S. A., Olsen, O., Tessier-Lavigne, M., & Gilbert, C. D. (2016). Physiological role for amyloid precursor protein in adult experience-dependent plasticity. *Proceedings of the National Academy of Sciences of the U S A, 113*(28), 7912–7917. http://dx.doi.org/10.1073/pnas.1604299113.
Marr, R. A., Guan, H., Rockenstein, E., Kindy, M., Gage, F. H., Verma, I., … Hersh, L. B. (2004). Neprilysin regulates amyloid Beta peptide levels. *Journal of Molecular Neuroscience, 22*(1–2), 5–11. http://dx.doi.org/10.1385/JMN:22:1-2:5.
Marttinen, M., Takalo, M., Natunen, T., Wittrahm, R., Gabbouj, S., Kemppainen, S., … Hiltunen, M. (2018). Molecular mechanisms of synaptotoxicity and neuroinflammation in Alzheimer's disease. *Frontiers in Neuroscience, 12*, 963. http://dx.doi.org/10.3389/fnins.2018.00963.
Masellis, M., Sherborn, K., Neto, P., Sadovnick, D. A., Hsiung, G. Y., Black, S. E., … Gauthier, S. (2013). Early-onset dementias: Diagnostic and etiological considerations. *Alzheimer's Research & Therapy, 5*(Suppl. 1), S7. http://dx.doi.org/10.1186/alzrt197.
Maynard, C. J., Bush, A. I., Masters, C. L., Cappai, R., & Li, Q. X. (2005). Metals and amyloid-beta in Alzheimer's disease. *International Journal of Experimental Pathology, 86*(3), 147–159. http://dx.doi.org/10.1111/j.0959-9673.2005.00434.x.
Mendez, M. F. (2017). Early-onset Alzheimer disease. *Neurologic Clinics, 35*(2), 263–281. http://dx.doi.org/10.1016/j.ncl.2017.01.005.
Millis, S. R., Rosenthal, M., Novack, T. A., Sherer, M., Nick, T. G., Kreutzer, J. S., … Ricker, J. H. (2001). Long-term neuropsychological outcome after traumatic brain injury. *The Journal of Head Trauma Rehabilitation, 16*(4), 343–355.
Morris, G. P., Clark, I. A., & Vissel, B. (2014). Inconsistencies and controversies surrounding the amyloid hypothesis of Alzheimer's disease. *Acta Neuropathologica Communications, 2*, 135. http://dx.doi.org/10.1186/s40478-014-0135-5.
Morris, J. K., Vidoni, E. D., Johnson, D. K., Van Sciver, A., Mahnken, J. D., Honea, R. A., … Burns, J. M. (2017). Aerobic exercise for Alzheimer's disease: A randomized controlled pilot trial. *PLoS One, 12*(2), e0170547. http://dx.doi.org/10.1371/journal.pone.0170547.
Moussa, C. E. (2017). Beta-secretase inhibitors in phase I and phase II clinical trials for Alzheimer's disease. *Expert Opinion on Investigational Drugs, 26*(10), 1131–1136. http://dx.doi.org/10.1080/13543784.2017.1369527.
Mucke, L., & Selkoe, D. J. (2012). Neurotoxicity of amyloid beta-protein: Synaptic and network dysfunction. *Cold Spring Harbor Perspectives in Medicine, 2*(7), a006338. http://dx.doi.org/10.1101/cshperspect.a006338.
Mudher, A., Colin, M., Dujardin, S., Medina, M., Dewachter, I., Alavi Naini, S. M., … Brion, J. P. (2017). What is the evidence that tau pathology spreads through prion-like propagation? *Acta Neuropathologica Communications, 5*(1), 99. http://dx.doi.org/10.1186/s40478-017-0488-7.
Murphy, M. P., & LeVine, H., 3rd (2010). Alzheimer's disease and the amyloid-beta peptide. *Journal of Alzheimer's Disease, 19*(1), 311–323. http://dx.doi.org/10.3233/JAD-2010-1221.
Murrell, J. R., Hake, A. M., Quaid, K. A., Farlow, M. R., & Ghetti, B. (2000). Early-onset Alzheimer disease caused by a new mutation (V717L) in the amyloid precursor protein gene. *Archives of Neurology, 57*(6), 885–887.

Nerelius, C., Sandegren, A., Sargsyan, H., Raunak, R., Leijonmarck, H., Chatterjee, U., ... Johansson, J. (2009). Alpha-helix targeting reduces amyloid-beta peptide toxicity. *Proceedings of the National Academy of Sciences of the U S A, 106*(23), 9191–9196. http://dx.doi.org/10.1073/pnas.0810364106.

Nomura, I., Takechi, H., & Kato, N. (2012). Intraneuronally injected amyloid beta inhibits long-term potentiation in rat hippocampal slices. *Journal of Neurophysiology, 107*(9), 2526–2531. http://dx.doi.org/10.1152/jn.00589.2011.

Nuttall, J. R., & Oteiza, P. I. (2014). Zinc and the aging brain. *Genes Nutr, 9*(1), 379. http://dx.doi.org/10.1007/s12263-013-0379-x.

O'Brien, R. J., & Wong, P. C. (2011). Amyloid precursor protein processing and Alzheimer's disease. *Annual Review of Neuroscience, 34*, 185–204. http://dx.doi.org/10.1146/annurev-neuro-061010-113613.

Oddo, S. (2008). The ubiquitin-proteasome system in Alzheimer's disease. *Journal of Cellular and Molecular Medicine, 12*(2), 363–373. http://dx.doi.org/10.1111/j.1582-4934.2008.00276.x.

Opazo, C., Ruiz, F. H., & Inestrosa, N. C. (2000). Amyloid-beta-peptide reduces copper (II) to copper (I) independent of its aggregation state. *Biological Research, 33*(2), 125–131. http://dx.doi.org/10.4067/s0716-97602000000200012.

Paglia, G., Miedico, O., Cristofano, A., Vitale, M., Angiolillo, A., Chiaravalle, A. E., ... Di Costanzo, A. (2016). Distinctive pattern of serum elements during the progression of Alzheimer's disease. *Scientific Reports, 6*, 22769. http://dx.doi.org/10.1038/srep22769.

Peeraer, E., Bottelbergs, A., Van Kolen, K., Stancu, I. C., Vasconcelos, B., Mahieu, M., ... Moechars, D. (2015). Intracerebral injection of preformed synthetic tau fibrils initiates widespread tauopathy and neuronal loss in the brains of tau transgenic mice. *Neurobiology of Disease, 73*, 83–95. http://dx.doi.org/10.1016/j.nbd.2014.08.032.

Penninkilampi, R., Brothers, H. M., & Eslick, G. D. (2016). Pharmacological agents targeting gamma-secretase increase risk of cancer and cognitive decline in Alzheimer's disease patients: A systematic review and meta-analysis. *Journal of Alzheimer's Disease, 53*(4), 1395–1404. http://dx.doi.org/10.3233/JAD-160275.

Peters, I., Igbavboa, U., Schutt, T., Haidari, S., Hartig, U., Rosello, X., ... Eckert, G. P. (2009). The interaction of beta-amyloid protein with cellular membranes stimulates its own production. *Biochimica et Biophysica Acta, 1788*(5), 964–972. http://dx.doi.org/10.1016/j.bbamem.2009.01.012.

Pini, L., Pievani, M., Bocchetta, M., Altomare, D., Bosco, P., Cavedo, E., ... Frisoni, G. B. (2016). Brain atrophy in Alzheimer's Disease and aging. *Ageing Research Reviews, 30*, 25–48. http://dx.doi.org/10.1016/j.arr.2016.01.002.

Poojari, C., Kukol, A., & Strodel, B. (2013). How the amyloid-beta peptide and membranes affect each other: An extensive simulation study. *Biochimica et Biophysica Acta, 1828*(2), 327–339. http://dx.doi.org/10.1016/j.bbamem.2012.09.001.

Price, J. L., McKeel, D. W., Jr., Buckles, V. D., Roe, C. M., Xiong, C., Grundman, M., ... Morris, J. C. (2009). Neuropathology of nondemented aging: Presumptive evidence for preclinical Alzheimer disease. *Neurobiology of Aging, 30*(7), 1026–1036. http://dx.doi.org/10.1016/j.neurobiolaging.2009.04.002.

Price, D. L., Sisodia, S. S., & Gandy, S. E. (1995). Amyloid beta amyloidosis in Alzheimer's disease. *Current Opinion in Neurology, 8*(4), 268–274.

Probst, A., Tolnay, M., Langui, D., Goedert, M., & Spillantini, M. G. (1996). Pick's disease: Hyperphosphorylated tau protein segregates to the somatoaxonal compartment. *Acta Neuropathologica, 92*(6), 588–596.

Rajasekhar, K., Chakrabarti, M., & Govindaraju, T. (2015). Function and toxicity of amyloid beta and recent therapeutic interventions targeting amyloid beta in Alzheimer's disease. *Chemical Communications, 51*(70), 13434–13450. http://dx.doi.org/10.1039/c5cc05264e.

Ramos-Cejudo, J., Wisniewski, T., Marmar, C., Zetterberg, H., Blennow, K., de Leon, M. J., et al. (2018). Traumatic brain injury and Alzheimer's disease: The cerebrovascular link. *EBioMedicine, 28*, 21–30. http://dx.doi.org/10.1016/j.ebiom.2018.01.021.

Ries, M., & Sastre, M. (2016). Mechanisms of abeta clearance and degradation by glial cells. *Frontiers in Aging Neuroscience, 8*, 160. http://dx.doi.org/10.3389/fnagi.2016.00160.

Rosenmann, H., Blum, D., Kayed, R., & Ittner, L. M. (2012). Tau protein: Function and pathology. *International Journal of Alzheimer's Disease, 2012*, 707482. http://dx.doi.org/10.1155/2012/707482.

Savage, M. J., Kalinina, J., Wolfe, A., Tugusheva, K., Korn, R., Cash-Mason, T., … McCampbell, A. (2014). A sensitive abeta oligomer assay discriminates Alzheimer's and aged control cerebrospinal fluid. *Journal of Neuroscience, 34*(8), 2884–2897. http://dx.doi.org/10.1523/JNEUROSCI.1675-13.2014.

Schultz, S. A., Larson, J., Oh, J., Koscik, R., Dowling, M. N., Gallagher, C. L., … Okonkwo, O. C. (2015). Participation in cognitively-stimulating activities is associated with brain structure and cognitive function in preclinical Alzheimer's disease. *Brain Imaging Behav, 9*(4), 729–736. http://dx.doi.org/10.1007/s11682-014-9329-5.

Sengupta, U., Nilson, A. N., & Kayed, R. (2016). The role of amyloid-beta oligomers in toxicity, propagation, and immunotherapy. *EBioMedicine, 6*, 42–49. http://dx.doi.org/10.1016/j.ebiom.2016.03.035.

Sheng, M., Sabatini, B. L., & Sudhof, T. C. (2012). Synapses and Alzheimer's disease. *Cold Spring Harbor Perspectives in Biology, 4*(5). http://dx.doi.org/10.1101/cshperspect.a005777.

Sherrington, R., Rogaev, E. I., Liang, Y., Rogaeva, E. A., Levesque, G., Ikeda, M., … St George-Hyslop, P. H. (1995). Cloning of a gene bearing missense mutations in early-onset familial Alzheimer's disease. *Nature, 375*(6534), 754–760. http://dx.doi.org/10.1038/375754a0.

Smith, M. A., Harris, P. L., Sayre, L. M., & Perry, G. (1997). Iron accumulation in Alzheimer disease is a source of redox-generated free radicals. *Proceedings of the National Academy of Sciences of the U S A, 94*(18), 9866–9868. http://dx.doi.org/10.1073/pnas.94.18.9866.

Sochocka, M., Zwolinska, K., & Leszek, J. (2017). The infectious etiology of Alzheimer's disease. *Current Neuropharmacology, 15*(7), 996–1009. http://dx.doi.org/10.2174/1570159X15666170313122937.

Solarski, M., Wang, H., Wille, H., & Schmitt-Ulms, G. (2018). Somatostatin in Alzheimer's disease: A new role for an old player. *Prion, 12*(1), 1–8. http://dx.doi.org/10.1080/19336896.2017.1405207.

Soscia, S. J., Kirby, J. E., Washicosky, K. J., Tucker, S. M., Ingelsson, M., Hyman, B., … Moir, R. D. (2010). The Alzheimer's disease-associated amyloid beta-protein is an antimicrobial peptide. *PLoS One, 5*(3), e9505. http://dx.doi.org/10.1371/journal.pone.0009505.

Soto, C., Castano, E. M., Frangione, B., & Inestrosa, N. C. (1995). The alpha-helical to beta-strand transition in the amino-terminal fragment of the amyloid beta-peptide modulates amyloid formation. *Journal of Biological Chemistry, 270*(7), 3063–3067. http://dx.doi.org/10.1074/jbc.270.7.3063.

Spires-Jones, T. L., Kopeikina, K. J., Koffie, R. M., de Calignon, A., & Hyman, B. T. (2011). Are tangles as toxic as they look? *Journal of Molecular Neuroscience, 45*(3), 438–444. http://dx.doi.org/10.1007/s12031-011-9566-7.

Steward, K. A., Kennedy, R., Novack, T. A., Crowe, M., Marson, D. C., & Triebel, K. L. (2018). The role of cognitive reserve in recovery from traumatic brain injury. *The Journal of Head Trauma Rehabilitation, 33*(1), E18–E27. http://dx.doi.org/10.1097/HTR.0000000000000325.

Stockley, J. H., & O'Neill, C. (2007). The proteins BACE1 and BACE2 and beta-secretase activity in normal and Alzheimer's disease brain. *Biochemical Society Transactions, 35*(Pt 3), 574–576. http://dx.doi.org/10.1042/BST0350574.

Swaminathan, A., & Jicha, G. A. (2014). Nutrition and prevention of Alzheimer's dementia. *Frontiers in Aging Neuroscience, 6*, 282. http://dx.doi.org/10.3389/fnagi.2014.00282.

Swerdlow, R. H., Burns, J. M., & Khan, S. M. (2010). The Alzheimer's disease mitochondrial cascade hypothesis. *Journal of Alzheimer's Disease, 20*(Suppl. 2), S265–S279. http://dx.doi.org/10.3233/JAD-2010-100339.

Terwel, D., Dewachter, I., & Van Leuven, F. (2002). Axonal transport, tau protein, and neurodegeneration in Alzheimer's disease. *NeuroMolecular Medicine, 2*(2), 151–165. http://dx.doi.org/10.1385/NMM:2:2:151.

Texido, L., Martin-Satue, M., Alberdi, E., Solsona, C., & Matute, C. (2011). Amyloid beta peptide oligomers directly activate NMDA receptors. *Cell Calcium, 49*(3), 184–190. http://dx.doi.org/10.1016/j.ceca.2011.02.001.

Tharp, W. G., & Sarkar, I. N. (2013). Origins of amyloid-beta. *BMC Genomics, 14*, 290. http://dx.doi.org/10.1186/1471-2164-14-290.

Tracy, T. E., & Gan, L. (2018). Tau-mediated synaptic and neuronal dysfunction in neurodegenerative disease. *Current Opinion in Neurobiology, 51*, 134–138. http://dx.doi.org/10.1016/j.conb.2018.04.027.

Trinczek, B., Biernat, J., Baumann, K., Mandelkow, E. M., & Mandelkow, E. (1995). Domains of tau protein, differential phosphorylation, and dynamic instability of microtubules. *Molecular Biology of the Cell, 6*(12), 1887–1902. http://dx.doi.org/10.1091/mbc.6.12.1887.

Tsai, H. H., Lee, J. B., Tseng, S. S., Pan, X. A., & Shih, Y. C. (2010). Folding and membrane insertion of amyloid-beta (25-35) peptide and its mutants: Implications for aggregation and neurotoxicity. *Proteins, 78*(8), 1909–1925. http://dx.doi.org/10.1002/prot.22705.

Tucker, H. M., Kihiko, M., Caldwell, J. N., Wright, S., Kawarabayashi, T., Price, D., … Estus, S. (2000). The plasmin system is induced by and degrades amyloid-beta aggregates. *Journal of Neuroscience, 20*(11), 3937–3946.

Tundo, G., Ciaccio, C., Sbardella, D., Boraso, M., Viviani, B., Coletta, M., et al. (2012). Somatostatin modulates insulin-degrading-enzyme metabolism: Implications for the regulation of microglia activity in AD. *PLoS One, 7*(4), e34376. http://dx.doi.org/10.1371/journal.pone.0034376.

Urbanc, B., Cruz, L., Yun, S., Buldyrev, S. V., Bitan, G., Teplow, D. B., et al. (2004). In silico study of amyloid beta-protein folding and oligomerization. *Proceedings of the National Academy of Sciences of the U S A, 101*(50), 17345–17350. http://dx.doi.org/10.1073/pnas.0408153101.

Valencia, A., & Moran, J. (2004). Reactive oxygen species induce different cell death mechanisms in cultured neurons. *Free Radical Biology and Medicine, 36*(9), 1112–1125. http://dx.doi.org/10.1016/j.freeradbiomed.2004.02.013.

Veeraraghavalu, K., Choi, S. H., Zhang, X., & Sisodia, S. S. (2010). Presenilin 1 mutants impair the self-renewal and differentiation of adult murine subventricular zone-neuronal progenitors via cell-autonomous mechanisms involving notch signaling. *Journal of Neuroscience, 30*(20), 6903–6915. http://dx.doi.org/10.1523/JNEUROSCI.0527-10.2010.

Venkat, P., Chopp, M., & Chen, J. (2015). Models and mechanisms of vascular dementia. *Experimental Neurology, 272*, 97–108. http://dx.doi.org/10.1016/j.expneurol.2015.05.006.

Verma, M., Vats, A., & Taneja, V. (2015). Toxic species in amyloid disorders: Oligomers or mature fibrils. *Annals of Indian Academy of Neurology, 18*(2), 138–145. http://dx.doi.org/10.4103/0972-2327.144284.

Vugmeyster, L., Clark, M. A., Falconer, I. B., Ostrovsky, D., Gantz, D., Qiang, W., et al. (2016). Flexibility and solvation of amyloid-beta hydrophobic core. *Journal of Biological Chemistry, 291*(35), 18484–18495. http://dx.doi.org/10.1074/jbc.M116.740530.

Wainaina, M. N., Chen, Z., & Zhong, C. (2014). Environmental factors in the development and progression of late-onset Alzheimer's disease. *Neuroscience Bulletin, 30*(2), 253–270. http://dx.doi.org/10.1007/s12264-013-1425-9.

Wang, G. (2014). Human antimicrobial peptides and proteins. *Pharmaceuticals,* 7(5), 545–594. http://dx.doi.org/10.3390/ph7050545.

Wang, J. Z., Grundke-Iqbal, I., & Iqbal, K. (2007). Kinases and phosphatases and tau sites involved in Alzheimer neurofibrillary degeneration. *European Journal of Neuroscience, 25*(1), 59–68. http://dx.doi.org/10.1111/j.1460-9568.2006.05226.x.

Wang, C., Sun, B., Zhou, Y., Grubb, A., & Gan, L. (2012). Cathepsin B degrades amyloid-beta in mice expressing wild-type human amyloid precursor protein. *Journal of Biological Chemistry, 287*(47), 39834–39841. http://dx.doi.org/10.1074/jbc.M112.371641.

Wang, J., Tan, L., Wang, H. F., Tan, C. C., Meng, X. F., Wang, C., … Yu, J. T. (2015). Anti-inflammatory drugs and risk of Alzheimer's disease: An updated systematic review and meta-analysis. *Journal of Alzheimer's Disease, 44*(2), 385–396. http://dx.doi.org/10.3233/JAD-141506.

Weingarten, M. D., Lockwood, A. H., Hwo, S. Y., & Kirschner, M. W. (1975). A protein factor essential for microtubule assembly. *Proceedings of the National Academy of Sciences of the U S A, 72*(5), 1858–1862. http://dx.doi.org/10.1073/pnas.72.5.1858.

Weller, J., & Budson, A. (2018). Current understanding of Alzheimer's disease diagnosis and treatment. *F1000Res,* 7. http://dx.doi.org/10.12688/f1000research.14506.1.

Wolfe, M. S. (2019). Dysfunctional gamma-secretase in familial Alzheimer's disease. *Neurochemical Research, 44*(1), 5–11. http://dx.doi.org/10.1007/s11064-018-2511-1.

Wollmer, M. A. (2010). Cholesterol-related genes in Alzheimer's disease. *Biochimica et Biophysica Acta, 1801*(8), 762–773. http://dx.doi.org/10.1016/j.bbalip.2010.05.009.

Xia, D., Watanabe, H., Wu, B., Lee, S. H., Li, Y., Tsvetkov, E., … Kelleher, R. J., 3rd (2015). Presenilin-1 knockin mice reveal loss-of-function mechanism for familial Alzheimer's disease. *Neuron, 85*(5), 967–981. http://dx.doi.org/10.1016/j.neuron.2015.02.010.

Xiong, H., Callaghan, D., Jones, A., Walker, D. G., Lue, L. F., Beach, T. G., … Zhang, W. (2008). Cholesterol retention in Alzheimer's brain is responsible for high beta- and gamma-secretase activities and Abeta production. *Neurobiology of Disease, 29*(3), 422–437. http://dx.doi.org/10.1016/j.nbd.2007.10.005.

Yiannopoulou, K. G., & Papageorgiou, S. G. (2013). Current and future treatments for Alzheimer's disease. *Therapeutic Advances in Neurological Disorders, 6*(1), 19–33. http://dx.doi.org/10.1177/1756285612461679.

Yoon, S. S., & Jo, S. A. (2012). Mechanisms of amyloid-beta peptide clearance: Potential therapeutic targets for Alzheimer's disease. *Biomolecules & Therapeutics, 20*(3), 245–255. http://dx.doi.org/10.4062/biomolther.2012.20.3.245.

Zanetti, O., Solerte, S. B., & Cantoni, F. (2009). Life expectancy in Alzheimer's disease (AD). *Archives of Gerontology and Geriatrics, 49*(Suppl. 1), 237–243. http://dx.doi.org/10.1016/j.archger.2009.09.035.

Zhan, S. S., Veerhuis, R., Kamphorst, W., & Eikelenboom, P. (1995). Distribution of beta amyloid associated proteins in plaques in Alzheimer's disease and in the non-demented elderly. *Neurodegeneration, 4*(3), 291–297.

Zhao, L. N., Long, H., Mu, Y., & Chew, L. Y. (2012). The toxicity of amyloid beta oligomers. *International Journal of Molecular Sciences, 13*(6), 7303–7327. http://dx.doi.org/10.3390/ijms13067303.

CHAPTER TWELVE

Influence of the reducing environment in the misfolding of wine proteins

Paolo Ruzza[a,*], Claudia Honisch[a,b], Matteo Marangon[c,d], Andrea Curioni[c,d], Alan Bakalinsky[e] and Simone Vincenzi[c,d]
[a]Institute of Biomolecular Chemistry of CNR, Padua Unit, Padova, Italy
[b]Department of Chemical Sciences, University of Padua, Padova, Italy
[c]Department of Agronomy, Food, Natural Resources, Animals and Environment (DAFNAE), University of Padua, Legnaro (PD), Italy
[d]Centre for Research in Viticulture and Enology (CIRVE), Conegliano (TV), Italy
[e]Department of Food Science and Technology, Oregon State University, Corvallis, Oregon, United States
*Corresponding author: E-mail: paolo.ruzza@unipd.it

Contents

Abstract

While proteins are present in wine at low concentration, and are largely associated with undesirable haze formation in white wines, certain types or fractions make direct and indirect contributions to sensory quality and physical stability. The proteins found in wine represent a small subclass of the total pool of grape proteins that remain soluble in the non-physiological conditions of the wine matrix which is characterised by the presence of alcohol, high acidity, and relatively high levels of phenolic compounds. Although initially stable in these conditions, during storage of white and rosé wines proteins undergo changes leading to haze formation which is considered one of the most relevant non-microbiological defects, and which makes the wine commercially unacceptable. This phenomenon involves the two most abundant proteins present in wines: thaumatin-like proteins and chitinases, both belonging to pathogenesis-related proteins of the grape berry. Haze formation is often triggered by thermal fluctuations occurring during storage of white wines, although the presence of other

Advances in Protein Chemistry and Structural Biology, Volume 118
ISSN 1876-1623
https://doi.org/10.1016/bs.apcsb.2019.08.004
© 2020 Elsevier Inc.
All rights reserved.

non-protein-related factors seems to be necessary. Here, we review the characteristics of these two protein families and the factors that influence their solubility with a focus on the disulfide bonds reduction as a possible trigger for the onset of their aggregation.

1. Introduction

Proteins are present in grape juice at concentrations ranging from 15 to 700 mg/L, and have considerable importance because they greatly affect wine stability. Nonetheless, most grape proteins that are extracted into wine during vinification are readily lost due to exposure to non-native conditions: a significant drop in pH, the presence of alcohol, and the accessibility to the interactions with phenolic compounds, all factors leading to protein precipitation. Formation of insoluble tannin-protein complexes, occurring mainly in red wines, accounts for additional losses (Somers & Ziemelis, 1973). Moreover, also the action of grape and/or yeast proteases can affect the grape proteins (Mamo & Assefa, 2018; Younes et al., 2011). Thus, the proteins that remain soluble in bottled wine represent a small subclass of the pool initially present in grapes (Van Sluyter et al., 2015).

The total quantity of proteins present in wine depends on the initial protein content of the grapes, on the amount extracted from the yeast (and malolactic bacteria) during fermentation and post-fermentation contact, and on the aforementioned factors that greatly reduce their stability. In some wines, a protein content close to 1 g/L has been reported, whereas values less than 1 mg/L have been measured in others. Some of the reported variability is likely due to the analytical method used because wine contains several factors interfering with the quantification methods (Gazzola, Vincenzi, Pasini, Lomolino, & Curioni, 2015; Vincenzi, Polesani, & Curioni, 2005). A more representative value for protein content in white wines is in the range 40–100 mg/L (Vincenzi et al., 2005).

The major proteins implicated in haze formation in white and rosé wines were initially identified as belonging to the pathogenesis-related (PR) proteins of the grape berry (Waters, Shirley, & Williams, 1996). Plant PR proteins comprise 17 families whose main function is related to the defense of the plant from pathogens and/or parasites (Table 1).

To exploit this function, most of the PR-proteins have to be intrinsically resistant to proteolysis and denaturation and, generally, are expressed in higher quantities when the plant is subjected to biotic, but also abiotic, stresses (Ali et al., 2018).

Table 1 Classification of pathogenesis-related proteins.

Families	Proteins
PR1	Role still unknown
PR2	β-1,3-Glucanase
PR5	Thaumatin-Like Protein (TLP)
PR3, PR4, PR8, PR11	Chitinase ed Endochitinase
PR6	Protease inhibitor
PR7	Endoprotease
PR9	Peroxidase
PR10	Ribonuclease
PR12, PR13, PR14	Difensine, Tionin, Lipid Transfer Protein (LTP)
PR15, PR16	Oxalate oxidase, Oxalate oxidase-like
PR17	Role still unknown

In an illustrative study of proteins present in the juice of 19 different *Vitis vinifera* varieties, two major classes of about 21 and 25 kDa were detected over a range from 13 to 33 kDa. LC-MS analysis identified the 21 kDa class as thaumatin-like proteins (TLP) while chitinases were identified in the 25 kDa class (Hayasaka et al., 2001). The TLPs and chitinases families account for about 80%–100% of the total (Mierczynska-Vasilev, Boyer, Vasilev, & Smith, 2017) which is in reasonable agreement with a previous study of grape proteins (Waters, Hayasaka, Tattersall, Adams, & Williams, 1998). Additionally, many authors reported that TLP concentrations in wines generally exceeds that of chitinases (Mierczynska-Vasilev et al., 2017; Tian et al., 2017; Van Sluyter et al., 2013), which is an important fact that has significant implications for protein haze formation (see below).

TLPs and chitinases, which are constitutively expressed in grapes, have been reported to accumulate rapidly in ripening berries (Bin Tian, Harrison, Morton, & Jaspers, 2019) and to be absent in juices obtained from unripe fruit (prior to veraison) (Koch & Sajak, 1959; Tattersall, van Heeswijck, & Hoj, 1997).

1.1 Chitinases

According to their defense function in plants, chitinases are able to hydrolyze the glycosidic bonds of chitin, a constituent of fungal cell walls (Oyeleye & Normi, 2018). Among the different types of chitinases, the most frequently described in wines are those belonging to class IV (Gazzola, Pasini, Tolin, Curioni, & Vincenzi, 2017; Robinson, Jacobs, & Dry, 1997). Class IV chitinases, as those belonging to class I, are characterised by having a cysteine-rich N-terminal region able to bind chitin (chitin-binding domain), which is absent from those of the other types (classes II and III chitinases). Classes I and III are expressed in roots and leaves, but not in the berries. On the contrary, class IV is expressed in berries but not in roots or leaves, and an isoform of this class IV chitinase is the one constantly present during berry ripening (Robinson et al., 1997).

Moreover, grape class IV chitinases have been shown to contain a sequence which is able to bind flavonoids and in particular quercetin and catechin, which in this way are involved in the regulation of the chitinases activity in the berry (Filippi et al., 2016).

However, the 3D molecular structure of the grape berry chitinase has not been determined yet, while this information could be crucial for the elucidation of the mechanism involved in its denaturation and aggregation in wine. Nevertheless, given that the genome of *Vitis vinifera* has been sequenced (Velasco et al., 2007), the 3D structure of Class IV grape chitinases can be modeled by using its sequence and building the structure in silico by taking advantage of the 3D structure of a chitinase from *Ozyza sativa*, that however presents a degree of similarity of only 45% (Fig. 1).

In its native form grape class IV chitinase appears to be a globular protein (Fig. 1) that, however, presents a domain which is sticking out from the main part of the molecule. This section corresponds to the above-mentioned chitin-binding domain which also in wine has been shown to maintain its capacity to bind chitin from both shellfish (Vincenzi et al., 2005) and yeast (Ndlovu, Divol, & Bauer, 2018).

The domain that binds flavonoids is located in the C-terminal region of the protein, aa 250-259 (Filippi et al., 2016) that, according to the predicted 3D structure shown in Fig. 2, seems to be in the external part of the molecule. This suggests that the binding of flavonoids such as quercetin and catechin to this highly accessible region has the ability to affect the conformational flexibility of chitinases thus modulating its enzymatic activity.

Grape chitinase contains 14 cysteine residues, but only 10 of them are involved in forming disulfide bridges according to the predicted structure.

Two of these disulfide bonds are located in the chitin-binding domain. It has been shown that in the reduced form grape chitinases lose the ability to bind to chitin (Vincenzi & Curioni, 2005), thus suggesting that the integrity of the SS-bonds is necessary for the functionality, as shown also in Rye seed class I chitinase-a (Takeshi & Gunki, 1996).

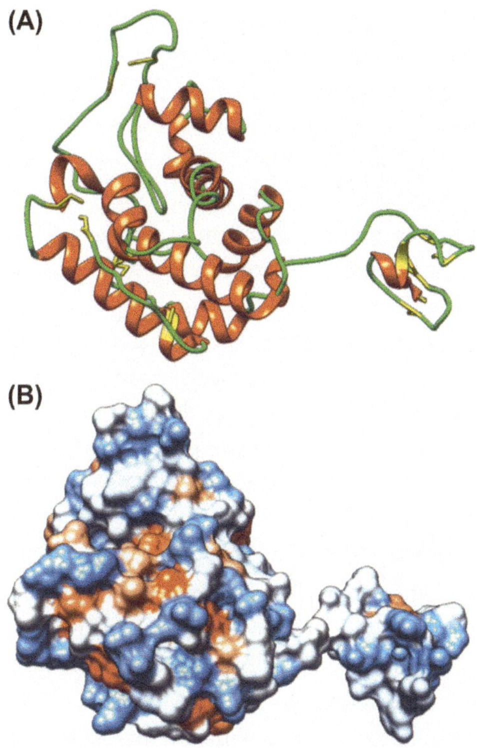

Fig. 1 (A) Predicted 3D structure of grape class IV chitinases (UniProt Q7XAU6) based on the alignment with the *Oryza sativa* counterpart (PDB file: 2DKV). The position of the cysteine residues are indicated in yellow. (B) Mapping of the surface hydrophobicity of grape class IV chitinases. The amino acid hydrophobicity is based on the Kyte-Doolittle scale from dodger blue for the most hydrophilic, to white, to orange/red for the most hydrophobic (Kyte & Doolittle, 1982; Sanner, Olson, & Spehner, 1996). The images were generated using the Chimera 1.13.1 software (Pettersen et al., 2004).

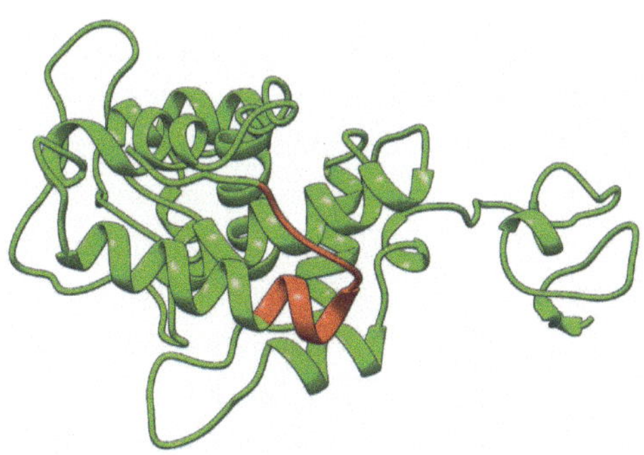

Fig. 2 Flavonoid binding domain (aa 250–259, coil depicted in red) of grape class IV chitinases next to the C-terminus. The 3D structure is based on the alignment with the *Oryza sativa* counterpart (PDB file:2DKV). The image was generated using the Chimera 1.13.1 software (Pettersen et al., 2004).

The other three SS-bridges seem to be located at the turns of different α-helices, thus blocking the core of the protein in a compact form (see Fig. 1B). This seems to be demonstrated by the fact that without reduction the SDS-PAGE mobility increases when compared to that of the reduced form (Vincenzi & Curioni, 2005). The active site of the enzyme is surrounded by three SS-bridges. From an evolutionary point of view this could be seen as a way of protecting the function of the enzyme in the plant. At the same time, this rigidity could also result in the well-known resistance of the enzyme against proteolysis.

However, it has to be noted that the sole temperature has been able to denature irreversibly grape chitinases in both model and real wines (Falconer et al., 2010; Marangon, Sauvage, Waters, & Vernhet, 2011) causing its aggregation and subsequent precipitation in juice (Marangon et al., 2012). The possible mechanisms involved in aggregation will be discussed in section 2.

1.2 Thaumatin-like proteins

TLPs are characterized by a high degree of sequence similarity to Thaumatin, a sweet-tasting protein isolated from the seed arils of *Thaumatococcus daniellii* (van der Wel & Loeve, 1972). As is true for other PR proteins, TLPs accumulate in fruits in response to stress of biotic and/or abiotic origin, in particular as a consequence of wounding by insects (Sudisha, Sharathchandra, Amruthesh, Kumar, & Shetty, 2012).

Five different TLPs that share a high degree of sequence similarity have been identified in *Vitis vinifera* for which the major isoform has been designated *Vitis vinifera* thaumatin-like protein 1 or VVTL1 (Pocock, Hayasaka, McCarthy, & Waters, 2000). Recently, Marangon et al. reported the crystallographic structure of three different *Vitis vinifera* TLPs called F2, H2, and I, corresponding to PDB files 4JRU, 4MBT, and 4L5H, respectively (Marangon, Van Sluyter, Waters, & Menz, 2014).

These proteins share a high degree of sequence similarity with a banana TLP (PBD file: 1Z3Q, UniProt O22322), whose crystallographic structure has been solved. The amino acid alignment shown in Table 2 indicates a 79.6% overlap between the banana TLP and one of the TLP isoforms found in grape juice (F2/4JRU), and a 73.3% overlap with other two grape juice TLP isoforms (I/4L5H and H2/4MBT). These latter, despite sharing identical sequences (Marangon et al., 2014) showed different chromatographic properties (Van Sluyter et al., 2009), suggesting the possibility for these differences to be related to their different stability behavior (see below). In particular, it has been shown that F2/4JRU was a TLP isoform unstable

Table 2 Alignment of thaumatin-like protein sequences from banana (UniProt O22322) with those from *Vitis vinifera* called F2/4JRU (UniProt F6HUG9), I/4L5H and H2/4MBT (UniProt O04708 for both). Table was generated by Chimera 1.13.1 software (Meng et al., 2006; Thompson et al., 1997).

	1	11	21	31	41
Consensus	mrrssiLsis	--ifllfsLl	fTpshAATFx	IqNhCsYTVW	AAAvPGGGrq
Conservation					
O04708 V. vinif.	MRFTTTLPIL	--IPLLLSLL	FTSTHAATFD	ILNKCTYTVW	AAASPGGGRR
F6HUG9 V. Vinif.	MGLCKILSIS	--SFLLTALF	FTPSYAATFN	IQNHCSYTVW	AAAVPGGGMQ
O22322 Banana	GTRSSXLSLS	FFFFFFFSLL	VTLSHAATFX	IVNRCSYTVW	AAAVPGGGRQ
	51	61	71	81	91
Consensus	LdsGQSWtin	VNaGTTggRi	WgRTsCsFDa	sGrGkCeTGD	CgGlLqCtaY
Conservation					
O04708 V. vinif.	LDSGQSWTIT	VNPGTTNARI	WGRTSCTFDA	NGRGKCETGD	CNGLLECQGY
F6HUG9 V. Vinif.	LGSGQSWSLN	VNAGTTGGRV	WARTNCNFDA	SGNGKCETGD	CGGLLQCTAY
O22322 Banana	LNQGQSWTIN	VNAGTTGGRI	WGRTGCSFDG	SGRGRCQTGD	CGGVLSCTAY
	101	111	121	131	141
Consensus	GsPPNTLAEF	ALNQfnNLDf	fDISLVDGFN	vPMdFspts-	GC-RGIqCaa
Conservation					
O04708 V. vinif.	GSPPNTLAEF	ALNQPNNLDY	IDISLVDGFN	IPMDFS----	GC-RGIQCSV
F6HUG9 V. Vinif.	GTPPNTLAEF	ALNQFSNLDF	FDISLVDGFN	VPMAFNPTSN	GCTRGISCTA
O22322 Banana	GNPPNTLAEF	ALNQFNNLDF	FDISLVDGFN	VPMDFSPTSG	GC-RGIRCAA
	151	161	171	181	191
Consensus	DInGqCPaaL	kapGGCnNPC	TVFkTdeYCC	n--sGSCspT	dYSrFFk-rc
Conservation					
O04708 V. vinif.	DINGQCPSEL	KAPGGCNNPC	TVFKTNEYCC	TDGPGSCGPT	TYSKFFKDRC
F6HUG9 V. Vinif.	DIVGECPAAL	KTTGGCNNPC	TVFKTDEYCC	N--SGSCSAT	DYSRFFKTRC
O22322 Banana	DINGQCPGAL	XAPGGCXNPC	TVFXTDXYCC	N--SGSCSPT	DYSRFF----
	201	211	221		
Consensus	pdaysyp-dd	-ts-ftc--g	tny-v-fcp		
Conservation					
O04708 V. vinif.	PDAYSYPQDD	KTSLFTCPSG	TNYKVTFCP		
F6HUG9 V. Vinif.	PDAYSYPKDD	QTSTFTCTAG	TNYEVVFCP		
O22322 Banana	----------	----------	---------		

upon heating, whereas I/4L5H and H2/4MBT were not. The match between stable and unstable TLPs was found to be about 69% (Marangon et al., 2014). The differences in protein sequence, essentially in β-sheets 9 and 10 and in α-helices 2 and 3, allow two functional families belonging to the Thaumatin superfamily fold to be distinguished: a *thaumatin-like protein-like domain* (1Z3Q and F2/4JRU) and a *pathogenesis-related protein-like domain* (I/4L5H and H2/4MBT) (Marangon et al., 2014).

The thaumatin superfamily fold possesses three protein domains stabilized by the presence of eight highly conserved disulfide bonds (Fig. 3). Domain I is the biggest and represents the central core of the molecule. The two β-sheets of domain I are stabilised by four disulfide bridges. Domain II is composed of a long α-helix (α5) and four short helicoidal segments (α1-α4), and contains four disulfide bonds, that stabilise the domain and link domain II to domain I. The last domain, consisting of only 32 amino acids, is characterized by the presence of a long loop and two

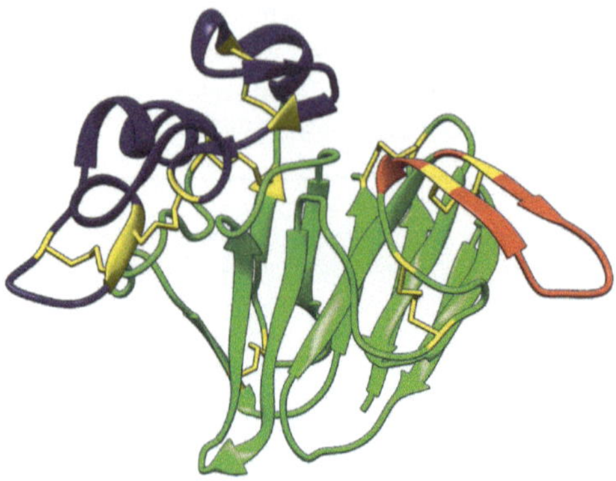

Fig. 3 Three dimensional structure of the F2/4JRU protein (PDB file: 4JRU). Thaumatin-like protein-like domains are indicated in green (Domain I), blue (Domain II), red (Domain III), while disufide bridges are showed in yellow. The image was generated using the Chimera 1.13.1 software (Pettersen et al., 2004).

β-strands, and contains two disulfide bridges, with one of them linking domain III with domain I. The surface of the F2/4JRU protein shows a cleft between domains I and II, a common feature in TLPs, containing four highly conserved acidic residues (Glu107, Asp120, Asp125 and Asp206) (highlighted in orange in Fig. 4). This acidic cleft is likely to be associated with the antifungal properties toward *Phomopsis viticola* and *Botrytis cinerea* as shown by Monteiro et al. (2003).

Fig. 5 shows that several areas of the protein have a hydrophobic character (Fig. 5, orange/red areas). These hydrophobic areas could represent

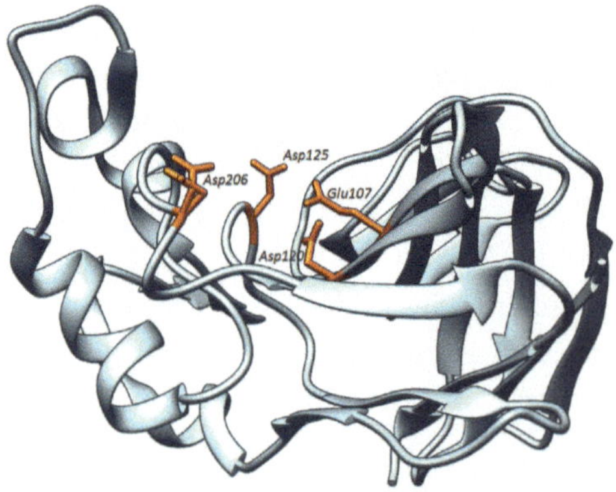

Fig. 4 Three dimensional structure of the protein 4JRU where the residues involved in the formation of the acidic left (Glu107, Asp120, Asp125 and Asp206) are shown in orange. The image was generated using the Chimera 1.13.1 software (Pettersen et al., 2004).

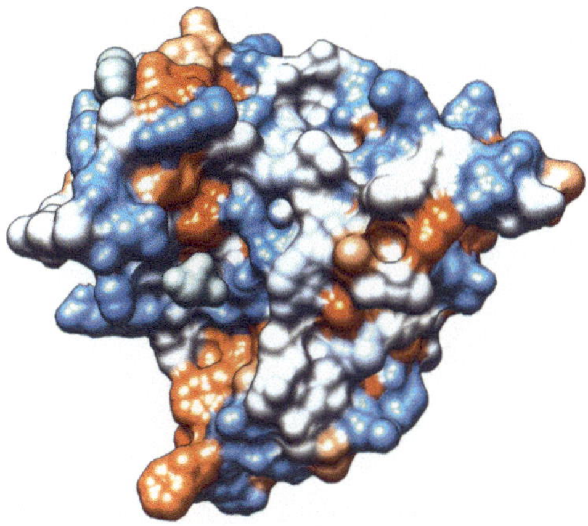

Fig. 5 Mapping of the surface hydrophobicity of the grape TLP F2/4JRU. The molecular surface is based on the Kyte-Doolittle scale from dodger blue for the most hydrophilic, to white, to orange/red for the most hydrophobic (Kyte & Doolittle, 1982; Sanner, Olson, & Spehner, 1996). The images were generated using the Chimera 1.13.1 software (Pettersen et al., 2004).

binding sites for tannins present in wines, that have been indicated as main factors involved in protein aggregation leading to haze formation in white wines (Marangon, Vincenzi, Lucchetta, & Curioni, 2010). Moreover, it has been demonstrated by computational docking calculations that quercetin can bind to TLPs by entering in the cleft, and this binding mechanism has been suggested being potentially involved in the formation of haze in white wines (Toledo, Salazar, & Aquino, 2017). Given that catechin is structurally very similar to quercetin, it is possible that also this molecule could enter in the TLP cleft as the terminal unit of a polymeric tannin chain thus supporting the idea that tannins are involved in the mechanism of protein aggregation in wine. Therefore it seems that both the main protein components found in wine, TLPs and chitinases, have the tendency to specifically bind flavonoid molecules (Filippi et al., 2016; Toledo et al., 2017). This notion should be kept in mind when studying the formation of aggregates in wines.

In particular, it could be hypothesised that grape proteins in their native state bind to flavonoids as soon as they are released from the grape cells. Despite the lack of analytical evidence supporting this theory, during protein purification from juices and wines it is common to obtain protein solutions with a yellowish color, which is attributable to the likely presence of tannins bound to the proteins. This is further supported by the fact that treating juices and wines with PVPP results in the removal of the phenolics but also to a certain percentage of proteins, thus implying that protein and tannins are associated. This protein-tannin association occurring at the very beginning of grape processing can have implications for the mechanism of haze formation that will be discussed below in section 2.

2. Protein instability

Protein instability is typically considered to be the most common non-microbiological defect in white wines (Bayly & Berg, 1967; Hsu & Heatherbell, 1987; Waters, Wallace, & Williams, 1992) and involves soluble proteins which can aggregate over time in the bottle to form haze and amorphous sediments. Three classes of grape proteins with low conformational stability have been implicated: β-glucanases, chitinases, and thaumatin-like proteins (Dufrechou, Poncet-Legrand, Sauvage, & Vernhet, 2012; Dufrechou, Vernhet, Roblin, Sauvage, & Poncet-Legrand, 2013; Marangon et al., 2010; Marangon et al., 2014). The first stage of haze formation is thought to involve protein denaturation. Subsequent interactions between the denatured proteins or with other wine components, such as polyphenols, then lead to aggregation and flocculation (Falconer et al., 2010; Pocock & Waters, 2006).

In general, three physical mechanism of protein/peptide aggregation are possible: (1) a sequential process where monomeric units are added individually to a growing chain, (2) a multimeric interaction characterized by the association of multimers of any size, and (3) a nucleation-dependent event involving slow formation of a critical aggregate size followed by rapid polymerization (Fig. 6).

However, the mechanism depicted in Fig. 6 are unlikely to be applicable to wine proteins as other wine components (e.g. sulfate) must be present for wine proteins to aggregate as when tested pure in model wine they have been shown not to be able to form aggregation even upon heating (Pocock, Alexander, Hayasaka, Jones, & Waters, 2007). Moreover, temperature and pH also play crucial roles. While an increase in temperature is typically associated with aggregates' formation, greater instability was observed at pH 2.5 than at higher pH (Dufrechou et al., 2012). This pH value is quite far from the isoelectric point of wine proteins (4.5−5.0), where they have the lowest solubility, so the increased instability can be explained by the increase of the degree of misfolding due to the repulsive interactions between charged

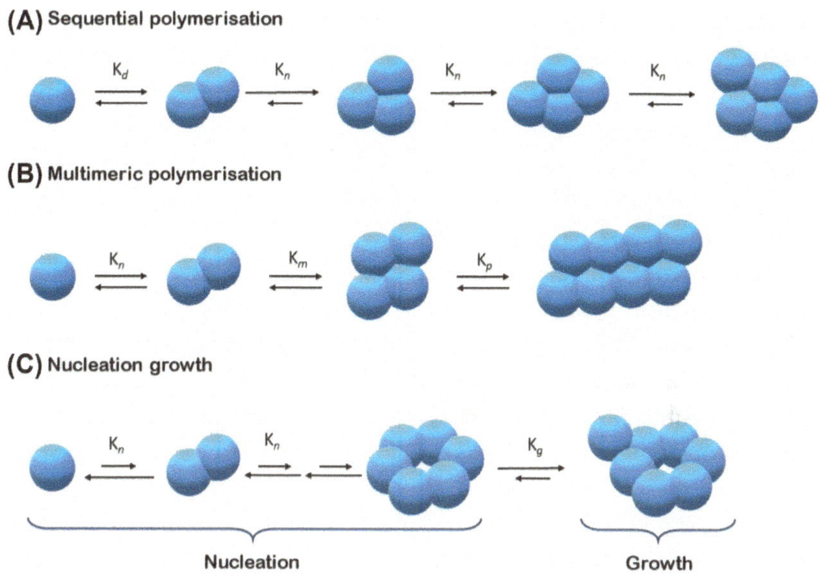

Fig. 6 Models depicting the three proposed polymerisation mechanism of aggregation.

groups that destabilize the fold conformation (Dufrechou, Sauvage, Bach, & Vernhet, 2010).

The thermal stability of wine proteins, as assessed by their melting temperature (T_M) values, and their hydrophobicity were analyzed by Falconer et al. (2010) who found that proteins with lower T_M values (55–56 °C compared to 62–63 °C) had a greater tendency to aggregate and to form precipitates. Chitinases were found to have lower thermal stability and a great tendency to denature irreversibly, whereas TLPs were prone to denature reversibly, adopting the native structure upon cooling, with the exception of the unstable isoform F2/4JRU (Marangon et al., 2014). As a matter of fact the comparison of the 3D structures of the unstable TLP versus the stable one indicates noteworthy differences only in the loop between β-strands β9 and β10 in domain I, which was longer and therefore more exposed in the unstable TLP (Marangon et al., 2014).

In addition to the intrinsic protein characteristics, heat tests carried out dissolving wine proteins in a model wine solution revealed that additional factors, such as sulfate, ionic strength, polyphenols and polysaccharides, all contribute to the aggregation of denatured proteins (Pocock & Waters, 2006). The formation of haze was described as a multifactorial process where some elements, including protein concentration, pH and ionic strength are well defined, whereas others remain unknown (Batista, Monteiro, Loureiro, Teixeira, & Ferreira, 2009, 2010; Dufrechou et al., 2012; Pocock et al., 2007). The unknown factor(s), along with proteins, seems to be a fundamental prerequisite in order to have haze formation. Pocock et al. (2007) suggested that sulfate ions may be this non-proteic factor essential for the formation of turbidity. Different hypothesis on the mechanism of aggregation of wine proteins have been proposed.

A model proposed by Stranks and coauthors (2009), described the amorphous aggregation process of TLPs in a model wine solution containing sulfate. Initially TLPs aggregated via a three-dimensional tree pathway, followed by bi-dimensional cylindrical behavior. The authors reasoned that hydrophobic interactions take place between aggregates and monomers, such that consecutive addition of monomers to the aggregate leads to stereochemical restrictions due to the decrease in exposed hydrophobic sites. Large aggregates form even larger clusters reducing the surface available for interaction with other monomers.

A more comprehensive model that takes into account also chitinases, different isoforms of TLPs and other wine components has been proposed to explain what leads to haze formation in wines in real conditions (Van Sluyter et al., 2015). This model does not distinguish the wine proteins based on their type but on the reversibility of their heat-mediated misfolding. Starting from the shared step of heat denaturation, all soluble wine proteins are to be found in an misfolded state. At this stage the mechanism can take two different routes: the proteins able to refold upon cooling (stable TLPs) do not form aggregates, while those unable to refold undergo self-aggregation, a step favored by the presence of salts. This leads to the formation of small protein aggregates invisible to the naked eye. Upon cooling, the proteins continue to interact and the resulting aggregates become cross-linked by non-protein wine components such as sulfate and phenolics to a point in which their size is sufficiently large to be visible as haze (Fig. 7).

2.1 Influence of reducing agents on protein aggregation

An additional factor influencing the aggregation and precipitation of proteins is the presence of reducing agents (Chagas, Ferreira, Laia, Monteiro, & Ferreira, 2016; Pocock et al., 2007). Reducing agents present in wine include sulfur dioxide and glutathione. The first has been used for a long time by winemakers as a preservative due to its antimicrobial and antioxidant properties. The maximum concentration of sulfur dioxide allowed in wine is

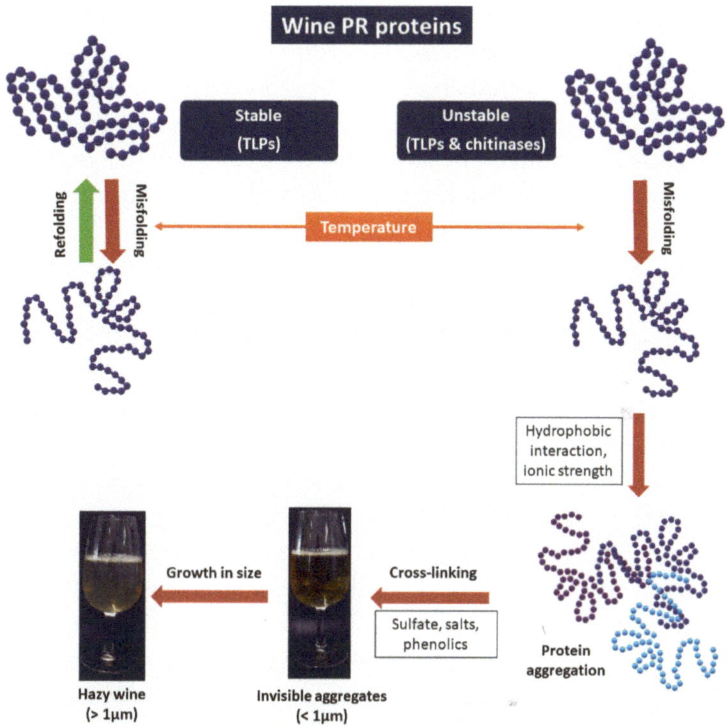

Fig. 7 Behavior of stable and unstable proteins in wine.

established by laws. In the European Union the allowed limit is 160 mg/L in red wines and 210 mg/L in white and rosé wines. Sulfur dioxide is also produced by yeast as a normal metabolic intermediate. The amount excreted into wine is a function of strain and vinification conditions, but generally does not exceed 10–20 mg/L.

Glutathione (L-γ-glutamyl-L-cysteinyl-glycine, GSH) is the most abundant non-proteic sulfur compound found in most living organisms including the wine yeast *Saccharomyces cerevisiae* (Bachhawat et al., 2009; Kritzinger, Bauer, & du Toit, 2013). It is also an important constituent of grapes, must and wine where it has antioxidant properties. Thanks to this property glutathione can be used in winemaking, and its use has been recently allowed by the OIV (International Organization of Vine and Wine) (Resolution OIV-OENO 445–2015 and OIV-OENO 446–2015).

Therefore reducing agents are always present in wine, thus determining a reducing environment that can affect protein 3D structure by reducing their disulphide bonds.

The suggestion of an involvement of the reduction mechanism on wine protein stability and protein reactivity with tannins was firstly reported by Marangon et al. (2010). In that project, the reducing agent used was dithiothreitol (DTT) which was shown to increase the reactivity between tannins and heat-misfolded proteins by facilitating the opening of the protein molecule so to have more hydrophobic binding sites exposed to react with the wine tannins. Therefore it was hypothesised that the reduction of disulfide bridges could be a phenomenon potentially involved in irreversible haze formation.

This idea was further considered when the 3D structure of the wine TLPs was solved (Marangon et al., 2014). Indeed it was shown that the section of the TLP structure that distinguish stable versus unstable isoforms was an exposed loop containing a disulfide bond whose ease of reduction could explain the differences in heat stability observed. The authors suggested that sulfur dioxide could be a candidate for these actions.

This idea was later experimentally confirmed by thermal stability tests performed in the presence of increasing concentrations of sulfur dioxide on proteins isolated from wine. It was shown that increasing the concentration of sulfur dioxide led to an increased haze formed upon heating (Chagas et al., 2016; Chagas, Laia, Ferreira, & Ferreira, 2018). The results obtained by Chagas et al. indicate that the addition of sulfur dioxide can induce reversible cleavage of TLPs disulphide bridges via sulfitolysis according to the reaction described in Fig. 8 (Würfel, Häberlein, & Follmann, 1993).

Based on this, a new model for TLPs aggregation mechanism in wine was proposed in which the cysteine residues could not reform the disulphide bridges as they were bound to SO_2. When this occurs hydrophobic interactions and formation of new pattern of covalent bonds by disulfide bridges scrambling occur (Chagas et al., 2018).

We have to consider, however, that during winemaking large quantities of antioxidants (mainly sulfur dioxide, but also glutathione, tannins, ascorbic acid, etc.) can be added to wine. The authors further suggested that in real wine-like conditions the decrease of redox potential occurring after bottling could be sufficient to trigger this denaturation upon consumption of free oxygen by the wine polyphenols.

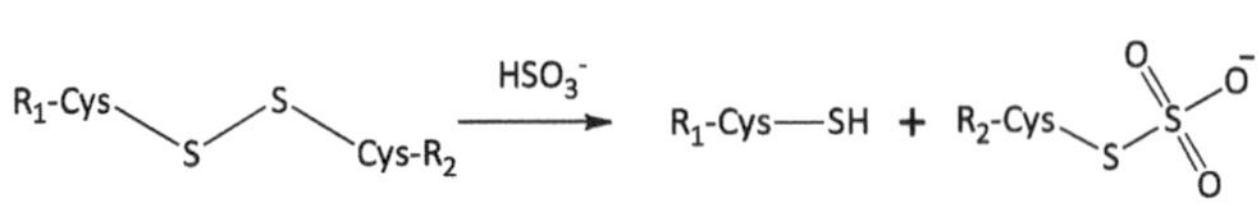

Fig. 8 Proposed reaction of SO_2 on disulfide bond. *Adapted from Chagas et al., 2016.*

The influence of a reducing environment, able to reduce disulfide bonds, thus promoting the exposure of hydrophobic sites that favors protein aggregation, has been evaluated using dithiothreitol and tris(2-carboxyethyl) phosphine (TCEP). Unfortunately, DTT is scarcely reactive at acidic pH values, while TCEP is reactive in a wider pH range (Getz, Xiao, Chakrabarty, Cooke, & Selvin, 1999). Surprisingly, in a model wine solution TCEP did not lead to the expected remarkable turbidity, negligible with respect to that induced by $NaHSO_3$, suggesting a different mechanism in the two reducing agents (Marangon et al., 2010).

3. A case study: effects of reducing agent on VVTL1

The effects of reducing agents on the tertiary structure of a TLP protein isolated from Manzoni Bianco grape, identified as VVTL1 (Di Gaspero et al., 2017), were evaluated by different biophysical techniques as fluorescence, circular dichroism and NMR spectroscopies.

In order to see the effects of these agents on protein hydrophobicity variations, 8-anilinonaphthalene-1-sulfonic acid (ANS) was used as fluorescent probe (Gasymov & Glasgow, 2007). The wavelength of fluorescence emission of ANS is highly sensitive to the polarity of the environment and it is shifted to higher wavelengths with a decrease in the fluorescence emission intensity with the increasing of the polarity of the medium. These properties make ANS an excellent probe to detect the presence of hydrophobic sites which may appear on the protein surface following either denaturation or reduction of disulfide bridges with a proportional increase in the fluorescence emission intensity.

VVTL1 protein was incubated in a model wine solution at pH 3.5 either at room temperature or at 45 °C for 22 days with or without equimolar amounts of different reducing agents, including SO_2, DTT and TCEP. Samples were then analyzed at different incubation times by fluorescence spectroscopy after the addition of ANS. An example of the normalized fluorescence emission spectra of ANS added to VVTL1-TCEP solution is shown in Fig. 9.

The time-courses of the intensity of normalized ANS fluorescence emission (Fig. 10, left and right) show a significant increase in the fluorescence intensity only in the presence of TCEP. On the contrary, samples added with DTT or SO_2 showed similar fluorescence values as the control without reducing agent. This indicates that at room temperature neither DTT nor SO_2 were able to modify the tertiary structure of VVTL1 with exposure

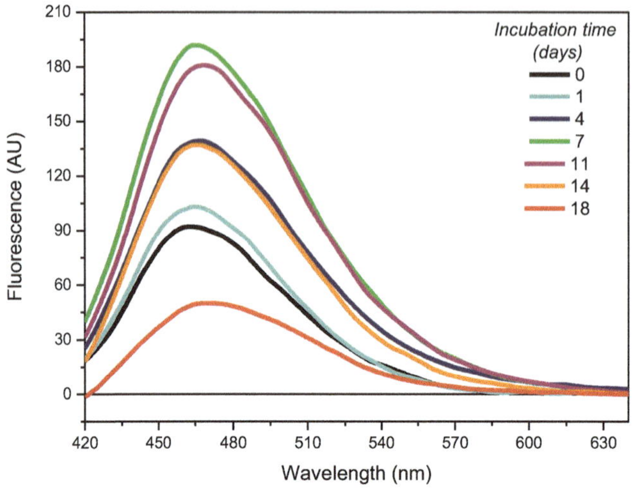

Fig. 9 Fluorescence emission spectra of ANS added to a VVTL1 protein solution (0.21 mg/mL) in model wine, pH 3.5, in presence of TCEP (0.895 mg/mL) at different incubation times (indicated). Spectra were the average of 4 scans; λ_{exc} 350 nm, emission and excitation slits 3.5 nm; scan speed 500 nm/min.

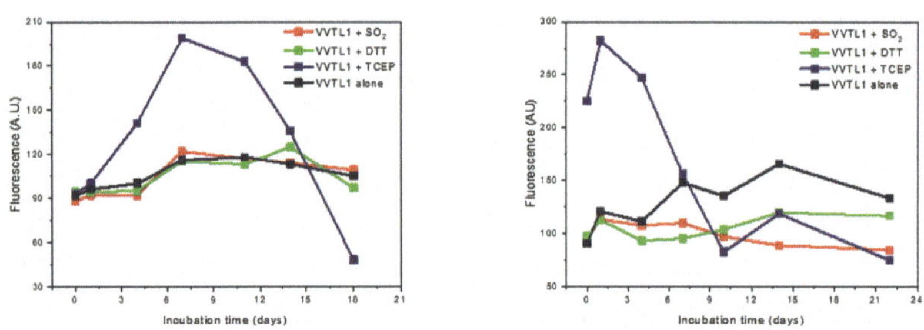

Fig. 10 Time-courses of the ANS fluorescence emission after addition to VVTL1 protein solutions (0.21 mg/mL) in model wine alone or in the presence of equimolar amounts of different reducing agents (indicated) incubated (**left**) at room temperature, and (**right**) at 45 °C.

of hydrophobic sites, which would represent the first step required for the onset of protein aggregation. The increase of fluorescence due to the presence of TCEP occurs both at room temperature and at 45 °C, although at different times. At higher temperature, TCEP maximum fluorescence is reached earlier, indicating that the whole kinetic is faster in these conditions. However, at both temperatures but at different times, the fluorescence of TCEP samples drops rapidly. This suggests a decrease in the content of exposed hydrophobic sites, most likely due to the protein aggregation process. Moreover, a slight generalized increase in the fluorescence emission, more evident in the VVTL1 alone, was detected in the samples incubated at 45 °C, suggesting that temperature contributes to the misfolding process.

Further information on the conformational effect of a reducing environment on the VVTL1 protein was obtained by circular dichroism (CD) spectroscopy. Unlike macromolecular crystallography and NMR where detailed structural models of the proteins at atomic resolution can be achieved, CD spectroscopy is the ideal technique to characterize and monitor the folding of protein as a function of environmental conditions as temperature, pH, solvent polarity, presence of salts, detergents, lipids and ligand interactions (Johnson, 1990). The far-UV CD spectra of VVTL1 are characterized by the presence of a negative band at about 210 nm and a positive band at 195 nm (Di Gaspero et al., 2017). When the above mentioned reducing agents were added, the far-UV CD spectrum resulted almost unchanged when compared to the control, with only a small change in the intensity of the positive band at 195 nm (data not shown).

The near-UV CD spectrum of VVTL1 yields information on the local tertiary structure around aromatic residues and disulfide bonds. In this region, the CD spectra show a negative band at 290 nm with a shoulder at about 300 nm, and a positive ellipticity at wavelengths lower than 260 nm (Fig. 11).

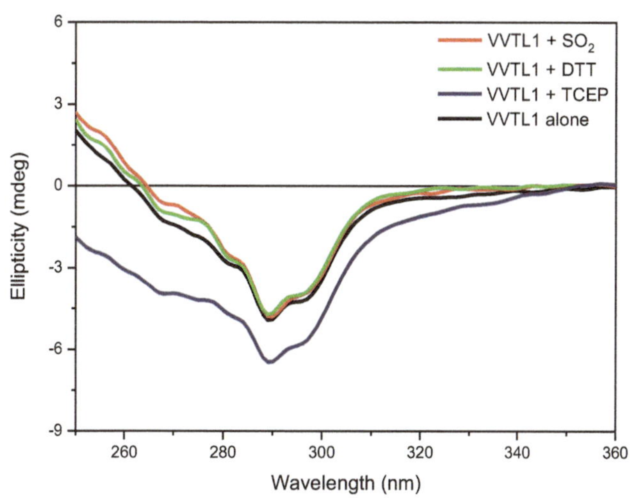

Fig. 11 Near-UV CD spectra of VVTL1 in model wine (0.7 mg/mL), pH 3.5, alone or in the presence of equimolar amounts of different reducing agents (indicated). The spectra were recorded using a 1 cm path length Suprasil cell (Hellma) filled with 900 μL of solution, 1.0 mm data pitch, bandwidth 2 nm, integration time 4 s, scanning speed 50 nm/min, average of 9 accumulations.

After the addition of the examined reducing agents at room temperature, only the CD spectra of VVTL1 in the presence of TCEP was modified, confirming that only TCEP was able to modify the protein structure, in accordance with the results shown in Fig. 10.

Our near-UV CD experiment did not indicate that SO_2 possessed the ability to affect protein structure even after 14 days of incubation (Figs. 11 and 12). These findings are in partial agreement with those by other authors that using CD experiments showed no effect of SO_2 on wine TLPs (Chagas et al. 2018).

Moreover, the thermal stability experiments showed that the melting temperature of VVTL1 is not influenced by the presence of DTT (T_M 54.4 °C vs. 53.1 °C of VVTL1 alone), while it increases in the presence of SO_2 (T_M 57.1 °C vs. 53.1 °C of VVTL1 alone).

Recently, the ^{15}N labeled VVTL1 protein has been expressed by one of the Authors (unpublished data). The availability of a labeled protein makes it possible to carry out Heteronuclear Single Quantum Coherence (HSQC) NMR experiments (Ozawa, Wu, Dixon, & Otting, 2006). NMR spectroscopy, permitting to characterize proteins at the level of single amino acid residues, has significantly contributed to comprehending the mechanism of protein folding and misfolding, and NMR techniques can be used to

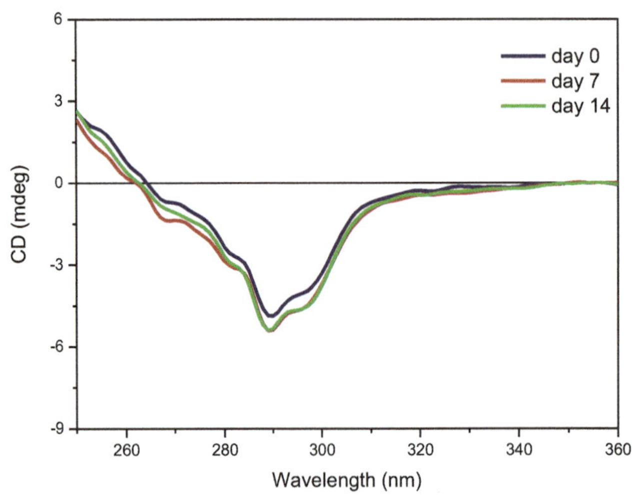

Fig. 12 Near-UV CD spectra of VVTL1in model wine (0.7 mg/mL), pH 3.5, in presence of SO_2 at different time of incubation at room temperature. The spectra were recorded using a 1 cm path length Suprasil cell (Hellma) filled with 900 μL of solution, 1.0 mm data pitch, bandwidth 2 nm, integration time 4 s, scanning speed 50 nm/min, average of 9 accumulations.

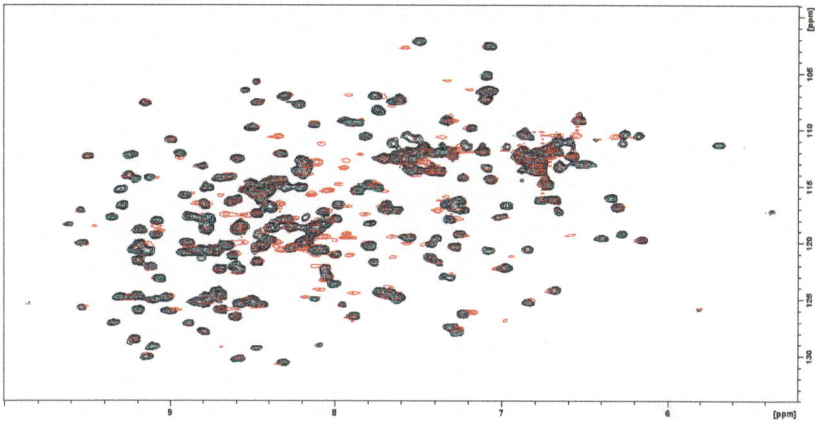

Fig. 13 $^1H,^{15}N$–HSQC NMR spectrum of ^{15}N-labelled VVTL1 in model wine solution alone (black), in presence of SO_2 after 35 days of incubation at room temperature (light blue) and following addition of TCEP (red).

monitor the effects of temperature, pH, denaturants, and solvent on protein conformation and folding.

The addition of SO_2 did not modify the amide resonance in the $^1H,^{15}N$–HSQC NMR spectrum of labeled VVTL1 protein in model wine indicating the absence of any immediate effect of SO_2 on the tertiary structure of the protein, confirming the results above reported. The sample containing SO_2 was incubated at room temperature for 35 days recording the $^1H,^{15}N$–HSQC at different times. As showed in Fig. 13, the HSQC spectrum recorded after 35 days is still superimposable to that of VVTL1 alone, confirming the inability of SO_2 to affect the protein structure.

Using an equimolar amount of TCEP which showed to be effective in modifying protein structure (see Figs. 10 and 11), the HSQC spectrum of a protein sample was recorded after 24 h of incubation (Fig. 13). Conversely from what observed for SO_2, the addition of TCEP produced a large modification in the HSQC spectrum of the VVTL1, revealing a substantial alteration in the structure of the protein that can be attributed to the reduction of its disulfide bridges.

4. Outlook

This chapter examined the instability of wine proteins, the most common non-microbiological defect in white wines related to protein misfolding, and the role of reducing agents in catalyzing this process. It is evident that appreciation of both protein structures and protein interactions is required to fully understand which factors lead to the misfolding of wine proteins and how small molecules present in wine may influence this process. A gap remains to be filled in comprehending the conformational changes correlated with the transition from native to misfolded states and the influence/interaction of other wine components, small or macro-molecules. Indeed, as showed by our studies, the presence of only one component in wine, as for example SO_2, is not sufficient to cause the denaturation of the VVTL1 protein and the formation of insoluble aggregates through exposure of hydrophobic sites. Most likely other factors, including heating, have an impact on the protein misfolding and subsequent aggregation of proteins with haze formation. Despite the challenges mentioned in this chapter, the achievability of controlling protein misfolding and haze formation is outstanding and continues to advance. However, a careful thoughtfulness is required when considering the interactions among the different protein components and those with other wine constituents.

References

Ali, S., Ganai, B. A., Kamili, A. N., Bhat, A. A., Mir, Z. A., Bhat, J. A., et al. (2018). Pathogenesis-related proteins and peptides as promising tools for engineering plants with multiple stress tolerance. *Microbiological Research, 212–213*, 29–37. http://dx.doi.org/10.1016/j.micres.2018.04.008.

Bachhawat, A. K., Ganguli, D., Kaur, J., Kasturia, N., Thakur, A., Kaur, H., et al. (2009). Glutathione production in yeast. In Satyanarayana, Tulasi, Kunze, & Gotthard (Eds.), *Yeast biotechnology: Diversity and applications* (1st ed., pp. 259–280). Netherlands: Springer, Dordrecht.

Batista, L., Monteiro, S., Loureiro, V. B., Teixeira, A. R., & Ferreira, R. B. (2009). The complexity of protein haze formation in wines. *Food Chemistry, 112*(1), 169–177. http://dx.doi.org/10.1016/j.foodchem.2008.05.070.

Batista, L., Monteiro, S., Loureiro, V. B., Teixeira, A. R., & Ferreira, R. B. (2010). Protein haze formation in wines revisited. The stabilising effect of organic acids. *Food Chemistry, 122*(4), 1067–1075. http://dx.doi.org/10.1016/j.foodchem.2010.03.076.

Bayly, F. C., & Berg, H. W. (1967). Grape and wine proteins of white wine varietals. *American Journal of Enology and Viticulture, 18*(1), 18–32.

Chagas, R., Ferreira, L. M., Laia, C. A., Monteiro, S., & Ferreira, R. B. (2016). The challenging SO_2-mediated chemical build-up of protein aggregates in wines. *Food Chemistry, 192*, 460–469. http://dx.doi.org/10.1016/j.foodchem.2015.07.052.

Chagas, R., Laia, C., Ferreira, R., & Ferreira, L. (2018). Sulfur dioxide induced aggregation of wine thaumatin-like proteins: Role of disulfide bonds. *Food Chemistry, 259*, 166–174. http://dx.doi.org/10.1016/j.foodchem.2018.03.115.

Di Gaspero, M., Ruzza, P., Hussain, R., Vincenzi, S., Biondi, B., Gazzola, D., et al. (2017). Spectroscopy reveals that ethyl esters interact with proteins in wine. *Food Chemistry, 217*, 373–378. http://dx.doi.org/10.1016/j.foodchem.2016.08.133.

Dufrechou, M., Poncet-Legrand, C., Sauvage, F. X., & Vernhet, A. (2012). Stability of white wine proteins: Combined effect of pH, ionic strength, and temperature on their aggregation. *Journal of Agricultural and Food Chemistry, 60*(5), 1308–1319. http://dx.doi.org/10.1021/jf204048j.

Dufrechou, M., Sauvage, F. X., Bach, B., & Vernhet, A. (2010). Protein aggregation in white wines: Influence of the temperature on aggregation kinetics and mechanisms. *Journal of Agricultural and Food Chemistry, 58*(18), 10209–10218. http://dx.doi.org/10.1021/jf1017687.

Dufrechou, M., Vernhet, A., Roblin, P., Sauvage, F. X., & Poncet-Legrand, C. (2013). White wine proteins: How does the pH affect their conformation at room temperature? *Langmuir, 29*(33), 10475–10482. http://dx.doi.org/10.1021/la401524w.

Falconer, R. J., Marangon, M., Van Sluyter, S. C., Neilson, K. A., Chan, C., & Waters, E. J. (2010). Thermal stability of thaumatin-like protein, chitinase, and invertase isolated from Sauvignon blanc and Semillon juice and their role in haze formation in wine. *Journal of Agricultural and Food Chemistry, 58*(2), 975–980. http://dx.doi.org/10.1021/jf902843b.

Filippi, A., Petrussa, E., Rajcevic, U., Čurin Šerbec, V., Passamonti, S., Renzone, G., et al. (2016). Flavonoid interaction with a chitinase from grape berry skin: Protein identification and modulation of the enzymatic activity. *Molecules, 21*(10), 1300.

Gasymov, O. K., & Glasgow, B. J. (2007). ANS fluorescence: Potential to augment the identification of the external binding sites of proteins. *Biochimica et Biophysica Acta, 1774*(3), 403–411. http://dx.doi.org/10.1016/j.bbapap.2007.01.002.

Gazzola, D., Pasini, G., Tolin, S., Curioni, A., & Vincenzi, S. (2017). Characterisation of chitinase isoforms from grape juice. *Italian Journal of Food Science, 29.*

Gazzola, D., Vincenzi, S., Pasini, G., Lomolino, G., & Curioni, A. (2015). Advantages of the KDS/BCA assay over the bradford assay for protein quantification in white wine and grape juice. *American Journal of Enology and Viticulture, 66*(2), 227–233. http://dx.doi.org/10.5344/ajev.2014.14076.

Getz, E. B., Xiao, M., Chakrabarty, T., Cooke, R., & Selvin, P. R. (1999). A comparison between the sulfhydryl reductants tris(2-carboxyethyl)phosphine and dithiothreitol for use in protein biochemistry. *Analitical Biochemistry, 273*(1), 73–80. http://dx.doi.org/10.1006/abio.1999.4203.

Hayasaka, Y., Adams, K. S., Pocock, K. F., Baldock, G. A., Waters, E. J., & Hoj, P. B. (2001). Use of electrospray mass spectrometry for mass determination of grape (*Vitis vinifera*) juice pathogenesis-related proteins: A potential tool for varietal differentiation. *Journal of Agricultural and Food Chemistry, 49*(4), 1830–1839.

Hsu, J.-C., & Heatherbell, D. A. (1987). Isolation and characterization of soluble proteins in grapes, grape juice, and wine. *American Journal of Enology and Viticulture, 38*(1), 6–10.

Johnson, W. C., Jr. (1990). Protein secondary structure and circular dichroism: A practical guide. *Proteins,* 7(3), 205–214. http://dx.doi.org/10.1002/prot.340070302.

Koch, J., & Sajak, E. (1959). A review and some studies on grape protein. *American Journal of Enology and Viticulture, 10*(3), 114–123.

Kritzinger, E. C., Bauer, F. F., & du Toit, W. J. (2013). Role of glutathione in winemaking: A review. *Journal of Agricultural and Food Chemistry, 61*(2), 269–277. http://dx.doi.org/10.1021/jf303665z.

Kyte, J., & Doolittle, R. F. (1982). A simple method for displaying the hydropathic character of a protein. *Journal of Molecular Biology, 157*(1), 105–132. http://dx.doi.org/10.1016/0022-2836(82)90515-0.

Mamo, J., & Assefa, F. (2018). The role of microbial aspartic protease enzyme in food and beverage industries. *Journal of Food Quality, 2018,* 15. http://dx.doi.org/10.1155/2018/7957269.

Marangon, M., Gazzola, D., Van Sluyter, S. C., Curioni, A., Waters, E. J., Vernhet, A., et al. (2012). Study of the heat induced aggregation behavior of wine macromolecules by means of scanning ion occlusion sensing and dynamic light scattering techniques. *Abstracts of Papers of the American Chemical Society, 244.*

Marangon, M., Sauvage, F.-X., Waters, E. J., & Vernhet, A. (2011). Effects of ionic strength and sulfate upon thermal aggregation of grape chitinases and thaumatin-like proteins in a model system. *Journal of Agricultural and Food Chemistry, 59*(6), 2652–2662. http://dx.doi.org/10.1021/jf104334v.

Marangon, M., Van Sluyter, S. C., Waters, E. J., & Menz, R. I. (2014). Structure of haze forming proteins in white wines: *Vitis vinifera* thaumatin-like proteins. *PLoS One, 9*(12), e113757. http://dx.doi.org/10.1371/journal.pone.0113757.

Marangon, M., Vincenzi, S., Lucchetta, M., & Curioni, A. (2010). Heating and reduction affect the reaction with tannins of wine protein fractions differing in hydrophobicity. *Analitica Chimica Acta, 660*(1–2), 110–118. http://dx.doi.org/10.1016/j.aca.2009.10.038.

Meng, E. C., Pettersen, E. F., Couch, G. S., Huang, C. C., & Ferrin, T. E. (2006). Tools for integrated sequence-structure analysis with UCSF Chimera. *BMC Bioinformatics, 7*, 339. http://dx.doi.org/10.1186/1471-2105-7-339.

Mierczynska-Vasilev, A., Boyer, P., Vasilev, K., & Smith, P. A. (2017). A novel technology for the rapid, selective, magnetic removal of pathogenesis-related proteins from wines. *Food Chemistry, 232*, 508–514. http://dx.doi.org/10.1016/j.foodchem.2017.04.050.

Monteiro, S., Piçarra-Pereira, M. A., Teixeira, A. R., Loureiro, V. B., & Ferreira, R. B. (2003). Environmental conditions during vegetative growth determine the major proteins that accumulate in mature grapes. *Journal of Agricultural and Food Chemistry, 51*(14), 4046–4053. http://dx.doi.org/10.1021/jf020456v.

Ndlovu, T., Divol, B., & Bauer, F. F. (2018). Yeast cell wall chitin reduces wine haze formation. *Applied and Environmental Microbiology, 84*(13). http://dx.doi.org/10.1128/aem.00668-18. e00668-00618.

Oyeleye, A., & Normi, Y. M. (2018). Chitinase: Diversity, limitations, and trends in engineering for suitable applications. *Bioscience Reports, 38*(4). http://dx.doi.org/10.1042/bsr20180323. BSR2018032300.

Ozawa, K., Wu, P. S. C., Dixon, N. E., & Otting, G. (2006). 15N-Labelled proteins by cell-free protein synthesis. *FEBS Journal, 273*(18), 4154–4159. http://dx.doi.org/10.1111/j.1742-4658.2006.05433.x.

Pettersen, E. F., Goddard, T. D., Huang, C. C., Couch, G. S., Greenblatt, D. M., Meng, E. C., et al. (2004). UCSF Chimera-a visualization system for exploratory research and analysis. *Journal of Computational Chemistry, 25*(13), 1605–1612. http://dx.doi.org/10.1002/jcc.20084.

Pocock, K. F., Alexander, G. M., Hayasaka, Y., Jones, P. R., & Waters, E. J. (2007). Sulfate-a candidate for the missing essential factor that is required for the formation of protein haze in white wine. *Journal of Agricultural and Food Chemistry, 55*(5), 1799–1807. http://dx.doi.org/10.1021/jf062658n.

Pocock, K. F., Hayasaka, Y., McCarthy, M. G., & Waters, E. J. (2000). Thaumatin-like proteins and chitinases, the haze-forming proteins of wine, accumulate during ripening of grape (*Vitis vinifera*) berries and drought stress does not affect the final levels per berry at maturity. *Journal of Agricultural and Food Chemistry, 48*(5), 1637–1643. http://dx.doi.org/10.1021/jf9905626.

Pocock, K. F., & Waters, E. J. (2006). Protein haze in bottled white wines: How well do stability tests and bentonite fining trials predict haze formation during storage and transport? *Australian Journal of Grape and Wine Research, 12*(3), 212–220. http://dx.doi.org/10.1111/j.1755-0238.2006.tb00061.x.

Robinson, S. P., Jacobs, A. K., & Dry, I. B. (1997). A class IV chitinase is highly expressed in grape berries during ripening. *Plant Physiology, 114*(3), 771–778. http://dx.doi.org/10.1104/pp.114.3.771.

Sanner, M. F., Olson, A. J., & Spehner, J. C. (1996). Reduced surface: An efficient way to compute molecular surfaces. *Biopolymers, 38*(3), 305–320. http://dx.doi.org/10.1002/(sici)1097-0282(199603)38:3%3c305::aid-bip4%3e3.0.co;2-y.

Somers, T. C., & Ziemelis, G. (1973). Direct determination of wine proteins. *American Journal of Enology and Viticulture, 24*(2), 47–50.

Stranks, S. D., Ecroyd, H., Van Sluyter, S., Waters, E. J., Carver, J. A., & von Smekal, L. (2009). Model for amorphous aggregation processes. *Physical Review, 80*(5), 051907. http://dx.doi.org/10.1103/PhysRevE.80.051907.

Sudisha, J., Sharathchandra, R. G., Amruthesh, K. N., Kumar, A., & Shetty, H. S. (2012). Pathogenesis related proteins in plant defense response. In J. M. Mérillon, & K. G. Ramawat (Eds.), *Plant defence: Biological control* (1st ed., Vol. 12, pp. 379–403). Villenave d'Ornon, France: Springer Nature.

Takeshi, Y., & Gunki, F. (1996). Limited proteolysis and reduction-carboxymethylation of Rye seed chitinase-a: Role of the chitin-binding domain in its chitinase action. *Bioscience Biotechnology and Biochemistry, 60*(7), 1081–1086. http://dx.doi.org/10.1271/bbb.60.1081.

Tattersall, D. B., van Heeswijck, R., & Hoj, P. B. (1997). Identification and characterization of a fruit-specific, thaumatin-like protein that accumulates at very high levels in conjunction with the onset of sugar accumulation and berry softening in grapes. *Plant Physiology, 114*(3), 759–769. http://dx.doi.org/10.1104/pp.114.3.759.

Thompson, J. D., Gibson, T. J., Plewniak, F., Jeanmougin, F., & Higgins, D. G. (1997). The CLUSTAL_X windows interface: Flexible strategies for multiple sequence alignment aided by quality analysis tools. *Nucleic Acids Research, 25*(24), 4876–4882. http://dx.doi.org/10.1093/nar/25.24.4876.

Tian, B., Harrison, R., Morton, J., & Jaspers, M. (2019). Changes in pathogenesis-related proteins and phenolics in *Vitis vinifera* L. cv. 'Sauvignon Blanc' grape skin and pulp during ripening. *Scientia Horticulturae, 243*, 78–83. http://dx.doi.org/10.1016/j.scienta.2018.08.018.

Tian, B., Harrison, R., Morton, J., Jaspers, M., Hodge, S., Grose, C., et al. (2017). Extraction of pathogenesis-related proteins and phenolics in Sauvignon blanc as affected by grape harvesting and processing conditions. *Molecules, 22*(7). http://dx.doi.org/10.3390/molecules22071164.

Toledo, L. N., Salazar, F. N., & Aquino, A. J. A. (2017). A theoretical approach for understanding the haze phenomenon in bottled white wines at molecular level. *South African Journal for Enology & Viticulture, 38*(1). http://dx.doi.org/10.21548/38-1-837.

Van Sluyter, S. C., Marangon, M., Stranks, S. D., Neilson, K. A., Hayasaka, Y., Haynes, P. A., & Waters, E. J. (2009). Two-step purification of pathogenesis-related proteins from grape juice and crystallization of thaumatin-like proteins. *Journal of Agricultural and Food Chemistry, 57*(23), 11376–11382. http://dx.doi.org/10.1021/jf902365r.

Van Sluyter, S. C., McRae, J. M., Falconer, R. J., Smith, P. A., Bacic, A., Waters, E. J., et al. (2015). Wine protein haze: Mechanisms of formation and advances in prevention. *Journal of Agricultural and Food Chemistry, 63*(16), 4020–4030. http://dx.doi.org/10.1021/acs.jafc.5b00047.

Van Sluyter, S. C., Warnock, N. I., Schmidt, S., Anderson, P., van Kan, J. A. L., Bacic, A., et al. (2013). Aspartic acid protease from *Botrytis cinerea* removes haze-forming proteins during white winemaking. *Journal of Agricultural and Food Chemistry, 61*(40), 9705–9711. http://dx.doi.org/10.1021/jf402762k.

Velasco, R., Zharkikh, A., Troggio, M., Cartwright, D. A., Cestaro, A., Pruss, D., et al. (2007). A high quality draft consensus sequence of the genome of a heterozygous grapevine variety. *PLoS One, 2*(12), e1326. http://dx.doi.org/10.1371/journal.pone.0001326.

Vincenzi, S., & Curioni, A. (2005). Anomalous electrophoretic behavior of a chitinase isoform from grape berries and wine in glycol chitin containing SDS-PAGE gels. *Electrophoresis, 26*(1), 60–63.

Vincenzi, S., Mosconi, S., Zoccatelli, G., Pellegrina, C. D., Veneri, G., Chignola, R., et al. (2005). Development of a new procedure for protein recovery and quantification in wine. *American Journal of Enology and Viticulture, 56*(2), 182–187.

Vincenzi, S., Polesani, M., & Curioni, A. (2005). Removal of specific protein components by chitin enhances protein stability in a white wine. *American Journal of Enology and Viticulture, 56*(3), 246–254.

Waters, E. J., Hayasaka, Y., Tattersall, D. B., Adams, K. S., & Williams, P. J. (1998). Sequence analysis of grape (*Vitis vinifera*) berry chitinases that cause haze formation in wines. *Journal of Agricultural and Food Chemistry, 46*(12), 4950–4957. http://dx.doi.org/10.1021/jf980421o.

Waters, E. J., Shirley, N., & Williams, P. (1996). *Nuisance proteins of wine are grape pathogenesis-related proteins* (Vol. 44).

Waters, E. J., Wallace, W., & Williams, P. J. (1992). Identification of heat-unstable wine proteins and their resistance to peptidases. *Journal of Agricultural and Food Chemistry, 40*(9), 1514–1519. http://dx.doi.org/10.1021/jf00021a008.

van der Wel, H., & Loeve, K. (1972). Isolation and characterization of thaumatin I and II, the sweet-tasting proteins from *Thaumatococcus daniellii* benth. *European Journal of Biochemistry, 31*(2), 221–225. http://dx.doi.org/10.1111/j.1432-1033.1972.tb02522.x.

Würfel, M., Häberlein, I., & Follmann, H. (1993). Facile sulfitolysis of the disulfide bonds in oxidized thioredoxin and glutaredoxin. *European Journal of Biochemistry, 211*(3), 609–614. http://dx.doi.org/10.1111/j.1432-1033.1993.tb17588.x.

Younes, B., Cilindre, C., Villaume, S., Parmentier, M., Jeandet, P., & Vasserot, Y. (2011). Evidence for an extracellular acid proteolytic activity secreted by living cells of *Saccharomyces cerevisiae* PlR1: Impact on grape proteins. *Journal of Agricultural and Food Chemistry, 59*(11), 6239–6246. http://dx.doi.org/10.1021/jf200348n.

FSC
www.fsc.org
MIX
Papier | Fördert gute Waldnutzung
FSC® C083411

Kolibri 360 GmbH
Ferdinand-Jühlke-Straße 7
99095 Erfurt, Deutschland
produktsicherheit@kolibri360.de